Electromagnetic Optics

Online at: https://doi.org/10.1088/978-0-7503-6064-7

Electromagnetic Optics

Gregory J Gbur
*Department of Physics and Optical Science, University of North Carolina at Charlotte,
Charlotte, NC, USA*

IOP Publishing, Bristol, UK

ISBN   978-0-7503-6064-7 (ebook)
ISBN   978-0-7503-6062-3 (print)
ISBN   978-0-7503-6065-4 (myPrint)
ISBN   978-0-7503-6063-0 (mobi)

DOI   10.1088/978-0-7503-6064-7

Version: 20250201

IOP ebooks

British Library Cataloguing-in-Publication Data: A catalogue record for this book is available from the British Library.

Published by IOP Publishing, wholly owned by The Institute of Physics, London

IOP Publishing, No.2 The Distillery, Glassfields, Avon Street, Bristol, BS2 0GR, UK

US Office: IOP Publishing, Inc., 190 North Independence Mall West, Suite 601, Philadelphia, PA 19106, USA

For my father John Gbur

who has always inspired me to be curious

and for Zoe

the best feline friend anyone could wish for.

Contents

Preface

It has been known since the introduction of Maxwell's equations in the 1860s that light is an electromagnetic wave, but the teaching of introductory optics often avoids the discussion of electromagnetic theory until it is absolutely necessary. This avoidance is, in part, historical: the wave nature of light was first demonstrated by Thomas Young in the early 1800s, and thus many of the significant discoveries and theoretical ideas involving interference and diffraction were made without reference to the electromagnetic nature of light. The avoidance is also, of course, partly pedagogical: when working with highly directional (paraxial) light waves with a simple polarization state, the full electromagnetic description of light is an unnecessary complication in understanding key phenomena.

But many novel effects and applications cannot be described adequately, or at all, with a paraxial scalar model. One example of this is so-called optical tweezing, a technique in which microscopic particles are trapped in a single beam of light. The particles can only be adequately trapped in the focus of a high numerical aperture optical system, and in the description of such fields, the transverse wave nature of light cannot be neglected.

Even for highly directional fields, novel effects can be found by considering the full electromagnetic properties of light. In recent years, there has been intense research into what is now known as structured light: beams of light that possess nontrivial phase, amplitude, and polarization properties. Important classes of structured light beams include radially and azimuthally polarized beams, in which the state of polarization varies within the cross section of the beam. Among other interesting properties, radially polarized beams have been shown to possess a smaller focal spot than uniformly polarized beams, making them potentially useful in imaging.

At optical wavelengths, most natural materials respond directly to the electric field of an illuminating light wave but not the magnetic field. Traditional optics is based on the assumption that materials are nonmagnetic. But over the last two decades, researchers have demonstrated that it is possible to manipulate the structure of matter on a subwavelength scale to produce optical responses not seen in nature. These artificial materials are now known as metamaterials, and they can be designed to have a number of rare properties that are useful in unusual applications. They can be made to possess a magnetic response, which in turn allows the creation of materials with a negative refractive index. This, in turn, allows for the creation of lenses which are, in principle, 'perfect,' creating a perfect image of an object. In 2006, researchers made worldwide news by showing that, theoretically, it is possible to use metamaterials to make an invisibility cloak.

These examples show that there are great advantages in looking at the full electromagnetic nature of light, and many surprises have been found by doing so. This book is intended to explore those electromagnetic properties in detail, including discussions of the applications mentioned above. The field of physical optics is broadly divided into the study of the interference, diffraction, coherence, and polarization of light, and it could be said that this book focuses on the latter part.

Electromagnetic Optics is based on a one-semester course that I have been teaching at the University of North Carolina at Charlotte for the last decade. It is one of the core graduate courses, along with mathematical methods, physical optics, geometrical optics, optical sources and detectors, and the optical properties of materials. Electromagnetic waves have long been considered an important part of an optics education, but there has not been a suitable book for the subject. The textbooks on electromagnetic waves I investigated while preparing my course were either far too theoretical or aimed at describing long-wavelength microwave and radio waves.

As with all of my academic books, I attempt here to thread a very fine needle in making *Electromagnetic Optics* both a textbook for students to learn from and a reference book that will provide useful insight to optics researchers. To that end, I endeavor to include plenty of references on the subjects at hand to guide further investigations. I have also included tables of optical constants, with original references, throughout the book; I have kept the significant figures of the original references, no matter how overly precise they may seem.

One note: there is some unavoidable overlap of topics between this book and my *Singular Optics* text; I have significantly revised my descriptions of the topics for this text, but the reader will note similarities in the figures and the presentation. There are only so many ways to explain things.

It is my hope that this book will provide a comprehensive and practical introduction to the electromagnetic nature of visible light and an idea of how that nature is key to understanding modern optics.

Each textbook I have written has been a learning experience for me, and one thing I have gotten much better at is writing interesting and insightful exercises that are intended to genuinely provide insight into the relevant physics. To support these, I have written a complete solutions manual for the exercises in the book to help instructors who might be interested in using the book for a course. Instructors can request the manual from me at http://emopticssolutions@gmail.com.

Acknowledgments

Every book takes a village, in terms of friends, colleagues, and family members who offered advice, support, or just generally helped keep me sane.

I extend a special thanks to Professor Taco Visser for reading the entire book draft and giving me suggestions and corrections; his help was invaluable!

I would like to acknowledge the Air Force Office of Scientific Research and Program Manager Dr. Arje Nachman, who supported much of my original research that informed and occasionally appears in this book.

I would like to thank my family for their support: my father John Gbur, my mother Patricia Gbur, my sister Gina Darby, and my nieces Fern and Ceci.

I would also like to thank some of my friends for helping keep me sane and happy while I was writing! This includes my guitar instructor Toby Watson, my skating coach Tappie Dellinger, and my regular online gaming friends: Donna Lanclos, Mindy Weisberger, Dani Marzano, Karyn Murphy, Lali DeRosier, Lisa Manglass, Al Houghton, Josh Witten, Nathan Taylor, and Hugo González.

Author biography

Gregory J Gbur

Gregory J Gbur is a full professor of Physics and Optical Science at the University of North Carolina at Charlotte with over 150 peer-reviewed scientific papers. He is the author of three textbooks—*Mathematical Methods for Optical Physics and Engineering, Singular Optics*, and *Electromagnetic Optics*—as well as two popular science books—*Falling Felines and Fundamental Physics and Invisibility*. He maintains a long-running blog about physics, the history of science, and horror fiction called 'skulls in the stars.'

IOP Publishing

Electromagnetic Optics

Gregory J Gbur

Chapter 1

Introduction: the electromagnetic spectrum

This text focuses primarily on the electromagnetic properties of visible light. Visible light, which has frequencies between roughly 430 THz (terahertz) and 750 THz, is only a small fraction of the electromagnetic spectrum that is relevant to our lives. The lowest-frequency radio waves measured have a frequency on the order of 8 Hz, while the highest-energy gamma rays ever detected had frequencies of 10^{29} Hz, or 10^{17} THz! In between these extremes, there are many different parts of the electromagnetic spectrum that each have their own distinguishing features and physics, and it is worth our time to review them to get a sense of their similarities and differences.

The electromagnetic spectrum divides electromagnetic waves by their frequency of oscillation. As we have already indicated, frequency is measured in cycles per second (hertz) and is typically denoted by the symbol ν. The inverse of this quantity is the period $T = 1/\nu$, which indicates the duration of a single cycle of the wave. In theoretical work, we often prefer to use angular frequency, or radians per second, which is given by $\omega = 2\pi\nu$. In vacuum, the speed of light is $c = 3 \times 10^8$ m s^{-1}, or 186 282 miles per second. (We never use the latter units in calculations but provide them for those who more naturally think in terms of miles instead of meters.) We may then write the wavelength—the spatial length of one oscillation of the wave—in terms of this speed as $\lambda = c/\nu$, or $\lambda = 2\pi c/\omega$. Wavelength is the spatial analogy to temporal period, and we may also introduce $k = 2\pi/\lambda$ as the wavenumber, which is the spatial frequency of oscillation, or radians per meter. The wavelength may be roughly viewed as a measure of the 'size' of the photon when considering light–matter interactions.

In ordinary situations, the frequency of an electromagnetic wave does not change when it enters a medium, but its wavelength does. If the refractive index of a medium at frequency ν is n, then the wavelength within the medium is λ/n, as we will later discuss in more detail. When we talk about the wavelength of the electromagnetic spectrum, we are almost always talking about the vacuum wavelength. Wavelength

doi:10.1088/978-0-7503-6064-7ch1

and frequency are inversely related—a larger wavelength corresponds to a smaller frequency, and vice versa.

Though this book discusses the classical wave properties of electromagnetism, it is also helpful to think of the electromagnetic spectrum in terms of individual light particles, called photons. The energy of a photon is given by $E = h\nu$, where $h = 4.1357 \times 10^{-15}$ eV Hz^{-1} is Planck's constant. The photon energy is typically measured in eV, or electron volts, which is the kinetic energy acquired by an electron upon being accelerated through a one-volt potential difference. It is worth noting that we may also write $E = hc/\lambda$ or $E = \hbar\omega$, where $\hbar = h/2\pi = 6.5821 \times 10^{-16}$ eV s. The energy of a photon is directly proportional to its frequency and inversely proportional to its wavelength—short-wavelength photons have the highest energy. In vacuum, we may also describe the momentum of a photon by the expression $p = h/\lambda$, which indicates that short-wavelength photons also have the highest momentum.

Photon energy, wavelength, and frequency are all key parameters for describing the interaction of electromagnetic waves with matter. Differences in the observed behavior of different parts of the electromagnetic spectrum may be traced to how the frequency of the electromagnetic wave compares to the natural frequencies of matter and how the wavelength compares to the size of the components of matter. With this in mind, let us consider in brief the different components of the electromagnetic spectrum.

The complete electromagnetic spectrum is illustrated in figure 1.1. Starting at the lowest frequencies, we have radio waves, followed by microwaves, infrared waves, visible light, ultraviolet light, x-rays, and then gamma rays. It may be said that the

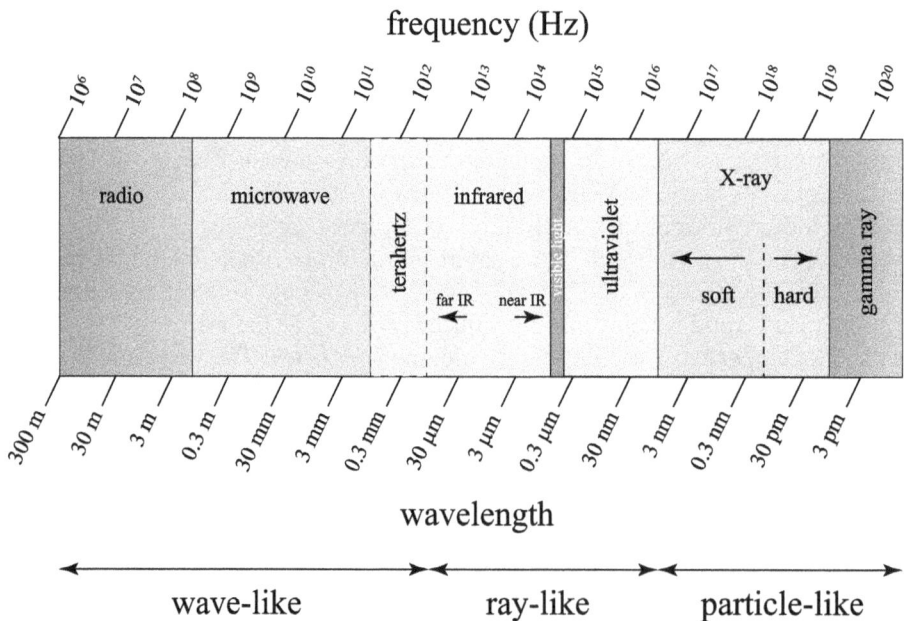

Figure 1.1. Illustration of the wavelengths and frequencies of the electromagnetic spectrum.

boundaries between these different wave bands are inherently fuzzy—there is no clear distinction between the highest-energy x-rays and the lowest energy gamma rays, for example—and the boundaries are conventions (mostly) agreed upon by researchers.

I have also demarcated the electromagnetic spectrum roughly in terms of its observable behavior on a terrestrial scale, dividing it into wave-like, ray-like, and particle-like behaviors. This demarcation is based on the relative size of the wavelength compared to ordinary day-to-day objects, as wave-like behavior generally only arises when an electromagnetic wave interacts with an object comparable to or smaller than the wavelength. Thus electromagnetic waves become 'wave-like' around the terahertz regime, where the wavelength becomes larger than 1 mm. The 'ray-like' regime, which includes visible light, involves electromagnetic waves that typically look particle-like on a terrestrial scale and which only show wave-like properties with careful preparation. 'Particle-like' electromagnetic waves are those with a wavelength so short (smaller than 1 nm) that their wave nature is almost never significant (with x-ray crystallography being a notable exception).

Let us start our survey with the most obvious part of the spectrum: visible light. As noted above, visible light covers the frequencies from 430 THz to 750 THz. This corresponds to a wavelength range from 700 nm to 400 nm, or photon energies from 1.77 eV to 3.10 eV. The Sun emits radiation roughly as a blackbody with a peak emission around 500 nm, and the human eye has evidently evolved to take advantage of the abundance of photons around that wavelength. The energy of visible-light photons is comparable to the energy levels of molecular and atomic electronic states, so the absorption of photons by matter is more or less the norm (with obvious exceptions such as glass, air, and water).

The wavelength of light is significantly larger than the size of an atom, which is on the order of 0.1 nm, or one angstrom. One consequence of this is that visible light does not 'see' the individual atoms in a bulk material, but only a spatial average that is characterized by properties like the refractive index. Visible light cannot, in general, be used to study the atomic structure of matter, but there have been curious exceptions. In 1912, Jean Baptiste Perrin used light interference measurements of colored soap bubbles to estimate the size of soap molecules [1]; this was successful because the molecules are extremely long chains (4.5 nm in length) that tend to align normal to the surface of a soap film.

Visible light is but a tiny fraction of the electromagnetic spectrum that affects our lives. Moving on to lower frequencies, we come to infrared (IR) light, which lies below the red end of the visible spectrum, with frequencies from 300 GHz to 430 THz. This corresponds to wavelengths from 1 mm to 700 nm and photon energies from 1.2×10^{-3} eV to 1.77 eV. Infrared radiation was discovered by William Herschel in 1800 [2]. While performing experiments measuring both the brightness and the heating power of the Sun's rays, Herschel noticed that the heating power of light increased towards the red end of the spectrum. Herschel had been using a prism to separate the colors of light and been situating a thermometer in each color to measure this heating; in a flash of inspiration, he placed the thermometer beyond the red end of the spectrum, where no light was apparently falling. Herschel found that the heating increased even further, demonstrating that invisible heating rays were being produced by the Sun.

Heating is one of the most noteworthy aspects of infrared radiation; photon energies correspond to the vibrational energies of molecules, making infrared exceptionally effective at heating matter. Infrared radiation from the Sun accounts for about 50% of the heating of the Earth, with the rest of the heating coming from visible light that is absorbed and then reemitted at longer wavelengths.

Infrared radiation is often categorized as near-, mid-, or far-infrared (NIR, MIR, or FIR) depending on its wavelength; again, the definitions of these categories vary somewhat between different engineering and scientific groups. The ISO 20 473 scheme specifies NIR as the wavelength range from 780 nm to 3 μm, MIR as ranging from 3 μm to 50 μm, and FIR as ranging from 50 μm to 1 mm. NIR light is often used in fiber optic cables, as silica has low absorption in this region. NIR light, being very close to the visible range, is also applied in night vision devices. These devices convert NIR light to visible light, increasing the number of photons that are seen at night by the user.

All objects emit thermal radiation depending on temperature, and objects at room temperature emit in the 8 μm to 25 μm wavelength range. Thermal imaging cameras can directly measure this MIR radiation and convert it to a visible-light image.

Infrared radiation crosses a number of significant absorption bands of water, with only a limited number of 'windows' available for transmission. Many applications are tailored to the frequencies of these windows. Because of its strong interactions, infrared radiation typically does not penetrate far into matter, though it can in many cases penetrate deeper than visible light. For this reason, infrared imaging has been used to look at details hidden in the subsurface layers of classic works of art [3].

At the end of the far-infrared, the next frequency band that is of significance is so-called terahertz radiation. This radiation has frequencies from 0.3 THz to 3 THz, with wavelengths between 1 mm and 100 μm and energies between 1.2×10^{-3} eV and 1.2×10^{-2} eV. One can see that terahertz radiation technically falls within the already-defined infrared band, but it has been given special attention due to its recently discovered usefulness in applications. Terahertz radiation is strongly absorbed by water but can penetrate fabrics and plastics, making it useful radiation for security screening. Some frequencies can also penetrate several millimeters of human tissue, giving terahertz radiation great potential for medical imaging. In contrast with x-rays, terahertz radiation is nonionizing and poses little risk to a human. It has also been used in archaeology for imaging the wrapping layers of mummies [4]. A review of the current state of terahertz imaging can be found in Li et al [5]. Researchers have been aware of terahertz radiation for a long time, but the study of its practical applications only took off with the 1995 paper of Hu and Nuss [6] that demonstrated a practical terahertz imaging system.

Below infrared and terahertz radiation, we enter the microwave band, with frequencies between 300 MHz and 300 GHz and corresponding wavelengths between 1 m and 1 mm; the photon energies are between 1.24×10^{-6} eV and 1.24×10^{-3} eV. In this band, the wavelength is now comparable to the size of ordinary objects, and microwaves exhibit significant wave properties. At the low end of the band, microwaves are absorbed only weakly by the atmosphere and water within it; a major application of microwaves is thus line-of-sight communications. Included in these applications are short-range devices such as Bluetooth devices, which operate at around 2.40 GHz and WiFi connections, which operate at standard frequencies of 2.4 GHz and 5 GHz. The higher frequency provides greater bandwidth but is more strongly absorbed and propagates over a shorter distance. 5G cellular networks operate in two ranges, 450 MHz to 6 GHz, which is the long-term evolution (LTE) range, and 24.25 GHz to 52.5 GHz.

Microwave ovens typically operate at 2.45 GHz, with a wavelength of 12 cm. At this frequency, electromagnetic waves are particularly good at exciting rotational modes of water molecules, producing heat. The longer wavelength of microwaves provides significant penetration depth into food, resulting in uniform heating. Microwave ovens are also a fun way to estimate the speed of light. By heating a chocolate bar for a few seconds in a microwave, one can deduce the wavelength from the melted spots on the bar, and from the listed frequency of operation, one can calculate the speed using $c = \nu\lambda$. This demonstration is in keeping with the spirit of the original invention of the microwave oven. In 1945, it is said, engineer Percy Spencer was standing in front of a microwave transmitter when he noticed that the candy bar in his pocket had melted. This motivated him to investigate the effect of microwaves on food, eventually leading to the commercial microwave oven.

The lowest-frequency electromagnetic waves are, of course, radio waves. Radio waves encompass all waves with a frequency lower than 300 MHz, or a wavelength greater than 1 m, with photon energies lower than 1.24×10^{-6} eV. Talking about photons in this frequency band is usually unnecessary, as the wavelengths are so long that radio waves can be described almost entirely by classical wave physics. Radio waves were discovered by Heinrich Hertz in a series of experiments between 1886 and 1889, as he sought to confirm Maxwell's prediction of electromagnetic waves [7]. Hertz developed an oscillating spark-gap generator—essentially a crude radio transmitter—and measured the nodes and antinodes of the standing waves produced by interference between incident and reflected waves. His experiments demonstrated a speed of wave propagation comparable to the speed of light and proved the existence of electromagnetic waves.

By virtue of their long wavelength, the generation of radio waves can be understood directly from classical considerations. Oscillating currents in a metal antenna produce time-varying electric and magnetic fields, resulting in an electromagnetic wave. We will discuss the physics of this mechanism in chapter 14. The detection of radio waves also proceeds similarly: an electromagnetic wave excites currents in a conducting metal receiver, which can then be measured.

The sources of radio waves must typically be as large as at least a quarter wavelength of the desired radiation. In North America, the amplitude modulation (AM) radio band is from 530 to 1700 kHz, which corresponds to a wavelength of 566 m to 176 m. The human ear has a maximum frequency sensitivity around 28 kHz, and AM radio stations within range of each other must have a frequency difference of at least 10 kHz in order to produce fair-quality audio signals without crosstalk. Frequency modulation (FM) radio uses a range from 87.5 to 108.0 MHz, with wavelengths from 3.43 m to 2.78 m. Because of their higher carrier frequencies, FM radio channels can have larger bandwidths and carry more information. AM radio waves, with their long wavelength, can diffract over obstacles and follow the curvature of the Earth for long distances; FM radio waves, with their shorter wavelengths, are typically limited to propagation up to the horizon.

The lowest-frequency radio waves, with a frequency on the order of 8 Hz, are naturally occurring Schumann resonances. These are ultralong wavelength radio waves that are excited by the continuous lightning strikes that occur around the Earth; our atmosphere acts as a waveguide that conducts these waves around the globe. Schumann resonances were first predicted theoretically by Winfried Otto Schumann in 1952 [8] and were experimentally observed for the first time in 1960 by Balser and Wagner [9]. As resonant phenomena, Schumann resonances have distinct resonant peaks, the first of which is at 7.83 Hz and the second at 14.3 Hz. The wavelength of the lowest Schumann resonance is a truly astounding 38 300 km, which is comparable to the circumference of the Earth at 40 000 km. These resonances can be put to practical scientific use, for example, in measuring global lightning activity [10].

Let us now move on to the electromagnetic spectrum at frequencies above those of visible light. Just beyond the violet end of the visible spectrum, we have ultraviolet (UV) light, with frequencies from 750 THz to 3×10^4 THz. This range corresponds to wavelengths from 400 nm to 10 nm and photon energies from 3.10 eV to 124 eV. The low end of the UV spectrum coincides with the limit of human vision; it also roughly corresponds with the beginning of photon energy levels that can damage organic molecules and cause chemical reactions.

UV light was discovered by Johann Wilhelm Ritter in 1801, who naturally concluded from Herschel's discovery of infrared light that there must be invisible rays beyond the violet end of the visible spectrum. Ritter initially believed, from philosophical reasoning, that the invisible rays on the violet end must cause cooling in contrast with infrared; however, his experiments found no such effect. He then investigated whether unseen rays could cause chemical reactions, and indeed found this to be the case [11]. The UV he had discovered could darken a solution of silver chloride even more strongly than violet light. As already noted, UV photons have sufficient energy to break chemical bonds and ionize atoms, causing permanent chemical changes. This is problematic for outdoor activities, as UV radiation can damage DNA and cause skin cancer.

Ultraviolet light is further divided into subbands with different effects and significance. According to the ISO standard 21 348, the three longest wavelength bands are UV-A, with wavelengths of 315–400 nm, UV-B, with wavelengths of 280–315 nm, and UV-C, with wavelengths of 100–280 nm. UV-A is the band used in black lights to induce fluorescence; it is the least damaging of the UV bands and does not directly break DNA bonds but can nevertheless cause indirect damage through the creation of chemical free radicals. UV-B is needed by humans for the production of vitamin D, but too much can cause sunburn and skin cancer. UV-C can cause the most damage to biological organisms, and UV-C lamps have become common for sterilizing tools and workspaces in medical and biological laboratories. Most UV-B and UV-C radiation is blocked by the ozone layer, and higher-energy UV radiation is almost entirely absorbed by the atmosphere.

Ultraviolet light has wavelengths that are still significantly larger than the sizes of atoms, so materials can still be modeled as bulk materials rather than arrangements of individual atoms.

Beyond ultraviolet light, the next electromagnetic band is x-rays, which have frequencies from 3×10^4 THz to 3×10^7 THz and corresponding wavelengths of 10 nm to 10 pm (picometers). Their photon energies range from 124 eV to 124 keV. At these wavelengths, which are comparable to or smaller than atoms themselves, x-rays can 'see' the structure of bulk matter, forming the basis of x-ray crystallography, where x-ray scattering measurements are used to deduce the structure of a bulk crystal.

X-rays were discovered serendipitously by Wilhelm Conrad Röntgen in 1895 during his studies of cathode rays (accelerated electrons) in cathode ray tubes [12]. During his experiments, Röntgen noticed that a nearby fluorescent screen was glowing, even though no cathode rays could be hitting it; evidently, a new, mysterious type of ray was being produced in the tube. These rays were dubbed 'x'-rays due to their unknown nature.

Many researchers suspected that x-rays were another type of electromagnetic radiation, but it turned out to be difficult to prove. Because of their short wavelength, x-rays do not reflect like visible light and, in fact, scatter when encountering a surface. Because their photon energies are much higher than most resonances of matter, they pass easily through most materials and do not exhibit appreciable refraction. Furthermore, most x-rays are produced in unpolarized form, and so directly detecting any polarization effects was not possible. The demonstration of their electromagnetic properties was finally achieved by Charles Barkla in 1904 [13], when he used scattering from one gas cell to produce polarized x-rays and scattering from a second gas cell to measure this polarization.

Because their frequencies are far from the ordinary resonances of atoms, x-rays tend to travel in straight lines through soft matter and are only weakly absorbed. The discovery of x-rays immediately led to their use for medical imaging. Much later, in the 1970s, Godfrey Hounsfield demonstrated that multiple x-rays of a patient, taken from multiple directions, could be used to construct a three-dimensional image of a patient's interior, inventing the technique of computed tomography [14].

X-rays are broadly divided into 'hard' and 'soft' x-rays, with soft x-rays being those below 5 keV, and hard ones above this. Hard x-rays are the ones that can easily penetrate matter and are used in applications; soft x-rays are absorbed in air over short propagation distances.

X-rays are generated through the rapid acceleration or deceleration of charged particles, typically electrons. In cathode ray tubes, the collision of electrons with the anode—and their sudden deceleration—causes an emission of x-rays. Large circular particle accelerators such as synchrotrons can produce high-energy x-rays, which can be used for a variety of experimental purposes. However, a large accelerator is not needed to produce x-rays; in 2008, researchers demonstrated that x-rays are generated when simple adhesive tape is peeled in a vacuum [15]. Peeling tape causes charge separation between the two parts of the tape, and in vacuum, electrons can accelerate to high speed when crossing the gap to recombine.

Finally, at the high 'end' of the electromagnetic spectrum, we have gamma rays. Gamma rays are defined as any electromagnetic wave with a frequency above 3×10^7 THz, or a wavelength below 10 pm. The first gamma rays were observed in radioactive decay, in which a high-energy photon is emitted from the nucleus. These photons sometimes have energies that lie within the x-ray band, so the term 'gamma ray' is often used generically to refer to any high-energy photons from the nucleus.

The electron has a rest mass of 0.511 MeV; when the energy of a photon exceeds twice this amount, the photon can interact with matter and be converted into an electron-antielectron pair. At even higher energies, photons can create showers of particles upon interaction with matter.

The highest-energy gamma ray ever detected originated in the Crab Nebula. It was detected in 2019 and had an approximate frequency of 10^{29} Hz, or a wavelength of 3.0×10^{-12} nm [16]. This is an energy of 4.14×10^{14} eV, or 6.63×10^{-5} joules! For comparison, the kinetic energy of a mosquito in flight is estimated to be 10^{-7} joules, so this single photon had the energy of hundreds of mosquitoes. It seems likely that even higher-energy photons will eventually be detected, meaning that we have not yet reached the end of the electromagnetic spectrum.

For all these bands of electromagnetic waves, from radio waves all the way up to cosmic rays, Maxwell's equations can be used in some form to characterize their behavior. Even if we restrict our attention to the visible spectrum and perhaps the bands immediately adjacent to it, we find that there is a wealth of phenomena and unexpected surprises to be found and understood. This book is an attempt to highlight those phenomena and capture some of the beauty inherent in electromagnetic waves. Even though Maxwell's equations were developed over 150 years ago, we have yet to fully understand the implications of these equations. Hopefully, this text will serve as an excellent launching point for the efforts of future researchers.

References

[1] Perrin J 1913 Observations sur les lames minces *Arch. Sci. Phys. Naturelles.* **35** 384–5

[2] Herschel W 1800 Investigation of the powers of the prismatic colours to heat and illuminate objects; with remarks, that prove the different refrangibility of radiant heat. To which is added an inquiry into the method of viewing the Sun advantageously, with telescopes of large apertures and high magnifying powers *Phil. Trans. R. Soc.* **90** 255–83

[3] Delaney J K, Thoury M and Zeibel J 2016 Visible and infrared imaging spectroscopy of paintings and improved reflectography *Heritage Sci.* **4** 6

[4] Fukunaga K, Cortes E, Cosentino A, Stünkel I, Leona M, Duling I N III and Mininberg D T 2011 Investigating the use of terahertz pulsed time domain reflection imaging for the study of fabric layers of an Egyptian mummy *J. Eur. Opt. Soc. Rapid Publ.* **6** 11040

[5] Li X, Li J, Li Y, Ozcan A and Jarrahi M 2023 High-throughput terahertz imaging: progress and challenges *Light: Sci. Appl.* **12** 233

[6] Hu B B and Nuss M C 1995 Imaging with terahertz waves *Opt. Lett.* **20** 1716–8

[7] Hertz H 1962 *Electric Waves* (New York: Dover) pp 95–106

[8] Schumann W O 1952 Über die strahlungslosen Eigenschwingungen einer leitenden Kugel, die von einer Luftschicht und einer Ionosphärenhülle umgeben ist *Z. Nat.forsch.* A **7** 149–54

[9] Balser M and Wagner C A 1960 Observations of earth-ionosphere cavity resonances *Nature* **188** 638–41

[10] Heckman S J, Williams E and Boldi B 1998 Total global lightning inferred from Schumann resonance measurements *J. Geophys. Res. Atmos.* **103** 31775–9

[11] Ritter J W 1801 [Am 22sten Febr ...] *Ann. Phys., Lpz.* **7** 527

[12] Röntgen W C 1896 On a new kind of rays *Science* **3** 227–31

[13] Barkla C G 1904 Polarisation in Röntgen rays *Nature* **69** 463–3

[14] Hounsfield G N 1973 Computerized transverse axial scanning (tomography): part I. Description of system *Br. J. Radiol.* **46** 1016–22

[15] Camara C G, Escobar J V, Hird J R and Putterman S J 2008 Correlation between nanosecond X-ray flashes and stick-slip friction in peeling tape *Nature* **455** 1089–92

[16] Amenomori M, Bao Y W and Bi X 2019 First detection of photons with energy beyond 100 TeV from an astrophysical source *Phys. Rev. Lett.* **123** 051101

IOP Publishing

Electromagnetic Optics

Gregory J Gbur

Chapter 2

Maxwell's equations

Maxwell's equations form the foundation of the electromagnetic theory of light, and in this chapter we discuss the history of each of the four equations, their physical interpretation, and their alternate representations.

We begin our story in the middle, so to speak. The complete set of Maxwell's equations, in their modern and most general differential form, is listed below:

$$\nabla \cdot \mathbf{D}(\mathbf{r}, t) = \rho_f(\mathbf{r}, t), \quad \text{(Gauss's law)} \tag{2.1}$$

$$\nabla \cdot \mathbf{B}(\mathbf{r}, t) = 0, \quad \text{('No magnetic monopoles')} \tag{2.2}$$

$$\nabla \times \mathbf{E}(\mathbf{r}, t) = -\frac{\partial \mathbf{B}(\mathbf{r}, t)}{\partial t}, \quad \text{(Faraday's law)} \tag{2.3}$$

$$\nabla \times \mathbf{H}(\mathbf{r}, t) = \mathbf{J}_f(\mathbf{r}, t) + \frac{\partial \mathbf{D}(\mathbf{r}, t)}{\partial t}. \quad \text{(Ampère–Maxwell law)} \tag{2.4}$$

There are two vector equations and two scalar equations, and each quantity in the equations depends in general on spatial position $\mathbf{r}$ and time t. Because each vector equation consists of three component equations in three-dimensional space, there are eight equations in total.

The quantity $\mathbf{E}(\mathbf{r}, t)$ is the *electric field*, one of the fundamental electromagnetic fields; it has units of V m^{-1}. The quantity $\mathbf{D}(\mathbf{r}, t)$ is known as the *electric displacement* and has units of C m^{-2}; it is a derived quantity that is used to simplify calculations when working in dielectric materials.

The quantity $\mathbf{B}(\mathbf{r}, t)$ is the *magnetic induction*, and it is the other fundamental electromagnetic field; it has units of Wb m^{-2}. The quantity $\mathbf{H}(\mathbf{r}, t)$ is the *magnetic field* and is a derived quantity that is used to simplify calculations when working in magnetic materials; it has units of A m^{-1}. We will see how this derived quantity $\mathbf{H}(\mathbf{r}, t)$ and its electric analogue $\mathbf{D}(\mathbf{r}, t)$ are used in a later chapter.

doi:10.1088/978-0-7503-6064-7ch2   2-1

The term 'magnetic induction' is a historical holdover and a rather clunky term. We will generally refer to both **B** and **H** as the 'magnetic field,' when the context makes it clear which we are referring to. We will also use the shorthand 'B-field' and 'H-field' at times to discuss them, as well as 'E-field' and 'D-field.'

We also have source terms in Maxwell's equations: those quantities that produce electric and magnetic fields. Electric fields are created by electric charges, and the *charge density* $\rho(\mathbf{r}, t)$ describes the local charge per unit volume, with units of C m^{-3}. We also have *current density* $\mathbf{J}(\mathbf{r}, t)$, and magnetic fields are created by electric currents; the units of current density are A m^{-2}. In the complete form of Maxwell's equations given above, these quantities have a subscript f which indicates 'free' charges and currents. When we discuss electromagnetism in materials, we will distinguish between the charges bound to materials in the form of dipoles and the charges that are unbound—free—of the material.

Though they do not appear explicitly in Maxwell's equations, we will often need to refer to the net charge Q and net current I of a system, which have units of C and A = C s^{-1}, respectively.

As can be seen from the description above, Maxwell's equations in their most general form involve a dizzying number of quantities and equations. A big part of the strategy in solving them, and a strategy that will be used throughout the book, is to simplify the equations for a particular case of interest and look at the behavior of the solution in that simplified system.

With this strategy in mind, we note that the relationships between **D** and **E** and between **H** and **B** depend on the materials the electromagnetic fields are interacting with. In general, this relationship can be extremely complicated, and much of our future work will involve exploring this relationship for various situations. For now, we consider electromagnetic fields in vacuum, where the relationship is exceedingly simple:

$$\mathbf{D}(\mathbf{r}, t) = \varepsilon_0\mathbf{E}(\mathbf{r}, t), \quad \mathbf{B}(\mathbf{r}, t) = \mu_0\mathbf{H}(\mathbf{r}, t). \tag{2.5}$$

The quantity ε_0 is the free-space *permittivity*, which has the value $\varepsilon_0 = 8.85 \times 10^{-12}$ F m^{-1}. The quantity μ_0 is the free-space *permeability*, with the value $\mu_0 = 1.26 \times 10^{-6}$ N A^{-2}.

In vacuum, we do not need the D-field and the H-field, so we substitute from equation (2.5) into Maxwell's equations to remove those derived fields. The simplified form of Maxwell's equations in vacuum is then:

$$\nabla \cdot \mathbf{E}(\mathbf{r}, t) = \rho(\mathbf{r}, t)/\varepsilon_0, \text{ (Gauss's law)} \tag{2.6}$$

$$\nabla \cdot \mathbf{B}(\mathbf{r}, t) = 0, \text{ ('No magnetic monopoles')} \tag{2.7}$$

$$\nabla \times \mathbf{E}(\mathbf{r}, t) = -\frac{\partial \mathbf{B}(\mathbf{r}, t)}{\partial t}, \text{ (Faraday's law)} \tag{2.8}$$

$$\nabla \times \mathbf{B}(\mathbf{r}, t) = \mu_0\mathbf{J}(\mathbf{r}, t) + \mu_0\varepsilon_0\frac{\partial \mathbf{E}(\mathbf{r}, t)}{\partial t}. \text{ (Ampère–Maxwell law)} \tag{2.9}$$

We have dropped the subscript f for the moment because, in the absence of any materials containing dipoles, all charges may be considered free.

There is one more important piece of this complicated puzzle to be added. Maxwell's equations describe how electric charges and currents produce electric and magnetic fields, but those electric and magnetic fields themselves exert forces on electric charges. For a point charge, this force $\mathbf{F}$ is called the *Lorentz force* and satisfies the equation

$$\mathbf{F} = q(\mathbf{E} + \mathbf{v} \times \mathbf{B}), \tag{2.10}$$

where q is the value of the point charge, $\mathbf{v}$ is the velocity of the charge, and $\mathbf{E}$ and $\mathbf{B}$ are the electric and magnetic fields at the location of the charge.

So the above description tells us what Maxwell's equations *are*, but does not tell us what they *mean* and where they came from. We now look at the history of each of these equations in turn and discuss what they tell us about the relationship between fields and charges. We will rely heavily on the vector calculus foundations discussed in appendix A.

2.1 Gauss's law

In vacuum, Gauss's law has the simple form:

$$\nabla \cdot \mathbf{E}(\mathbf{r}, t) = \rho(\mathbf{r}, t)/\varepsilon_0. \tag{2.11}$$

The origin of Gauss's law goes back to the experimental work of the French physicist Charles-Augustin de Coulomb. In 1785, he first published the law that we now know as *Coulomb's law* [1], which describes the force between two point-like charges:

$$\mathbf{F}_{1 \to 2} = \frac{1}{4\pi\varepsilon_0} \frac{q_1 q_2}{R^2} \hat{\mathbf{R}}. \tag{2.12}$$

Here, $1 \to 2$ indicates that the equation describes the force exerted on charge 2 by charge 1; q_1 and q_2 are the strengths of the two charges. As illustrated in figure 2.1, the quantity R is the distance between the two charges and $\hat{\mathbf{R}}$ is the unit vector pointing from charge 1 to charge 2.

Coulomb's law tells us three important things about the electric force between two charges. The first is that the force always acts along the line between the two charges. The second is the familiar statement that 'like charges repel, opposite

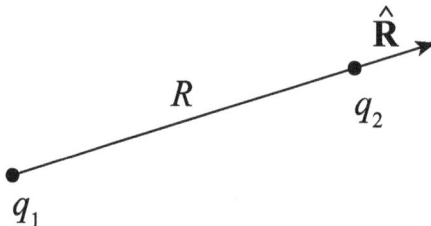

Figure 2.1. Illustration of the notation for Coulomb's law.

charges attract.' Charges with the same sign repel each other, while charges of opposite sign attract each other. The third observation is that the force between electric charges satisfies an inverse square law, varying with the inverse square of the distance between the charges.

Coulomb was not the first to propose an inverse square law for the force between charges. Other researchers also performed experiments suggesting such a law, notably the English physicist Joseph Priestley in 1767, but Coulomb performed the most delicate and conclusive tests of the idea. His experimental arrangement is illustrated in figure 2.2, showing the torsion balance he used in his research. When a hanging fiber is twisted, it experiences a restoring force that attempts to untwist it. This force can be calculated and measured quite precisely, and Coulomb used this torsion force to measure the electric force between charges. One static charge was placed on the arm of the hanging fiber, and another on a fixed sphere a close distance away. By varying the distance between the two static charges, he could measure the force of repulsion as a function of distance and found it agreed with an inverse square law.

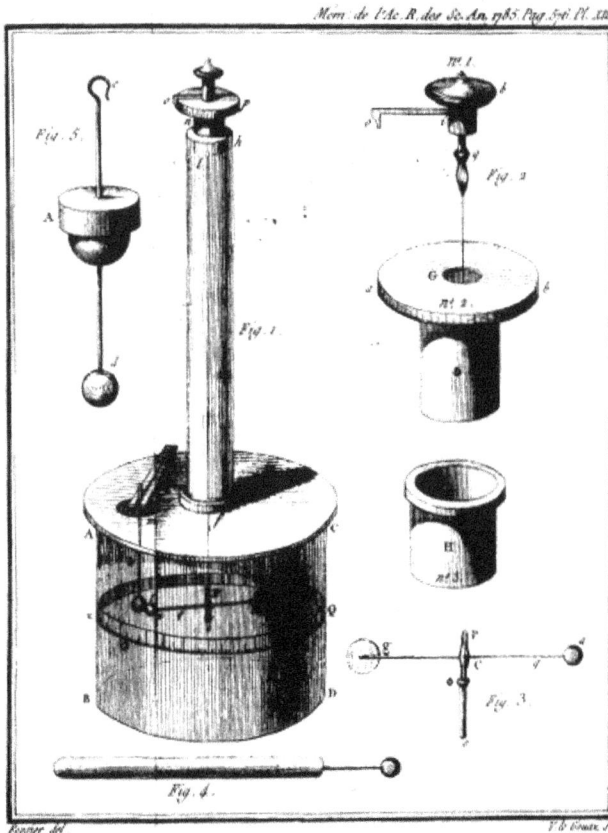

Figure 2.2. Coulomb's drawing of his torsion balance experiment, from [1].

We now derive Gauss's law by building upon this fundamental description by Coulomb. Our first step is to introduce the electric field of a given point charge. This is done by noting that the force produced by charge 1 acting on charge 2 may be written in the form

$$\mathbf{F}_{1\to2} = q_2\mathbf{E}_1, \tag{2.13}$$

where $\mathbf{E}_1$ is defined as the electric field produced by charge 1. It may be readily seen that it can be written as

$$\mathbf{E}_1 = \frac{1}{4\pi\varepsilon_0}\frac{q_1}{R^2}\hat{\mathbf{R}}, \tag{2.14}$$

and does not depend upon the properties of charge 2 at all. We may then interpret the force between two charges as follows: charge 1 creates an electric field that interacts with charge 2, producing an electric force.

Historically, the force between two charges was viewed as a direct 'action at a distance,' with no intermediate cause. In 1852, at age 61, Michael Faraday introduced the concept of fields as a better way to visualize and calculate the interactions between charges and between currents [2]. Faraday's work was largely ignored by the scientific community at the time, but James Clerk Maxwell would take Faraday's lessons to heart and use them to derive what we now know as Maxwell's equations.

In general, we may describe the electric field at a point $\mathbf{r}$, produced by a charge q located at point $\mathbf{r}'$, by the expression

$$\mathbf{E}(\mathbf{r}) = \frac{1}{4\pi\varepsilon_0}\frac{q}{R^2}\hat{\mathbf{R}}, \tag{2.15}$$

where $\mathbf{R} = \mathbf{r} - \mathbf{r}'$ and $R = |\mathbf{r} - \mathbf{r}'|$. The next logical question to ask is: what is the electric field produced by a *collection* of point charges?

We make one important fundamental assumption in answering this question: the *principle of superposition*. If we have a collection of N point charges q_1, q_2, q_3, and so on, which produce electric fields $\mathbf{E}_1$, $\mathbf{E}_2$, $\mathbf{E}_3$, and so on, respectively, then the principle of superposition states that the total electric field is the vector sum of the individual fields:

$$\mathbf{E}_{tot}(\mathbf{r}) = \mathbf{E}_1(\mathbf{r}) + \mathbf{E}_2(\mathbf{r}) + \mathbf{E}_3(\mathbf{r}) + \dots + \mathbf{E}_N(\mathbf{r}) = \sum_{i=1}^{N}\mathbf{E}_i(\mathbf{r}), \tag{2.16}$$

where we have introduced a summation notation at the end of the equation simply to get used to using this notation in future work. The total field, in terms of the fields of the individual point charges, is therefore given by

$$\mathbf{E}_{tot}(\mathbf{r}) = \frac{1}{4\pi\varepsilon_0}\sum_{i=1}^{N}\frac{q_i}{|\mathbf{r} - \mathbf{r}_i|^2}\hat{\mathbf{R}}_i, \tag{2.17}$$

where $\mathbf{r}_i$ is the position of the ith point charge and $\hat{\mathbf{R}}_i$ is the unit vector pointing from $\mathbf{r}_i$ to $\mathbf{r}$.

It is mathematically more convenient to work with a continuous distribution of charge with charge density $\rho(\mathbf{r}')$. We may make this transition for equation (2.17) by viewing each infinitesimal volume d^3r of a charge distribution as a point charge q_i, with the relationship $q_i = \rho(\mathbf{r}_i)d^3r$. We replace q_i in equation (2.17) using this expression and, assuming we are summing over a large number of point charges, convert the summation into an integral,

$$\mathbf{E}(\mathbf{r}) = \frac{1}{4\pi\varepsilon_0}\int_V \frac{\rho(\mathbf{r}')\hat{\mathbf{R}}}{R^2}d^3r'. \tag{2.18}$$

Since we are integrating over $\mathbf{r}'$, we write the differential element as d^3r'. The position labeled by $\mathbf{r}'$ is referred to as the *source point*, as it is the source of electric fields; the position labeled by $\mathbf{r}$ is referred to as the *field point*, as it is the location where we are calculating the electric field.

We are now in a position to derive the differential form of Gauss's law with the help of some vector calculus. We first take the divergence of both sides of equation (2.18),

$$\nabla \cdot \mathbf{E}(\mathbf{r}) = \frac{1}{4\pi\varepsilon_0}\nabla \cdot \int_V \frac{\rho(\mathbf{r}')\hat{\mathbf{R}}}{R^2}d^3r'. \tag{2.19}$$

We note that the integral is over $\mathbf{r}'$, and the divergence is taken with respect to $\mathbf{r}$—they are independent operations. We therefore make the assumption that we can interchange the order of integration and differentiation,

$$\nabla \cdot \mathbf{E}(\mathbf{r}) = \frac{1}{4\pi\varepsilon_0}\int_V \nabla \cdot \left(\frac{\hat{\mathbf{R}}}{R^2}\right)\rho(\mathbf{r}')d^3r'. \tag{2.20}$$

This assumption is potentially risky because there are cases where this interchange alters the value of the integral—we will see one of those cases much later in the book. Proving it is acceptable in our case takes a lot of effort and provides little insight, so we take it as a given.

We now take advantage of the following identity,

$$\nabla \cdot \left(\frac{\hat{\mathbf{R}}}{R^2}\right) = 4\pi\delta^{(3)}(\mathbf{r} - \mathbf{r}'), \tag{2.21}$$

where $\delta^{(3)}$ is the three-dimensional Dirac delta function, which exhibits the 'sifting' property:

$$\int \delta^{(3)}(\mathbf{r} - \mathbf{r}')f(\mathbf{r}')d^3r' = f(\mathbf{r}). \tag{2.22}$$

With this property of the Dirac delta, we may immediately simplify equation (2.20) to the form

$$\nabla \cdot \mathbf{E}(\mathbf{r}) = \rho(\mathbf{r})/\varepsilon_0. \tag{2.23}$$

This is the differential form of Gauss's law. It should be noted that we derive this under the assumption that all the charges are fixed in position and that the formula is therefore independent of time; it has been confirmed, however, that it applies even for time-varying charge distributions.

We may also derive an integral form of Gauss's law by integrating both sides of equation (2.23) over a volume V. We then have

$$\int_V \nabla \cdot \mathbf{E}(\mathbf{r})d^3r = \frac{1}{\varepsilon_0} \int_V \rho(\mathbf{r})d^3r. \tag{2.24}$$

The integral on the right-hand side is an integral of the charge density over the volume V; it represents the net charge Q_{enc} enclosed by the volume. With this definition, we may write

$$\int_V \nabla \cdot \mathbf{E}(\mathbf{r})d^3r = \frac{Q_{enc}}{\varepsilon_0}. \tag{2.25}$$

We now use Gauss's theorem (equation (A.45)) on the left-hand side of this expression:

$$\int_V \nabla \cdot \mathbf{E}(\mathbf{r})d^3r = \oint_S \mathbf{E}(\mathbf{r}) \cdot d\mathbf{a}, \tag{2.26}$$

where S is the surface bounding the volume V and $d\mathbf{a}$ is a vector infinitesimal surface element. Replacing the volume integral with the surface integral, Gauss's law becomes

$$\oint_S \mathbf{E}(\mathbf{r}) \cdot d\mathbf{a} = \frac{Q_{enc}}{\varepsilon_0}. \tag{2.27}$$

This is the integral form of Gauss's law and the familiar one that is used in introductory electromagnetism classes to calculate the electric fields of symmetric charge distributions such as spheres and cylinders. Gauss first derived a version of this formula in 1835, though in a form not easily shown to be equivalent to its modern counterpart.

We may physically interpret the differential form of Gauss's law with the help of our understanding of the meaning of the divergence. The divergence of a vector field indicates the amount of vector field being created or destroyed at that point. Since the divergence of the electric field is proportional to the charge density, Gauss's law tells us that electric charges are the sources of electric fields, and that electric field lines begin or end at electric charges.

2.1.1 Uniformly charged sphere

It is worth looking at a couple of examples of the derivation of electric fields using Gauss's law, as this is an approach that we will find useful later in other contexts. We begin with the classic example of a uniformly charged spherical volume of radius a and total charge Q, and calculate the electric field inside and outside the sphere.

Symmetry plays a key role in these problems. Because the charge density is uniform throughout the sphere, the entire problem is spherically symmetric. We may conclude from this that the electric field itself must be a spherically symmetric distribution and can therefore only point in the radial direction. We consider two spherical shells, one of radius $r < a$ and one of radius $r > a$, as illustrated in figure 2.3.

Regardless of whether we consider the interior or the exterior, the integral form of Gauss's law applies:

$$\oint_S \mathbf{E}(\mathbf{r}) \cdot d\mathbf{a} = \frac{Q_{enc}}{\varepsilon_0}. \tag{2.28}$$

In this case, we expect that the electric field points only in the radial direction, i.e. $\mathbf{E}(\mathbf{r}) = E(r)\hat{\mathbf{r}}$, and that the surface element is, in spherical coordinates, $r^2 \sin\theta d\theta d\phi \hat{\mathbf{r}}$. Upon substitution, we obtain

$$\oint_S E(r)r^2 \sin\theta d\theta d\phi = \frac{Q_{enc}}{\varepsilon_0}. \tag{2.29}$$

The electric field is independent of the angles θ and ϕ, so we integrate over the solid angle of the sphere to get 4π. We are then left with

$$E(r)r^2 = \frac{Q_{enc}}{4\pi\varepsilon_0}. \tag{2.30}$$

For $r > a$, the total charge enclosed is simply Q, so we have

$$E_{out}(r) = \frac{Q}{4\pi\varepsilon_0 r^2}, \tag{2.31}$$

exactly what we would expect for a point charge of strength Q. For $r < a$, only the charge enclosed plays a role, and that enclosed charge is given by

$$Q_{enc} = \frac{Q}{4\pi a^3/3}(4\pi r^3/3) = \frac{Qr^3}{a^3}. \tag{2.32}$$

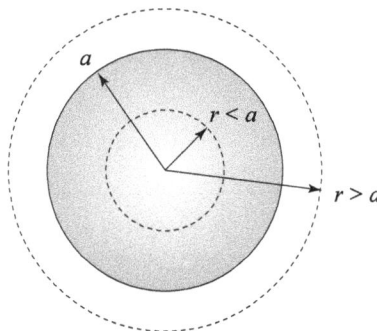

Figure 2.3. Gaussian shell used to calculate the electric field of a uniformly charged sphere.

This leaves us with

$$E_{in}(r) = \frac{Qr}{4\pi a^3}.$$

(2.33)

Of course, this is the magnitude of the electric field only; the vector electric field points in the $\hat{r}$ direction. We note that the electric field is continuous throughout space; it increases linearly from the center of the sphere up to its boundary and then decreases according to the inverse square law.

2.1.2 Two infinite sheets of electric charge

As a second example, we consider two infinite sheets of uniform electric charge density, as illustrated in figure 2.4. The upper sheet has a surface charge density $+\sigma$, and the lower sheet has a surface charge density $+2\sigma$. Again, we rely on symmetry to solve the problem; because the sheets are uniformly charged, the system is completely invariant in the x- and y-directions, and therefore the only direction the electric field can point in is the z-direction.

The ideal Gaussian surface to determine the strength of this field is a cylindrical pillbox with endcap area A. A natural instinct is to create a pillbox that is centered between the two surfaces and consider the cases where the pillbox lies entirely between the sheets and where it stretches out and encompasses both sheets. However, this does not give the correct result, because the system as a whole is asymmetric around the plane $z = a/2$. This flawed approach would suggest that the field between the plates is zero, which cannot be true because the charges on the plates are not equal. The proper approach is to instead calculate the field of each sheet separately using a pillbox centered on it, which possesses the proper symmetry, and then combine the fields of the two sheets using the superposition principle.

Let us imagine we have a pillbox centered on a single sheet at $z = 0$ with surface charge density $+\sigma'$. Because the electric field is in the z-direction, the quantity $\mathbf{E} \cdot d\mathbf{a} = 0$ on the entire cylinder of the pillbox. On the upper endcap, $d\mathbf{a}$ points in the $+\hat{z}$ direction, and on the lower endcap, $d\mathbf{a}$ points in the $-\hat{z}$ direction; the electric field is constant on both caps. Assuming $\sigma' > 0$, the electric field always points away from the sheet. Gauss's law for this single sheet is thus

$$\oint_S \mathbf{E}(\mathbf{r}) \cdot d\mathbf{a} = 2E(z)A = \sigma'A/\varepsilon_0,$$

(2.34)

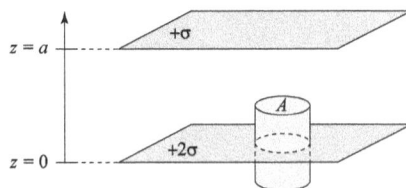

Figure 2.4. Two-sheet example for Gauss's law.

where the factor of two comes from the combination of both endcaps. We therefore have

$$E(z) = \frac{\sigma'}{2\varepsilon_0}. \tag{2.35}$$

The *direction* of this field is important. It is defined as $+\hat{\mathbf{z}}$ above the sheet and $-\hat{\mathbf{z}}$ below the sheet. The vector field should thus be written

$$\mathbf{E}(z) = \begin{cases} \dfrac{\sigma'}{2\varepsilon_0}\hat{\mathbf{z}}, & z > 0, \\[2ex] -\dfrac{\sigma'}{2\varepsilon_0}\hat{\mathbf{z}}, & z < 0. \end{cases} \tag{2.36}$$

We now use this result twice to determine the total field of our original system. There are three regions: above the two sheets, between the two sheets, and below the two sheets. Above the two sheets, we use the 'upper' value of the electric field for both. Between the two sheets, we use the 'lower' value of the field for the upper sheet and the 'upper' value of the lower sheet. Below the two sheets, we use the 'lower' value for each field. Our results can be summarized in a piecewise form as

$$\mathbf{E}(z) = \begin{cases} \dfrac{3\sigma}{2\varepsilon_0}\hat{\mathbf{z}}, & z > a, \\[2ex] \dfrac{\sigma}{2\varepsilon_0}\hat{\mathbf{z}}, & 0 < z < a, \\[2ex] -\dfrac{3\sigma}{2\varepsilon_0}\hat{\mathbf{z}}, & z < 0. \end{cases} \tag{2.37}$$

2.2 'No magnetic monopoles'

The next of Maxwell's equations has no formal name, and we refer to it as the 'no magnetic monopoles' law:

$$\nabla \cdot \mathbf{B}(\mathbf{r}, t) = 0. \tag{2.38}$$

Again, we note that the divergence of a vector field is a measure of how much the vector field is created or destroyed at that point. The fact that the divergence of the magnetic field is always zero tells us that there are no sources of magnetic fields—magnetic field lines always form closed loops or stretch out to infinity. Two classic examples, illustrated in figure 2.5, are a long wire carrying a constant current and a bar magnet.

Ordinary magnets, including the Earth as a whole, have two magnetic poles, usually labeled north and south. The absence of sources of magnetic fields indicates that every magnet must have two equal and opposite poles, i.e. that isolated magnetic monopoles (single poles) do not exist. It is a common laboratory demonstration to cut a thin bar magnet in half, thus separating the two poles,

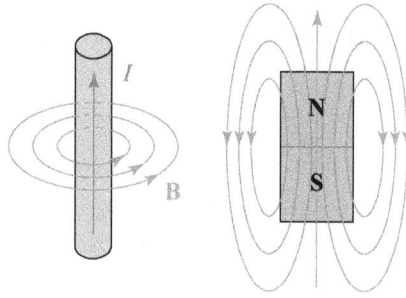

Figure 2.5. Rough illustrations of the magnetic field lines of a long wire and a bar magnet.

only to find that, in fact, there are now two new bar magnets, each possessing its own north and south poles.

No magnetic monopoles have ever been seen, but there have been active searches for them in high-energy physics experiments [3], where it is thought that monopoles might be massive particles that have an extremely short lifetime. The motivation for these searches can be traced back to a theoretical paper by Dirac published in 1931 [4], in which he showed that the existence of even a single magnetic monopole, combined with the rules of quantum physics, would require electric charge to be quantized. We know that electric charge is in fact quantized, with the charge of the electron being the smallest isolated charge possible, and so Dirac's theory has been considered a serious possible explanation for this quantization.

Though no fundamental magnetic monopoles have been discovered, it is worth noting that virtual monopoles have been created in the laboratory through a number of clever techniques. In 2015, researchers experimentally demonstrated a metamaterial structure that acts like a magnetic wormhole [5]; if one end of a bar magnet is placed in this wormhole, it is virtually moved to the other end of the wormhole, allowing separation of the poles from each other. This strategy of 'stretching' a magnet is curiously analogous to the theoretical method that Dirac used to derive his monopole results, and the idea of magnetic wormholes was introduced theoretically as a means of making a virtual monopole [6]. In condensed-matter physics, 'magnetic monopoles' have been predicted [7] and observed in exotic materials such as spin ice [8] and Bose–Einstein condensates [9]. These are monopoles of derived fields such as the H-field or other quantities and thus do not violate the 'no magnetic monopoles' rule, but they can provide insight into what a true magnetic monopole would look like.

An integral form of the 'no magnetic monopoles' rule can be derived by simply integrating equation (2.38) over a volume V and using Gauss's theorem to convert it to a surface integral. The result is

$$\oint_S \mathbf{B}(\mathbf{r}, t) \cdot d\mathbf{a} = 0. \tag{2.39}$$

This indicates that any magnetic field line that enters the surface must exit the surface at some other point. It is, in essence, another way of saying that magnetic field lines never begin or end in a finite volume.

2.3 Faraday's law

Though electricity and magnetism have been observed and studied for millennia, for most of that history they had been considered two separate phenomena. Electricity was associated with shocks on cold, dry days, with lightning, and with weak attractive forces caused by rubbing the right materials together. Magnetism was associated with bar magnets, lodestones, and compasses. Electricity was seen to come in positive and negative types, while magnetism was seen to always come in pairs of poles—a dipole—that could not be separated.

By the early 1800s, however, researchers began to suspect, and investigate, the possibility that there might be a link between electricity and magnetism. The crucial experiment was finally made by the Danish philosopher Hans Christian Ørsted in the year 1820. In his experiment, Ørsted placed an ordinary magnetic compass above a wire connected to a battery. He found that when a current was allowed to flow, the compass needle was deflected, thus showing that an electrical current can produce a magnetic field. This was the first definitive demonstration of a link between electricity and magnetism.

Incredibly, Ørsted performed this experiment for the first time in a crowded lecture hall, making it likely the only major discovery in history to have been made in front of an audience. It is often reported that Ørsted's discovery was accidental, but this is not correct. Ørsted was a proponent of a German philosophical movement called Naturphilosophie, which looked at all of nature as a unified whole. Within this philosophy, it was natural to look for connections between different phenomena in nature, and Ørsted was convinced that there must be a link between electricity and magnetism. He was originally planning to test his experimental apparatus before the lecture but ran out of time; when he did it before the audience, he saw a small twitch of the compass needle that heralded a new era of physics. Ørsted later recounted, however, that 'the experiment made no strong impression on the audience' [10], likely because they had no idea of the significance of what they were seeing.

With the recognition that electricity could produce magnetism, it was natural for scientists to ask whether the reverse was true: could magnetism produce electricity? In 1832, the English scientist Michael Faraday demonstrated that this is possible, though in an unexpected way [11]. Through an exceedingly thorough series of experiments, Faraday demonstrated that a *change* in a magnetic field over time produces a circulating electric field; in integral form, the law derived from Faraday's observation may be written as

$$\oint_C \mathbf{E}(\mathbf{r}, t) \cdot d\mathbf{r} = -\frac{\partial}{\partial t} \int_S \mathbf{B}(\mathbf{r}, t) \cdot d\mathbf{a}, \qquad (2.40)$$

where the left side of the integral is a path integral over a closed curve C with vector path element $d\mathbf{r}$ and the right side is an integral over an open surface S that is bounded by the curve C. The relationship between $d\mathbf{a}$ and the direction of integration over C is, of course, dictated by the right-hand rule.

The integral on the left is traditionally referred to as the *electromotive force*; thus Faraday's law may be stated as follows:

> The electromotive force around a closed path is the opposite of the time rate of change of magnetic flux through the path.

The term 'electromotive force' is a misnomer in modern times, as the quantity has dimensions of potential per charge and not force. The term originated in an era before the meaning of the words 'force' and 'energy' had solidified in the physics community.

The key experiment that demonstrated Faraday's law is illustrated in figure 2.6. Faraday created two wire helices, wound in the same sense and electrically insulated from each other, looped around the same block of wood. He attached one helix to a galvanometer (current meter) and the other to a battery. We quote Faraday for the result of his experiment:

> When the contact was made, there was a sudden and very slight effect at the galvanometer, and there was also a similar slight effect when the contact with the battery was broken. But whilst the voltaic current was continuing to pass through the one helix, no galvanometrical appearances of any effect like induction upon the other helix could be perceived,

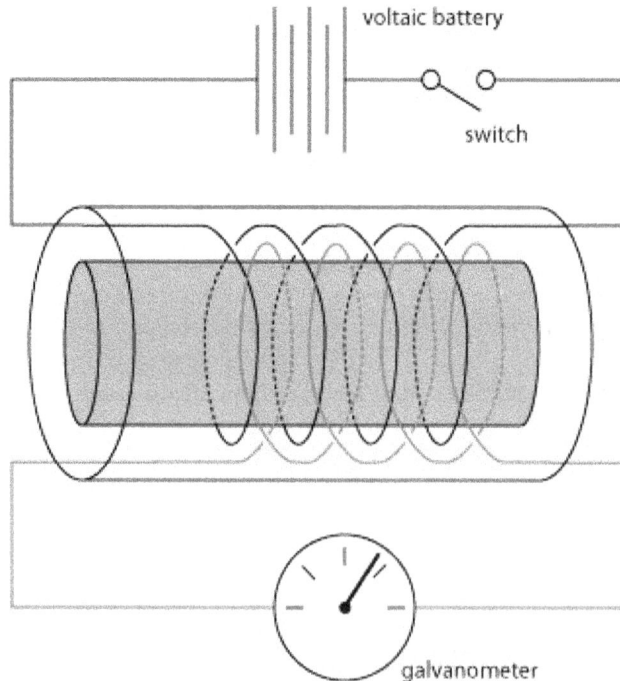

Figure 2.6. Faraday's key experiment, showing how an electric field can be induced in one coil of wire by a changing magnetic field in another coil.

although the active power of the battery was proved to be great, by its heating the whole of its own helix, and by the brilliancy of the discharge when made through charcoal.

In other words, a constant magnetic field flowing through one helix produced no effect in the other helix, but a changing magnetic field flowing through the first helix induced a circulating electric field, and consequently a current, in the second.

It was later established that any current produced by the Faraday effect generates a magnetic field that *opposes* the change in the original magnetic field; this is often referred to as Lenz's Law, after the physicist Emil Lenz, who extensively studied the Faraday effect in 1834. This effect forms the basis of operation of inductors in electrical circuits.

Faraday's law in integral form, given by equation (2.40), can be converted to differential form by the use of Stokes's theorem (equation (A.46)),

$$\oint_C \mathbf{E}(\mathbf{r},\, t) \cdot d\mathbf{r} = \int_S \nabla \times \mathbf{E}(\mathbf{r},\, t) \cdot d\mathbf{a}. \tag{2.41}$$

Using this formula to replace the path integral of Faraday's law with a surface integral, we have

$$\int_S \nabla \times \mathbf{E}(\mathbf{r},\, t) \cdot d\mathbf{a} = -\frac{\partial}{\partial t} \int_S \mathbf{B}(\mathbf{r},\, t) \cdot d\mathbf{a}. \tag{2.42}$$

We finally note that this equation must hold true regardless of the specific choice of the surface S. This can only happen if the integrands on both sides of the equation are the same, or

$$\nabla \times \mathbf{E}(\mathbf{r},\, t) = -\frac{\partial \mathbf{B}(\mathbf{r},\, t)}{\partial t}, \tag{2.43}$$

which is Faraday's law.

We know that the curl of a vector field at a point mathematically represents the amount of circulation the vector field has at that point. Faraday's law in differential form thus tells us that a time-varying magnetic field produces a circulating electric field; this circulating electric field can induce currents in current loops.

Faraday's discovery secured his reputation as one of the most important scientists of the 19th century and showed that there is a dynamic relationship between electricity and magnetism: interesting connections arise between the seemingly different phenomena when there are variations over time.

2.4 Ampère–Maxwell law

Ørsted's 1820 discovery that an electric current produces magnetic fields led others to attempt to quantify this effect. In 1823, the French physicist André-Marie Ampère found a mathematical formula that describes the force between two current-carrying wires, which is often known as Ampère's force law. What later became known as *Ampère's circuital law*, or simply *Ampère's law*, was curiously not derived by him, but by James Clerk Maxwell in 1861 [12]; it has apparently been named in honor of Ampère's contributions to the field.

In integral form, Ampère's law is the magnetic analogy to Gauss's law: it states that the integrated magnetic flux around a closed loop C is proportional to the current I_{enc} passing through the surface S bounded by C,

$$\oint_C \mathbf{B}(\mathbf{r}) \cdot d\mathbf{r} = \mu_0 I_{enc}. \tag{2.44}$$

Much as Gauss's law can be used to determine the electric field in systems with high symmetry, Ampère's law can be used to determine the magnetic field in systems with high symmetry through the use of an 'Ampèrian loop.'

We can convert Ampère's law to a differential form by first noting that we may write the enclosed current as

$$I_{enc} = \int_S \mathbf{J}(\mathbf{r}) \cdot d\mathbf{a}, \tag{2.45}$$

where $\mathbf{J}(\mathbf{r})$ is the current density. We then have

$$\oint_C \mathbf{B}(\mathbf{r}) \cdot d\mathbf{r} = \mu_0 \int_S \mathbf{J}(\mathbf{r}) \cdot d\mathbf{a}. \tag{2.46}$$

We now use Stokes's theorem to rewrite the left-hand side of this equation as a surface integral,

$$\oint_C \mathbf{B}(\mathbf{r}) \cdot d\mathbf{r} = \int_S \nabla \times \mathbf{B}(\mathbf{r}) \cdot d\mathbf{a}, \tag{2.47}$$

and upon using this, equation (2.46) takes on the form

$$\int_S \nabla \times \mathbf{B}(\mathbf{r}, t) \cdot d\mathbf{a} = \mu_0 \int_S \mathbf{J}(\mathbf{r}) \cdot d\mathbf{a}. \tag{2.48}$$

Again, we argue that this equation must be independent of the choice of integration domain, which means the integrands must be equal, or

$$\nabla \times \mathbf{B}(\mathbf{r}) = \mu_0 \mathbf{J}(\mathbf{r}). \tag{2.49}$$

This is the differential form of Ampère's law. It should be noted that we have implicitly only considered static (i.e. time-independent) fields and currents, which was historically the only case that was considered.

But when one looks closely, one finds that there are mathematical and physical inconsistencies in Ampère's law. For example, let us take the divergence of equation (2.49); because the divergence of a curl is always zero, the equation reduces to

$$\nabla \cdot \mathbf{J}(\mathbf{r}) = 0. \tag{2.50}$$

This is not necessarily wrong, but it is puzzling; in principle, we have great freedom to design current distributions, and there does not seem to be an obvious reason why the current density must always be free of divergence.

More troubling: let us consider the simple electrical circuit of figure 2.7(a), in which a battery is being used to charge a capacitor. With the Ampère loop shown,

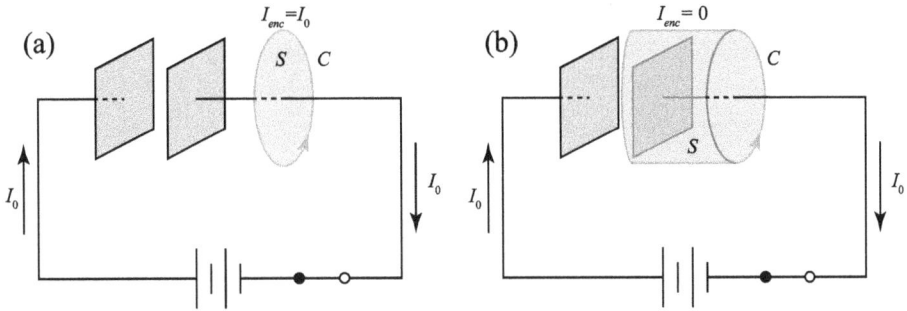

Figure 2.7. Illustration of two contradictory results obtained using (a) an Ampère loop with the current enclosed and (b) an Ampère loop with the current excluded.

the enclosed current is nonzero and therefore the integrated magnetic field on the loop is nonzero.

But in the case of a surface integral such as that in Ampère's law, we are free to distort the shape of the surface as long as the bounding path remains unchanged. This means that we may also use the surface of figure 2.7(b), i.e. a surface that does not intersect the wire at all; in this case, the enclosed current is zero and the integrated magnetic field on the loop is zero.

Clearly, both of these conclusions cannot be true, which suggests that something is missing from Ampère's law. To see what is missing, let us return and look more closely at the quantity $\nabla \cdot \mathbf{J}(\mathbf{r})$. First, we note that the dimensions of $\mathbf{J}$ are of the form

$$\mathbf{J} = \frac{\text{vector current}}{\text{unit area}} = \frac{\text{Coulomb}}{\text{s} \cdot \text{m}^2}. \tag{2.51}$$

If we integrate this current density over a surface, we get the total current I passing through the surface,

$$I = \int_S \mathbf{J}(\mathbf{r},\, t) \cdot d\mathbf{a}. \tag{2.52}$$

Let us consider a closed surface S, as illustrated in figure 2.8. Because S is a closed surface, we may use Gauss's theorem to convert the surface integral to a volume integral, so that

$$I = \int_V \nabla \cdot \mathbf{J}(\mathbf{r},\, t) d^3r. \tag{2.53}$$

Let us assume that electric charge is conserved. (Of course, we all know it is conserved, but as far as our theoretical calculations are concerned, it is an assumption.) The total current I represents the rate at which charge leaves the volume. If the total charge in the volume is Q, then we may write

$$\int_V \nabla \cdot \mathbf{J}(\mathbf{r},\, t) d^3r = -\frac{\partial Q(t)}{\partial t}. \tag{2.54}$$

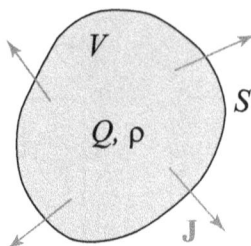

Figure 2.8. Illustration of the volume, its bounding surface, and the charge and current densities contained within it.

We may also write the total charge Q as

$$Q(t) = \int_V \rho(\mathbf{r}, t) d^3 r. \tag{2.55}$$

On substituting this expression into equation (2.54), we obtain

$$\int_V \nabla \cdot \mathbf{J}(\mathbf{r}, t) d^3 r = -\int_V \frac{\partial \rho(\mathbf{r}, t)}{\partial t} d^3 r. \tag{2.56}$$

Again, such an integral relation is independent of the choice of V, which means the integrands must be equal. We arrive at the result

$$\nabla \cdot \mathbf{J}(\mathbf{r}, t) = -\frac{\partial \rho(\mathbf{r}, t)}{\partial t}. \tag{2.57}$$

This is known as the *continuity equation,* and it is a mathematical representation of charge conservation. We again recall that the divergence represents the amount of vector field being created or destroyed at a given point; the continuity equation says that a net flow of current from a point must represent a net change in the charge at that point.

We may use this continuity equation to determine what is missing from Ampère's law. We solve equation (2.11), Gauss's law, for $\rho(\mathbf{r}, t)$ and then substitute it into equation (2.57). We then have

$$\nabla \cdot \mathbf{J}(\mathbf{r}, t) = -\frac{\partial}{\partial t}[\varepsilon_0 \nabla \cdot \mathbf{E}(\mathbf{r}, t)]. \tag{2.58}$$

We interchange the order of differentiation on the right-hand side of this expression (there is no obvious problem with doing so) and get the expression

$$\nabla \cdot \mathbf{J}(\mathbf{r}, t) = \nabla \cdot \left[-\varepsilon_0 \frac{\partial \mathbf{E}(\mathbf{r}, t)}{\partial t} \right]. \tag{2.59}$$

We may move the right-hand term of the equation to the left side and group together the terms under the same divergence,

$$\nabla \cdot \left[\mathbf{J}(\mathbf{r}, \, t) + \varepsilon_0 \frac{\partial \mathbf{E}(\mathbf{r}, \, t)}{\partial t} \right] = 0. \tag{2.60}$$

We recall from equation (2.50) that one of the concerns with Ampère's law is the zero divergence of $\mathbf{J}$. Let us replace the $\mathbf{J}$ in Ampère's law with the zero divergence term of equation (2.60). If we do so, the new form of Ampère's law is

$$\nabla \times \mathbf{B}(\mathbf{r}, \, t) = \mu_0 \mathbf{J}(\mathbf{r}, \, t) + \mu_0 \varepsilon_0 \frac{\partial \mathbf{E}(\mathbf{r}, \, t)}{\partial t}. \tag{2.61}$$

This equation is the vacuum version of what is known as the Ampère–Maxwell law, which was first derived by James Clerk Maxwell in the 1860s. It introduces a new term into Ampère's law that serves the same mathematical purpose as an electric current and is therefore called the *displacement current*.

$$\mathbf{J}_{disp}(\mathbf{r}, \, t) \equiv \varepsilon_0 \frac{\partial \mathbf{E}(\mathbf{r}, \, t)}{\partial t}. \tag{2.62}$$

The displacement current depends on the time derivative of $\mathbf{E}(\mathbf{r}, \, t)$; in the case of a static system, this current is zero, and then the Ampère–Maxwell law reduces to Ampère's law.

The introduction of the displacement current resolves the paradox illustrated by figure 2.7. While the capacitor is charging, the electric field between the capacitor plates is increasing over time. This means that there is a displacement current passing between the plates, and further calculation can show that this displacement current is equal to the actual current flowing in the wire. Therefore, the two choices of surface give the same result.

If we compare the Ampère–Maxwell law, equation (2.61), to Faraday's law, equation (2.43), we find that there is a striking similarity and complementarity between them. Faraday's law shows that a time-varying magnetic field creates a circulating electric field, and the Ampère–Maxwell law shows that a time-varying electric field creates a circulating magnetic field. This, in turn, suggests that creating a time-varying electric field induces a time-varying magnetic field, and vice versa. If we excite a time-varying field, say with an oscillating electric current, then the fields continue to maintain each other. This is the origin of electromagnetic waves: mutually maintained electric and magnetic fields.

We begin our discussion of these electromagnetic waves properly in the next chapter. Before doing so, however, it is worth reviewing a few simple applications of Ampère's law to gain familiarity with the properties of static magnetic fields. These calculations will also be useful in later chapters.

2.4.1 Magnetic field of a thick wire

Just as we use Gauss's law to determine the electric field of static charge distributions of sufficient symmetry, we may use Ampère's law to determine the magnetic field of static current distributions of sufficient symmetry. We consider a thick straight wire of radius a and infinite length, with a uniform current I passing through it, as illustrated in figure 2.9.

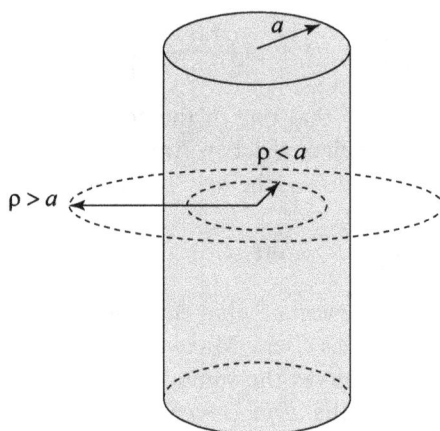

Figure 2.9. Geometry for determining the magnetic field inside and outside a thick wire.

With magnetic fields, taking advantage of the symmetry of the system requires a bit more thought. The current flows entirely in the $\hat{\mathbf{z}}$-direction, and the system is rotationally symmetric around the z-axis. This implies that the magnetic field must be constant on a circular path centered on the axis of the wire. We thus choose such a circular path for our Ampère loop, with the path of integration along $+\hat{\phi}$. The path element is $d\mathbf{l} = \rho \, d\phi \hat{\phi}$, and for the path completely encircling the wire, we have

$$\oint_C \mathbf{B}(\mathbf{r}) \cdot d\mathbf{r} = 2\pi \rho B(\rho) = \mu_0 I, \tag{2.63}$$

or

$$B_{out}(\rho) = \frac{\mu_0 I}{2\pi\rho}. \tag{2.64}$$

Inside the wire, the enclosed current is determined by

$$I_{\text{enc}} = \frac{I}{\pi a^2} (\pi \rho^2), \tag{2.65}$$

and the magnetic field in the wire is

$$B_{in}(\rho) = \frac{\mu_0 I \rho}{2\pi a^2}. \tag{2.66}$$

This field, from our calculation, circulates in the $+\hat{\phi}$ direction if the current I is positive.

2.4.2 Magnetic field of a current sheet

Next, let us calculate the magnetic field of an infinite sheet of uniform current. We assume that the sheet is in the $z = 0$ plane, and the surface current density travels in the $\hat{\mathbf{y}}$-direction, i.e. $\sigma = \sigma_0 \hat{\mathbf{y}}$. The geometry is illustrated in figure 2.10.

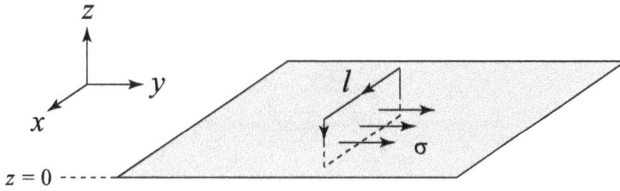

Figure 2.10. Geometry for an infinite sheet of uniform electric current.

We first consider the natural symmetry of the system. If we treat the current sheet as an infinite set of parallel wires running along $\hat{\mathbf{y}}$, we expect that all the vertical ($\hat{\mathbf{z}}$) components of the magnetic field of adjacent wires cancel out. We therefore expect that the magnetic field travels only in the $\hat{\mathbf{x}}$ direction, $+\hat{\mathbf{x}}$ above the sheet and $-\hat{\mathbf{x}}$ below the sheet, according to the right-hand rule. At a constant z, the field should be constant as well. We therefore create a loop to integrate over, as shown in the figure. With the assumed field directions, the vertical components of the path contribute nothing, and Ampère's law takes the form

$$2Bl = \mu_0 I_{\text{enc}} = \mu_0 l \sigma. \tag{2.67}$$

We may readily solve for

$$B = \frac{\mu_0 \sigma}{2}, \tag{2.68}$$

with a direction of $+\hat{\mathbf{x}}$ for $z > 0$ and $-\hat{\mathbf{x}}$ for $z < 0$.

2.4.3 Magnetic field of a solenoid

There is one more example worth considering: the magnetic field produced by an infinite solenoid of radius a. A solenoid is a cylinder with a single continuous wire wound around its circumference from one end to the other. We assume that there are N turns per unit length in the axial direction of the solenoid. If the wires are sufficiently thin and dense, we may approximate the current by a pure azimuthal surface current density $\mathbf{K} = NI\hat{\phi}$, where I is the current running through the wire.

The symmetry argument here is a bit more subtle than in the previous examples. We may view our solenoid as an infinite stack of current loops. The magnetic field passes through the center of each loop and circles around the outside of it; with an increasing number of stacked loops, however, this circulation is pushed further and further away, and in the limit of an infinite solenoid, the magnetic field outside must be zero. Inside, we expect that the field can only point in the $\hat{\mathbf{z}}$ direction, due to the symmetry of the system and the fact that there is nowhere for a field to 'go' to form a closed loop if it points in any other direction.

We therefore choose an Ampèrian loop as shown in figure 2.11. The loop penetrates a distance $w/2$ outside and inside the solenoid and has a length h along the length of the solenoid. Because there is no field in the azimuthal or radial directions, the w part of the loop contributes nothing to the integral; also, there is no

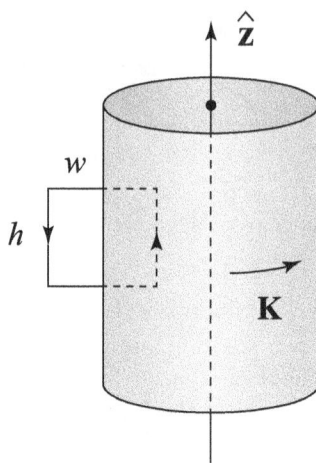

Figure 2.11. Geometry for an infinite solenoid.

magnetic field outside the loop, so only the $\hat{\mathbf{z}}$ component of the field within the solenoid contributes. Ampère's law then reduces to the form

$$B_z h = \mu_0 NIh, \tag{2.69}$$

or we find that the field inside the solenoid is

$$\mathbf{B} = \mu_0 NI\hat{\mathbf{z}}. \tag{2.70}$$

The examples we have considered in this chapter have been for electrostatic and magnetostatic fields, though it is to be noted that, in certain circumstances, it is possible to solve time-varying problems using an Ampèrian loop for Faraday's law and the Ampère–Maxwell law.

2.5 Exercises

These exercises are mostly a review of electrostatics and magnetostatics problems, to get the reader ready for more complicated manipulations of fields.

1. Suppose there is a region containing an unknown constant electric and magnetic field. We perform three experiments with a test charge of 1 C moving at $1\ \mathrm{m\ s^{-1}}$ passing through the region, and determine the following resultant forces: (i) $\mathbf{f} = 3\hat{\mathbf{x}} - \hat{\mathbf{y}} + 2\hat{\mathbf{z}}$ when $\mathbf{v} = \hat{\mathbf{x}}$, (ii) $\mathbf{f} = 2\hat{\mathbf{x}} - 2\hat{\mathbf{y}} - \hat{\mathbf{z}}$ when $\mathbf{v} = \hat{\mathbf{y}}$, and (iii) $\mathbf{f} = 2\hat{\mathbf{x}} + \hat{\mathbf{z}}$ when $\mathbf{v} = \hat{\mathbf{z}}$. What are the electric and magnetic fields $\mathbf{E}$ and $\mathbf{B}$ causing these forces? IMPORTANT: we are talking about the same fields for each case, with the charges being sent through the fields in different directions. In short: using the Lorentz force law, you should be able to get nine self-consistent equations for the six fields.

2. Suppose there is a region containing an unknown constant electric and magnetic field. We perform three experiments with a test charge of 1 C moving at $1\ \mathrm{m\ s^{-1}}$ passing through the region, and determine the following resultant forces: (i) $\mathbf{f} = \hat{\mathbf{x}} + \hat{\mathbf{z}}$ when $\mathbf{v} = \hat{\mathbf{x}}$, (ii) $\mathbf{f} = \hat{\mathbf{x}}$ when $\mathbf{v} = \hat{\mathbf{y}}$, and

(iii) $\mathbf{f} = \hat{\mathbf{x}} + \hat{\mathbf{y}} + \hat{\mathbf{z}}$ when $\mathbf{v} = \hat{\mathbf{z}}$. What are the electric and magnetic fields $\mathbf{E}$ and $\mathbf{B}$ causing these forces? IMPORTANT: we are talking about the same fields for each case, with the charges being sent through the same fields in different directions. In short: using the Lorentz force law, you should be able to get nine self-consistent equations for the six fields.

3. Consider a pair of infinite sheets of electric charge of thickness d, one with volume charge density $+\rho$ and one with volume charge density $-\rho$, arranged as shown in figure 2.12. Determine the electric field for all values of z. NOTE: you should determine the electric field of each sheet separately, then determine the total electric field piecewise in each region.

4. A sphere of radius a has a charge density given by $\rho(\mathbf{r}) = Br^3$, with B a constant, for $r < a$, and no charge outside. Find the electric field both inside and outside the sphere, as a function of r, using Gauss's law.

5. A sphere with inner radius a and outer radius b has a charge density given by $\rho(\mathbf{r}) = A/r$, with A a constant, for $a \leqslant r \leqslant b$, and no charge elsewhere. Find the electric field both inside and outside the sphere as a function of r.

6. An infinitely long hollow cylinder with an inner radius of a and an outer radius of b possesses a total charge per unit length of Λ within these two boundaries. Calculate the electric field everywhere inside and outside the cylinder as a function of distance ρ from the axis.

7. Let us suppose we have an infinite slab of current of total thickness d, centered on the plane $z = 0$. The current density is constant within the slab and has the value $\mathbf{J} = J_0\hat{\mathbf{x}}$. Find the magnetic field everywhere inside and outside the slab.

8. Let us suppose that we have a solenoid of radius a in which the current is increasing linearly, i.e. $I(t) = Ct$. First use Ampère's law to find the magnetic field within the solenoid (assuming a quasistatic approximation where the increase of current is so slow that Ampère's law can still be used). Now, with this time-varying magnetic field, use Faraday's law to determine the electric field induced within the wires of the solenoid. Finally, assuming that the vector current induced in the wire by an electric field is $\mathbf{I}_{ind} = \sigma\mathbf{E}$, where σ is a conductivity, determine the induced current. (This is a very rough illustration of how an inductor works.)

9. In a coaxial metal waveguide, waves propagate between cylindrical metal surfaces with radii a and b, with $b > a$. The magnetic field of a monochromatic wave propagating in such a waveguide is

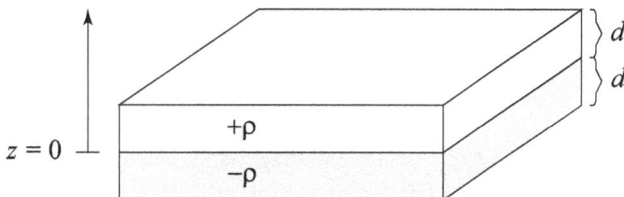

Figure 2.12. Geometry for an infinite sheet of uniform electric current.

$$\mathbf{B}(x, y, z) = \hat{\phi}\frac{A}{cr} \cos(kz - \omega t),$$

where ω is the frequency of the wave and $k = \omega/c$ is the wavenumber. Use the Ampère–Maxwell law to determine the electric field, and then use Faraday's law to confirm that this electric field is consistent with Maxwell's equations.

10. We can roughly demonstrate that the displacement current resolves the Ampère's law paradox of figure 2.7. Assume that the current in the wire is of the form $I(t) = I_0 \exp[-t/\sigma]$, where σ is a characteristic charging time. (a) Assuming the capacitor is uncharged at $t = 0$, determine the charge on it at time t. (b) Assuming that the capacitor plates are circular and of area A and that the plates are close enough together to approximate the fields between them by infinite sheets of charge, calculate the electric field between the plates. (c) Show that the Ampère–Maxwell law gives the same result if the surface is a circle of area A between the capacitor plates or surrounding the charging wire of the capacitor.

11. How would Maxwell's equations change if a magnetic charge density ρ_m were to exist? Assume that the 'no magnetic monopoles' law becomes the 'yes, magnetic monopoles!' law,

$$\nabla \cdot \mathbf{B}(\mathbf{r}, t) = \mu_0 \rho_m(\mathbf{r}, t).$$

Assuming that $\rho_m(\mathbf{r}, t)$ satisfies a continuity equation with a magnetic current density $\mathbf{J}_m(\mathbf{r}, t)$, determine how Faraday's law would change due to the presence of magnetic charge. (Hint: add an unknown vector to Faraday's law and take the divergence; figure out from this what the vector must be.) What are the SI units of magnetic charge in this formalism?

References

[1] de Coulomb C A 1785 Premier mémoire sur l'électricité et le magnétisme *Histoire de l'Académie R. Sci.* 569–77

[2] Faraday M 1852 Experimental researches in electricity *Twenty-Eighth Series. Phil. Trans. R. Soc. Lond.* **142** 25–56

[3] Milton K A 2006 Theoretical and experimental status of magnetic monopoles *Rep. Prog. Phys.* **69** 1637–711

[4] Dirac P A M 1931 Quantised singularities in the electromagnetic field *Proc R. Soc.* A **133** 60–72

[5] Prat-Camps J, Navau C and Sanchez A 2015 A magnetic wormhole *Sci. Rep.* **5** 12488

[6] Greenleaf A, Kurylev Y, Lassas M and Uhlmann G 2007 Electromagnetic wormholes and virtual magnetic monopoles from metamaterials *Phys. Rev. Lett.* **99** 183901

[7] Castelnovo C, Moessner R and Sondhi S 2008 Magnetic monopoles in spin ice *Nature* **451** 42–5

[8] Morris D J P *et al* 2009 Dirac strings and magnetic monopoles in the spin ice $Dy_2Ti_2O_7$ *Science* **326** 411–4

[9] Ray M, Ruokokoski E, Kandel S, Möttönen M and Hall D S 2014 Observation of Dirac monopoles in a synthetic magnetic field *Nature* **505** 657–60

[10] Oersted H C 1832 Thermo-electricity *The Edinburgh Encyclopedia* **vol 17** ed D Brewster (Philadelphia, PA: Joseph Parker) pp 715–32

[11] Faraday M 1832 Experimental researches in electricity *Phil. Trans. R. Soc. Lond.* **122** 125–62

[12] Maxwell J C 1861 XXV. On physical lines of force *London, Edinburgh Dublin, Phil. Mag. J. Sci.* **21** 161–75

Chapter 3

Electromagnetic waves

3.1 The wave equation

To explore the properties of electromagnetic waves, we now simplify the vacuum forms of Maxwell's equations further and assume that the charge density $\rho(\mathbf{r}, t) = 0$ and the current density $\mathbf{J}(\mathbf{r}, t) = 0$. We look for nontrivial solutions of Maxwell's equations in the absence of source terms because one of the defining characteristics of waves is their ability to propagate beyond the region where they are excited. Under these conditions, Maxwell's equations may be written as

$$\nabla \cdot \mathbf{E}(\mathbf{r}, t) = 0, \quad \text{(Gauss's law)} \tag{3.1}$$

$$\nabla \cdot \mathbf{B}(\mathbf{r}, t) = 0, \quad \text{('No magnetic monopoles')} \tag{3.2}$$

$$\nabla \times \mathbf{E}(\mathbf{r}, t) = -\frac{\partial \mathbf{B}(\mathbf{r}, t)}{\partial t}, \quad \text{(Faraday's law)} \tag{3.3}$$

$$\nabla \times \mathbf{B}(\mathbf{r}, t) = \mu_0 \varepsilon_0 \frac{\partial \mathbf{E}(\mathbf{r}, t)}{\partial t}. \quad \text{(Ampère–Maxwell law).} \tag{3.4}$$

In this form, Maxwell's equations are coupled (i.e. they mix $\mathbf{E}$ and $\mathbf{B}$) and it is difficult to see the nature of the solutions at a glance. We therefore attempt to combine the equations in such a way as to find a single equation for the electric or magnetic field. To do this, we first take the curl of equation (3.3),

$$\nabla \times [\nabla \times \mathbf{E}(\mathbf{r}, t)] = \nabla \times \left[\frac{\partial \mathbf{B}(\mathbf{r}, t)}{\partial t} \right] = -\frac{\partial}{\partial t} [\nabla \times \mathbf{B}(\mathbf{r}, t)]. \tag{3.5}$$

We have changed the order of the derivatives in the last term and may now substitute from equation (3.4) into this one. We then have

doi:10.1088/978-0-7503-6064-7ch3

$$\nabla \times [\nabla \times \mathbf{E}(\mathbf{r},\, t)] = -\mu_0 \varepsilon_0 \frac{\partial^2 \mathbf{E}(\mathbf{r},\, t)}{\partial t^2}. \tag{3.6}$$

We next use the familiar vector identity (equation (A.29)),

$$\nabla \times (\nabla \times \mathbf{A}) = \nabla(\nabla \cdot \mathbf{A}) - \nabla^2 \mathbf{A}, \tag{3.7}$$

to simplify the left-hand side of equation (3.6). The result has the form

$$\nabla[\nabla \cdot \mathbf{E}(\mathbf{r},\, t)] - \nabla^2 \mathbf{E}(\mathbf{r},\, t) = -\mu_0 \varepsilon_0 \frac{\partial^2 \mathbf{E}(\mathbf{r},\, t)}{\partial t^2}. \tag{3.8}$$

We note that the leftmost term of the above equation is zero, according to equation (3.1). With some slight rearrangement, we finally obtain

$$\nabla^2 \mathbf{E}(\mathbf{r},\, t) - \mu_0 \varepsilon_0 \frac{\partial^2 \mathbf{E}(\mathbf{r},\, t)}{\partial t^2} = 0. \tag{3.9}$$

This is a three-dimensional vector wave equation for the electric field; each component of the electric field satisfies an identical three-dimensional scalar wave equation of the form

$$\nabla^2 U(\mathbf{r},\, t) - \frac{1}{c^2} \frac{\partial^2 U(\mathbf{r},\, t)}{\partial t^2} = 0, \tag{3.10}$$

where c represents the speed of the wave. On comparison with equation (3.9), we find that the speed of electromagnetic waves is given by

$$c = \frac{1}{\sqrt{\varepsilon_0 \mu_0}} = 3 \times 10^8 \text{ m s}^{-1}, \tag{3.11}$$

which is the vacuum speed of light.

Maxwell first derived this wave equation in 1865 [1]. However, he had already estimated the speed of electromagnetic waves several years earlier using a hydro-dynamical analysis of his equations [2], finding the same result as equation (3.11). He concluded at that early stage,

> …we can scarcely avoid the inference that light consists in the transverse undulations of the same medium which is the cause of electric and magnetic phenomena.

A similar derivation may be performed for the magnetic field. We begin by taking the curl of equation (3.4) and then apply equation (3.2). We end up with the expression

$$\nabla^2 \mathbf{B}(\mathbf{r},\, t) - \mu_0 \varepsilon_0 \frac{\partial^2 \mathbf{B}(\mathbf{r},\, t)}{\partial t^2}, \tag{3.12}$$

which shows that the magnetic field also satisfies a vector wave equation.

Maxwell's work was theoretical. He showed that electromagnetic waves exist and that they propagate at the speed of light, suggesting that light is an electromagnetic wave. However, experimental confirmation was needed for all of these hypotheses. The first steps, showing that electromagnetic waves exist and that they propagate at the speed of light, were undertaken by the German physicist Heinrich Hertz between 1886 and 1889. He created the first artificial radio waves, with a wavelength of 9.3 meters, and reflected them from a mirror to produce standing waves. He knew the frequency ν of the waves and could measure the wavelength λ, which allowed him to determine the speed v of the waves using the expression $v = \lambda\nu$. He indeed found that $v = c$, confirming most of Maxwell's predictions; few doubted that light was an electromagnetic wave after that. (We will discuss the experimental confirmation that light is an electromagnetic wave later in this book.)

3.2 Solutions of the wave equation

It is one thing to demonstrate that electric and magnetic fields satisfy a wave equation; it is another thing entirely to find solutions to those wave equations; we now look for those solutions. We first try to find a simple solution to the wave equation, which is known as a plane wave, and we then go further and show that all free-propagating electromagnetic waves can be constructed from a combination of plane waves.

For the simple solution, let us make an educated guess and look for solutions $U(z, t)$ of equation (3.10) that depend only on time and one spatial variable, which we take to be z in a Cartesian coordinate system. (We will worry about the vector part of the wave equation in a little while.) Then, for this special case, we find that the scalar wave equation reduces to the form

$$\frac{\partial^2 U(z, t)}{\partial z^2} - \frac{1}{c^2}\frac{\partial^2 U(\mathbf{r}, t)}{\partial t^2} = 0. \tag{3.13}$$

Let us go further and assume that the field oscillates at a single frequency ω, so that its time dependence can be written in a complex exponential form,

$$U(z, t) = U(z)e^{-i\omega t}, \tag{3.14}$$

where $U(z)$ is now generally a complex function. Although physical electromagnetic waves have real values, we have adopted a complex solution for mathematical convenience. Once we have a complete solution, we can always take the real part of our result to get a real-valued quantity. This is an approach we will use quite regularly throughout the book.

On substituting our complex plane wave into equation (3.13), we find that our equation reduces to the ordinary differential equation

$$\frac{d^2 U(z)}{dz^2} + k^2 U(z) = 0, \tag{3.15}$$

where we have divided out a common term $\exp[-i\omega t]$ and have also defined $k \equiv \omega/c$ as the wavenumber of the wave. But this is the simple harmonic oscillator equation, and we can write the solution in complex form as

$$U(z) = A_+ e^{ikz} + A_- e^{-ikz}, \tag{3.16}$$

where A_+ and A_- are constants determined by the specific conditions of the problem. Our solution to the wave equation may thus be written as

$$U(z, t) = A_+ e^{i(kz-\omega t)} + A_- e^{-i(kz+\omega t)}. \tag{3.17}$$

This is an example of a *plane wave* solution to the wave equation. We see that the arguments of the exponentials are constant in every xy-plane of constant z value. These planes where the argument is constant are referred to as the *wave fronts* of the plane wave. As time passes, the wave fronts of the A_+ exponential travel in the positive z-direction at a speed $v = \omega/k = c$, and the wave fronts of the A_- exponential travel in the negative z-direction at speed c. These two solutions are independent, as we are free to independently choose the values of A_+ and A_-; we will often keep only the positive-going plane wave solution and write

$$U(z, t) = A e^{i(kz-\omega t)}. \tag{3.18}$$

The preceding calculation shows that a plane wave is one type of solution for the wave equation; next, we show that every solution of a vector wave equation in unbounded space can be written in terms of plane waves. We try a slightly unconventional approach and apply the three-dimensional spatial Fourier transform to equation (3.9); this transform is defined as

$$\tilde{\mathbf{E}}(\mathbf{K}, t) = \frac{1}{(2\pi)^3} \int \mathbf{E}(\mathbf{r}, t) e^{-i\mathbf{K}\cdot\mathbf{r}} d^3r, \tag{3.19}$$

where d^3r is the infinitesimal volume element and the integral is over all of three-dimensional space. The corresponding inverse transform has the form

$$\mathbf{E}(\mathbf{r}, t) = \int \tilde{\mathbf{E}}(\mathbf{K}, t) e^{i\mathbf{K}\cdot\mathbf{r}} d^3K, \tag{3.20}$$

where the integral is now over the infinite volume of spatial frequency space and d^3K is the infinitesimal spatial frequency volume element.

We take the Fourier transform of both sides of equation (3.9); for the leftmost term, this requires simplifying the integral of

$$\frac{1}{(2\pi)^3} \int \nabla^2 \mathbf{E}(\mathbf{r}, t) e^{-i\mathbf{K}\cdot\mathbf{r}} d^3r. \tag{3.21}$$

This may be done with a bit of vector calculus trickery. First, we note that we may write

$$\nabla \cdot [\nabla v(\mathbf{r}) e^{-i\mathbf{K}\cdot\mathbf{r}}] = \nabla^2 v(\mathbf{r}) e^{-i\mathbf{K}\cdot\mathbf{r}} - i\mathbf{K} \cdot \nabla v(\mathbf{r}) e^{-i\mathbf{K}\cdot\mathbf{r}}, \tag{3.22}$$

which is essentially the product rule applied to the divergence operation. We rewrite this as

$$\nabla^2 v(\mathbf{r}) e^{-i\mathbf{K}\cdot\mathbf{r}} = \nabla \cdot [\nabla v(\mathbf{r}) e^{-i\mathbf{K}\cdot\mathbf{r}}] + i\mathbf{K} \cdot \nabla v(\mathbf{r}) e^{-i\mathbf{K}\cdot\mathbf{r}}. \tag{3.23}$$

We can use this identity for each component of the electric field separately and substitute from it into expression (3.21). We then have

$$\frac{1}{(2\pi)^3} \int \nabla^2 \mathbf{E}(\mathbf{r},\,t) e^{-i\mathbf{K}\cdot\mathbf{r}} d^3 r = \frac{1}{(2\pi)^3} \int \nabla \cdot [\nabla \mathbf{E}(\mathbf{r},\,t) e^{-i\mathbf{K}\cdot\mathbf{r}}] d^3 r + \frac{i}{(2\pi)^3} \mathbf{K} \cdot \int \nabla \mathbf{E}(\mathbf{r},\,t) e^{-i\mathbf{K}\cdot\mathbf{r}} d^3 r. \quad (3.24)$$

As a result of doing the manipulation in this way, both integrals on the right-hand side now have a direct product of vector-like quantities in the form $\nabla \mathbf{E}(\mathbf{r},\,t)$. The resulting quantity is a dyadic or can equivalently be viewed as a matrix or a second-rank tensor. We will discuss dyadics in detail in chapter 14; for now, they will disappear by the end of our calculation.

The first integral on the right-hand side of equation (3.24) has an integrand which is, in effect, the divergence of a vector. We may therefore use Gauss's theorem (equation (A.45)) to convert it into a surface integral. Because we are integrating over all of three-dimensional space, the surface is of infinite size; if we assume that our field $\mathbf{E}(\mathbf{r},\,t)$ vanishes at infinity, as all physical fields should, then we may assume that the surface integral goes to zero. We are left with

$$\frac{1}{(2\pi)^3} \int \nabla^2 \mathbf{E}(\mathbf{r},\,t) e^{-i\mathbf{K}\cdot\mathbf{r}} d^3 r = \frac{i}{(2\pi)^3} \mathbf{K} \cdot \int \nabla \mathbf{E}(\mathbf{r},\,t) e^{-i\mathbf{K}\cdot\mathbf{r}} d^3 r. \quad (3.25)$$

We may then repeat the above process, now using the identity

$$\nabla[v(\mathbf{r}) e^{-i\mathbf{K}\cdot\mathbf{r}}] = \nabla v(\mathbf{r}) e^{-i\mathbf{K}\cdot\mathbf{r}} - i\mathbf{K} v(\mathbf{r}) e^{-i\mathbf{K}\cdot\mathbf{r}}. \quad (3.26)$$

The result, after similar manipulations, is

$$\frac{1}{(2\pi)^3} \int \nabla^2 \mathbf{E}(\mathbf{r},\,t) e^{-i\mathbf{K}\cdot\mathbf{r}} d^3 r = -|\mathbf{K}|^2 \frac{1}{(2\pi)^3} \int \mathbf{E}(\mathbf{r},\,t) e^{-i\mathbf{K}\cdot\mathbf{r}} d^3 r. \quad (3.27)$$

In short, the Fourier transform has replaced each ∇ in the wave equation by a factor of $i\mathbf{K}$. This is the usual effect of Fourier transforms on derivatives, and as a shorthand for three-dimensional spatial Fourier transforms, we can simply replace any ∇ in a transform by $i\mathbf{K}$ without explicitly doing all the math to prove it.

The Fourier transform of our wave equation therefore has the form

$$\frac{1}{c^2} \frac{\partial^2 \tilde{\mathbf{E}}(\mathbf{K},\,t)}{\partial t^2} + |\mathbf{K}|^2 \tilde{\mathbf{E}}(\mathbf{K},\,t) = 0. \quad (3.28)$$

This expression is simply the harmonic oscillator equation in time; if we define $\omega \equiv |\mathbf{K}|c$, we may make this more obvious and write the preceding expression as

$$\frac{\partial^2 \tilde{\mathbf{E}}(\mathbf{K},\,t)}{\partial t^2} + \omega^2 \tilde{\mathbf{E}}(\mathbf{K},\,t) = 0. \quad (3.29)$$

We express the solution of this equation in terms of complex exponentials,

$$\tilde{\mathbf{E}}(\mathbf{K},\,t) = \mathbf{A}(\mathbf{K}) e^{-i\omega t} + \mathbf{B}(\mathbf{K}) e^{i\omega t}, \quad (3.30)$$

where $\mathbf{A}(\mathbf{K})$ and $\mathbf{B}(\mathbf{K})$ are undetermined functions that depend on $\mathbf{K}$ but are constant with respect to t. These functions depend on the initial conditions that set up the wave; we do not concern ourselves with such specific solutions but focus on the behavior of the general solution.

We may substitute from equation (3.30) into our inverse Fourier transform, equation (3.20), to get an expression for the general solution to the wave equation in unbounded space. It has the form

$$\mathbf{E}(\mathbf{r},\, t) = \int [\mathbf{A}(\mathbf{K})e^{i[\mathbf{K}\cdot\mathbf{r}-\omega t]} + \mathbf{B}(\mathbf{K})e^{i[\mathbf{K}\cdot\mathbf{r}+\omega t]}]d^3K. \qquad (3.31)$$

Again, we have $\omega \equiv |\mathbf{K}|c$; this is an early example of what we will later refer to as a *dispersion relation*, which connects the wavenumber $K = |\mathbf{K}|$ of light to the angular frequency ω.

Before we discuss this solution further, we note again that electric fields are real-valued quantities, but our solution in equation (3.31) is complex-valued. It can be shown that the requirement that our solution is real-valued results in the condition $\mathbf{A}(-\mathbf{K}) = \mathbf{B}^*(\mathbf{K})$. However, it is very convenient to work with complex waves, so we will often use the complex representation with the understanding that we are to take the real part at the end of the calculation to get the 'physical' field.

The part of equation (3.31) that depends on $\mathbf{A}(\mathbf{K})$ represents a plane wave traveling in the direction $+\mathbf{K}$ at speed c. This can be seen because the wave is unchanging when $\mathbf{K} \cdot \mathbf{r} - \omega t$ is constant. Similarly, the part of the solution that depends on $\mathbf{B}(\mathbf{K})$ is a plane wave traveling in the direction $-\mathbf{K}$ at speed c. These two solutions are a bit redundant, as we are integrating over all positive and negative values of $\mathbf{K}$; generally, we may say that the solution to the electric field wave equation is the superposition of a set of plane waves of the form

$$\mathbf{E}_{\mathbf{K}}(\mathbf{r},\, t) = \mathbf{E}_0(\mathbf{K})e^{i[\mathbf{K}\cdot\mathbf{r}-\omega t]}, \qquad (3.32)$$

where $\mathbf{E}_0(\mathbf{K})$ is a vector that is constant with respect to $\mathbf{r}$ and t.

This lengthy derivation shows that plane waves are natural solutions to the electromagnetic wave equation and that any electromagnetic waves in unbounded space can be expressed in terms of a superposition of plane waves. Because of this, we will use plane waves extensively throughout the book as a simple model of a highly directional electromagnetic wave.

There is one significant limitation to our calculation: we assumed that our waves exist in an unbounded, empty three-dimensional space. This automatically excludes common and important situations such as waves at the interface between two media, where so-called inhomogeneous plane waves, or evanescent waves, can appear. Our solution is therefore not the most general solution for a light wave, but it is an excellent model when we are looking at waves propagating in a uniform medium far away from any boundaries. We discuss inhomogeneous plane waves at the end of the chapter.

3.3 Plane waves

Much of the time, we will study electromagnetic waves by considering plane waves of a single frequency ω, i.e. monochromatic waves. In such a case, we have a single wave vector $\mathbf{k}$ for which $|\mathbf{k}| = k = \omega/c$,

$$\mathbf{E_k}(\mathbf{r},\ t) = \mathbf{E}_0 e^{i[\mathbf{k}\cdot\mathbf{r}-\omega t]}. \tag{3.33}$$

The quantity $\mathbf{E}_0$ is now simply a constant complex vector. Again we note that physical fields are real-valued, so we work with this complex field with the understanding that we must take the real part of the equation at the end to get the physical result.

Waves of this form are known as 'plane waves' because the argument of the exponential (the phase), and therefore the entire function, is constant on planar surfaces such that $\mathbf{k}\cdot\mathbf{r} - \omega t = $ constant, i.e. the wave fronts. This idea is illustrated in figure 3.1. To visualize plane waves, we often draw those surfaces where the phase is equal to a multiple of 2π, i.e. $\mathbf{k}\cdot\mathbf{r} - \omega t = 2\pi m$, where m is an integer.

Curiously, though equation (3.33) represents a solution to the vector wave equation, it is not necessarily a solution of Maxwell's equations. To derive the vector wave equation, we combined several of Maxwell's equations together, and the result does not include all the constraints of Maxwell's original formulas.

We can readily see this by plugging our plane wave solution into Gauss's law, equation (3.1). We then have

$$\nabla \cdot \mathbf{E_k}(\mathbf{r},\ t) = \mathbf{E}_0 \cdot \nabla e^{i[\mathbf{k}\cdot\mathbf{r}-\omega t]} = 0. \tag{3.34}$$

If we take the derivative and divide out the nonzero complex exponential, we find that

$$\mathbf{k} \cdot \mathbf{E}_0 = 0. \tag{3.35}$$

This expression indicates that the electric field vector $\mathbf{E}_0$ is perpendicular to the wave vector and direction of propagation $\mathbf{k}$; in other words, the electric field is *transverse* to the direction of wave propagation. This is the origin of the often-repeated statement that 'light is a transverse wave,' though we will note momentarily that this is not entirely accurate.

Next, we substitute our plane wave solution into Faraday's law, equation (3.3), to get

$$\nabla \times \mathbf{E_k}(\mathbf{r},\ t) = i\mathbf{k} \times \mathbf{E}_0 e^{i[\mathbf{k}\cdot\mathbf{r}-\omega t]} = -\frac{\partial \mathbf{B}(\mathbf{r},\ t)}{\partial t}. \tag{3.36}$$

It should be noted that the time derivative of a complex exponential $\exp[-i\omega t]$ is proportional to that same complex exponential; in order for Faraday's law to be satisfied, it is necessary for the magnetic field to have the same functional form:

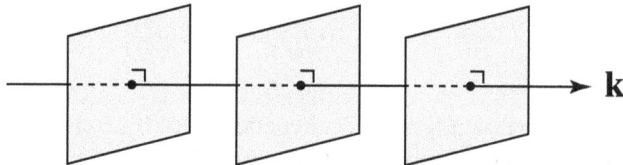

Figure 3.1. Illustration of surfaces of constant phase for a plane wave. The wave vector $\mathbf{k}$ is perpendicular to the surfaces.

$$\mathbf{B_k}(\mathbf{r},\, t) = \mathbf{B}_0 e^{i[\mathbf{k}\cdot\mathbf{r}-\omega t]}. \tag{3.37}$$

We substitute this into the above expression, take the derivative, and quickly find the relation

$$\mathbf{B}_0 = \frac{1}{\omega}\mathbf{k} \times \mathbf{E}_0. \tag{3.38}$$

Because the cross product of two vectors is automatically perpendicular to both of them, we see that the magnetic field vector is perpendicular to both the direction of propagation and the electric field vector; the magnetic field is also transverse. Because of this observation, we don't have to plug our expression for $\mathbf{B_k}(\mathbf{r},\, t)$ into the 'no magnetic monopoles' law, equation (3.2); the result is obviously $\mathbf{k} \cdot \mathbf{B}_0 = 0$.

If we finally substitute our expressions for the electric and magnetic fields into the Ampère–Maxwell law, we find that

$$\mathbf{k} \times \mathbf{B}_0 = -\frac{\omega}{c^2}\mathbf{E}_0. \tag{3.39}$$

To check whether this is consistent with our existing expressions, we insert equation (3.38) into this formula:

$$\frac{1}{\omega}\mathbf{k} \times (\mathbf{k} \times \mathbf{E}_0) = \frac{1}{\omega}[\mathbf{k}(\mathbf{k} \cdot \mathbf{E}_0) - \mathbf{E}_0|\mathbf{k}|^2] = -\frac{\omega}{c^2}\mathbf{E}_0, \tag{3.40}$$

where we have used $\mathbf{k} \cdot \mathbf{E}_0 = 0$ and $|\mathbf{k}| = \omega/c$. The result is consistent, and therefore our plane wave solution is fully consistent with Maxwell's equations.

Looking at equations (3.38) and (3.39), it becomes clear that the vectors $(\mathbf{E}_0, \mathbf{B}_0, \mathbf{k})$—in that order—form what we can call a right-handed triplet. If we take the cross product of any two of those vectors in cyclic order—$\mathbf{E}_0$ times $\mathbf{B}_0$, $\mathbf{B}_0$ times $\mathbf{k}$, $\mathbf{k}$ times $\mathbf{E}_0$—the result is proportional to the third. Equations (3.38) and (3.39) already confirm two of these products; we may directly confirm the third:

$$\mathbf{E}_0 \times \mathbf{B}_0 = \frac{1}{\omega}\mathbf{E}_0 \times (\mathbf{k} \times \mathbf{E}_0) = \frac{1}{\omega}[\mathbf{k}|\mathbf{E}_0|^2 - \mathbf{E}_0(\mathbf{E}_0 \cdot \mathbf{k})] = \frac{|\mathbf{E}_0|^2}{\omega}\mathbf{k}, \tag{3.41}$$

which is indeed proportional to $\mathbf{k}$.

We note this right-handed nature of electromagnetic waves in vacuum, where the right-hand rule for cross products applies, because we will later see that negative-refractive-index materials have a left-handed nature and are often called 'left-handed materials' because of it. This is one of many curious properties of negative-refractive-index materials.

So what does one of these electromagnetic plane waves look like? If we assume that $\mathbf{E}_0$ is a real-valued vector, then a picture of an electromagnetic plane wave is shown in figure 3.2. It should be noted that the fields themselves do not extend out into space; the illustration shows the magnitude and direction of the fields along a line of propagation. The electric and magnetic fields oscillate in phase with each other and are perpendicular to each other and the direction of propagation.

We are often told that light is a transverse wave, but this is only strictly true for the plane waves we have considered here and a few other simple forms of

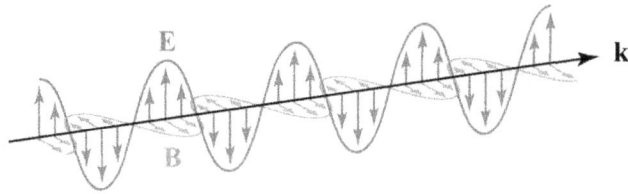

Figure 3.2. Illustration of the fields of an electromagnetic plane wave.

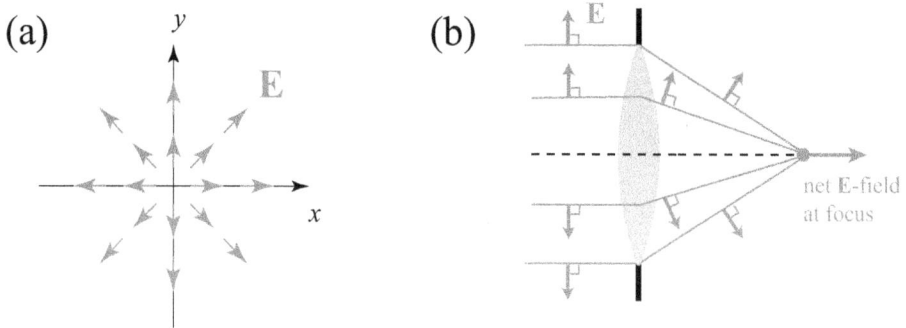

Figure 3.3. Illustration of (a) the polarization of a radially polarized beam in its cross section and (b) the effect of tight focusing of a radially polarized beam.

electromagnetic waves. More complicated waves may have a longitudinal component of the field in the overall direction of propagation. The word 'overall' is important here; whenever we look at complicated fields, such as a Gaussian beam emitted from a laser or a light wave traveling through a focusing system, we can assign an overall direction of propagation but, as equation (3.31) shows, the field consists of a collection of transverse plane waves traveling in different directions. Though each plane wave has fields perpendicular to its individual direction of propagation, those fields are not necessarily perpendicular to the overall direction, which is, in a sense, the 'average' direction of the collection of plane waves.

An example will help illustrate this. In recent years, researchers have extensively studied the properties of beams with a nonuniform state of polarization in their cross section, such as beams where the electric field points radially away from the beam center [3, 4]. An illustration of the electric field vector of a radially polarized beam is shown in figure 3.3(a). If such a beam is tightly focused, we may imagine it as a collection of plane waves, each of which is bent towards the geometrical focus while remaining transverse, as illustrated in figure 3.3(b).

All of these focused plane waves have a component of the electric field that lies along the axis of the focusing system, which serves as the overall direction of travel of the light wave. If the field is tightly focused, i.e. the plane waves are greatly deflected, the longitudinal components of the individual waves combine to form an overall longitudinal component of the propagating wave. This longitudinal component has been suggested for use in particle acceleration and imaging.

This example shows that the description of light as a transverse wave applies strictly to the simplest types of waves. It is perhaps better to say that the fundamental solutions of Maxwell's equations are transverse electromagnetic plane waves. In optics, where we are usually dealing with highly directional beams, the longitudinal component of the field is typically quite small, and we can say that the beam is effectively transverse.

The introduction of nonuniformly polarized light shows that the behavior of the electric field vector can, in general, be more complicated than the simple plane wave picture shown in figure 3.2. In the next chapter, we look at polarization in detail.

3.4 Waves in a half space

Our solution of the wave equation for the electric field, equation (3.31), is quite general but leaves out important cases, such as the behavior of waves near a planar interface. Such a situation includes problems such as diffraction, where an electric field is propagating from the space $z < 0$ onto the plane $z = 0$ and the solution is to be determined in the space $z > 0$. It is worth seeing how our plane wave solution for the electric field is changed when we consider waves restricted to propagate into the positive half-space $z > 0$.

This calculation is not essential for the rest of the book, so it can be safely skipped on a first reading; we include it to highlight some of the big assumptions that we made in our solution of the wave equation in unbounded space.

This time, we begin immediately with the assumption that our waves are monochromatic, i.e. their time dependence may be written as $\exp[-i\omega t]$. Our vector wave equation reduces to a vector Helmholtz equation of the form

$$[\nabla^2 + k^2]\mathbf{E}(\mathbf{r}, \omega) = 0, \tag{3.42}$$

where $k = \omega/c$ and $\mathbf{E}(\mathbf{r}, \omega)$ represents the electric field at frequency ω. We cannot now take a complete three-dimensional Fourier transform because of our restriction to $z > 0$; instead, we look at the Fourier transform with respect to x and y alone,

$$\tilde{\mathbf{E}}(K_x, K_y, z) = \frac{1}{(2\pi)^2} \int \mathbf{E}(x, y, z) e^{-i\mathbf{K}_\perp \cdot \rho} d^2\rho, \tag{3.43}$$

where $\rho \equiv (x, y)$ and $\mathbf{K}_\perp \equiv (K_x, K_y)$. If we take the transform of the vector Helmholtz equation, we have

$$\frac{\partial^2 \tilde{\mathbf{E}}(K_x, K_y, z)}{\partial z^2} + \left[k^2 - K_\perp^2 \right] \tilde{\mathbf{E}}(K_x, K_y, z) = 0. \tag{3.44}$$

This is again a harmonic oscillator equation, and we can write the solution for $\tilde{\mathbf{E}}(K_x, K_y, z)$ as

$$\tilde{\mathbf{E}}(K_x, K_y, z) = \mathbf{A}_+(\mathbf{K}_\perp) e^{ik_z z} + \mathbf{A}_+(\mathbf{K}_\perp) e^{-ik_z z}, \tag{3.45}$$

where

$$k_z \equiv \sqrt{k^2 - K_\perp^2}. \tag{3.46}$$

Let us introduce the inverse Fourier transform

$$\mathbf{E}(x, y, z) = \int \tilde{\mathbf{E}}(K_x, K_y, z)e^{-i\mathbf{K}_\perp \cdot \rho}d^2K_\perp, \tag{3.47}$$

which allows us to write

$$\mathbf{E}(x, y, z) = \int [\mathbf{A}_+(\mathbf{K}_\perp)e^{ik_z z} + \mathbf{A}_+(\mathbf{K}_\perp)e^{-ik_z z}]e^{i\mathbf{K}_\perp \cdot \rho}d^2K_\perp. \tag{3.48}$$

We therefore again find that the solutions appear in plane wave form, just as they did in unbounded space. In particular, we have waves $\mathbf{E}_+(x, y, z)$ going in the positive z-direction and waves $\mathbf{E}_-(x, y, z)$ going in the negative z-direction:

$$\mathbf{E}_+(x, y, z) = \mathbf{A}_+(\mathbf{K}_\perp)e^{i(\mathbf{K}_\perp \cdot \rho + k_z z)}, \tag{3.49}$$

$$\mathbf{E}_-(x, y, z) = \mathbf{A}_-(\mathbf{K}_\perp)e^{i(\mathbf{K}_\perp \cdot \rho - k_z z)}. \tag{3.50}$$

There are a few subtle but significant differences from unbounded space, however. First, the z-direction has been singled out as special in our result as a consequence of the way we derived the solution with a z-centric approach. Second, and even more significantly, we note that there is no limit on the magnitude of $\mathbf{K}_\perp$; therefore, two classes of solutions arise:

$$k_z = \begin{cases} \sqrt{k^2 - K_\perp^2}, & K_\perp < k, \\ i\sqrt{K_\perp^2 - k^2}, & K_\perp > k. \end{cases} \tag{3.51}$$

For $K_\perp < k$ we get a real-valued k_z, and the resulting solution is an ordinary plane wave. For $K_\perp > k$, we get an imaginary k_z, and the wave exponentially grows or decays in the z-direction; these waves are referred to as *inhomogeneous plane waves* or *evanescent waves*.

Let us recall that we are interested in solutions that are valid in the half-space $z > 0$ and have propagated from the plane $z = 0$. This means that not only must the plane waves propagate in the positive z-direction but also that the evanescent waves must exponentially decay in that direction. Solutions where the field exponentially grows are unphysical and discarded. Both of these conditions can be satisfied by setting $\mathbf{A}_- = 0$, and we finally get a solution for waves propagating into the half-space $z > 0$, namely

$$\mathbf{E}(x, y, z) = \int \mathbf{A}(\mathbf{K}_\perp)e^{i(\mathbf{K}_\perp \cdot \rho + k_z z)}d^2K_\perp. \tag{3.52}$$

This sort of solution for the electric field propagating into the positive half-space is known as the *angular spectrum representation* of the solution; it represents the electric field as a space-frequency 'spectrum' of plane waves propagating at different 'angles.'

Evanescent waves are waves that can appear at an interface under the right conditions, and we will see a number of these conditions throughout this book.

The most common is total internal reflection of light at a planar interface, to be discussed in section 9.4.4. We will have much more to say about such waves in the future, but for the moment, we note that equation (3.51) indicates that evanescent waves have a transverse wavenumber $K_\perp$ that is greater than the total wavenumber k; this has implications for imaging optics.

Evanescent waves do not appear in equation (3.31) for unbounded space, simply because we have taken the space to be unbounded; it is implicit in the solution that any evanescent waves must have already decayed away to nothing.

Our approach here has been somewhat different from the approach for unbounded space, in that we used a two-dimensional Fourier transform for the half-space and a three-dimensional Fourier transform for unbounded space. You might wonder whether it is possible to derive the half-space solution using a three-dimensional transform, and the answer is yes! If we take a two-dimensional Fourier transform with respect to x and y, where the limits of integration are $-\infty$ to ∞, and a Laplace transform with respect to z, where the limits of integration are 0 to ∞, we can derive equation (3.52). This approach is much more involved, and we do not consider it here.

The angular spectrum representation is most often used for scalar waves in optics, where it is assumed that the wave is highly directional, i.e. $K_\perp \ll k$ for all significant components of the field. This is one way to define the *paraxial approximation* in optics, which assumes that the wave is 'mostly' propagating along the z-axis. In the paraxial limit, the electric field of the wave is approximately perpendicular to z; if we further assume that the electric field direction is constant, we can represent the electric field by a scalar wave $U(x, y, z)$. Our angular spectrum representation of the field then becomes

$$U(x, y, z) = \int A(\mathbf{K}_\perp)e^{i(\mathbf{K}_\perp \cdot \rho + k_z z)}d^2 K_\perp, \tag{3.53}$$

where $A(\mathbf{K}_\perp)$ is now a scalar angular spectrum.

We can derive one very important result for this scalar representation. If we let $z = 0$, we find that the field $U_0(x, y) \equiv U(x, y, 0)$ satisfies the expression

$$U_0(x, y) = \int A(\mathbf{K}_\perp)e^{i\mathbf{K}_\perp \cdot \rho}d^2 K_\perp. \tag{3.54}$$

This equation indicates that the field in the plane $z = 0$ is related by an inverse Fourier transform to the angular spectrum. We may take the Fourier transform of both sides of this equation to determine that

$$A(\mathbf{K}_\perp) = \tilde{U}_0(\mathbf{K}_\perp), \tag{3.55}$$

where $\tilde{U}_0(\mathbf{K}_\perp)$ represents the two-dimensional Fourier transform of the field in the plane $z = 0$. This expression gives us a very straightforward way to calculate, analytically or computationally, the propagation of the field from the $z = 0$ plane. If we know $U_0(x, y)$, we may take the Fourier transform to determine $A(\mathbf{K}_\perp)$. We may then plug this angular spectrum into equation (3.53) to determine the field at any z distance.

We may take a similar approach to finding the angular spectrum of the vector electric field $\mathbf{E}(x, y, z)$ from equation (3.52), but it requires a bit more work. Recall that $\mathbf{A}(\mathbf{K}_\perp)$ must be transverse to the direction of propagation in order to satisfy Maxwell's equations, so to find a proper electromagnetic solution we must decompose $\mathbf{A}(\mathbf{K}_\perp)$ into an appropriate orthogonal pair of transverse vectors. Except in very special cases, this is more trouble than it is worth, so we do not pursue it further here.

3.5 Paraxial waves and Gaussian beams

The paraxial approximation introduced in the previous section simplifies the mathematics of wave propagation immensely and is therefore frequently used to analyze optical problems. It helps that lasers produce highly directional beams that automatically satisfy the paraxial condition, and many optics problems can therefore be studied using it. In fact, we can readily calculate the propagation integrals for Gaussian beams and more general Hermite–Gauss and Laguerre–Gauss beams, and it is worth taking some time to do so.

The derivation of Gaussian beams is usually done by solving a paraxial form of the wave equation directly; here, we take an alternative approach and derive them using the angular spectrum. Like the previous section, the discussion here is not essential for the rest of the book and can be safely skipped if the math gets a little overwhelming.

To begin, let us assume that the plane $z = 0$ is the waist plane of a Gaussian beam, where the scalar beam amplitude is of the form

$$U_0(x, y) = U_0 e^{-(x^2 + y^2)/w_0^2}, \tag{3.56}$$

where w_0 represents the effective width of the beam and U_0 is a constant. From equation (3.55), we know that the angular spectrum $A(k_x, k_y)$ of the beam is given by the Fourier transform of $U_0(x, y)$, i.e.

$$A(\mathbf{K}_\perp) = \frac{1}{(2\pi)^2} \int U_0(\boldsymbol{\rho}) e^{-i\mathbf{K}_\perp \cdot \boldsymbol{\rho}} d^2\rho. \tag{3.57}$$

In the case of our Gaussian, this is readily found to have the form

$$A(\mathbf{K}_\perp) = \frac{w_0^2 U_0}{4\pi} e^{-K_\perp^2 w_0^2 / 4}, \tag{3.58}$$

using the standard result that the Fourier transform of a Gaussian is a Gaussian.

To propagate this Gaussian beam, we would like to substitute this angular spectrum into equation (3.53); however, due to the nontrivial form of k_z given by equation (3.51), the integration cannot be performed analytically. Let us now apply a paraxial approximation, which involves assuming the beam is highly directional. This, in turn, means that the angular spectrum must be a very narrow function of $\mathbf{K}_\perp$ peaked at $\mathbf{K}_\perp = 0$; we may quantify this by the inequality $kw_0 \gg 1$, or

$$w_0 \gg \frac{\lambda}{2\pi}. \tag{3.59}$$

This is a way of saying that at the plane wave/evanescent wave boundary, $|\mathbf{K}_\perp| = k$, the angular spectrum is insignificant and that the angular spectrum is very concentrated around $\mathbf{K}_\perp = 0$.

We can compare equation (3.59) to common laser beams, for which the width is usually on the order of a few millimeters. With a wavelength in the visible range, e.g. 500 nm, the inequality is easily satisfied.

In this paraxial limit, we can approximate k_z by the first two terms of its Taylor series expansion,

$$k_z = \sqrt{k^2 - K_\perp^2} \approx k - \frac{K_\perp^2}{2k}. \tag{3.60}$$

Let us use this in equation (3.53); we then have

$$U(x, y, z) = e^{ikz} \int A(\mathbf{K}_\perp) e^{i\mathbf{K}_\perp \cdot \rho} e^{-i\frac{K_\perp^2 z}{2k}} d^2 K_\perp. \tag{3.61}$$

If we put our Gaussian angular spectrum into this equation, we have

$$U(x, y, z) = e^{ikz} \frac{w_0^2 U_0}{4\pi} \int e^{-\frac{K_\perp^2}{4}\left[w_0^2 + \frac{2iz}{k}\right]} e^{i\mathbf{K}_\perp \cdot \rho} d^2 K_\perp. \tag{3.62}$$

The integral is now again the Fourier transform of a Gaussian function, albeit a Gaussian with complex width. Assuming that we can use the standard Gaussian integral formulas to evaluate this (and yes, we can), we get

$$U(x, y, z) = e^{ikz} \frac{U_0}{\beta^2} e^{-\rho^2/w_0^2 \beta^2}, \tag{3.63}$$

where we have introduced

$$\beta^2 \equiv 1 + \frac{2iz}{kw_0^2}. \tag{3.64}$$

This is an extremely compact and elegant form of the solution, which shows that a Gaussian beam keeps its Gaussian structure upon propagation. We may also write it in a more physically suggestive form by first defining the Rayleigh range z_0,

$$z_0 \equiv \frac{kw_0^2}{2} = \frac{\pi w_0^2}{\lambda}. \tag{3.65}$$

We furthermore simplify our expression for the Gaussian beam by introducing the functions

$$w(z) = w_0 \sqrt{1 + z^2/z_0^2}, \tag{3.66}$$

$$R(z) = \frac{z_0^2}{z} + z, \text{ and} \tag{3.67}$$

$$\Phi(z) = \arctan(z/z_0), \tag{3.68}$$

and with some manipulation, we may write

$$U(x, y, z) = e^{ikz} U_0 e^{-i\Phi(z)} \left[\frac{w_0}{w(z)} e^{ik\rho^2/2R(z)} \right] e^{-\rho^2/w^2(z)}. \tag{3.69}$$

The quantity $w(z)$ represents the z-dependent width of the Gaussian beam, while $R(z)$ represents the wavefront curvature. The quantity $\Phi(z)$ represents the so-called Gouy phase of the beam, usually described as the difference of phase that a Gaussian beam exhibits relative to a spherical wave.

An illustration of the intensity of the beam and its width upon propagation is shown in figure 3.4. The Rayleigh range represents the distance at which the beam width has increased by a factor of $\sqrt{2}$ relative to its waist width.

An important practical aspect of Gaussian beams is that they are shape invariant, i.e. the field maintains a Gaussian intensity profile upon propagation. We now go a step further and demonstrate that there are entire classes of beams, namely the Hermite–Gaussian and Laguerre–Gaussian beams, that also exhibit shape invariance.

Let us first consider the set of one-dimensional Hermite–Gaussian functions, which are defined as

$$\psi_n(u) = e^{-u^2/2} H_n(u), \tag{3.70}$$

where $H_n(u)$ is the Hermite polynomial of the nth order, n is a nonnegative integer, and u is taken as a dimensionless variable. The Hermite polynomials can be formally defined by the relation [5]

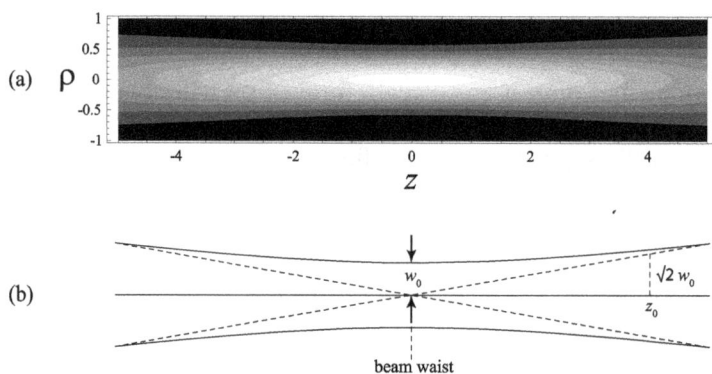

Figure 3.4. Illustration of (a) the intensity and (b) the effective width $w(z)$ of a Gaussian beam in cross section. The width of the beam is 1 μm, the wavelength is 500 nm, and the corresponding Rayleigh range is $z_0 = 6.28$m.

$$H_n(u) = (-1)^n e^{u^2} \frac{d^n}{du^n}[e^{-u^2}]. \tag{3.71}$$

The first few polynomials are given in table 3.1.

To construct Hermite–Gaussian beams, we take advantage of a property of the Hermite functions $\psi_n(x)$: they are eigenfunctions of the Fourier transform operator. This can be represented mathematically as

$$\frac{1}{\sqrt{2\pi}} \int_{-\infty}^{\infty} \psi_n(u) e^{-iuv} du = (-i)^n \psi_n(v). \tag{3.72}$$

Looking at equation (3.70) and the definition of the Gaussian width in equation (3.56), we let $u_x = \sqrt{2}\,x/w_0$ and $u_y = \sqrt{2}\,y/w_0$, and we define a Hermite–Gaussian beam of order m, n in the waist plane $z = 0$ as follows:

$$HG_{mn}(x, y) = H_m(\sqrt{2}\,x/w_0) H_n(\sqrt{2}\,y/w_0) e^{-(x^2 + y^2)/w_0^2}. \tag{3.73}$$

If we calculate the angular spectrum for this beam, we can let $v_x = k_x w_0/\sqrt{2}$ and $v_y = k_y w_0/\sqrt{2}$, and through the use of equation (3.72) for x and y, we can readily find that

$$A_{mn}^{HG}(k_x, k_y) = \left(\frac{w_0}{\sqrt{2}}\right)^2 \frac{1}{2\pi} (-i)^{n+m} H_m(k_x w_0/\sqrt{2}) H_n(k_y w_0/\sqrt{2}) e^{-(k_x^2 + k_y^2)w_0^2/4}. \tag{3.74}$$

We now substitute this angular spectrum into equation (3.61). With some rearrangement, we may write

$$U(x, y, z) = \left(\frac{w_0}{\sqrt{2}}\right)^2 \frac{1}{2\pi} (-i)^{n+m} e^{ikz} \int H_m\left(\frac{k_x w_0}{\sqrt{2}}\right) H_n\left(\frac{k_y w_0}{\sqrt{2}}\right) e^{-(k_x^2 + k_y^2)w_0^2\beta^2/4} d^2K_\perp, \tag{3.75}$$

with β again defined as in equation (3.64). We note that the integrals can be completely factorized into integrals in k_x and k_y. We now take advantage of a Hermite function integral given in section 7.374 of [6]:

$$\int_{-\infty}^{\infty} e^{-(x-y)^2/2\alpha} H_n(x) dx = (2\pi\alpha)^{1/2}(1 - 2\alpha)^{n/2} H_n[y(1 - 2\alpha)^{-1/2}]. \tag{3.76}$$

Table 3.1. The first few Hermite polynomials.

$H_0(u) = 1$
$H_1(u) = 2u$
$H_2(u) = 4u^2 - 2$
$H_3(u) = 8u^3 - 12u$
$H_4(u) = 16u^4 - 48u^2 + 12$
$H_5(u) = 32u^5 - 160u^3 + 120u$
$H_6(u) = 64u^6 - 480u^4 + 720u^2 - 120$

By assigning the identities $x \to k_x w_0/\sqrt{2}$, $y \to ix\sqrt{2}/w_0\beta^2$, and $\alpha \to 1/\beta^2$, we can evaluate the integrals for k_x and k_y. With a significant amount of effort (no exaggeration), we find that

$$HG_{mn}(x, y, z) = \frac{w_0}{w(z)}e^{ikz}e^{-(x^2+y^2)/w^2(z)}e^{ik(x^2+y^2)/2R(z)}e^{-i(n+m+1)\Phi(z)}H_m\left(\frac{\sqrt{2}x}{w(z)}\right)H_n\left(\frac{\sqrt{2}y}{w(z)}\right), \quad (3.77)$$

where $w(z)$ and $R(z)$ are defined as for the fundamental Gaussian beam. We find that the Hermite–Gaussian beams are shape invariant, retaining their Hermite–Gaussian form upon propagation, and that they all have the same propagation-dependent wavefront curvature and widths. Their Gouy phases depend on the order m, n of the beam, however.

It can be shown that these Hermite–Gaussian beams form a complete orthogonal set of paraxial beams. Therefore, any general paraxial beam can be written as a linear superposition of Hermite–Gaussian beams. The intensities of some low-order Hermite–Gaussian beams are shown in figure 3.5. The orders m and n of the beams equal the number of zeros of the beam in the x and y directions, respectively.

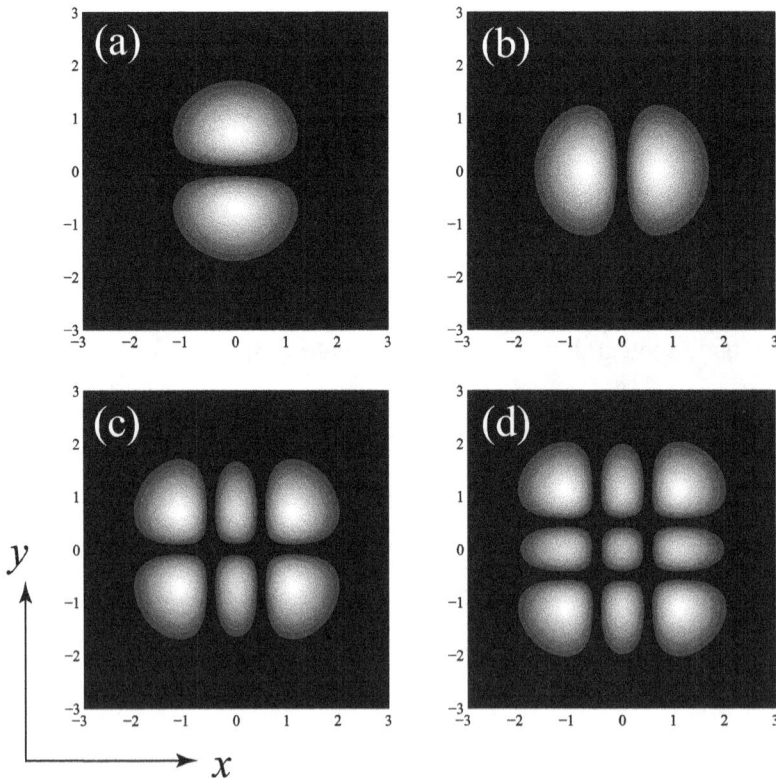

Figure 3.5. Intensity profiles of four Hermite–Gaussian beams in the plane $z = 0$. (a) HG_{01}, (b) HG_{10}, (c) HG_{32}, (d) HG_{23}. In all figures, $w_0 = 1$ cm.

A similar approach can be taken to derive the propagation characteristics of Laguerre–Gaussian beams, which in the source plane in cylindrical coordinates have the form

$$LG_{lm}(\rho, \phi) = \left(\frac{\sqrt{2}\rho}{w_0}\right)^{|m|} L_l^{|m|}\left(\frac{2\rho^2}{w_0^2}\right) \exp[im\phi]e^{-\rho^2/w_0^2}, \tag{3.78}$$

where the function $L_l^{|m|}(x)$ is an associated Laguerre function of order l and $|m|$. In this case, $l \geqslant 0$ but m can be any integer. As we are starting to wander too far into math and away from electromagnetism, we leave out the propagation calculation this time and simply state the result:

$$LG_{lm}(\rho, \phi) = \left[\frac{w_0}{w(z)}e^{ikz}e^{ik\rho^2/2R(z)}\right]e^{-\rho^2/w^2(z)}$$

$$\times \left(\frac{\sqrt{2}\rho}{w(z)}\right)^{|m|} L_l^{|m|}\left(\frac{2\rho^2}{w^2(z)}\right) \exp[im\phi]\exp\left[-i(2l + |m| + 1)\Phi(z)\right]. \tag{3.79}$$

The associated Laguerre functions are derivatives of the ordinary Laguerre polynomials that satisfy the equation

$$L_l^m(x) = (-1)^m \frac{d^m}{dx^m} L_{l+m}(x). \tag{3.80}$$

The first few ordinary Laguerre polynomials are tabulated in table 3.2. The order l of the associated Laguerre function indicates the number of zeros that the function possesses, which is equal to the number of rings of zero intensity in a Laguerre–Gaussian beam.

The meaning of the order m of a Laguerre–Gaussian beam can be illustrated by noting that we may write

$$\rho^{|m|}e^{im\phi} = (x \pm iy)^{|m|}, \tag{3.81}$$

where the $\pm$ sign corresponds to the sign of m. The order m represents the number of 2π 'twists' the phase of the beam undergoes in a circular path around the central beam axis, often called the *topological charge*. This phase structure is referred to as an *optical vortex* because it represents a circulation of phase around the axis, and the

Table 3.2. The first few Laguerre polynomials.

$L_0(x) = 1$
$L_1(x) = -x + 1$
$2!L_2(x) = x^2 - 4x + 2$
$3!L_3(x) = -x^3 + 9x^2 - 18x + 6$
$4!L_4(x) = x^4 - 16x^3 + 72x^2 - 96x + 24$
$5!L_5(x) = -x^5 + 25x^4 - 200x^3 + 600x^2 - 600x + 120$

study of beams possessing optical vortices falls into the field known as *singular optics* [7–9]. The intensities and phases of some low-order Laguerre–Gaussian beams are shown in figure 3.6. The zero rings in intensity, based on l, and the phase twist, based on m, are clearly visible in the figure.

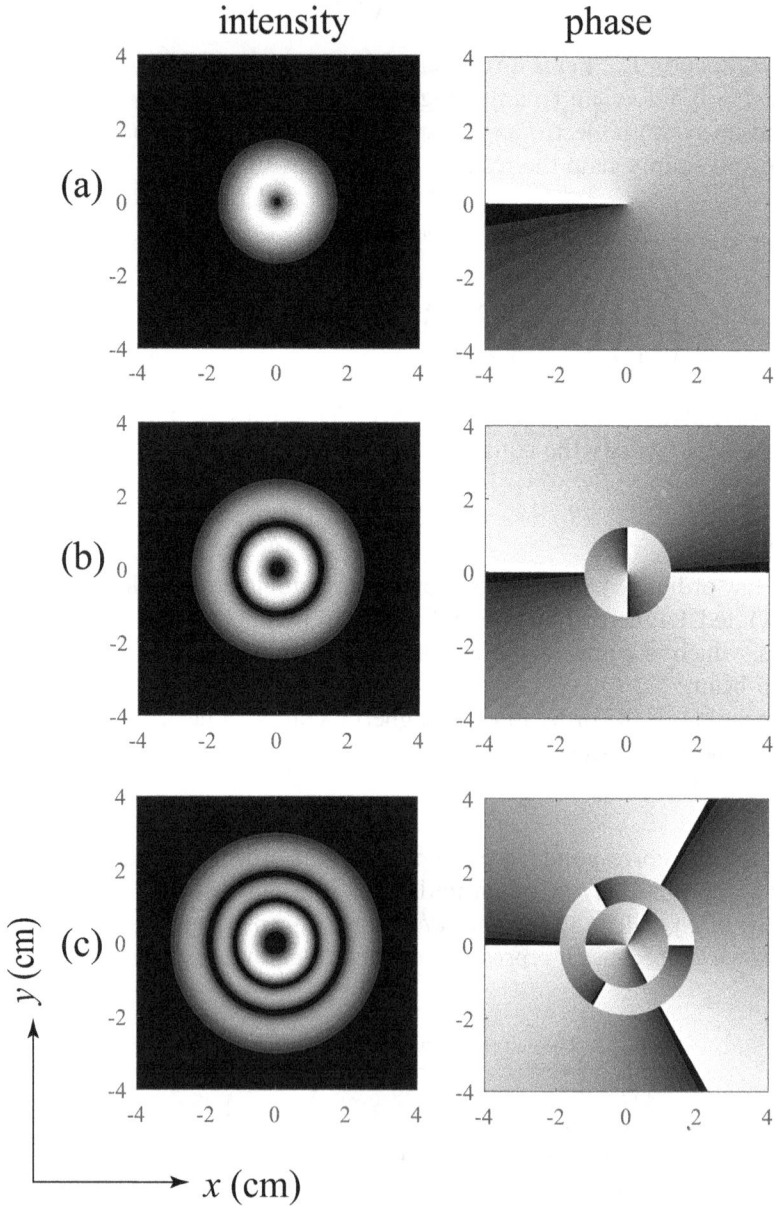

Figure 3.6. The intensity and phase of Laguerre–Gaussian beams in the plane $z = 0$ with (a) $l = 0$, $m = 1$, (b) $l = 1$, $m = 2$, and (c) $l = 2$, $m = -3$. The phase ranges from $-\pi$ (black) to π (white), and $w_0 = 1$ cm in each case.

Like the Hermite–Gaussian beams, the Laguerre–Gaussian beams form a complete orthogonal set of paraxial beams.

You may have noticed that we used Cartesian coordinates to derive the Hermite–Gaussian beams and cylindrical coordinates to derive the Laguerre–Gaussian beams. These two sets represent the shape-invariant modes with respect to their particular coordinate systems. There are other coordinate systems that can be used to derive beam-like fields. By solving the wave equation in elliptical cylindrical coordinates, one can derive so-called Mathieu beams, which are in their ideal case propagation-invariant [10]. By solving the wave equation in parabolic cylindrical coordinates, one can derive propagation-invariant parabolic beams [11]. These examples largely exhaust the possible beam classes that can be solved for analytically.

We have endeavored to provide a description of Gaussian beam classes because they play such a fundamental role in many optical experiments. Due to their high directionality, fundamental Gaussian beams are as close to an ideal plane wave as we can create in the laboratory.

We have derived these results for scalar fields. Because the beams are highly directional, we can introduce a constant state of polarization and an electric field simply by multiplying by a constant transverse unit vector, e.g. $\hat{\mathbf{x}}$ or $\hat{\mathbf{y}}$. We may also make more sophisticated beams where the state of polarization varies within the beam's cross section by a sum of beams of different orders with different polarization states; we discuss this briefly in section 4.8.

3.6 Exercises

1. From Maxwell's equations, derive the wave equation for the magnetic field, i.e.

$$\nabla^2 \mathbf{B}(\mathbf{r},\, t) - \mu_0 \varepsilon_0 \frac{\partial^2 \mathbf{B}(\mathbf{r},\, t)}{\partial t^2}.$$

2. For each of the following monochromatic plane waves in vacuum (time dependence $\exp[-i\omega t]$), determine the direction of propagation of the wave (give the unit vector of propagation), the wavelength λ, and the frequency ν. Position variables x, y, and z should be interpreted as having units of meters.
   (a) $\mathbf{E}(\mathbf{r}) = E_0 \hat{\mathbf{y}} e^{-i2\pi z}$,
   (b) $\mathbf{E}(\mathbf{r}) = E_0(\hat{\mathbf{x}} + \hat{\mathbf{y}}) e^{i4\pi(x-y)}$,
   (c) $\mathbf{E}(\mathbf{r}) = E_0(\hat{\mathbf{x}} + \hat{\mathbf{y}} + \hat{\mathbf{z}}) e^{i8\pi(-x-y+2z)}$.
3. For each monochromatic (time dependence $\exp[-i\omega t]$) electric/magnetic field given, determine the unit vector of the direction of propagation and the corresponding magnetic/electric field. Confirm that $\mathrm{Re}\{\mathbf{E} \cdot \mathbf{B}^*\} = 0$ for each case, i.e. that the fields are orthogonal. In all cases, fields are in vacuum, and $k = \omega/c$.
   (a) $\mathbf{E}(\mathbf{r}) = E_0(\hat{\mathbf{z}} + i\hat{\mathbf{y}}) e^{-ikx}$,
   (b) $\mathbf{B}(\mathbf{r}) = B_0 \hat{\mathbf{y}} e^{ik(x-z)/\sqrt{2}}$,
   (c) $\mathbf{E}(\mathbf{r}) = E_0(\hat{\mathbf{x}} + \hat{\mathbf{y}} + 2\hat{\mathbf{z}}) e^{ik(x+y-z)/\sqrt{3}}$.

4. For each of the following pairs of monochromatic electric and magnetic fields (time dependence $\exp[-i\omega t]$), determine whether the fields satisfy Maxwell's equations. If they do, determine the relationship between E_0 and B_0. In all cases, $k = \omega/c$.
   (a) $\mathbf{E}(\mathbf{r}) = (\hat{\mathbf{x}} + i\hat{\mathbf{y}})E_0 \exp[ikz]$, $\mathbf{B}(z) = (\hat{\mathbf{x}} - i\hat{\mathbf{y}})B_0 \exp[ikz]$,
   (b) $\mathbf{E}(\mathbf{r}) = (\hat{\mathbf{x}} - \hat{\mathbf{y}})E_0 \exp[ik(x + z)/\sqrt{2}]$,
       $\mathbf{B}(z) = (\hat{\mathbf{x}} + \hat{\mathbf{y}} - \hat{\mathbf{z}})B_0 \exp[ik(x + z)/\sqrt{2}]$,
   (c) $\mathbf{E}(\mathbf{r}) = (-2\hat{\mathbf{x}} + i\hat{\mathbf{y}} + \hat{\mathbf{z}})E_0 \exp[ik(x + 2z)/\sqrt{5}]$,
       $\mathbf{B}(z) = (-2\hat{\mathbf{x}} + 5i\hat{\mathbf{y}} + \hat{\mathbf{z}})B_0 \exp[ik(x + 2z)\sqrt{5}]$.

5. An example of a monochromatic evanescent wave in the space $z > 0$ is given by the expression

$$\mathbf{E}(\mathbf{r}) = \frac{E_0(k_z\hat{\mathbf{y}} + ik_y\hat{\mathbf{z}})}{k_0} e^{ik_y y} e^{-k_z z} e^{-i\omega t},$$

where k_y and k_z are real-valued. By substituting into the wave equation for $\mathbf{E}$, determine what condition k_y and k_z must satisfy for the wave equation to be satisfied. (You may want to use $k_0 \equiv \omega/c$ for notational convenience.) How does the value of k_y compare to the value of k_0? Show that this electric field is consistent with Maxwell's equations (i.e. consistent with $\nabla \cdot \mathbf{E} = 0$).

6. A two-dimensional electromagnetic field is completely independent of a single coordinate, say the z-coordinate. Show that Maxwell's equations uncouple for fields of this form and become two independent sets: the first relating E_z, B_x, and B_y to J_z, and the second relating B_z, E_x, and E_y to J_x and J_y.

7. In the early 1900s, Paul Ehrenfest argued that it is possible to have oscillating charges that do not produce radiation. He gave two illustrative examples: an infinite plane of uniform electric charge that oscillates in a direction normal to the plane, and a uniform spherical shell of charge that pulsates radially. Demonstrate, using symmetry arguments alone, that neither of these two examples can produce electromagnetic waves. (Think of the directions of $\mathbf{E}$ and $\mathbf{B}$ needed for a propagating wave.)

8. Another fundamental class of electromagnetic waves is spherical waves. Assume there is a monochromatic electric dipole at the origin, pointing in the $\hat{\mathbf{z}}$-direction with dipole moment $p_0\hat{\mathbf{z}}$. Far from the origin (when r is very large), the electric field takes on the approximate form

$$\mathbf{E}(r, \theta, t) \approx -\frac{k^2 p_0}{\varepsilon_0}\hat{\boldsymbol{\theta}} \sin(\theta)\frac{e^{i(kr-\omega t)}}{r}.$$

Calculate the magnetic field $\mathbf{B}(r, \theta, t)$ of this wave. From this magnetic field, use Maxwell's equations to calculate the electric field and determine whether it is consistent with the original function.

9. Though we generally refer to electromagnetic waves as transverse waves, there are many situations where the field can have longitudinal components. Let us consider a wave propagating in a hollow rectangular metal waveguide

of widths a and b at frequency ω with transverse electric field components of the form

$$E_x(x, y, z, t) = - E_0 \cos(m\pi x/a)\sin(n\pi y/b)e^{ik_l z}e^{-i\omega t},$$

$$E_y(x, y, z, t) = E_0 \frac{bm}{an} \sin(m\pi x/a)\cos(n\pi y/b)e^{ik_l z}e^{-i\omega t},$$

where m and n are positive integers and k_l is a longitudinal propagation constant. Calculate the magnetic field $\mathbf{B}(x, y, z, t)$ for this waveguide mode. Use Maxwell's equations to calculate the electric field $\mathbf{E}(x, y, z, t)$ from the magnetic field, and use the result to determine the form of k_l.

References

[1] Maxwell J C 1865 A dynamical theory of the electromagnetic field *Phil. Trans. R. Soc. Lond.* **155** 459–512

[2] Maxwell J C 1862 III. On physical lines of force *London, Edinburgh Dublin Phil. Mag. J. Sci.* **23** 12–24

[3] Youngworth K S and Brown T G 2000 Focusing of high numerical aperture cylindrical vector beams *Opt. Exp.* **7** 77–87

[4] Brown T G 2008 Unconventional polarization states: beam propagation, focusing, and imaging *Progress in Optics* **vol 56** ed E Wolf (Amsterdam: Elsevier) 81–129

[5] Gbur G J 2011 *Mathematical Methods for Optical Physics and Engineering* (Cambridge: Cambridge University Press)

[6] Gradshteyn I S and Ryzhik I M 2007 *Table of Integrals, Series, and Products* 7th edn (Burlington, MA: Academic)

[7] Soskin M S and Vasnetsov M V 2001 Singular optics *Progress in Optics* **vol 42** ed E Wolf (Amsterdam: Elsevier) 219

[8] Dennis M R, O'Holleran K and Padgett M J 2009 Singular optics: optical vortices and polarization singularities *Progress in Optics* **vol 53** ed E Wolf (Amsterdam: Elsevier) 293

[9] Gbur G J 2016 *Singular Optics* (Boca Raton, FL: CRC Press)

[10] Gutiérrez-Vega J C, Iturbe-Castillo M D and Chávez-Cerda S 2000 Alternative formulation for invariant optical fields: Mathieu beams *Opt. Lett.* **25** 1493–5

[11] Bandres M A, Gutiérrez-Vega J C and Chávez-Cerda S 2004 Parabolic nondiffracting optical wave fields *Opt. Lett.* **29** 44–6

IOP Publishing

Electromagnetic Optics

Gregory J Gbur

Chapter 4

The polarization of light

The concept of polarization predates the discovery of the electromagnetic nature of light. For centuries, people had noticed that light passing through a piece of Iceland spar (optical calcite) produces two images of the object behind it. This phenomenon, called double refraction, baffled scientists, who could not explain it using the existing wave and particle theories of light. In 1808, however, the French physicist Étienne-Louis Malus happened to look through a piece of Iceland spar at the windows of the Luxembourg Palace, and to his surprise realized that one of the images of reflected sunlight was brighter than the other—and he could change which image was brighter by rotating the crystal. He found that in certain circumstances only a single image is produced in Iceland spar using reflected light, and he called such single-image light 'polarized.'

The preliminary explanation of the double image came from Thomas Young, who in 1817 penned a letter to his colleague François Arago speculating that light is a transverse wave. Because there are two possible orthogonal oscillation directions of a transverse wave, each of the images formed in Iceland spar comes from one of these transverse oscillations. Maxwell's theory confirmed that the electric and magnetic fields oscillate transverse to the wave direction; since then, the term 'polarization' has become synonymous with 'the behavior of the electric field of a light wave.' Why do we associate polarization with the electric field and not the magnetic field? We will answer that question in an upcoming chapter.

In this chapter, we study in detail the polarization of monochromatic light waves, looking at the physics of light polarization as well as a number of ways to describe, model, measure, and manipulate it.

4.1 Polarization basics

We first derive the most general state of polarization of a monochromatic electric field. Let us begin by giving the answer: we will find that, in general, a monochromatic electric field traces out an ellipse at every point in space. Proving this in a

doi:10.1088/978-0-7503-6064-7ch4

systematic manner, however, takes significant effort; this is what we will spend our time on in this section. Our discussion follows closely the mathematics laid out by Nye [1], which is the clearest and most rigorous derivation I have encountered.

We start our calculation by assuming a general, real-valued electric field at a given point in space, written in Cartesian coordinates as

$$E_x(\mathbf{r}, t) = a_x(\mathbf{r})\cos[\omega t - \delta_x(\mathbf{r})], \tag{4.1a}$$

$$E_y(\mathbf{r}, t) = a_y(\mathbf{r})\cos[\omega t - \delta_y(\mathbf{r})], \tag{4.1b}$$

$$E_z(\mathbf{r}, t) = a_z(\mathbf{r})\cos[\omega t - \delta_z(\mathbf{r})], \tag{4.1c}$$

where $a_i(\mathbf{r})$ is the amplitude of the ith component of the field and $\delta_i(\mathbf{r})$ is the corresponding phase, with $i = x, y, z$. It is to be noted that there is nothing about this definition that depends on Maxwell's equations in particular; the results we find apply to any monochromatic vector wave in three-dimensional space.

We may expand each of the cosine terms using the familiar trigonometric identity,

$$\cos[\omega t - \delta_i] = \cos(\omega t)\cos(\delta_i) + \sin(\omega t)\sin(\delta_i), \tag{4.2}$$

and then group the $\cos(\omega t)$ and $\sin(\omega t)$ terms together to write the total electric field vector in the compact form,

$$\mathbf{E}(\mathbf{r}, t) = \mathbf{p}(\mathbf{r})\cos(\omega t) + \mathbf{q}(\mathbf{r})\sin(\omega t), \tag{4.3}$$

where we have defined

$$p_i(\mathbf{r}) \equiv a_i(\mathbf{r})\cos[\delta_i(\mathbf{r})], \quad q_i \equiv a_i(\mathbf{r})\sin[\delta_i(\mathbf{r})]. \tag{4.4}$$

We may immediately make one observation: the electric field may be written in terms of two real-valued vectors $\mathbf{p}$ and $\mathbf{q}$ that are constant in time, which in general define a plane. (If the vectors are parallel, then they do not define a unique plane, but we can always choose one.) The electric field vector, therefore, always oscillates within a plane. Furthermore, because we are dealing with a periodic function, the electric field vector must trace out a closed path in this plane.

But what is the shape of this closed path? For simplicity, and without loss of generality, let us choose a new coordinate system so that the path lies in the xy-plane. We now write the electric field as

$$\begin{aligned} E_x &= a_x \cos[\omega t - \delta_x], \\ E_y &= a_y \cos[\omega t - \delta_y], \end{aligned} \tag{4.5}$$

and focus on the behavior at a single point in space, dropping the functional dependence on $\mathbf{r}$. We again introduce the vectors $\mathbf{p}$ and $\mathbf{q}$ and can write the electric field in terms of a two-dimensional vector-matrix equation,

$$|\mathbf{E}\rangle = \mathbf{P}|\mathbf{C}\rangle, \tag{4.6}$$

where

$$|\mathbf{E}\rangle = \begin{bmatrix} E_x \\ E_y \end{bmatrix}, \quad |\mathbf{C}\rangle = \begin{bmatrix} \cos(\omega t) \\ \sin(\omega t) \end{bmatrix}, \tag{4.7}$$

and

$$\mathbf{P} = \begin{bmatrix} p_x & q_x \\ p_y & q_y \end{bmatrix}. \tag{4.8}$$

The clever trick to finding the path that the electric field traces out is to invert equation (4.6) for the vector $|\mathbf{C}\rangle$, i.e.

$$|\mathbf{C}\rangle = \mathbf{P}^{-1}|\mathbf{E}\rangle. \tag{4.9}$$

It is straightforward to invert the 2×2 matrix $\mathbf{P}$, and the inverse takes the form

$$\mathbf{P}^{-1} = \frac{1}{\det[\mathbf{P}]} \begin{bmatrix} q_y & -q_x \\ -p_y & p_x \end{bmatrix}. \tag{4.10}$$

Here, $\det[\mathbf{P}] = p_x q_y - q_x p_y$ is the determinant of $\mathbf{P}$, which can readily be shown to have the explicit form

$$\det[\mathbf{P}] = a_x a_y \sin \delta, \tag{4.11}$$

and $\delta \equiv \delta_y - \delta_x$ is the phase difference between the field components.

We can eliminate the time dependence completely by taking the matrix product

$$\langle \mathbf{C}|\mathbf{C}\rangle = (\mathbf{P}^{-1}|\mathbf{E}\rangle)^T (\mathbf{P}^{-1}|\mathbf{E}\rangle), \tag{4.12}$$

where the superscript T represents the transpose. We note that

$$\langle \mathbf{C}|\mathbf{C}\rangle = \cos^2(\omega t) + \sin^2(\omega t) = 1, \tag{4.13}$$

which means we may write

$$\langle \mathbf{E}|(\mathbf{P}^{-1})^T \mathbf{P}^{-1}|\mathbf{E}\rangle = \langle \mathbf{E}|\mathbf{Q}|\mathbf{E}\rangle = 1, \tag{4.14}$$

where

$$\mathbf{Q} \equiv (\mathbf{P}^{-1})^T \mathbf{P}^{-1} = \frac{1}{[\det(\mathbf{P})]^2} \begin{bmatrix} p_y^2 + q_y^2 & -(p_x p_y + q_x q_y) \\ -(p_x p_y + q_x q_y) & p_x^2 + q_x^2 \end{bmatrix}. \tag{4.15}$$

We now use equation (4.4) to write the components of $\mathbf{p}$ and $\mathbf{q}$ in terms of the field parameters. With a little effort, we find that

$$\mathbf{Q} = \frac{1}{(a_x a_y \sin \delta)^2} \begin{bmatrix} a_y^2 & -a_x a_y \cos \delta \\ -a_x a_y \cos \delta & a_x^2 \end{bmatrix}. \tag{4.16}$$

Equation (4.14) is the key result of this derivation. When multiplied out, it represents a quadratic equation for the two electric field components E_x and E_y, and the curve that it represents must be a conic section: a hyperbola, parabola, or ellipse. Now we borrow from the theory of conic sections: if the matrix $\mathbf{M}$ of a quadratic equation is given by

$$\mathbf{M} = \begin{bmatrix} A & B/2 \\ B/2 & C \end{bmatrix}, \tag{4.17}$$

then the type of conic section is determined by the sign of the discriminant $D \equiv B^2 - 4AC$. If D is positive, the result is a hyperbola; if it is negative, the result is an ellipse. A zero discriminant represents a parabola. For our electric field matrix, we have

$$D = \frac{4a_x^2 a_y^2 \cos^2 \delta - 4a_x^2 a_y^2}{(a_x a_y \sin \delta)^4}. \tag{4.18}$$

Because $\cos^2 \delta \leqslant 1$, this quantity is negative. We therefore conclude that the electric field vector generally traces out an elliptical path in E_x, E_y space.

An ellipse is defined by three independent geometric parameters: the length of the major axis a_M, the length of the minor axis a_m, and the orientation angle of the ellipse ψ with respect to the E_x axis. We next determine how these geometric parameters of the polarization ellipse are related to the field amplitudes a_x, a_y and the phase difference δ. The relevant geometry is illustrated in figure 4.1.

The strategy we take is to diagonalize the matrix $\mathbf{Q}$; as a real symmetric matrix, it is always diagonalizable by rotation. The angle of rotation that diagonalizes the matrix is the orientation angle ψ, while the eigenvalues are directly related to the major and minor axes of the ellipse. We introduce a matrix $\mathbf{\Theta}$ that rotates the coordinate system by the angle ψ,

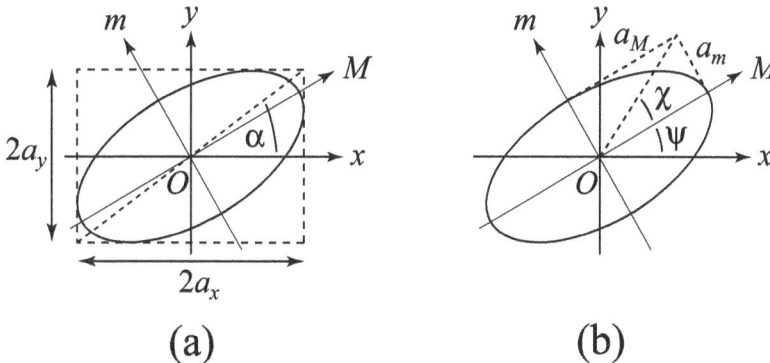

Figure 4.1. Illustration of the polarization ellipse in terms of (a) its field properties and (b) its geometric properties. The angle α is the angle of the diagonal of the box of side $2a_x$, $2a_y$; it is generally different from the orientation angle ψ. Reprinted from [5] by permission of the publisher (Taylor & Francis Ltd, http://www.tandfonline.com).

$$\Theta = \begin{bmatrix} \cos\psi & \sin\psi \\ -\sin\psi & \cos\psi \end{bmatrix}. \tag{4.19}$$

The form of $\mathbf{Q}$ in this new coordinate system is found by a similarity transformation,

$$\mathbf{Q}' = \Theta \mathbf{Q} \Theta^T$$
$$= \begin{bmatrix} Q_{xx}\cos^2(\psi) + Q_{xy}\sin(2\psi) + Q_{yy}\sin^2(\psi) & (Q_{yy} - Q_{xx})\cos(\psi)\sin(\psi) + Q_{xy}\cos(2\psi) \\ (Q_{yy} - Q_{xx})\cos(\psi)\sin(\psi) + Q_{xy}\cos(2\psi) & Q_{yy}\cos^2(\psi) - Q_{xy}\sin(2\psi) + Q_{xx}\sin^2(\psi) \end{bmatrix}, \tag{4.20}$$

where Q_{ij} is the i, jth component of $\mathbf{Q}$. The matrix $\mathbf{Q}'$ is diagonal when the off-diagonal elements are zero, or

$$(Q_{yy} - Q_{xx})\cos(\psi)\sin(\psi) + Q_{xy}\cos(2\psi) = (Q_{yy} - Q_{xx})\sin(2\psi)/2 + Q_{xy}\cos(2\psi) = 0. \tag{4.21}$$

We may solve the latter part of the equation to determine an expression for the orientation angle ψ,

$$\tan(2\psi) = \frac{2Q_{xy}}{Q_{xx} - Q_{yy}} = \frac{2a_x a_y \cos\delta}{a_x^2 - a_y^2}. \tag{4.22}$$

It should be noted that this equation is startlingly ambiguous. The tangent function is π-periodic, and because we have 2ψ in the argument, there are *four* possible angles that diagonalize our matrix. This is readily understandable: an ellipse can be rotated 180 degrees and look the same, which accounts for two possible angles, and the matrix can be diagonalized in two ways: by choosing ψ to align a coordinate axis along the major or minor axis of the ellipse.

To remove part of the ambiguity, we instead introduce the four-quadrant arctangent atan2(y, x), which takes as arguments a pair of x, y values in the xy-plane and returns the polar angle $-\pi \leqslant \theta < \pi$ associated with them.

We define the sine and cosine of 2ψ as

$$\sin(2\psi) \equiv \frac{-2Q_{xy}}{\sqrt{4Q_{xy}^2 + (Q_{xx} - Q_{yy})^2}} = \frac{2a_x a_y \cos\delta}{\sqrt{4a_x^2 a_y^2 \cos^2\delta + (a_y^2 - a_x^2)^2}}, \tag{4.23a}$$

$$\cos(2\psi) \equiv \frac{Q_{yy} - Q_{xx}}{\sqrt{4Q_{xy}^2 + (Q_{xx} - Q_{yy})^2}} = \frac{a_x^2 - a_y^2}{\sqrt{4a_x^2 a_y^2 \cos^2\delta + (a_y^2 - a_x^2)^2}}, \tag{4.23b}$$

and then reintroduce the angle ψ as

$$\psi = \text{atan2}[\sin(2\psi), \cos(2\psi)]/2, \tag{4.24}$$

which is defined over the range $-\pi/2 \leqslant \psi < \pi/2$. There is still some ambiguity in these definitions for ψ. If we flip the sign of the sine and cosine functions, the value of the tangent is unchanged, but this flip is equivalent to a rotation by $\pi/2$. This transformation amounts to choosing the angle ψ to align with either the major or minor axis, but which is which? For now, we make a lucky guess and assume that the positive sign aligns with the major axis. We will need to explore further to confirm this.

In the diagonal representation, equation (4.14) may be written as

$$\frac{|E_M|^2}{a_M^2} + \frac{|E_m|^2}{a_m^2} = 1, \tag{4.25}$$

where we have made the association, based on equation (4.20),

$$\frac{1}{a_M^2} = Q'_{xx} = Q_{yy} \sin^2(\psi) + Q_{xy} \sin(2\psi) + Q_{xx} \cos^2(\psi), \tag{4.26a}$$

$$\frac{1}{a_m^2} = Q'_{yy} = Q_{yy} \cos^2(\psi) - Q_{xy} \sin(2\psi) + Q_{xx} \sin^2(\psi). \tag{4.26b}$$

We may write the elements of the matrix $\mathbf{Q}$ in terms of the field variables and then solve for a_M^2 and a_m^2. This is a rather challenging calculation, and we leave it as an advanced exercise! The final result is

$$a_M^2 = a_y^2 \sin^2 \psi + a_x^2 \cos^2 \psi + a_x a_y \sin(2\psi)\cos \delta, \tag{4.27a}$$

$$a_m^2 = a_y^2 \cos^2 \psi + a_x^2 \sin^2 \psi - a_x a_y \sin(2\psi)\cos \delta. \tag{4.27b}$$

If a_M and a_m represent the major and minor axes, respectively, we should then find that

$$\Delta \equiv a_M^2 - a_m^2 \geqslant 0. \tag{4.28}$$

This quantity Δ is easy to evaluate, and we find that

$$\Delta = (a_x^2 - a_y^2)\cos(2\psi) + 2a_x a_y \sin(2\psi)\cos \delta. \tag{4.29}$$

If we use equation (4.23), we readily find that $\Delta \geqslant 0$, and our definitions of a_M and a_m are consistent, provided we have taken the positive values of the sine and cosine functions, as mentioned earlier.

We have therefore shown that the path of the electric field vector is generally elliptical, and we have derived equations that relate the parameters of the field—a_x, a_y, and δ—to the geometric parameters of the ellipse—a_M, a_m, and ψ.

There is one important property of polarization that is not encompassed by the geometric parameters: the direction in which the electric field circulates around the ellipse. It is usually described in terms of 'handedness,' and the most commonly used convention is to call the polarization *left-handed* or *right-handed* depending on whether it appears to be circulating counterclockwise or clockwise with respect to an observer looking directly into the oncoming field, respectively. This is illustrated in figure 4.2. This view is often called the *receiver view* because one imagines looking from the receiver toward the source. If the viewer points their thumb toward the source, the curve of their fingers gives a sense of the direction of circulation. The opposite convention, the *source view*, is also sometimes used, in which one imagines

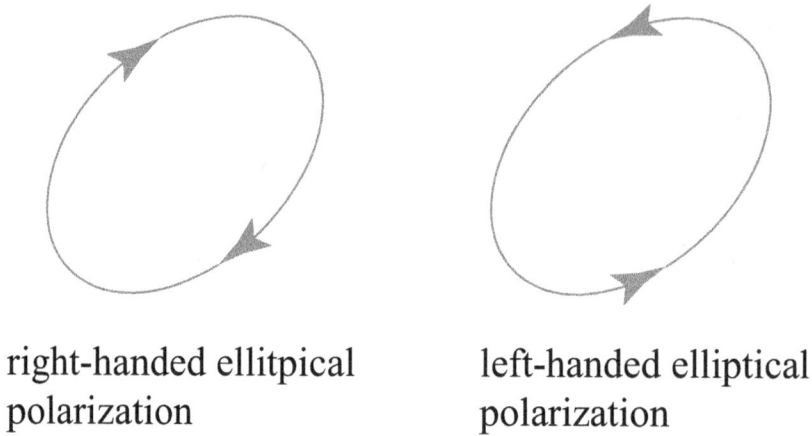

right-handed ellitpical
polarization

left-handed elliptical
polarization

Figure 4.2. Illustration of the definitions of polarization handedness. From the receiver view, left-handed is counterclockwise, while right-handed is clockwise.

instead looking from the source to the receiver. We will exclusively use the receiver view in all our discussions of handedness.

The handedness comes directly from the value of $\sin \delta$. For $\sin \delta > 0$, we have left-hand elliptically polarized light, and for $\sin \delta < 0$, we have right-hand elliptically polarized light. This will be easiest to see when we introduce the Poincaré sphere in section 4.5.

This handedness is typically included in a derived geometric parameter of the ellipse. The *ellipticity* describes the ratio of the minor and major axes and is defined as the angle χ satisfying

$$\tan \chi = \mp a_m / a_M. \tag{4.30}$$

The two signs are introduced to distinguish the two paths the electric field vector may traverse around the ellipse and may formally be considered the sign of a_m. Because $0 \leqslant |a_m/a_M| \leqslant 1$, the ellipticity has a total range of $-\pi/4 \leqslant \chi \leqslant \pi/4$. A negative ellipticity is right-handed, and a positive ellipticity is left-handed.

This derivation may have seemed excessively lengthy, but fortunately, it has given us very general results. We now know that the state of polarization of light is generally elliptical, and we have explicit formulas for calculating the properties of the polarization ellipse from the field quantities.

We have done this derivation entirely with real-valued fields. It should be noted that we can also write our fields in complex form,

$$E_x(t) = a_x e^{-i\omega t} e^{i\delta_x}, \tag{4.31a}$$

$$E_y(t) = a_y e^{-i\omega t} e^{i\delta_y}, \tag{4.31b}$$

which will be convenient for calculations going forward. To reproduce the real fields, we simply take the real part of these expressions. The complex fields will be particularly useful, however, in calculating the average observable properties of the field, as we will see.

4.2 Special cases

Though the general state of polarization is elliptical, most optical experiments use two special and important cases: linear polarization and circular polarization. We briefly consider the properties of both of these states.

4.2.1 Linear polarization

Linear polarization arises when the minor axis of the ellipse is zero, resulting in an electric field that oscillates back and forth along a line. This occurs when $\delta = n\pi$, where n is an integer. With some effort, one can show from equations (4.27) and equation (4.22) that

$$a_M^2 = a_x^2 + a_y^2, \tag{4.32}$$

$$a_m^2 = 0, \tag{4.33}$$

$$\tan(2\psi) = (-1)^n \frac{2a_x a_y}{a_x^2 - a_y^2}. \tag{4.34}$$

It should be noted that the handedness is undefined for linear polarization, with $\sin \delta = 0$. There is no sense of circulation of the field along a linear path. A special case of linear polarization is when either $a_x = 0$ or $a_y = 0$; then, the corresponding δ_x or δ_y is undefined (phase has no meaning when the amplitude is zero).

Linear polarization is the most commonly used state of polarization in optics, as it is easy to produce using a simple linear polarizer.

4.2.2 Circular polarization

The other important special case is circular polarization, which arises when $a_x = a_y = a$ and

$$\delta = \delta_y - \delta_x = (2n + 1)\pi/2, \quad (n = 0, \pm 1, \pm 2, ...). \tag{4.35}$$

With this requirement, equation (4.14) for the shape of the polarization ellipse reduces to the form

$$\left(\frac{E_x}{a}\right)^2 + \left(\frac{E_y}{a}\right)^2 = 1, \tag{4.36}$$

which is the equation for a circle.

It can be seen that the angle of orientation of the polarization ellipse, equation (4.22), is undefined for circular polarization, as it reduces to the indeterminate form 0/0. Because a circle has no major axis, there is no definition for the direction of the major axis.

Let us specifically consider the case $\delta_x = 0$, $\delta_y = \pm\pi/2$, so that $\delta = \pm\pi/2$. It is then straightforward to show that the electric field has the form

$$\mathbf{E}(t) = a\hat{\mathbf{x}}\cos(\omega t) \pm a\hat{\mathbf{y}}\sin(\omega t). \tag{4.37}$$

From this expression, it is clear that $\delta > 0$ represents left-hand circular polarization and $\delta < 0$ represents right-hand circular polarization.

Circular polarization states are regularly used in optics. It is convenient to note the complex form of the field for such states,

$$\mathbf{E}(t) = E_0(\hat{\mathbf{x}} \pm i\hat{\mathbf{y}})e^{-i\omega t}, \tag{4.38}$$

where E_0 is a complex constant, and the positive and negative signs represent left-hand and right-hand circular polarization, respectively. By taking the real part of this expression, one can easily see that the circulation of the electric field behaves as described.

4.3 Polarization-sensitive optical elements

Though we have introduced two ways to characterize the polarization of light (the field and geometric parameters), we have yet to describe how one can experimentally measure these parameters. Doing such measurements requires the use of polarization-sensitive optical elements, and we now briefly describe the behavior of the most important of these elements: the linear polarizer, the wave plate, and the rotator.

All of these are assumed to be transmission-based elements, i.e. the element changes the state of the light that passes directly through it. It is also possible to have reflection-based elements, as we will see later.

Let us consider a monochromatic plane wave traveling in the $+z$-direction, i.e.

$$\mathbf{E}(z, t) = \mathbf{E}e^{i(kz - \omega t)}, \tag{4.39}$$

where the complex electric field vector is written as

$$\mathbf{E} = E_x\hat{\mathbf{x}} + E_y\hat{\mathbf{y}}. \tag{4.40}$$

We do not explicitly use the time and space dependence of the field in the following calculations, but we want to be clear about the direction of light propagation and the convention used for the phase of the time- and space-dependent parts of the wave.

We are interested in describing the output polarization state $\mathbf{E}'$ based on the input polarization state $\mathbf{E}$. We describe the effect of each element both by writing the electric field output in vector form as well as in component form.

An ideal polarizer completely blocks the component of the electric field in one transverse direction while perfectly transmitting the component in the orthogonal direction. Let us call the angle of orientation of the transmission direction with respect to the x-axis θ and label the unit vector in this direction $\hat{\mathbf{w}}$; this direction is called the *polarization axis*. We may then describe the effect of a polarizer quite simply in terms of vectors as

$$\mathbf{E}' = \hat{\mathbf{w}}(\mathbf{E} \cdot \hat{\mathbf{w}}). \tag{4.41}$$

In Cartesian coordinates, $\hat{\mathbf{w}} = \hat{\mathbf{x}} \cos \theta + \hat{\mathbf{y}} \sin \theta$; the output electric field may be written as

$$E'_x = \cos^2 \theta E_x + \cos \theta \sin \theta E_y, \tag{4.42}$$

$$E'_y = \sin \theta \cos \theta E_x + \sin^2 \theta E_y. \tag{4.43}$$

Clearly, if $\theta = 0$ we have $E'_x = E_x$ and $E'_y = 0$. There are several types of polarizing elements, the most common of which is an absorptive polarizer. Such a device is made of material that strongly absorbs light whose electric field is perpendicular to the polarization axis; it is an example of an anisotropic material that will be discussed in chapter 8. A standard Polaroid filter consists of chains of microscopic molecules all oriented in one direction, along which current can flow freely. An electric field incident on the filter parallel to one of these chains is blocked by the freely oscillating electrons, much like in a metal; an electric field perpendicular to the chains can pass without loss because the electrons cannot oscillate freely in that direction. Therefore, the polarization axis is perpendicular to the orientation of the molecular chains.

Another common optical element is a wave plate, which advances the phase of one transverse component of the field by α relative to the other. The origin of this effect is typically an anisotropic material, which has different refractive indices and therefore different speeds of light for different polarizations. The axis with the faster speed is called the 'fast' axis, and we take it to point in the direction $\hat{\mathbf{w}}$. In vector form, the effect of a wave plate may be described as

$$\mathbf{E}' = e^{-i\alpha}\hat{\mathbf{w}}(\mathbf{E} \cdot \hat{\mathbf{w}}) + (\hat{\mathbf{w}} \times \hat{\mathbf{z}})[\mathbf{E} \cdot (\hat{\mathbf{w}} \times \hat{\mathbf{z}})]. \tag{4.44}$$

Note the $\exp[-i\alpha]$ factor of the first term: because this is the 'fast' axis, the wavenumber is smaller and thus the phase advances less than it does over the slow axis. In the second term, we take advantage of the orthogonality properties of the cross product to describe the perpendicular axis. In Cartesian components,

$$E'_x = (e^{-i\alpha} \cos^2 \theta + \sin^2 \theta)E_x - (-e^{-i\alpha} + 1)\sin \theta \cos \theta E_y, \tag{4.45}$$

$$E'_y = -(-e^{-i\alpha} + 1)\sin \theta \cos \theta E_x + (e^{-i\alpha} \sin^2 \theta + \cos^2 \theta)E_y. \tag{4.46}$$

The two most common wave plates are the quarter-wave plate, for which $\alpha = \pi/2$ (one quarter of 2π), and the half-wave plate, for which $\alpha = \pi$. The output components for the quarter-wave plate are

$$E'_x = (-i \cos^2 \theta + \sin^2 \theta)E_x - (i + 1)\sin \theta \cos \theta E_y, \tag{4.47}$$

$$E'_y = -(i + 1)\sin \theta \cos \theta E_x + (-i \sin^2 \theta + \cos^2 \theta)E_y, \tag{4.48}$$

and for the half-wave plate are

$$E_x' = -\cos(2\theta)E_x - \sin(2\theta)E_y, \tag{4.49}$$

$$E_y' = -\sin(2\theta)E_x + \cos(2\theta)E_y. \tag{4.50}$$

To see the effect of these wave plates in their simplest form, let $\theta = 0$. The quarter-wave plate then has the effect

$$E_x' = -iE_x, \tag{4.51}$$

$$E_y' = E_y, \tag{4.52}$$

which is a relative phase shift of $\pi/2$ between the x- and y-components of the electric field. We will soon see how this can be used to convert linear polarization to circular polarization, and vice versa. For the half-wave plate, we have

$$E_x' = -E_x, \tag{4.53}$$

$$E_y' = E_y, \tag{4.54}$$

which is a relative phase shift of π. We will see that this can be used to change the state of polarization to its orthogonal complement.

The final optical element we discuss is a rotator, which simply rotates the direction of linear polarization by an angle of θ_0, regardless of the incident angle of linear polarization. This is represented by the formulas

$$E_x' = \cos\theta_0 E_x - \sin\theta_0 E_y, \tag{4.55}$$

$$E_y' = \sin\theta_0 E_x + \cos\theta_0 E_y. \tag{4.56}$$

Here, we note that the angle θ_0 is not the angle of orientation of the device, as was the case for the wave plate and polarizer, but represents the angle by which the polarization is rotated, regardless of device orientation.

There are several types of rotators, with subtly different physics. A Faraday rotator uses magnetism-induced Faraday rotation, and the rotation is always in the same sense with respect to the direction of the magnetic field passing through the device, regardless of whether the wave is moving forward or backward through the system. An optically active material is made of chiral molecules, and the rotation is always in the same sense with respect to the direction of propagation. This is discussed in more detail in section 8.7.

It should be noted that all of these devices typically introduce an overall phase shift resulting from the propagation of light through the finite thickness of the device. If we are not doing interferometry, i.e. interfering two different waves, this overall phase factor is unimportant, and we leave it out of the equations.

4.4 Stokes parameters

The field parameters a_x, a_y, δ and the geometric parameters a_M, a_m, ψ are both valid ways to describe the state of polarization of the field, but at optical frequencies—which are around 10^{14} cycles/second—they are not directly measurable in most experiments. (Though new techniques have appeared quite recently that do allow the direct measurement of the electric field [2].)

In 1852, George Gabriel Stokes introduced a set of *four* parameters that can be measured experimentally and are directly related to the field and geometric parameters [3]. These parameters are now the standard tool for characterizing polarization experimentally. In terms of the field properties, the so-called Stokes parameters are *defined* as

$$S_0 = a_x^2 + a_y^2, \tag{4.57a}$$

$$S_1 = a_x^2 - a_y^2, \tag{4.57b}$$

$$S_2 = 2a_x a_y \cos\delta, \tag{4.57c}$$

$$S_3 = 2a_x a_y \sin\delta. \tag{4.57d}$$

We can already see that these parameters convey the same information as the electric field parameters, as it is possible to extract a_x and a_y from S_0 and S_1 and δ from S_2 and S_3. However, we will have to do some work to show that these parameters are useful for experimentally characterizing the polarization.

The fact that there are four parameters, where three quantities can uniquely describe the state of polarization, suggests that the Stokes parameters are not independent. It can readily be shown that

$$S_0^2 = S_1^2 + S_2^2 + S_3^2, \tag{4.58}$$

which suggests that we can view S_1, S_2 and S_3 as components of some sort of three-dimensional vector whose length is S_0.

We note that the matrix $\mathbf{Q}$ of equation (4.16) can be written entirely in terms of the Stokes parameters, in the form

$$\mathbf{Q} = \frac{2}{S_3^2} \begin{bmatrix} S_0 - S_1 & - S_2 \\ - S_2 & S_0 + S_1 \end{bmatrix}. \tag{4.59}$$

The Stokes parameters are derived from time-averaged intensity measurements of the electromagnetic field. Let us explicitly label the real-valued field components with a superscript '(r),' as we will soon transition to working with complex fields. We then define the time average of two real-valued field components $E_i^{(r)}$, $E_j^{(r)}$, with $i, j = x, y$, as

$$\langle E_i^{(r)}(t) E_j^{(r)}(t) \rangle = \lim_{T \to \infty} \frac{1}{2T} \int_{-T}^{T} E_i^{(r)}(t) E_j^{(r)}(t)\,dt = \lim_{T \to \infty} \frac{1}{2T} \int_{-T}^{T} a_i a_j \cos(\omega t - \alpha_i)\cos(\omega t - \alpha_j)\,dt, \tag{4.60}$$

where, for future use, we introduce α_i as the phase of the field after passing through an optical element. In short, we average the product of fields over a finite time interval T and then see what happens to this average as T gets arbitrarily large; this process approximates the averaging performed by a physical detector when the detector response is much slower than the oscillations of the field.

We use the familiar trigonometric identity

$$\cos(A)\cos(B) = \frac{1}{2}[\cos(A + B) + \cos(A - B)] \tag{4.61}$$

to write

$$\langle E_i^{(r)}(t)E_j^{(r)}(t)\rangle = \lim_{T\to\infty}\frac{a_i a_j}{4T}\int_{-T}^{T}[\cos(2\omega t - \alpha_i - \alpha_j) + \cos(\alpha_j - \alpha_i)]dt. \tag{4.62}$$

The integral can be evaluated explicitly:

$$\langle E_i^{(r)}(t)E_j^{(r)}(t)\rangle = \lim_{T\to\infty}\frac{a_i a_j}{4T}\left[\frac{1}{2\omega}\sin(2\omega t - \alpha_i - \alpha_j)_{-T}^{T} + 2T\cos(\alpha_j - \alpha_i)\right] \tag{4.63}$$

In the limit $T \to \infty$, the first term on the right goes to zero, and we are left with:

$$\langle E_i^{(r)}(t)E_j^{(r)}(t)\rangle = \frac{a_i a_j}{2}\cos(\alpha), \tag{4.64}$$

where we have introduced $\alpha \equiv \alpha_j - \alpha_i$.

This averaging process can be made much simpler if we use the complex form of the field in equations (4.31). From those equations and equation (4.64), it is straightforward to see that we may then write that

$$\langle E_i^{(r)}(t)E_j^{(r)}(t)\rangle = \frac{1}{2}\text{Re}\{E_i^*(t)E_j(t)\}, \tag{4.65}$$

where $E_i^{(r)}$ and $E_j^{(r)}$ on the left-hand side of the expression are the *real-valued* fields, and E_i and E_j on the right-hand side are the *complex-valued* fields. The use of complex fields allows us to readily calculate the cycle average without having to perform the explicit time integrals, and we will use this approach quite often in our future work, so expect more complex fields to come!

We note for future reference that the Stokes parameters can be written in terms of the complex field components in the simple form

$$S_0 = |E_x|^2 + |E_y|^2, \tag{4.66a}$$

$$S_1 = |E_x|^2 - |E_y|^2, \tag{4.66b}$$

$$S_2 = 2\text{Re}\left\{E_x^* E_y\right\}, \tag{4.66c}$$

$$S_3 = 2\text{Im}\left\{E_x^* E_y\right\}. \tag{4.66d}$$

This can be readily proven by substituting from equations (4.31) into the above.

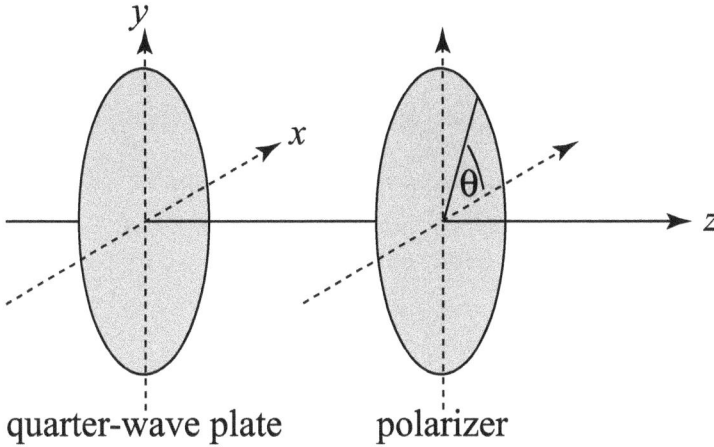

Figure 4.3. The experimental arrangement for measuring Stokes parameters.

We now ask: how can we determine the Stokes parameters from simple intensity measurements? We consider a uniformly polarized field that is sent through an optical system that can include a quarter-wave plate and a polarizer, as shown in figure 4.3.

Let us work with complex field components to evaluate the effect of these optical elements. The field emerging from this system is labeled $\mathbf{E}(\theta, q)$, where θ is the angle of transmission of the linear polarizer and $q = \pi/2, 0$ represents the presence or absence of the quarter-wave plate, oriented with its fast axis along the x-axis. We may use the results of section 4.3 to evaluate the effect of every element in these measurements.

The six measurements and the fields that emerge are listed below.

$$\mathbf{E}(0, 0) = \hat{\mathbf{x}} a_x e^{-i\omega t} e^{i\delta_x}, \tag{4.67}$$

$$\mathbf{E}(\pi/2, 0) = \hat{\mathbf{y}} a_y e^{-i\omega t} e^{i\delta_y}, \tag{4.68}$$

$$\mathbf{E}(\pi/4, 0) = \left(\frac{\hat{\mathbf{x}} + \hat{\mathbf{y}}}{\sqrt{2}}\right)\left[\frac{a_x}{\sqrt{2}} e^{-i\omega t} e^{i\delta_x} + \frac{a_y}{\sqrt{2}} e^{-i\omega t} e^{i\delta_y}\right], \tag{4.69}$$

$$\mathbf{E}(-\pi/4, 0) = \left(\frac{\hat{\mathbf{x}} - \hat{\mathbf{y}}}{\sqrt{2}}\right)\left[\frac{a_x}{\sqrt{2}} e^{-i\omega t} e^{i\delta_x} - \frac{a_y}{\sqrt{2}} e^{-i\omega t} e^{i\delta_y}\right], \tag{4.70}$$

$$\mathbf{E}(\pi/4, \pi/2) = \left(\frac{\hat{\mathbf{x}} + \hat{\mathbf{y}}}{\sqrt{2}}\right)\left[\frac{a_x}{\sqrt{2}} e^{-i\omega t} e^{i(\delta_x - \pi/2)} + \frac{a_y}{\sqrt{2}} e^{-i\omega t} e^{i\delta_y}\right], \tag{4.71}$$

$$\mathbf{E}(-\pi/4, \pi/2) = \left(\frac{\hat{\mathbf{x}} - \hat{\mathbf{y}}}{\sqrt{2}}\right)\left[\frac{a_x}{\sqrt{2}} e^{-i\omega t} e^{i(\delta_x - \pi/2)} - \frac{a_y}{\sqrt{2}} e^{-i\omega t} e^{i\delta_y}\right]. \tag{4.72}$$

We may calculate the cycle-averaged intensity of each of these cases using equation (4.64) to relate complex fields to time-averaged real fields,

$$\langle I(\theta, q) \rangle = \frac{1}{2} \mathbf{E}^*(\theta, q) \cdot \mathbf{E}(\theta, q). \tag{4.73}$$

The results of our measurements take the following forms:

$$\langle I(0, 0) \rangle = \frac{a_x^2}{2}, \tag{4.74a}$$

$$\langle I(\pi/2, 0) \rangle = \frac{a_y^2}{2}, \tag{4.74b}$$

$$\langle I(\pi/4, 0) \rangle = \frac{1}{4}\left[a_x^2 + a_y^2 + 2a_x a_y \cos \delta \right], \tag{4.74c}$$

$$\langle I(-\pi/4, 0) \rangle = \frac{1}{4}\left[a_x^2 + a_y^2 - 2a_x a_y \cos \delta \right], \tag{4.74d}$$

$$\langle I(\pi/4, \pi/2) \rangle = \frac{1}{4}\left[a_x^2 + a_y^2 - 2a_x a_y \sin \delta \right], \tag{4.74e}$$

$$\langle I(-\pi/4, \pi/2) \rangle = \frac{1}{4}\left[a_x^2 + a_y^2 + 2a_x a_y \sin \delta \right]. \tag{4.74f}$$

The Stokes parameters can easily be found from these measured intensities,

$$S_0 = 2\langle I(0, 0) \rangle + 2\langle I(\pi/2, 0) \rangle, \tag{4.75a}$$

$$S_1 = 2\langle I(0, 0) \rangle - 2\langle I(\pi/2, 0) \rangle, \tag{4.75b}$$

$$S_2 = 2\langle I(\pi/4, 0) \rangle - 2\langle I(-\pi/4, 0) \rangle, \tag{4.75c}$$

$$S_3 = 2\langle I(-\pi/4, \pi/2) \rangle - 2\langle I(\pi/4, \pi/2) \rangle. \tag{4.75d}$$

This set of Stokes measurements is clearly somewhat inefficient—we use *six* measurements to determine *four* Stokes parameters, which gives us information about *three* field parameters—but this particular arrangement weighs each measurement equally in the derivation of the parameters.

From equation (4.75a), it is clear that S_0 is proportional to the total intensity of the electromagnetic wave at that point, and that intensity does not play a role in the orientation and ellipticity of the polarization ellipse. We have also seen from equation (4.58) that S_0 is determined by the other Stokes parameters. It is convenient, therefore, to introduce a smaller set of normalized Stokes parameters of the form

$$s_1 = S_1/S_0, \tag{4.76a}$$

$$s_2 = S_2/S_0, \tag{4.76b}$$

$$s_3 = S_3/S_0. \tag{4.76c}$$

These normalized parameters satisfy the relation

$$s_1^2 + s_2^2 + s_3^2 = 1, \tag{4.77}$$

and we will find in the next section that they have special significance. Furthermore, this normalization eliminates the pesky factor of two difference between the calculation of field averages from real-valued quantities and the averages obtained using complex-valued quantities.

There is one final manipulation of the Stokes parameters that can prove useful in many cases. We introduce unit vectors $\hat{\mathbf{e}}_+$ and $\hat{\mathbf{e}}_-$ to represent a left- and right-circular polarization basis, i.e.

$$\hat{\mathbf{e}}_\pm = \frac{\hat{\mathbf{x}} \pm i\hat{\mathbf{y}}}{\sqrt{2}}. \tag{4.78}$$

In terms of $\hat{\mathbf{x}}$ and $\hat{\mathbf{y}}$, our electric field can be written as

$$\mathbf{E}(t) = [\hat{\mathbf{x}}a_x e^{i\delta_x} + \hat{\mathbf{y}}a_y e^{i\delta_y}]e^{-i\omega t}. \tag{4.79}$$

If we invert equations (4.78) to write the Cartesian unit vectors in terms of the circular ones, we have

$$\hat{\mathbf{x}} = \frac{\hat{\mathbf{e}}_+ + \hat{\mathbf{e}}_-}{\sqrt{2}}, \quad \hat{\mathbf{y}} = \frac{\hat{\mathbf{e}}_+ - \hat{\mathbf{e}}_-}{\sqrt{2}\,i}, \tag{4.80}$$

Then, in a circular polarization basis, the electric field can be written as

$$\mathbf{E}(t) = \left[\hat{\mathbf{e}}_+ \frac{a_x e^{i\delta_x} - ia_y e^{i\delta_y}}{\sqrt{2}} + \hat{\mathbf{e}}_- \frac{a_x e^{i\delta_x} + ia_y e^{i\delta_y}}{\sqrt{2}} \right]e^{-i\omega t}. \tag{4.81}$$

The components of the electric field in this circular basis are of the form,

$$E_+ \equiv \frac{a_x e^{i\delta_x} - ia_y e^{i\delta_y}}{\sqrt{2}}, \tag{4.82}$$

$$E_- \equiv \frac{a_x e^{i\delta_x} + ia_y e^{i\delta_y}}{\sqrt{2}}. \tag{4.83}$$

We have dropped the time dependence at this point since it is always canceled out in our complex representation. We can readily find that the Stokes parameters can be written in the circular polarization basis as

$$S_0 = |E_+|^2 + |E_-|^2, \tag{4.84}$$

$$S_1 = 2\mathrm{Re}\{E_+^* E_-\}, \tag{4.85}$$

$$S_2 = 2\text{Im}\{E_+^* E_-\}, \tag{4.86}$$

$$S_3 = |E_+|^2 - |E_-|^2. \tag{4.87}$$

We can confirm these results directly by plugging in the expressions for E_+ and E_-. On comparison with equations (4.66), we see that they are of a very similar form, but with the definitions of S_1, S_2, and S_3 being permutated among them. One gets the sense that our change of basis has amounted to some form of rotation, a sense that will be strengthened in our discussion of the Poincaré sphere.

4.5 The Poincaré sphere

The field parameters, geometric parameters, and Stokes parameters represent three equivalent ways to describe the state of polarization. There is an elegant geometric way to visualize the relationship between all these parameters, known as the Poincaré sphere, which we now derive. We follow a modern derivation of the properties of the sphere, as described by Jerrard [4].

Let us return to the complex form of the electric field components as defined in equations (4.31):

$$E_x = a_x e^{i\delta_x} e^{-i\omega t}, \tag{4.88}$$

$$E_y = a_y e^{i\delta_y} e^{-i\omega t}, \tag{4.89}$$

and define a complex number as the ratio of the field components

$$\zeta = u + iv \equiv \frac{E_y}{E_x} = \frac{a_y}{a_x} e^{i\delta}, \tag{4.90}$$

where u and v are real-valued.

It is readily apparent that each point in the complex plane represents a unique state of polarization. If we consider the position in polar coordinates, the ratio $\rho \equiv a_y/a_x$ is the distance from the origin and $\phi \equiv \delta$ is the azimuthal angle. It is well known that points on a plane can be mapped onto a sphere of unit radius through the use of what is known as a *stereographic projection*. For the polarization problem, the particular projection to be used is illustrated in figure 4.4.

The center of the sphere is placed at the origin of ζ, and the real and imaginary parts of ζ are oriented along the y- and z-axes, respectively, in three-dimensional space. Let us imagine a straight line drawn from the point where the $-x$-axis intersects the sphere to the point $u + iv$ in the complex plane. The sphere intersects this line at a point (x, y, z) which represents the mapping from the plane to the sphere.

The spatial variables x, y, and z can be determined in terms of u and v by solving the equations for the line and the sphere at their intersection point. The sphere obviously satisfies the equation $x^2 + y^2 + z^2 = 1$, while the line satisfies the equations

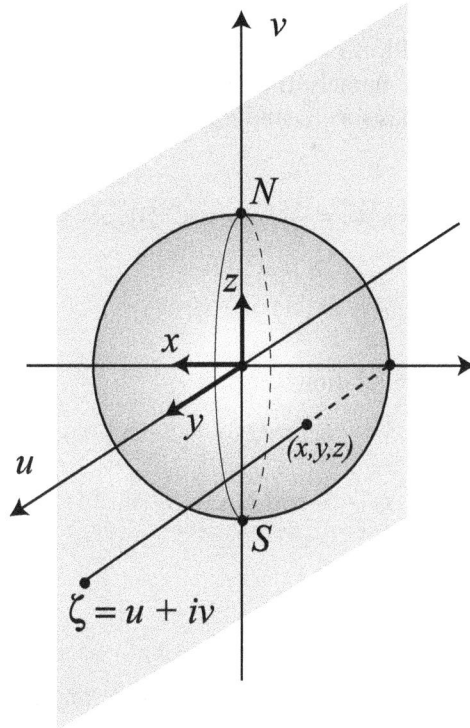

Figure 4.4. Stereographic projection from the complex plane onto the Poincaré sphere. Reprinted from [5] by permission of the publisher (Taylor & Francis Ltd, http://www.tandfonline.com).

$$x = \frac{y}{u} - 1, \quad z = \frac{v}{u}y. \tag{4.91}$$

There is a trivial solution to this equation at $(-1, 0, 0)$, which we ignore; the other solution satisfies the expressions

$$x = \frac{a_x^2 - a_y^2}{a_x^2 + a_y^2} = s_1, \tag{4.92a}$$

$$y = \frac{2a_x a_y \cos \delta}{a_x^2 + a_y^2} = s_2, \tag{4.92b}$$

$$z = \frac{2a_x a_y \sin \delta}{a_x^2 + a_y^2} = s_3, \tag{4.92c}$$

where we have already noted that (x, y, z) corresponds to (s_1, s_2, s_3)! We have shown that the coordinates on the sphere correspond to the normalized Stokes parameters of equations (4.76).

We may derive even more useful results for the Poincaré sphere, however, and show that the angular position on the sphere is directly related to the geometric parameters of polarization, namely the orientation angle ψ and the ellipticity χ. Looking back at the formulas for a_M and a_m, let us use the following trigonometric formulas:

$$\cos^2 \psi = \frac{1}{2}[1 + \cos(2\psi)], \tag{4.93}$$

$$\sin^2 \psi = \frac{1}{2}[1 - \cos(2\psi)], \tag{4.94}$$

which eventually lead to the relations

$$2a_M^2 = S_0 + S_1 \cos(2\psi) + S_2 \sin(2\psi), \tag{4.95}$$

$$2a_m^2 = S_0 - S_1 \cos(2\psi) - S_2 \sin(2\psi), \tag{4.96}$$

as well as

$$\tan(2\psi) = \frac{S_2}{S_1}. \tag{4.97}$$

Furthermore, let us apply a familiar tangent double-angle formula to the tangent of twice the ellipticity,

$$\tan(2\chi) = \frac{2 \tan(\chi)}{1 - \tan^2(\chi)} = \frac{2a_m a_M}{a_M^2 - a_m^2}. \tag{4.98}$$

We can deduce the form of $\sin(2\chi)$ and $\cos(2\chi)$ from this expression; it is then straightforward to show that

$$s_1 = \cos(2\chi)\cos(2\psi), \tag{4.99a}$$

$$s_2 = \cos(2\chi)\sin(2\psi), \tag{4.99b}$$

$$s_3 = \sin(2\chi). \tag{4.99c}$$

Here, 2χ represents the latitude on the sphere and 2ψ represents the longitude, so that the position on the sphere completely determines not only the Stokes parameters but also the geometric parameters. Furthermore, we note that 2χ is measured from the equator of the sphere; the handedness of the polarization ellipse is thus determined by the sign of s_3, which, as we can see from equation (4.92c), is determined by the sign of δ.

A point on the Poincaré sphere therefore fully characterizes all of the polarization characteristics in a very straightforward way. The Cartesian position on the sphere determines the normalized Stokes parameters and the angular position on the sphere determines the geometric parameters. The determination of the field parameters is less straightforward, but they can be found from either the Stokes or geometric

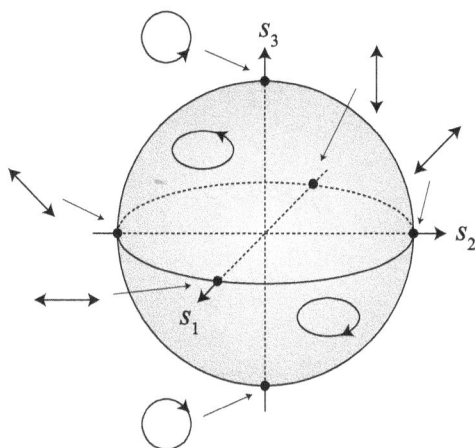

Figure 4.5. The relationship between the state of polarization, the Poincaré sphere, and the Stokes parameters. Reprinted from [5] by permission of the publisher (Taylor & Francis Ltd, http://www.tandfonline.com).

parameters. The relationship between the Stokes parameters, the state of polarization, and the Poincaré sphere is illustrated in figure 4.5. Each point on the sphere represents a distinct state of polarization, and every distinct state of polarization is represented by a single point on the sphere.

Points on the equator of the sphere are defined by $v = 0$, which indicates from equation (4.90) that either $\delta = 0$ or $\delta = \pi$. As we have seen in section 4.1, this corresponds to linear polarization. The north and south poles satisfy $u = 0$ and $v = \pm 1$; in terms of field parameters, they satisfy $a_x = a_y$ and $\delta = \pm \pi/2$, corresponding to left- and right-circular polarization, respectively. All other locations on the sphere represent elliptical polarization, with left-handed elliptical polarization above the equator and right-handed elliptical polarization below. The Poincaré sphere shows us that the elliptical states of polarization are far more common than linear polarization states, which in turn are far more common than circular polarization states.

4.6 The Pancharatnam phase

There is one polarization phenomenon for which the Poincaré sphere is a particularly useful representation. The phenomenon in question is an additional phase accrued when the state of polarization of light is changed; this phase is known as the *Pancharatnam phase*, after Shivaramakrishnan Pancharatnam, who first discovered the effect in 1956 [6].

When a monochromatic light wave propagates, it naturally accumulates a phase shift that is dependent upon the wavelength of light in the medium. This phase is referred to as a *dynamic phase*; the Pancharatnam phase is a *geometric phase* that arises purely from manipulations of the shape of the polarization ellipse.

We leave out the detailed mathematics and consider a simple illustration of how the Pancharatnam phase arises. Let us consider the experimental setup shown in figure 4.6. We start with left-hand circularly polarized light that is split into two

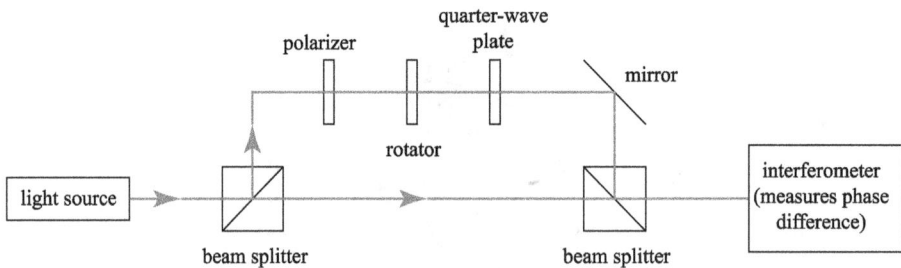

Figure 4.6. An experimental arrangement to illustrate the idea of the Pancharatnam phase.

beams. One beam takes a direct path through the system, while the other passes first through a linear polarizer that gives the light vertical polarization, then through a rotator that rotates the polarization 90 degrees, and then through a quarter-wave plate that returns the light to a circular polarization state.

In figure 4.7, we look exclusively at how the phase of the light—the position of the electric field around the polarization ellipse—changes as we change the shape of the ellipse. The black dot marks the state of the dynamic phase, which we leave unchanged. The only changes in phase come from the distortion of the ellipse.

The initial and final polarization states are both left-hand circular; however, the phase of the field has rotated 90 degrees from its starting position. This is the essence of the geometric phase: distortions of the polarization ellipse result in an additional phase shift. This phase shift can, in principle, be measured by interfering the two fields, as shown in figure 4.6.

The surprising thing about the geometric phase is that it can easily be calculated for any closed circuit of polarization using the Poincaré sphere. Any closed circuit of polarization changes can be represented by a solid angle Ω on the Poincaré sphere; it can be shown that the accumulated Pancharatnam phase is equal to $\Omega/2$. For example, for the state of changes we considered in figure 4.8, the solid angle is 1/4 of the sphere, or π; the corresponding Pancharatnam phase is $\pi/2$. The phase change is positive if the path is taken in a counterclockwise sense from the exterior perspective of the sphere and is negative if the clockwise sense is used.

The Pancharatnam phase is surprisingly difficult to isolate experimentally. For example, in the arrangement shown in figure 4.6, the dynamic phase of the upper path evolves on passing through the collection of free-space regions and regions of anisotropic materials. It is not clear how to define the dynamic phase in this case. One way to resolve the ambiguity is to change the polarization states in both arms of the interferometer in complementary ways, so that the dynamic phase is the same for both and the geometric phase is the solid angle between the two paths on the Poincaré sphere [7].

Pancharatnam's original paper spent many years in relative obscurity. In 1984, Michael Berry showed that certain quantum systems taken adiabatically around a circuit acquire a purely geometric phase [8]; Berry's colleagues in India pointed out the similarity between his discovery and Pancharatnam's own, and Berry wrote a paper highlighting the relationship soon after [9].

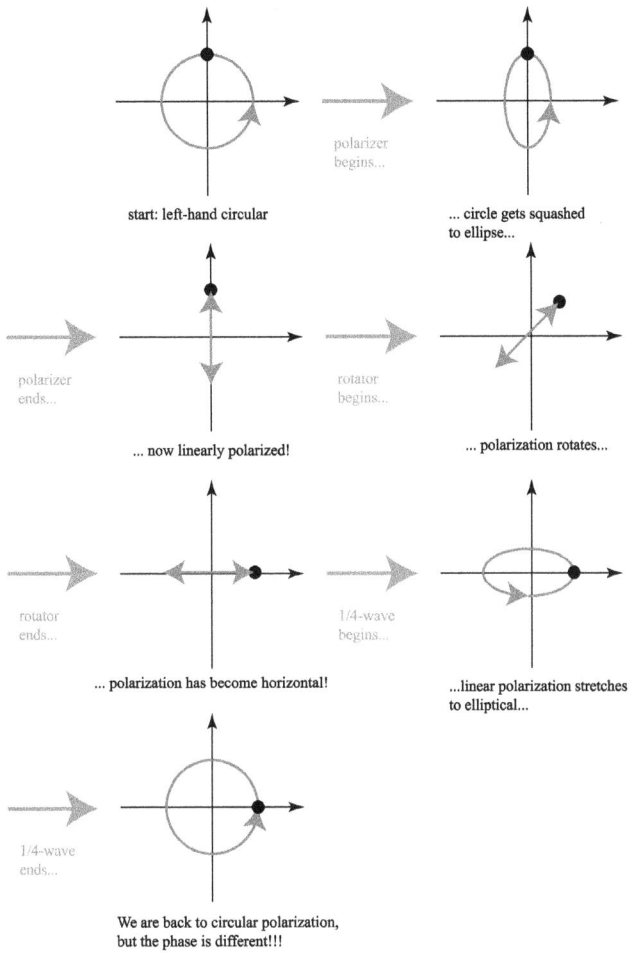

start: left-hand circular

polarizer
begins...

... circle gets squashed
to ellipse...

polarizer
ends...

... now linearly polarized!

rotator
begins...

... polarization rotates...

rotator
ends...

... polarization has become horizontal!

1/4-wave
begins...

...linear polarization stretches
to elliptical...

1/4-wave
ends...

We are back to circular polarization,
but the phase is different!!!

Figure 4.7. The accrued geometric phase is illustrated using a closed sequence of polarization changes.

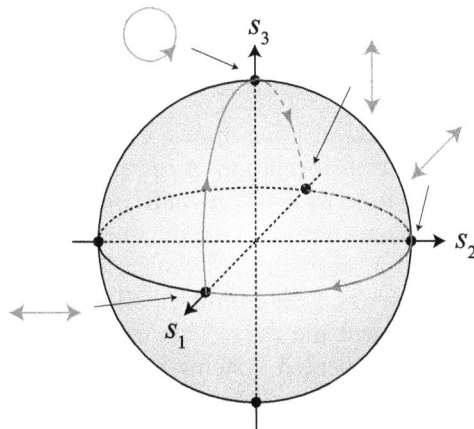

Figure 4.8. The path on the Poincaré sphere for the polarization cycle described in figure 4.7.

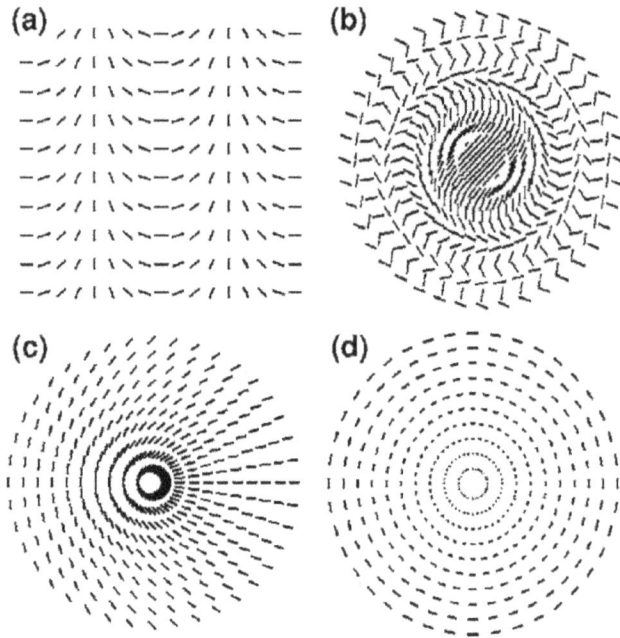

Figure 4.9. Crystal orientations of a Pancharatnam–Berry optical element used to produce (a) a circular polarizing beam splitter or switch, (b) a polarization-dependent lens, (c) a helical mode of order ±1, and (d) a helical mode of order ±2. Reprinted from [10] with permission of AIP Publishing.

In recent years, the Pancharatnam phase has been used to design optical devices for modifying the phase of a light wave. In 2006, Italian researchers used a liquid crystal display to create an inhomogeneous phase profile for a transmitted beam [10]. The device may be viewed as a collection of small anisotropic crystals with different orientations; circularly polarized light passing through the crystal accrues a different geometric phase that depends upon this orientation while emerging circularly polarized. An illustration of the orientations used to produce different phase elements is shown in figure 4.9. We explain the operation of this device using Jones vectors in the next section.

4.7 Jones vectors

The effect of every polarization-sensitive optical element introduced in section 4.3 can be described by a linear transformation of the electric field. This indicates that we may write the transformation as a vector-matrix equation, where the input and output electric fields are column vectors and the effect of the optical element is a matrix. If we are interested in designing an optical system where multiple polarization elements are used in series, their combined effect can be calculated by simple multiplication of all the element matrices. This method can be used not only to understand the effects of complicated systems but also to design new ones. This technique was first introduced by Jones [11] and is now known as the *Jones calculus*; the electric field vector is known as a *Jones vector*.

We represent the effect of an individual optical element on an electric field using the matrix equation

$$|\mathbf{E}'\rangle = \mathbf{M}|\mathbf{E}\rangle, \tag{4.100}$$

where $\mathbf{E}$ and $\mathbf{E}'$ are the input and output Jones vectors, which are column vectors, and $\mathbf{M}$ is the matrix representing the element. The input Jones vector has the form

$$|\mathbf{E}\rangle = \begin{bmatrix} E_x \\ E_y \end{bmatrix}. \tag{4.101}$$

The matrices for the various devices can be readily determined from the earlier equations. For a polarizer, we have

$$\mathbf{M}_{pol} = \begin{bmatrix} \cos^2\theta & \cos\theta\sin\theta \\ \cos\theta\sin\theta & \sin^2\theta \end{bmatrix}. \tag{4.102}$$

For a quarter-wave plate, we have

$$\mathbf{M}_{\lambda/4} = \begin{bmatrix} (-i\cos^2\theta + \sin^2\theta) & -(i+1)\sin\theta\cos\theta \\ -(i+1)\sin\theta\cos\theta & (-i\sin^2\theta + \cos^2\theta) \end{bmatrix}, \tag{4.103}$$

while for a half-wave plate, we have

$$\mathbf{M}_{\lambda/2} = \begin{bmatrix} -\cos(2\theta) & -\sin(2\theta) \\ -\sin(2\theta) & \cos(2\theta) \end{bmatrix}. \tag{4.104}$$

Finally, the rotator matrix is of the form

$$\mathbf{M}_{rot} = \begin{bmatrix} \cos\theta_0 & -\sin\theta_0 \\ \sin\theta_0 & \cos\theta_0 \end{bmatrix}. \tag{4.105}$$

So far, these matrix formulas simply reproduce the formulas derived from vector reasoning. The advantage of the Jones calculus is that we may combine the effects of multiple elements by vector multiplication. For example, if light passes through element 1, element 2, and then element 3 in turn, the final output field is of the form

$$|\mathbf{E}'\rangle = \mathbf{M}_3\mathbf{M}_2\mathbf{M}_1|\mathbf{E}\rangle. \tag{4.106}$$

The order of the multiplications is important: because the light encounters element 1 first, $\mathbf{M}_1$ is the first matrix that $|\mathbf{E}\rangle$ encounters. The combined product of matrices then provides the complete description of the system. This matrix can be used to analyze the properties of a system or to design new ones, though the design may still be nontrivial, as we will see.

We consider several simple examples to highlight the use of Jones vectors.

4.7.1 Example: linear to circular polarization

The most fundamental use of a quarter-wave plate is the conversion of linear to circular polarization and vice versa. We can see this effect quite readily using the

Jones calculus. Let us assume that we have a quarter-wave plate aligned along the y-axis ($\theta = \pi/2$), so that the matrix is of the form

$$\mathbf{M}_{\lambda/2} = \begin{bmatrix} 1 & 0 \\ 0 & -i \end{bmatrix}. \tag{4.107}$$

If the input is an arbitrary electric field with a Jones vector given by equation (4.101), the output is given by

$$|\mathbf{E}\rangle = \begin{bmatrix} E_x \\ -iE_y \end{bmatrix}. \tag{4.108}$$

In general, the output is elliptically polarized. But let us consider the special case where $E_x = E_0/\sqrt{2}$ and $E_y = \pm E_0/\sqrt{2}$, which represents linearly polarized light moving along one of the diagonals of the coordinate system. The output is then

$$|\mathbf{E}\rangle = \frac{E_0}{\sqrt{2}} \begin{bmatrix} 1 \\ \mp i \end{bmatrix}, \tag{4.109}$$

which is left (positive) or right (negative) circular polarization. This shows that, with the right orientation, a quarter-wave plate can convert linear polarization to circular polarization. Conversely, let us imagine we have an input field

$$|\mathbf{E}\rangle = \frac{E_0}{\sqrt{2}} \begin{bmatrix} 1 \\ \pm i \end{bmatrix}. \tag{4.110}$$

The output is then

$$|\mathbf{E}\rangle = \frac{E_0}{\sqrt{2}} \begin{bmatrix} 1 \\ \pm 1 \end{bmatrix}, \tag{4.111}$$

which represents linearly polarized light along one of the diagonals. Thus, a quarter-wave plate can also convert circularly polarized light into linearly polarized light.

4.7.2 Example: Pancharatnam–Berry optical element

In section 4.6, we noted that the Pancharatnam phase can be used to impose a transverse phase profile onto a light beam. Here, we show how this works, which is easiest to demonstrate with the Jones calculus.

Let us imagine that we have a thin optical element situated in the plane $z = 0$, and at each point within the plane, there is a small half-wave plate with an orientation $\alpha(\mathbf{r})$, i.e. the Jones matrix of the element is

$$\mathbf{M}_{\lambda/2}(\mathbf{r}) = \begin{bmatrix} -\cos[2\alpha(\mathbf{r})] & -\sin[2\alpha(\mathbf{r})] \\ -\sin[2\alpha(\mathbf{r})] & \cos[2\alpha(\mathbf{r})] \end{bmatrix}. \tag{4.112}$$

We now consider the effect of this element on light that is left-hand circularly polarized, i.e. the input Jones vector is

$$|\mathbf{E}\rangle = \frac{E_0}{\sqrt{2}}\begin{bmatrix}1\\i\end{bmatrix}. \tag{4.113}$$

We can readily find that the output Jones vector is

$$|\mathbf{E}'\rangle = \frac{E_0}{\sqrt{2}}\begin{bmatrix} -\cos[2\alpha(\mathbf{r})] & -\sin[2\alpha(\mathbf{r})] \\ -\sin[2\alpha(\mathbf{r})] & \cos[2\alpha(\mathbf{r})] \end{bmatrix}\begin{bmatrix}1\\i\end{bmatrix} = -\frac{E_0}{\sqrt{2}}e^{2i\alpha(\mathbf{r})}\begin{bmatrix}1\\-i\end{bmatrix}. \tag{4.114}$$

Remarkably, the output state of polarization is uniformly right-hand circularly polarized, but now with a spatially varying phase $2\alpha(\mathbf{r})$. The orientations of the small half-wave plates can therefore be chosen to impart a variety of useful phase structures on the field, such as wavefront curvature or an optical vortex.

4.7.3 Example: circular polarization filter

A polarizer can be used to isolate the orthogonal components E_x and E_y of the electric field by orienting the polarizer along each of the axes. One might wonder whether it is possible to do the same thing for circular polarization, i.e. we want to construct a device that isolates the amplitude of the left- or right-circularly polarized parts of the field. This turns out to be more subtle than it might first appear.

A straightforward attempt would simply use a polarizer to create a field polarized at 45° and then pass it through a quarter-wave plate, as was done to convert linear to circular polarization in subsection 4.7.1. We have as our system matrix

$$\mathbf{M} = \begin{bmatrix}1 & 0\\0 & -i\end{bmatrix}\begin{bmatrix}1/2 & 1/2\\1/2 & 1/2\end{bmatrix} = \frac{1}{2}\begin{bmatrix}1 & 1\\-i & -i\end{bmatrix}. \tag{4.115}$$

The output electric field vector $\mathbf{E}_o$ based on a general input vector $\mathbf{E}_i$ is given by

$$|\mathbf{E}_o\rangle = \mathbf{M}|\mathbf{E}_i\rangle = \frac{1}{2}\begin{bmatrix}1 & 1\\-i & -i\end{bmatrix}\begin{bmatrix}E_x\\E_y\end{bmatrix} = \frac{1}{2}\begin{bmatrix}E_x + E_y\\-i(E_x + E_y)\end{bmatrix}. \tag{4.116}$$

We can convert this output state into a circular polarization basis by means of the transformation

$$\begin{bmatrix}E_+\\E_-\end{bmatrix} = \frac{1}{\sqrt{2}}\begin{bmatrix}1 & -i\\1 & +i\end{bmatrix}\begin{bmatrix}E_x\\E_y\end{bmatrix}. \tag{4.117}$$

Let us apply this transformation to $|\mathbf{E}_o\rangle$; we readily find that, in a circular basis, we have

$$|\mathbf{E}_o\rangle = \begin{bmatrix}E_+\\E_-\end{bmatrix} = \begin{bmatrix}0\\\frac{1}{\sqrt{2}}(E_x + E_y)\end{bmatrix}, \tag{4.118}$$

which indeed is right-hand circular polarization. However, let us compare this to the circular components of the input field,

$$|\mathbf{E}_i\rangle = \begin{bmatrix} E_+ \\ E_- \end{bmatrix} = \frac{1}{\sqrt{2}} \begin{bmatrix} E_x - iE_y \\ E_x + iE_y \end{bmatrix}. \tag{4.119}$$

If we were truly making a filter that extracted the original right-hand circular polarization component, equation (4.118) should have given us a value $(E_x + iE_y)/\sqrt{2}$, which it did not. We must be more clever to extract circular polarization.

Let us instead start with a quarter-wave plate with its fast axis aligned in the x-direction. We have seen that circularly polarized light entering this device produces polarized light at $+45°$ or $-45°$, depending on whether the input was right or left circular polarization. So we follow our quarter-wave plate with a polarizer at $+45°$. We then pass the light through another quarter-wave plate oriented along y; the total system matrix is

$$\mathbf{M} = \begin{bmatrix} 1 & 0 \\ 0 & -i \end{bmatrix}\begin{bmatrix} 1/2 & 1/2 \\ 1/2 & 1/2 \end{bmatrix}\begin{bmatrix} -i & 0 \\ 0 & 1 \end{bmatrix} = \frac{1}{2}\begin{bmatrix} -i & 1 \\ -1 & -i \end{bmatrix}. \tag{4.120}$$

Let us examine the effect of this system matrix on a general input polarization state,

$$|\mathbf{E}_o\rangle = \frac{1}{2}\begin{bmatrix} -i & 1 \\ -1 & -i \end{bmatrix}\begin{bmatrix} E_x \\ E_y \end{bmatrix} = \frac{1}{2}\begin{bmatrix} -i(E_x + iE_y) \\ -(E_x + iE_y) \end{bmatrix}. \tag{4.121}$$

Finally, we look at this output vector in the circular polarization basis using equation (4.117):

$$|\mathbf{E}_o\rangle = \begin{bmatrix} E_+ \\ E_- \end{bmatrix} = \frac{1}{\sqrt{2}}\begin{bmatrix} 0 \\ -i(E_x + iE_y) \end{bmatrix}. \tag{4.122}$$

On comparison with equation (4.119), we can see that the output indeed gives us only the right-hand circularly polarized component of the initial field, with a trivial phase shift of $-i$.

If all we care about is the intensity of this circular component, and not preserving it in a circular polarization state, we can leave off the second quarter-wave plate. The output of the system is then linearly polarized, but its total intensity is the intensity of the right-hand circular component alone.

4.7.4 Example: optical attenuator

As a more complicated example, we consider the use of four polarization-sensitive elements to create an optical attenuator, which imposes a controlled amount of attenuation on an incident field. The four elements are shown in figure 4.10. We have a horizontal polarizer, a half-wave plate oriented at an angle θ_1, a vertical polarizer, and then a half-wave plate oriented at an angle θ_2.

The product of matrices that describes this system is given by

$$\mathbf{M} = \mathbf{H}_2\mathbf{P}_2\mathbf{H}_1\mathbf{P}_1 = \begin{bmatrix} -\cos(2\theta_2) & -\sin(2\theta_2) \\ -\sin(2\theta_2) & \cos(2\theta_2) \end{bmatrix}\begin{bmatrix} 0 & 0 \\ 0 & 1 \end{bmatrix}\begin{bmatrix} -\cos(2\theta_1) & -\sin(2\theta_1) \\ -\sin(2\theta_1) & \cos(2\theta_1) \end{bmatrix}\begin{bmatrix} 1 & 0 \\ 0 & 0 \end{bmatrix}. \tag{4.123}$$

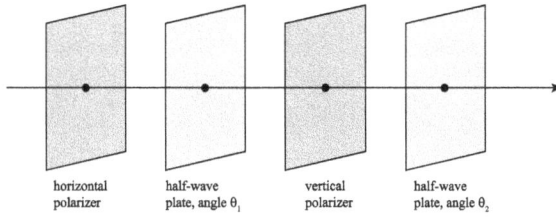

Figure 4.10. Arrangement of an optical attenuator, with light propagating from left to right.

The result of the multiplication is

$$\mathbf{M} = \begin{bmatrix} \sin(2\theta_2)\sin(2\theta_1) & 0 \\ -\cos(2\theta_2)\sin(2\theta_1) & 0 \end{bmatrix}. \tag{4.124}$$

We still have two free parameters, θ_1 and θ_2, and now we must use some creativity to design an isolator. Let us set $\theta_2 = \pi/4$; then we have

$$\mathbf{M} = \begin{bmatrix} \sin(2\theta_1) & 0 \\ 0 & 0 \end{bmatrix}. \tag{4.125}$$

This matrix looks like the matrix for a horizontally oriented linear polarizer, except that the amplitude of light transmitted is proportional to $\sin(2\theta_1)$. If we put in an arbitrary beam of light, the output Jones vector is

$$|\mathbf{E}\rangle = \begin{bmatrix} E_x \sin(2\theta_1) \\ 0 \end{bmatrix}. \tag{4.126}$$

Thus, we can control the output amplitude by changing θ_1. For $\theta_1 = 0$, there is no transmission, and for $\theta_1 = \pi/4$, the x-component is completely transmitted.

4.8 Nonuniform polarization

Traditionally, researchers have used beams of light which possess a uniform state of polarization: that is, the state of polarization is exactly the same at every point in the beam's cross section. In recent years, however, beams with a polarization that varies within the cross section—nonuniform polarization—have been extensively studied and applied, and it is worth describing the concepts briefly.

The two most familiar classes of nonuniformly polarized beams are radially polarized beams and azimuthally polarized beams; these beams can be mathematically expressed as the sum of two Hermite–Gaussian (HG) beams with orthogonal polarizations. As discussed in section 3.5, the amplitudes of the two lowest nontrivial HG beams are given by

$$HG_{01} = \frac{x}{w_0}e^{-(x^2 + y^2)/w_0^2}, \quad HG_{01} = \frac{y}{w_0}e^{-(x^2 + y^2)/2w_0^2}, \tag{4.127}$$

where w_0 is the width of the Gaussian. (These modes are often written in normalized form, but we leave off the normalization for simplicity.)

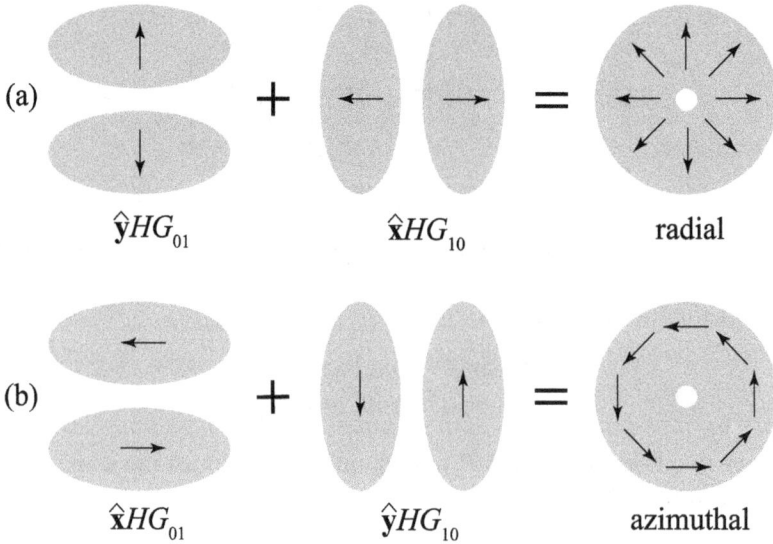

Figure 4.11. The construction of (a) a radially polarized beam and (b) an azimuthally polarized beam from HG modes of orthogonal polarization.

We can construct radially and azimuthally polarized beams by taking combinations of these HG beams, as shown in figure 4.11. For example, a radially polarized beam can be constructed using a vertically polarized HG_{01} mode and a horizontally polarized HG_{10} mode.

The resultant electric fields of the two cases are of the form

$$\mathbf{E}_{rad}(\mathbf{r}) = \frac{x\hat{\mathbf{x}} + y\hat{\mathbf{y}}}{w_0}e^{-(x^2 + y^2)/w_0^2}, \quad \mathbf{E}_{az}(\mathbf{r}) = \frac{-y\hat{\mathbf{x}} + x\hat{\mathbf{y}}}{w_0}e^{-(x^2 + y^2)/w_0^2}. \quad (4.128)$$

The corresponding polarization states are shown in figure 4.12, which illustrates why the radial and azimuthal states have their names. It should be noted that the state of polarization is linear and real-valued, so we show the electric field as a vector; for a general state of elliptical polarization, we would only be able to show the major axes of the polarization ellipse, as a complex electric field does not generally have a 'direction.'

From equation (3.77) given in section 3.5, it can readily be seen that the HG modes HG_{10} and HG_{01} have the same propagation characteristics, which means that their amplitudes and phases change in exactly the same manner during propagation. This, in turn, means that a radially or azimuthally polarized field remains radial or azimuthal at any propagation distance, at least within the paraxial approximation for beams.

Radial and azimuthal beams have a surprisingly long history, dating back to the creation of a quantum well semiconductor laser in 1992 that produced azimuthally polarized light [12]. In 2000, Youngworth and Brown [13] studied both classes extensively and demonstrated the strong longitudinal component of radially polarized light, which sparked intense interest in such beams and their properties. In 2003, Dorn, Qabis, and Leuchs experimentally demonstrated [14] that a radially

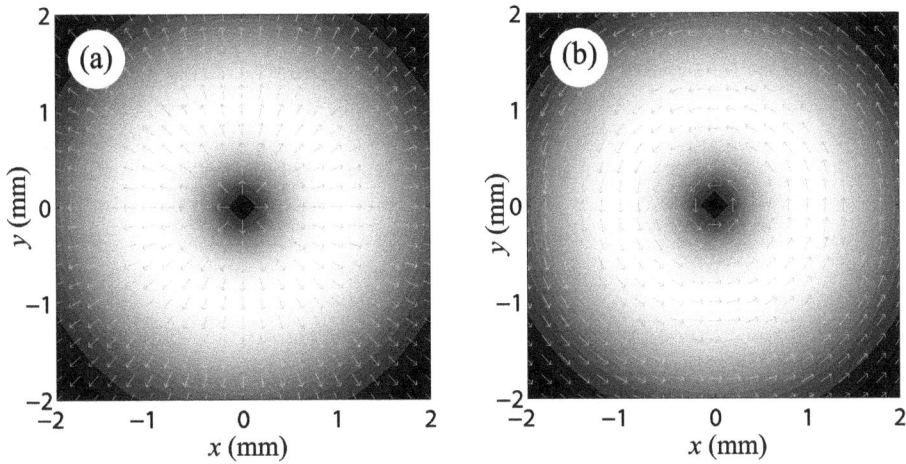

Figure 4.12. Illustration of the intensity and electric field directions of (a) a radially polarized beam and (b) an azimuthally polarized beam.

polarized field has a smaller focal spot than a uniformly polarized field, making such a field potentially interesting for imaging.

At the center of a radially or azimuthally polarized beam, the state of polarization is undefined, as all directions of polarization converge on that point. The center of the beam is an example of a *polarization singularity*, and the study of such singularities is a subset of the subfield of optics known as *singular optics* [15, 16].

One other class of nonuniformly polarized beams is worth noting here. It is possible to generate beams that possess nearly every state of polarization on the Poincaré sphere in their cross section. Such beams are known as *Poincaré beams* and were first introduced in 2010 [17].

These beams may be considered the superposition of an ordinary Gaussian beam with a Laguerre–Gaussian beam of order $(0, 1)$, which, from equation (3.78), have the unnormalized forms

$$LG_{00}(\mathbf{r}) = e^{-(x^2 + y^2)/w_0^2}, \quad LG_{01}(\mathbf{r}) = \frac{\rho}{w_0} e^{i\phi} e^{-(x^2 + y^2)/w_0^2}, \tag{4.129}$$

where ρ and ϕ are polar coordinates. We combine these fields with orthogonal polarizations $\mathbf{e}_1$ and $\mathbf{e}_2$, such that

$$\mathbf{E}(\mathbf{r}) = \cos\gamma\mathbf{e}_1 LG_{00}(\mathbf{r}) + \sin\gamma\mathbf{e}_2 LG_{01}, \tag{4.130}$$

where γ is a real-valued free parameter. If we factor out all common terms, we may write the state of polarization as

$$\mathbf{e} = \mathbf{e}_1 + \frac{\sqrt{2}\tan\gamma\rho}{w_0} e^{i\phi}\mathbf{e}_2. \tag{4.131}$$

It can now be noted that the value of ρ determines the relative amplitude ratio between the orthogonal polarizations, while the value of ϕ determines the relative phase. Over the entire transverse plane, every phase and amplitude ratio appears, which indicates that every state of polarization appears. During propagation, a constant phase factor arises between LG_{00} and LG_{01}, which changes the position of the different polarization states in the beam cross section, but all states still appear.

Of course, in practice, only a finite number of states appear because the beam is of finite width, and the intensity becomes too low to be measured outside a finite radius. By adjusting the value of γ, one can adjust the spatial extent of the polarization patterns, forcing more of the full Poincaré sphere to appear in a region of measurable intensity.

The creation of Poincaré beams is possible because the surface of a sphere can be mapped to a plane, just as the Poincaré sphere itself represents the reverse process: the mapping of all possible complex field states in the plane onto a sphere.

4.9 Exercises

1. For each of the uniform *monochromatic* plane waves in vacuum characterized below, describe the polarization ellipse (ellipticity, angle) and give the handedness, when appropriate:
   (a) $\mathbf{E} = (\hat{\mathbf{x}} + 2\hat{\mathbf{y}})e^{ikz}$,
   (b) $\mathbf{E} = (\hat{\mathbf{x}} - 2\hat{\mathbf{y}})e^{ikz}$,
   (c) $\mathbf{E} = (\hat{\mathbf{x}} + 2i\hat{\mathbf{y}})e^{ikz}$,
   (d) $\mathbf{E} = (i\hat{\mathbf{x}} + 2\hat{\mathbf{y}})e^{ikz}$.
   (Note: for these simple cases, you should be able to determine the properties of the ellipse without performing tedious calculations.)

2. A theory of optical activity due to Reusch and Sohncke states that a Faraday rotator can be constructed from an appropriate collection of phase-retarding plates. The simplest system of this form consists of a quarter-wave plate, a general wave plate with 'fast' axis phase α and orientation θ, followed by another quarter-wave plate perpendicular to the first. Show that, for an appropriate choice of α and θ, this system works as a Faraday rotator. (Hint: it turns out that α is the quantity that determines how much the polarization rotates.)

3. Fun with the Poincaré sphere! Using the definition of the Stokes parameters in terms of field quantities (a_x, a_y, δ), describe the state of polarization for light that lies on the sphere for (a) $s_1/s_0 = 1$, (b) $s_1/s_0 = -1$, (c) $s_2/s_0 = 1$, (d) $s_2/s_0 = -1$, (e) $s_3/s_0 = 1$, (f) $s_3/s_0 = -1$. Sketch the polarization ellipse in terms of E_x and E_y, showing the handedness.

4. Pancharatnam phase fun! Imagine we start with horizontally polarized light, use an optical element to rotate it to vertically polarized, use a quarter-wave plate to make it right-hand circularly polarized, and then another quarter-wave plate to take it directly back to horizontally polarized light. Using the graphical approach discussed in the chapter, determine the Pancharatnam phase shift produced. Confirm that this is consistent with what you would expect from the Poincaré sphere.

5. A polarization-dependent optical isolator consists of three elements: a vertically oriented polarizer, a Faraday rotator which rotates by 45°, and a second polarizer oriented at 45°. Using Jones vectors and the matrices for the different optical elements, determine the effect of this isolator on a polarized light wave propagating (a) forwards and (b) backwards through the device.

6. It is often helpful to work with a Jones vector defined in a circular polarization basis, such that

$$|E\rangle = \begin{bmatrix} E_+ \\ E_- \end{bmatrix},$$

where E_+ and E_- represent the right- and left-handed components of the field. Determine the matrices which describe a linear polarizer of angle θ, a quarter-wave plate of angle θ, and a rotator of angle θ_0 in this basis.

7. We can recreate the classic three-polarizer experiment using Jones vectors. First, consider a system that consists of a horizontally oriented polarizer ($\theta = 0$) followed by a vertically oriented polarizer ($\theta = 90°$). Show that no light passes through this system. Now imagine that a third polarizer is placed between the two previous polarizers at an angle θ. Determine the output electric field vector, and determine how the intensity of light, taken as $I = |\mathbf{E}|^2$, depends on the angle θ.

8. Suppose we have a monochromatic plane wave with a complex electric field of the form

$$\mathbf{E}(\mathbf{r}) = [\hat{\mathbf{x}} + 2i\hat{\mathbf{y}}]E_0 e^{ikz} e^{-i\omega t}.$$

By whatever means you deem appropriate, calculate the Stokes parameters S_1, S_2, S_3, and S_0 of this field. (Probably easiest to figure out a_x, a_y, and δ.)

9. Repeat the calculation of subsection 4.7.3 but design a system that extracts only the left-hand circular polarization component of a general field.

10. Complete the calculation to derive the major and minor axis values a_M and a_m of equations (4.27) from the previous formulas, equations (4.26).

11. The unit vector describing the state of polarization of a Poincaré beam is given by equation (4.131). Determine the locations in the transverse (x, y)-plane where the polarization is linear, left-hand circular, and right-hand circular, for the special cases
    (a) $\mathbf{e}_1 = \hat{\mathbf{x}}$, $\mathbf{e}_2 = \hat{\mathbf{y}}$,
    (b) $\mathbf{e}_1 = \mathbf{e}_+$, $\mathbf{e}_2 = \mathbf{e}_-$.
    Also, describe the regions where the polarization is left-hand elliptical and right-hand elliptical in each case.

References

[1] Nye J F 1983 Lines of circular polarization in electromagnetic wave fields *Proc. R. Soc.* A **389** 279–90

[2] Liu Y, Beetar J E, Nesper J, Gholam-Mirzaei S and Chini M 2021 Single-shot measurement of few-cycle optical waveforms on a chip *Nat. Photon.* **16** 109–12

[3] Stokes G G 1852 On the composition and resolution of streams of polarized light from different sources *Trans. Camb. Phil. Soc.* **9** 399–416

[4] Jerrard H G 1954 Transmission of light through birefringent and optically active media: the Poincaré sphere *J. Opt. Soc. Am.* **44** 634–40

[5] Gbur G J 2016 *Singular Optics* (Boca Raton, FL: CRC Press)

[6] Pancharatnam S 1956 Generalized theory of interference, and its applications *Proc. Indian Acad. Sci.* **44** 247–62

[7] Hariharan P, Mujumdar S and Ramachandran H 1999 A simple demonstration of the Pancharatnam phase as a geometric phase *J. Mod. Opt.* **46** 1443–6

[8] Berry M V 1984 Quantal phase factors accompanying adiabatic changes *Proc. R. Soc.* A **392** 45–57

[9] Berry M V 1987 The adiabatic phase and Pancharatnam's phase for polarized light *J. Mod. Opt.* **34** 1401–7

[10] Marrucci L, Manzo C and Paparo D 2006 Pancharatnam-Berry phase optical elements for wave front shaping in the visible domain: switchable helical mode generation *Appl. Phys. Lett.* **88** 221102

[11] Jones R C 1941 New calculus for the treatment of optical systems *J. Opt. Soc. Am.* **31** 488–93

[12] Erdogan T, King O, Hicks G W, Hall D G, Anderson E H and Rooks M J 1992 Circularly symmetric operation of a concentric-circle-grating, surface-emitting, AlGaAs/GaAs quantum-well semiconductor laser *Appl. Phys. Lett.* **60** 1921–3

[13] Youngworth K S and Brown T G 2000 Focusing of high numerical aperture cylindrical vector beams *Opt. Exp.* **7** 77–87

[14] Dorn R, Quabis S and Leuchs G 2003 Sharper focus for a radially polarized light beam *Phys. Rev. Lett.* **91** 233901

[15] Soskin M S and Vasnetsov M V 2001 Singular optics *Progress in Optics* **vol 42** ed E Wolf (Amsterdam: Elsevier) 219

[16] Dennis M R, O'Holleran K and Padgett M J 2009 Singular optics: optical vortices and polarization singularities *Progress in Optics* **vol 53** ed E Wolf (Amsterdam: Elsevier) 293

[17] Beckley A M, Brown T G and Alonso M A 2010 Full Poincaré beams *Opt. Exp.* **18** 10777–85

IOP Publishing

Electromagnetic Optics

Gregory J Gbur

Chapter 5

Maxwell's equations in matter

Up to this point, we have studied Maxwell's equations in vacuum, which has allowed us to work entirely with the **E** and **B** fields and neglect **D** and **H**. Of course, most of modern optical physics involves the interactions between light and matter; so, in this chapter, we explore the physics and formalism of basic light–matter interactions.

We have already noted that the **D** and **E** fields are related, as are the **B** and **H** fields. **E** and **B** are often referred to as the 'microscopic' fields, which represent the exact physical fields on all scales, and **D** and **H** are referred to as the 'macroscopic' fields, as they incorporate the microscopic interactions between light and matter that we cannot directly observe. The relationship between these fields in the time domain may be written generally as

$$\mathbf{D}(\mathbf{r}, t) = \mathbf{D}[\mathbf{E}(\mathbf{r}, t), \mathbf{B}(\mathbf{r}, t)], \tag{5.1}$$

$$\mathbf{H}(\mathbf{r}, t) = \mathbf{H}[\mathbf{E}(\mathbf{r}, t), \mathbf{B}(\mathbf{r}, t)]. \tag{5.2}$$

In the most general case, we note that **D** and **H** may both depend on the electric and magnetic fields, making them coupled. Furthermore, the following situations are possible:

- The relationship may be dispersive, in which the macroscopic fields at any given time depend on the microscopic fields over a range of times through an integral. This is almost always the case, and we will discuss dispersion in detail in chapter 6.
- The relationship may be anisotropic, in which the macroscopic fields depend on the microscopic fields through a matrix relation. This implies that the speed of light depends on the direction of wave propagation and the state of polarization. We will discuss anisotropy in chapter 8.
- The relationship may be nonlinear, in that **D** is not linearly proportional to **E**, and this may also be true for **H** and its relationship to **B**. Nonlinear optics is

doi:10.1088/978-0-7503-6064-7ch5

outside the scope of this book, but we refer the reader to Boyd's excellent *Nonlinear Optics* [1] for more information.

- The relationship may be nonlocal: the **D** and **H** fields at a point may depend on the **E** and **B** fields in a region around that point. This is a possibility typically considered only on the smallest scale, as a quantum effect; an electron can react to a field everywhere its quantum wave function extends. This is a very complicated subject, and we refer the reader to other texts; see, for instance, 'spatial dispersion' in Landau, Lifshitz, and Pitaevskii [2] and also Cho [3].

Though we cannot address all of these possibilities in this book, we mention them to stress that the relationship between light and matter can be exceedingly complicated. The use of **D** and **H** is intended to simplify related calculations.

5.1 Electric dipoles and the D-field

In general, all of the fields in Maxwell's equations are time-varying. The relationships between the **D** and **E** fields, however, are easiest to understand in the static limit, and so we start there, noting any differences in the time-dependent case as needed. We assume the reader has some basic familiarity with electrostatics and magnetostatics, though we derive many of the relevant equations as we go.

To begin, we take a microscopic view of matter. We view an atom as a positive nucleus surrounded by an electron cloud, as illustrated in figure 5.1(a). When an electric field is applied to this atom, the field distorts its shape: the nucleus is pulled to the right and the electron is pulled to the left, though typically the strongest effect arises from the less massive electrons. This is effectively a separation of charge, and the result is a dipole. The simplest approximation of the interaction of an electric field with a material is therefore that the electric field turns the material into a collection of dipoles.

Let us now take a look at a simple electric dipole. Its geometry is illustrated in figure 5.2. The dipole consists of a positive charge $+q$ and a negative charge $-q$ separated by a distance d. The dipole is assumed to be oriented along the z-axis for convenience; of course, we are free to choose our coordinate system such that this is true. For the moment, we also take our dipole to be centered on the origin of the coordinate system.

The electrostatic potential $\phi(\mathbf{r})$ of a dipole is given by

$$\phi(\mathbf{r}) = \frac{1}{4\pi\epsilon_0}\left[\frac{q}{R_+} - \frac{q}{R_-}\right], \tag{5.3}$$

where the electric field is given by $\mathbf{E}(\mathbf{r}) = -\nabla\phi(\mathbf{r})$. The two distances $R_\pm$ that satisfy

$$R_\pm^2 = r^2 + \left(\frac{d}{2}\right)^2 \mp rd\cos\theta = r^2\left[1 \mp \frac{d}{r}\cos\theta + \frac{d^2}{4r^2}\right] \tag{5.4}$$

(a)   (b)   E

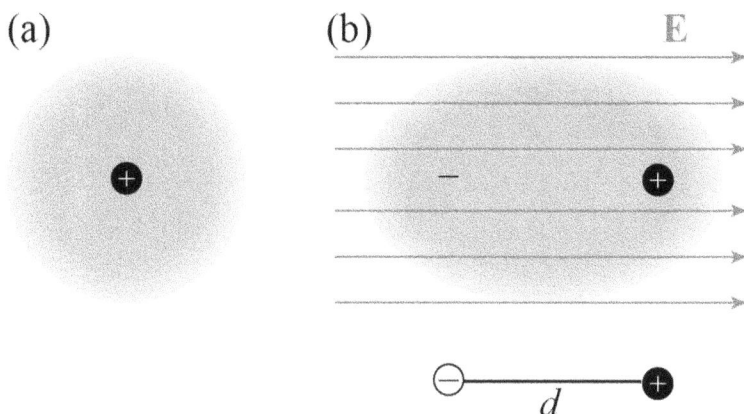

Figure 5.1. Illustration of (a) an isolated atom and (b) an atom under the influence of an electric field, with the equivalent dipole.

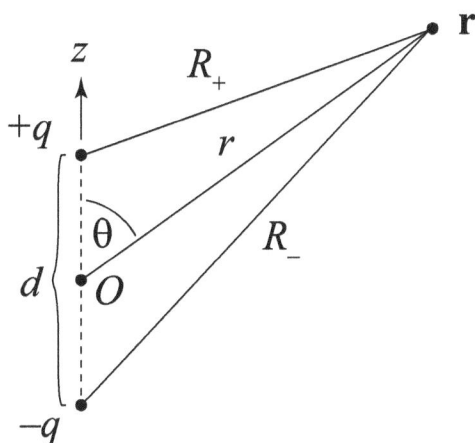

Figure 5.2. Illustration of the notation for an electric dipole.

may be determined using trigonometry. Let us consider the potential at a distance significantly larger than the dipole itself, i.e. $r \gg d$ or $r/d \gg 1$. This means that $d^2/4r^2 \ll d/r$, and we may neglect the third term of equation (5.4) and write

$$\frac{1}{R_{\pm}} \approx \frac{1}{r}\left(1 \mp \frac{d}{r}\cos\theta\right)^{-1/2}. \tag{5.5}$$

We may further approximate this square root using the first two terms of its binomial expansion,

$$(1 \mp x)^{-1/2} \approx 1 \pm \frac{x}{2}. \tag{5.6}$$

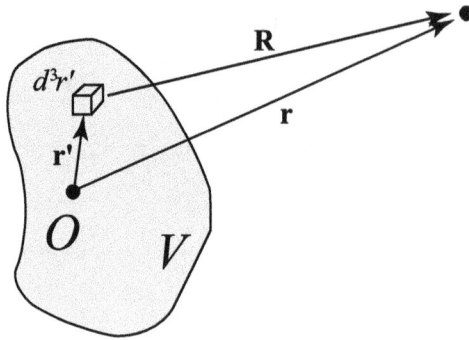

Figure 5.3. Illustration of the notation for the potential of a dipole distribution.

We thus have

$$\frac{1}{R_\pm} = \frac{1}{r}\left(1 \pm \frac{d}{2r}\cos\theta\right). \tag{5.7}$$

If we substitute this approximation into equation (5.3), we get the expression

$$\phi(\mathbf{r}) \approx \frac{1}{4\pi\epsilon_0}\frac{qd\cos\theta}{r^2}. \tag{5.8}$$

Let us finally make a point dipole approximation: we assume that $d \to 0$ and $q \to \infty$ such that the product $qd \to p$, a finite number. The quantity p is the magnitude of the dipole moment. Since θ is the angle between the vector $\mathbf{r}$ and the z-axis, and the dipole is oriented along the z-axis, we can define the vector dipole moment $\mathbf{p}$ as pointing from the negative to the positive charge. Our equation for the dipole potential can then be written in a coordinate-independent form as

$$\phi(\mathbf{r}) = \frac{1}{4\pi\epsilon_0}\frac{\mathbf{p}\cdot\mathbf{r}}{r^3}. \tag{5.9}$$

This is the potential for a single dipole; if we want to study the electrostatic potential of a continuous distribution of dipoles located in a volume V, we introduce the polarization density $\mathbf{P}(\mathbf{r})$, the dipole moment per unit volume. An infinitesimal volume then has a dipole moment $\mathbf{p}(\mathbf{r}') = \mathbf{P}(\mathbf{r}')d^3r'$, and the total potential is the sum of the potentials of all these dipole moments, converted into an integral over the volume. We write

$$\phi(\mathbf{r}) = \frac{1}{4\pi\epsilon_0}\int_V \frac{\mathbf{P}(\mathbf{r}')\cdot\mathbf{R}}{R^3}d^3r'. \tag{5.10}$$

Here, the vector $\mathbf{R} = \mathbf{r} - \mathbf{r}'$ represents the vector distance between $\mathbf{r}'$, called the source point (the point where a dipole lies), and $\mathbf{r}$, called the field point (the point at which we measure the total potential). The geometry is illustrated in figure 5.3.

We now use the simple identity

$$\nabla'\left(\frac{1}{R}\right) = \frac{\mathbf{R}}{R^3}, \tag{5.11}$$

where ∇' is the gradient with respect to the variable $\mathbf{r}'$. We substitute from this expression into equation (5.10) and get

$$\phi(\mathbf{r}) = \frac{1}{4\pi\epsilon_0} \int_V \mathbf{P}(\mathbf{r}') \cdot \nabla'\left(\frac{1}{R}\right) d^3r'. \tag{5.12}$$

We next use the vector identity,

$$\nabla' \cdot \left(\frac{\mathbf{P}}{R}\right) = \frac{1}{R}(\nabla' \cdot \mathbf{P}) + (\mathbf{P} \cdot \nabla')\left(\frac{1}{R}\right). \tag{5.13}$$

The rightmost term is equivalent to the term in the integrand, and we substitute this into the integrand to get the expression

$$\phi(\mathbf{r}) = \frac{1}{4\pi\epsilon_0}\left[\int_V \nabla' \cdot \left(\frac{\mathbf{P}}{R}\right) d^3r' - \int_V \frac{1}{R}(\nabla' \cdot \mathbf{P}) d^3r'\right]. \tag{5.14}$$

A brief note: you may have noticed that we have dropped the argument '$(\mathbf{r}')$' of the functions in the integrand above. This is done entirely for notational convenience in cases where the arguments of the relevant functions are obvious. We will do this throughout the book when appropriate, mainly in integrands and vector identities.

The first integral on the right-hand side of equation (5.14) is the integral of a divergence over a volume, which means we may use Gauss's theorem to convert it into a surface integral. The final result is of the form

$$\phi(\mathbf{r}) = \frac{1}{4\pi\epsilon_0}\left[\oint_S \frac{1}{R}\mathbf{P} \cdot d\mathbf{a}' - \int_V \frac{1}{R}(\nabla' \cdot \mathbf{P}) d^3r'\right]. \tag{5.15}$$

Our integral for the potential due to a distribution of electric dipoles therefore has two contributions: an integral over a surface and an integral over a volume. Each of these integrals has the form of an electric potential of a distribution of electric charges, and we may therefore introduce the following definitions:

$$\sigma_b(\mathbf{r}') \equiv \mathbf{P}(\mathbf{r}') \cdot \mathbf{n}'(\mathbf{r}'), \tag{5.16}$$

$$\rho_b(\mathbf{r}') \equiv -\nabla' \cdot \mathbf{P}(\mathbf{r}'), \tag{5.17}$$

where $\mathbf{n}'(\mathbf{r}')$ represents the unit normal to the surface at point $\mathbf{r}'$. The quantity σ_b is called the *bound surface charge density*, and ρ_b is called the *bound volume charge density*. They are referred to as *bound* charges because they are charges that are fixed to the dipoles and are 'bound' to atoms. This is in contrast to *free* charges, such as electrons in a conductor, which are free to move about within the conductor in response to applied electromagnetic forces.

Dipoles are electrically neutral, but a distribution of dipoles can result in unbalanced bound charges. The production of a surface charge density is illustrated

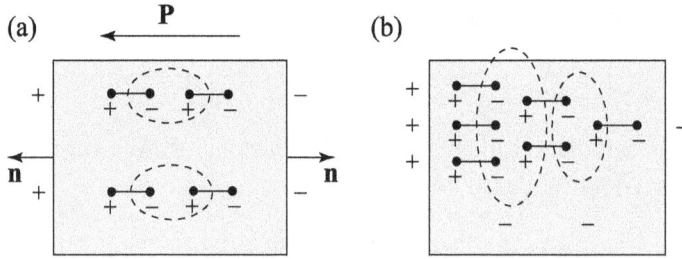

Figure 5.4. Illustrations of the concepts of (a) bound surface charge σ_b, (b) bound volume charge ρ_b.

in figure 5.4(a); we assume a uniform distribution of dipoles in a volume, all oriented in the same direction. Within the volume of the material, the charges of neighboring dipoles, indicated by the dashed loops, are effectively canceled. We are left with unbalanced charges at the surface. The production of a volume charge density is illustrated in figure 5.4(b), where we now assume that the density of dipoles varies with position. Now, within the volume, the negative charges outnumber the positive charges at every location, leaving a net negative charge within the volume.

Now we come to the point of why we did all this mathematics to develop bound charge densities. In a dielectric, the total charge is a combination of bound charges ρ_b and free charges ρ_f:

$$\rho = \rho_b + \rho_f. \tag{5.18}$$

We substitute from this equation into Gauss's law to get

$$\nabla \cdot \mathbf{E} = \rho/\epsilon_0 = (\rho_f - \nabla \cdot \mathbf{P})/\epsilon_0. \tag{5.19}$$

We now have a divergence on both sides of the equation; we may group these together to get the formula

$$\nabla \cdot (\epsilon_0 \mathbf{E} + \mathbf{P}) = \rho_f. \tag{5.20}$$

The quantity in parentheses apparently satisfies a form of Gauss's law, except it only depends upon the free charges; we define

$$\mathbf{D} \equiv \epsilon_0 \mathbf{E} + \mathbf{P} \tag{5.21}$$

as the displacement field, and we then have Gauss's law for $\mathbf{D}$:

$$\nabla \cdot \mathbf{D} = \rho_f. \tag{5.22}$$

By integrating both sides of this expression over a volume and using Gauss's theorem, we may write

$$\oint \mathbf{D} \cdot d\mathbf{a} = Q_{f,enc}, \tag{5.23}$$

where $Q_{f,enc}$ is the net free charge enclosed.

Gauss's law for the **D** field in integral form is extremely useful as a tool for finding the **D** field in a dielectric in cases of high symmetry, as **D** does not depend at all on the material properties of the dielectric. Getting the **E** field, however, requires knowledge of the material properties, which we will discuss in an upcoming section. Equation (5.21) is an exact formula, with the caveat that it assumes that the constituent parts of the material have a pure dipole response to an applied electric field. We will later discuss metamaterials, in which the constituent parts of the material are 'meta-atoms' that can be over ten times larger than natural atoms; in such a case, one may have to take into account the quadrupole response of the individual parts as well [4].

5.2 Magnetic dipoles and the H-field

We may also define the **H** field by looking at the magnetostatic case and considering a material that consists of a collection of *magnetic* dipoles. A magnetic dipole is typically viewed as a circular loop of radius a carrying current I, considered in the limit $a \rightarrow 0$ but with the product Ia^2 remaining finite. We start with the Biot–Savart law for magnetic fields and proceed to calculate the field of such a dipole, using

$$\mathbf{B}(\mathbf{r}) = \frac{\mu_0}{4\pi} \oint_C \frac{I d\mathbf{l}' \times \mathbf{R}}{R^3}. \tag{5.24}$$

We assume for the moment that our current loop lies in the xy-plane and is centered on the origin. Our path C is thus a circle of radius a in the xy-plane, and $d\mathbf{l}' = a d\phi' \hat{\phi}$; our notation is illustrated in figure 5.5. Before attempting to evaluate the integral, however, we note that we may write

$$\frac{\mathbf{R}}{R^3} = -\nabla \frac{1}{R}, \tag{5.25}$$

so that

$$\mathbf{B}(\mathbf{r}) = -\frac{\mu_0}{4\pi} \oint_C I d\mathbf{l}' \times \nabla \frac{1}{R} = \nabla \times \frac{\mu_0}{4\pi} \oint_C \frac{I d\mathbf{l}'}{R}, \tag{5.26}$$

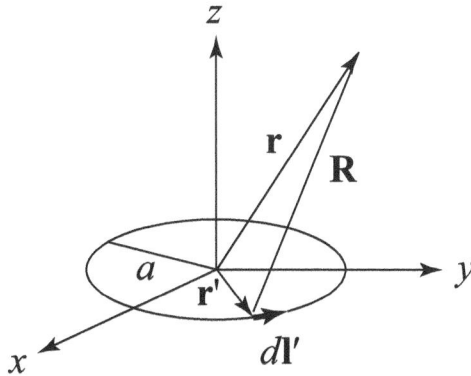

Figure 5.5. The geometry and notation used to derive the form of a magnetic dipole.

where we have taken the derivative out of the integral. We may therefore introduce a vector potential $\mathbf{A}$ that satisfies $\mathbf{B} = \nabla \times \mathbf{A}$; this has the form

$$\mathbf{A}(\mathbf{r}) = \frac{\mu_0}{4\pi} \oint_C \frac{I d\mathbf{l}'}{R}. \tag{5.27}$$

(We will discuss potentials in detail for the full electromagnetic case in a later chapter. For now, we just take them as useful mathematical tools.)

Before we attempt to evaluate the integral, we assume a small dipole, as we did in the electric case, i.e. $|\mathbf{r}'| \ll |\mathbf{r}|$. We may then write

$$\frac{1}{R} \approx \frac{1}{r}\left(1 + \frac{\hat{\mathbf{r}} \cdot \mathbf{r}'}{r}\right). \tag{5.28}$$

On substitution, we find that

$$\mathbf{A}(\mathbf{r}) \approx \frac{\mu_0}{4\pi r} \oint_C I d\mathbf{l}' + \frac{\mu_0}{4\pi r^2} \oint_C I(\hat{\mathbf{r}} \cdot \mathbf{r}') d\mathbf{l}'. \tag{5.29}$$

The first integral vanishes because of the symmetry of our loop; in essence, this is the statement that we do not have a magnetic monopole moment. To evaluate the second integral, we employ a modified version of Stokes's theorem, which we briefly outline. Stokes's theorem states that, for a well-behaved vector $\mathbf{v}$,

$$\int_S \nabla' \times \mathbf{v} \cdot d\mathbf{a}' = \oint_C \mathbf{v} \cdot d\mathbf{l}'. \tag{5.30}$$

Let us set $\mathbf{v} = \mathbf{c}V$, where V is a scalar function and $\mathbf{c}$ is a constant vector. We then have

$$\int_S \nabla' \times (\mathbf{c}V) \cdot d\mathbf{a}' = \mathbf{c} \cdot \oint_C V d\mathbf{l}'. \tag{5.31}$$

The derivative only acts on V because $\mathbf{c}$ is constant, and we may use the scalar triple product to rearrange terms. We thus obtain

$$\mathbf{c} \cdot \int_S d\mathbf{a} \times \nabla'V = \mathbf{c} \cdot \oint_C V d\mathbf{l}'. \tag{5.32}$$

Because $\mathbf{c}$ is an arbitrary constant vector, this expression must hold true for any $\mathbf{c}$. We must therefore have

$$\int_S \nabla'V \times d\mathbf{a} = -\oint_C V d\mathbf{l}'. \tag{5.33}$$

This is one modified form of Stokes's theorem.

Now let us set $V = [\hat{\mathbf{r}} \cdot \mathbf{r}']$. Then $\nabla'V = \hat{\mathbf{r}}$, leading to

$$\oint (\hat{\mathbf{r}} \cdot \mathbf{r}') d\mathbf{l}' = -\hat{\mathbf{r}} \times \int d\mathbf{a}', \tag{5.34}$$

and

$$\mathbf{A}(\mathbf{r}) \approx -\frac{\mu_0}{4\pi r^3}\mathbf{r} \times \int I d\mathbf{a}'. \tag{5.35}$$

The integral now only depends on the geometry of the loop, and we introduce this as the magnetic dipole moment:

$$\mathbf{m} \equiv I \int d\mathbf{a}' = \pi I a^2 \hat{\mathbf{z}}. \tag{5.36}$$

The magnetic dipole moment points normal to the loop, following the usual right-hand rule convention.

Before moving on, it is worth looking at the effective dipole moment of a localized current density $\mathbf{J}(\mathbf{r}')$, as this will be a useful quantity to be familiar with much, much later, in chapter 14. The generalized Biot–Savart law is now

$$\mathbf{B}(\mathbf{r}) = \frac{\mu_0}{4\pi} \int_V \frac{\mathbf{J}(\mathbf{r}') \times \mathbf{R}}{R^3} d^3 r', \tag{5.37}$$

which leads to a vector potential

$$\mathbf{A}(\mathbf{r}) = \frac{\mu_0}{4\pi} \int_V \frac{\mathbf{J}(\mathbf{r}')}{R} d^3 r'. \tag{5.38}$$

If we assume the condition $|\mathbf{r}'| \ll |\mathbf{r}|$, this may be written as

$$\mathbf{A}(\mathbf{r}) = \frac{\mu_0}{4\pi r} \int_V \mathbf{J}(\mathbf{r}') d^3 r' + \frac{\mu_0}{4\pi r^2} \int_V \mathbf{J}(\mathbf{r}')[\mathbf{r}' \cdot \hat{\mathbf{r}}] d^3 r'. \tag{5.39}$$

We now take advantage of the continuity equation, equation (2.57), which tells us that $\nabla \cdot \mathbf{J}(\mathbf{r}) = 0$ for a magnetostatic problem. Let us introduce two well-behaved functions $u(\mathbf{r}')$ and $v(\mathbf{r}')$ and consider the integral

$$\int_V \nabla' \cdot [u(\mathbf{r}')v(\mathbf{r}')\mathbf{J}(\mathbf{r}')] d^3 r' = \oint_S u(\mathbf{r}')v(\mathbf{r}')\mathbf{J}(\mathbf{r}') \cdot d\mathbf{a}' = 0, \tag{5.40}$$

where the zero value for the surface integral follows from assuming that our charge density is localized to a finite region. In such a case, we can always choose the surface to be outside the charge distribution to set the integral to zero. If we apply the product rule several times and use $\nabla' \cdot \mathbf{J}(\mathbf{r}') = 0$, we get the expression

$$\int_V [u(\mathbf{r}')\mathbf{J}(\mathbf{r}') \cdot \nabla'v(\mathbf{r}') + v(\mathbf{r}')\mathbf{J}(\mathbf{r}') \cdot \nabla'u(\mathbf{r}')] d^3 r' = 0. \tag{5.41}$$

For different choices of u and v, we can derive new integral identities related to the current density. First, if $u(\mathbf{r}') = 1$ and $v(\mathbf{r}') = x'_\alpha$, where x'_α is a component of the position vector, with $\alpha = 1, 2, 3$, we get

$$\int_V J_\alpha(\mathbf{r}') d^3 r' = 0, \tag{5.42}$$

which indicates that the first integral of equation (5.39) vanishes. This again shows that there is no monopole contribution to the magnetic field. If we let $u(\mathbf{r}') = x_\alpha'$ and $v(\mathbf{r}') = x_\beta'$, we get the result

$$\int_V x_\alpha' J_\beta(\mathbf{r}')d^3r' = -\int_V x_\beta' J_\alpha(\mathbf{r}')d^3r'. \qquad (5.43)$$

Let us now return to equation (5.39) and write the vectors in component form. We have

$$A_\alpha(\mathbf{r}) = \frac{\mu_0}{4\pi r^3}\sum_\beta x_\beta \int_V J_\alpha(\mathbf{r}')x_\beta' d^3r'. \qquad (5.44)$$

We now replace half of the integrand of this equation using equation (5.43), which gives us

$$A_\alpha(\mathbf{r}) = \frac{\mu_0}{4\pi r^3}\sum_\beta x_\beta \int_V \frac{1}{2}\Big[J_\alpha(\mathbf{r}')x_\beta' - x_\alpha' J_\beta(\mathbf{r}')\Big]d^3r'. \qquad (5.45)$$

This sum, however, can be associated with the vector triple product,

$$\mathbf{r} \times [\mathbf{r}' \times \mathbf{J}(\mathbf{r}')] = \mathbf{r}'[\mathbf{r} \cdot \mathbf{J}(\mathbf{r}')] - \mathbf{J}(\mathbf{r}')[\mathbf{r} \cdot \mathbf{r}']. \qquad (5.46)$$

We find that we may write

$$\mathbf{A}(\mathbf{r}) = -\frac{\mu_0}{4\pi r^3}\mathbf{r} \times \mathbf{m}, \qquad (5.47)$$

where we have defined

$$\mathbf{m} = \frac{1}{2}\int_V \mathbf{r}' \times \mathbf{J}(\mathbf{r}')d^3r'. \qquad (5.48)$$

This becomes our definition of the magnetic dipole moment for a system of currents. Though we have derived this for the magnetostatic case, we will see it is relevant for the time-varying case as well.

We now turn to the problem of characterizing a material that consists of a collection of magnetic dipoles. We start with the vector potential of a single magnetic dipole centered on the origin,

$$\mathbf{A}(\mathbf{r}) = \frac{\mu_0}{4\pi}\frac{\mathbf{m} \times \mathbf{r}}{r^3}. \qquad (5.49)$$

The calculation then follows in a similar manner to the electric dipole case. We introduce a magnetization $\mathbf{M}(\mathbf{r}')$, which is the magnetic dipole moment per unit volume, and thus obtain the total vector potential for a collection of dipoles,

$$\mathbf{A}(\mathbf{r}) = \frac{\mu_0}{4\pi}\int_V \frac{\mathbf{M} \times \mathbf{R}}{R^3}d^3r'. \qquad (5.50)$$

We use equation (5.11) to again rewrite the quantity $\mathbf{R}/R^3$ in the integrand,

$$\mathbf{A}(\mathbf{r}) = \frac{\mu_0}{4\pi} \int_V \mathbf{M} \times \nabla' \left(\frac{1}{R}\right) d^3r'. \tag{5.51}$$

The following curl identity is now used:

$$\nabla' \times \left[\frac{\mathbf{M}}{R}\right] = \frac{1}{R} \nabla' \times \mathbf{M} - \mathbf{M} \times \nabla' \left(\frac{1}{R}\right). \tag{5.52}$$

On substitution, the vector potential takes on the form,

$$\mathbf{A}(\mathbf{r}) = \frac{\mu_0}{4\pi} \left[\int_V \frac{1}{R} \nabla' \times \mathbf{M} d^3r' - \int_V \nabla' \times \left[\frac{\mathbf{M}}{R}\right] d^3r' \right]. \tag{5.53}$$

We may apply a modified version of the divergence theorem here, which we briefly derive. We use the divergence theorem but replace the vector $\mathbf{F} \rightarrow \mathbf{c} \times \mathbf{F}$, where $\mathbf{c}$ is an arbitrary constant vector:

$$\oint (\mathbf{c} \times \mathbf{F}) \cdot d\mathbf{a}' = \int_V \nabla' \cdot (\mathbf{c} \times \mathbf{F}) d^3r'. \tag{5.54}$$

We use the cyclic properties of the scalar triple product on both sides of the equation and then write

$$\mathbf{c} \cdot \oint \mathbf{F} \times d\mathbf{a}' = -\mathbf{c} \cdot \int_V (\nabla' \times \mathbf{F}) d^3r'. \tag{5.55}$$

Because this integral must be true for any choice of vector $\mathbf{c}$, we have

$$\oint \mathbf{F} \times d\mathbf{a}' = -\int_V (\nabla' \times \mathbf{F}) d^3r'. \tag{5.56}$$

We use this equation to replace the right-hand side of equation (5.53). We then have

$$\mathbf{A}(\mathbf{r}) = \frac{\mu_0}{4\pi} \left[\int_V \frac{1}{R} \nabla' \times \mathbf{M} d^3r' + \oint_S \frac{1}{R} \mathbf{M} \times d\mathbf{a}' \right]. \tag{5.57}$$

Similar to the electric case, we now have the magnetic vector potential expressed as the sum of two terms, one that is effectively a surface current and one that is a volume current. We call these the *bound surface current* $\mathbf{K}_b$ and the *bound volume current* $\mathbf{J}_b$, respectively, and they have the forms

$$\mathbf{K}_b(\mathbf{r}') = \mathbf{M}(\mathbf{r}') \times \mathbf{n}'(\mathbf{r}'), \tag{5.58}$$

$$\mathbf{J}_b(\mathbf{r}') = \nabla' \times \mathbf{M}(\mathbf{r}'), \tag{5.59}$$

where $\mathbf{n}'(\mathbf{r}')$ is the unit normal to the surface, which also, of course, depends on the position upon the surface.

We consider the total volume current to be of the form,

$$\mathbf{J} = \mathbf{J}_b + \mathbf{J}_f, \tag{5.60}$$

where $\mathbf{J}_f$ is the free current density. Using this in Ampère's law (remember, we are just doing magnetostatics right now), we have

$$\nabla \times \mathbf{B} = \mu_0(\nabla \times \mathbf{M}) + \mu_0 \mathbf{J}_f. \tag{5.61}$$

We group together the curl terms to find that

$$\nabla \times \left(\frac{\mathbf{B}}{\mu_0} - \mathbf{M} \right) = \mathbf{J}_f. \tag{5.62}$$

This suggests the introduction of an auxiliary field $\mathbf{H}$,

$$\mathbf{H} \equiv \left(\frac{\mathbf{B}}{\mu_0} - \mathbf{M} \right), \tag{5.63}$$

which leads to a new version of Ampère's law,

$$\nabla \times \mathbf{H} = \mathbf{J}_f. \tag{5.64}$$

The $\mathbf{H}$-field only depends on the free current density and not the bound current density; it does not 'see' the bound currents, and it is therefore often easier to derive in calculations.

In integral form, it may be written

$$\oint_C \mathbf{H} \cdot d\mathbf{l} = I_{f,enc}, \tag{5.65}$$

where $I_{f,enc}$ is the free current enclosed by the loop C.

We have therefore defined two new auxiliary fields for Maxwell's equations, $\mathbf{D}$ and $\mathbf{H}$. In systems of high symmetry, these fields can be calculated with knowledge of only the free currents and free charges, which makes them a useful intermediate step in many calculations.

However, to find the physical fields $\mathbf{B}$ and $\mathbf{E}$, which are directly related to the forces on charges, we must have some sort of model for the relationship between the physical fields and the auxiliary fields. We consider these relationships in section 5.4.

5.3 Closing the electromagnetic 'loop'

We have defined $\mathbf{D}$ and $\mathbf{H}$ for static fields, but we have not yet connected them back to our starting point and incorporated them into the complete form of Maxwell's equations. Let us look at each of the equations in turn and see how to incorporate these auxiliary fields into them.

First, we note that the 'no magnetic monopoles' law (2.2) and Faraday's law (2.3) are written entirely in terms of the microscopic vacuum fields $\mathbf{E}$ and $\mathbf{B}$ and need no adjustment.

Next, we argue that Gauss's law for the $\mathbf{D}$-field can be directly made into a time-varying equation without change. This is probably something that can only really be justified by experiment, so we take it as a given. We note that there is no explicit time derivative in the formula that would indicate anything needs to be revised for the time-varying case.

This leaves us with the Ampère–Maxwell law. In microscopic form, we have

$$\nabla \times \mathbf{B}(\mathbf{r}, t) = \mu_0 \mathbf{J}(\mathbf{r}, t) + \mu_0 \epsilon_0 \frac{\partial \mathbf{E}(\mathbf{r}, t)}{\partial t}. \tag{5.66}$$

We again break our current density into a free current $\mathbf{J}_f$ and a bound current $\mathbf{J}_b$. We might think from the magnetostatics discussion of section 5.2 that we should write $\mathbf{J}_b(\mathbf{r}, t) = \nabla \times \mathbf{M}(\mathbf{r}, t)$, but this does not include all the possible bound currents! To see this, let us consider the continuity equation, equation (2.57), applied to bound charges ρ_{ed} due to electric dipoles:

$$\nabla \cdot \mathbf{J}_{ed}(\mathbf{r}, t) = -\frac{\partial \rho_{ed}(\mathbf{r}, t)}{\partial t}, \tag{5.67}$$

where we have written $\mathbf{J}_{ed}$ to represent any currents associated with the time-varying electric dipoles. Using $\rho_{ed} = -\nabla \cdot \mathbf{P}$, we may write

$$\nabla \cdot \mathbf{J}_{ed}(\mathbf{r}, t) = \nabla \cdot \left[\frac{\partial \mathbf{P}(\mathbf{r}, t)}{\partial t} \right], \tag{5.68}$$

which implies that

$$\mathbf{J}_{ed}(\mathbf{r}, t) = \frac{\partial \mathbf{P}(\mathbf{r}, t)}{\partial t}. \tag{5.69}$$

In other words, when our fields are varying in time, we must add a second contribution to the bound current that arises from the electric dipoles. The total bound current is therefore

$$\mathbf{J}_b(\mathbf{r}, t) = \nabla \times \mathbf{M}(\mathbf{r}, t) + \frac{\partial \mathbf{P}(\mathbf{r}, t)}{\partial t}. \tag{5.70}$$

It should be noted that the magnetization part does not contribute to the continuity equation, because the divergence of a curl is identically zero, so there is no inconsistency in our treatment. Using this bound current in our microscopic Ampère–Maxwell law, we have

$$\nabla \times \mathbf{B}(\mathbf{r}, t) - \mu_0 \nabla \times \mathbf{M}(\mathbf{r}, t) = \mu_0 \mathbf{J}_f(\mathbf{r}, t) + \mu_0 \frac{\partial}{\partial t}[\epsilon_0 \mathbf{E}(\mathbf{r}, t) + \mathbf{P}(\mathbf{r}, t)]. \tag{5.71}$$

Using equations (5.21) and (5.63) that define $\mathbf{D}$ and $\mathbf{H}$, respectively, we finally have

$$\nabla \times \mathbf{H}(\mathbf{r}, t) = \mathbf{J}_f(\mathbf{r}, t) + \frac{\partial \mathbf{D}(\mathbf{r}, t)}{\partial t}, \tag{5.72}$$

which is the complete Ampère–Maxwell law in terms of $\mathbf{H}$ and $\mathbf{D}$.

With this, we have finally justified the complete form of Maxwell's equations that we introduced at the beginning of chapter 2. We are on firm ground to move forward in studying the interaction between electromagnetic waves and matter.

5.4 Permittivity, permeability, and the refractive index

Most materials of interest are, in their natural state, electrically neutral and nonmagnetic. It is possible to have materials with built-in static polarization and magnetization, but we are primarily interested in materials that have polarization and magnetization *induced* in them by electromagnetic fields. The optical response of a material comes from its electric and magnetic responses to an illuminating light wave.

The relationship between $\mathbf{D}$, $\mathbf{H}$ and $\mathbf{P}$, $\mathbf{M}$ can be quite complicated; we hypothesize that in ordinary circumstances, most materials satisfy the following equations,

$$\mathbf{P}(\mathbf{r}, \omega) = \epsilon_0 \chi_e(\mathbf{r}, \omega) \mathbf{E}(\mathbf{r}, \omega), \tag{5.73}$$

$$\mathbf{M}(\mathbf{r}, \omega) = \mu_0 \chi_m(\mathbf{r}, \omega) \mathbf{H}(\mathbf{r}, \omega). \tag{5.74}$$

These seemingly simple equations have a lot of information packed into them! First, we note that the quantities χ_e and χ_m are referred to as the *electric susceptibility* and *magnetic susceptibility*, respectively. They are inherent properties of a material that describe how strongly it responds to a locally applied field.

Next, we note that the equations are local, i.e. they are defined at a single point in space. Though there is a possibility that materials may have a nonlocal response, it is insignificant under most conditions.

Third, we note that if the material's response varies in space, the χ's may depend on space as well as time. We will generally look at the optical properties of bulk materials with constant χ; we will discuss how to deal with spatially varying responses when we look at scattering theory much later in chapter 15.

Fourth, we note that the equations are defined in the frequency domain. As a guess, we could have just as easily said that the fields and dipole densities are linearly proportional in the time domain; we will soon see that the frequency domain is the correct choice.

Fifth, we note that there is a curious asymmetry in the way the relationships are defined. $\mathbf{P}$ depends on the physical field $\mathbf{E}$, but $\mathbf{M}$ depends on the auxiliary field $\mathbf{H}$. This is a consequence of the way that we defined the fields $\mathbf{D}$ and $\mathbf{H}$, and it will not cause us any particular trouble, but it is worth noticing.

Finally, we note—and this is *crucial*—that $\mathbf{P}$ and $\mathbf{M}$ depend on the *total* fields $\mathbf{E}$ and $\mathbf{H}$ that the dipoles experience, and not just on the applied fields. If we put a dielectric in a charged capacitor, for example, the capacitor applies an electric field to the dielectric. But the dielectric, in turn, has dipoles induced in it, which produce their own electric fields, and these affect the dipoles in turn, and so on. When we solve for fields in matter, we must solve them self-consistently, taking into account the effect of the applied and induced fields simultaneously.

If we substitute our definitions for **P** and **M** into the expressions for **D** and **H**, we may write

$$\mathbf{D}(\mathbf{r}, \omega) = \epsilon(\mathbf{r}, \omega)\mathbf{E}(\mathbf{r}, \omega), \tag{5.75}$$

$$\mathbf{B}(\mathbf{r}, \omega) = \mu(\mathbf{r}, \omega)\mathbf{H}(\mathbf{r}, \omega). \tag{5.76}$$

Here, $\epsilon(\mathbf{r}, \omega)$ is the *permittivity* of the material, and $\mu(\mathbf{r}, \omega)$ is the *permeability*, and they can be written in terms of the susceptibilities as

$$\epsilon(\mathbf{r}, \omega) = [1 + \chi_e(\mathbf{r}, \omega)]\epsilon_0, \tag{5.77}$$

$$\mu(\mathbf{r}, \omega) = [1 + \chi_m(\mathbf{r}, \omega)]\mu_0. \tag{5.78}$$

At this point, we may finally introduce one of the most significant physical quantities in optics: the refractive index. We have noted that the speed of light in vacuum satisfies the relation $c^2 = 1/\epsilon_0\mu_0$; for a monochromatic wave propagating in a medium of uniform permittivity $\epsilon(\omega)$ and permeability $\mu(\omega)$, we may introduce a speed v of light in the medium at that frequency by the relation

$$v(\omega) = \frac{1}{\sqrt{\epsilon(\omega)\mu(\omega)}}. \tag{5.79}$$

As we will see in the next chapter, the frequency dependence has a significant effect on the propagation of time-dependent electromagnetic waves, so we should treat this simple expression for the speed of an electromagnetic wave with some caution unless we are looking at monochromatic fields. The speed of light is traditionally written in terms of a reduction of the speed of light by a factor $n(\omega)$, called the *refractive index*, in the form $v = c/n$. It immediately follows that we may write the refractive index as

$$n(\omega) = \sqrt{\frac{\epsilon(\omega)\mu(\omega)}{\epsilon_0\mu_0}}. \tag{5.80}$$

If the medium is nonmagnetic, i.e. it has a vacuum permeability $\mu = \mu_0$, then the expression for the refractive index takes on the form

$$n(\omega) = \sqrt{\frac{\epsilon(\omega)}{\epsilon_0}}. \tag{5.81}$$

The refractive index, as its name implies, affects the amount of refraction that light experiences when passing from one medium to another. We will explore this aspect of the refractive index in chapter 9. A few typical values of the refractive index are presented in table 5.1. For several materials, multiple wavelengths are given to provide a feeling for how strongly the refractive index depends on wavelength. It should be noted that the index also depends on temperature.

N-BK7 glass is a borosilicate crown glass that is a standard for making optical elements. Achromatic doublet lenses are often constructed from a crown glass lens and a flint glass lens to reduce chromatic aberration.

Table 5.1. Refractive indices of selected substances.

Material	Wavelength (nm)	Temperature (C)	Index n	[Ref]
Fused silica	546.07	26°	1.460 28	[5]
Fused silica	1128.66	26°	1.449 03	[5]
Aluminosilicate glass	546.07	26°	1.551 00	[5]
Aluminosilicate glass	1128.66	26°	1.536 99	[5]
Diamond	546.0	Unspecified	2.4237	[6]
Diamond	1200	Unspecified	2.390	[6]
Water	550	25°	1.333	[7]
Water	1200	25°	1.324	[7]
N-BK7 glass	546.1	22°	1.518 72	[8]
N-BK7 glass	1060.0	22°	1.506 69	[8]
F5 flint glass	546.1	22°	1.607 18	[8]
F5 flint glass	1060.0	22°	1.586 36	[8]
Air	546.227	15°	1.000 2779	[9]
Air	1129.05	15°	1.000 2738	[9]

The refractive index of air comes with significant caveats because it can depend on temperature, pressure, and chemical composition, as well as wavelength. The quoted value is usually 1.0003, which is good enough for applications short of high-precision metrology. Water comes with similar caveats, and books usually present a value of 1.33 for visible light.

One other material relation is worth noting here. If there are free charges in the system, they are moved under the action of an electric field. We may quantify this using the expression

$$\mathbf{J}_f(\mathbf{r}, \omega) = \sigma(\mathbf{r}, \omega)\mathbf{E}(\mathbf{r}, \omega), \tag{5.82}$$

where σ is the *electric conductivity* of the material. This equation is basically Ohm's law in vector form: the current is proportional to the applied field.

It is worth saying a bit more about Ohm's law, as it can cause some confusion when it is applied. Though all free currents are produced by electric fields, we note that we may conceptually divide them into two types: currents that are driven by some mechanism that is external to the system under consideration and those currents that are driven by fields within the system. In the former case, we have currents produced by chemical reactions or mechanical means such as electric generators; let us label these $\mathbf{J}_{ext}$. We do not usually model these current-generating mechanisms within Maxwell's equations, so the currents are treated as being fixed and independent of what the rest of our system is doing. The currents that are driven by fields within the system we label $\mathbf{J}_{int}$ and they satisfy Ohm's law.

Let us consider the Maxwell–Ampère law for monochromatic fields and sources. With an assumed time dependence $\exp[-i\omega t]$, this equation takes on the form

$$\nabla \times \mathbf{H}(\mathbf{r}, \omega) = \mathbf{J}_{ext}(\mathbf{r}, \omega) + \mathbf{J}_{int}(\mathbf{r}, \omega) - i\omega\epsilon(\mathbf{r}, \omega)\mathbf{E}(\mathbf{r}, \omega), \tag{5.83}$$

where we have used $\mathbf{J}_f = \mathbf{J}_{ext} + \mathbf{J}_{int}$ and equation (5.75). Writing $\mathbf{J}_{int}$ in terms of Ohm's law, we have

$$\nabla \times \mathbf{H}(\mathbf{r}, \omega) = \mathbf{J}_{ext}(\mathbf{r}, \omega) - i\omega \left[\epsilon(\mathbf{r}, \omega) - \frac{\sigma(\mathbf{r}, \omega)}{i\omega} \right] \mathbf{E}(\mathbf{r}, \omega). \qquad (5.84)$$

Here, we can see that, in the presence of electrical conductors with free electrons, the effective permittivity $\epsilon_{eff}(\mathbf{r}, \omega)$ at frequency ω can be written as

$$\epsilon_{eff}(\mathbf{r}, \omega) = \epsilon(\mathbf{r}, \omega) + \frac{i\sigma(\mathbf{r}, \omega)}{\omega}. \qquad (5.85)$$

The free electrons effectively modify the permittivity of the medium, adding an imaginary part to it and making the permittivity generally complex-valued. We will also see in the next chapter that the bound charges in physical models of the atom also produce a complex-valued permittivity, and so we will often combine the effects of the free charges and bound charges into one effective permittivity of the system.

To understand this rather artificial distinction between external and internal free currents, let us consider a broadcast radio antenna. The radio antenna is driven by external alternating currents that produce an electromagnetic wave. This electromagnetic wave can also induce currents in the metal of the broadcast antenna components or in the metal of a receiver antenna, which we can consider internal free currents.

In short, though all electromagnetic currents are driven by electric fields, in practice it is much easier to solve problems by separating those currents into ones generated externally to the system under consideration and ones generated within the system. The effects of currents generated within the system are largely indistinguishable from complex permittivity; therefore, going forward, we will usually just work with complex permittivity.

In closing this chapter, let us look again at equations (5.73) and (5.74) for $\mathbf{P}$ and $\mathbf{M}$. There are two big questions related to these equations: why did we assume a linear relationship between fields and dipole densities, and why did we choose to work in the frequency domain? We attempt to answer these questions in the next chapter.

5.5 Exercises

1. An infinitely long cylinder of radius a carries a fixed polarization $\mathbf{P} = P_0 \rho \sin \phi \hat{\phi} + P_1 \rho^2 \hat{\rho}$. Determine its surface and volume bound charge densities.

2. An infinitely long cylinder of radius a carries a fixed magnetization $\mathbf{M} = M_0 \rho^2 \cos^2 \phi \hat{\phi} + M_1 \rho^3 \hat{\rho} + M_2 \rho \hat{\mathbf{z}}$. Determine its surface and volume bound charge densities.

3. A cylindrical dielectric shell of inner radius a and outer radius b possesses a uniform positive permittivity ϵ_s. Along the axis of the cylinder is a line charge of charge per unit length Λ. (a) Find the displacement field $\mathbf{D}$ for all $\rho < b$. (b) Calculate the electric field $\mathbf{E}$ for all $\rho < b$. (c) Find the bound charge

densities for the dielectric. Do the bound charge densities enhance or reduce the field of the line charge within the dielectric?

4. A point charge Q is embedded in the center of a dielectric spherical shell of inner radius a, outer radius b, and permittivity $\epsilon(r) = \epsilon_0 r/a$. (a) Find the displacement field $\mathbf{D}$ for all values of r inside and outside the shell. (b) Calculate the electric field $\mathbf{E}$ for all r. (c) Determine the bound charge densities for the dielectric. (d) Which has a larger magnitude, the total bound surface charge or the total bound volume charge?

5. A parallel plate capacitor consists of two plates of area A separated by a distance d, where $d^2 \ll A$; the charge on the positive plate is $+Q$. The gap between the plates is filled with a dielectric with permittivity ϵ_d. (a) Calculate the $\mathbf{D}$ field and the $\mathbf{E}$ field between the plates, assuming we can approximate the fields as those due to an infinitely large capacitor. (b) The capacitance of a capacitor is defined as $C = Q/V$, where V is the potential difference between the plates; calculate the capacitance. (c) The stored energy E of a capacitor is given by $E = CV^2/2$. The energy of an electrostatic field in a volume V is given by

$$U_e = \frac{1}{2} \int_V \mathbf{D} \cdot \mathbf{E} d^3 r.$$

Taking the volume to be the volume between the plates, show that the two formulas predict the same total stored energy in the capacitor.

6. Let us suppose we have two thin sheets of current, one at $z = 0$ and one at $z = d$, with surface currents $\mathbf{K} = K_0 \hat{\mathbf{x}}$ and $\mathbf{K} = K_0 \hat{\mathbf{y}}$, respectively. Filling the space between the sheets is a material with permeability μ_s. (a) Calculate the $\mathbf{H}$ field between the sheets and outside the sheets. (b) Calculate the $\mathbf{B}$ field between the sheets and outside the sheets. (c) Calculate the bound current densities between the sheets.

7. Two coaxial cylindrical solenoids, with radius a and radius b, respectively, carry equal currents I but have N_a and N_b turns per unit length, respectively. The space between the two solenoids is filled with a magnetic material of permeability $\mu_s(\rho) = \mu_0 \rho^2/a^2$. Calculate the $\mathbf{H}$ fields everywhere, the $\mathbf{B}$ fields everywhere, and determine the bound currents of the system.

8. A conductive cylinder, with radius a running in the z direction, carries a uniform free current density $\mathbf{J}_j = J_0 \hat{\mathbf{z}}$. The permeability is μ_a inside the cylinder and μ_b outside. Determine the magnetic field and the bound current distribution inside and outside the cylinder.

9. A long cylindrical solenoid of radius a carries a current I; it has a length l and N total turns. Assume that the solenoid is filled with a material of permeability μ_s. (a) Calculate the fields $\mathbf{H}$ and $\mathbf{B}$ within the solenoid. (b) The inductance L of a solenoid is given by $L = N\phi_B/I$, where ϕ_B is the magnetic field flux passing through the solenoid. Calculate the inductance of the solenoid. (c) The stored energy of a solenoid is given by $E = LI^2/2$. The magnetostatic energy of a magnetic field in a volume V is given by

$$U_b = \frac{1}{2} \int_V \mathbf{H} \cdot \mathbf{B} d^3 r.$$

Assuming that the fields of the solenoid can be treated as those of an ideal (infinite) solenoid, show that the two formulas give the same result.

References

[1] Boyd R W 2008 *Nonlinear Optics* 3rd edn (Amsterdam: Academic)

[2] Landau L D, Lifshitz E M and Pitaevskii L P 1984 *Electrodynamics of Continuous Media* 2nd edn (Oxford: Pergamon)

[3] Cho K 2003 *Optical Response of Nanostructures* (Berlin: Springer)

[4] Cho D J, Wang F, Zhang X and Shen Y R 2008 Contribution of the electric quadrupole resonance in optical metamaterials *Phys. Rev.* **78** 121101

[5] Wray J H and Neu J T 1969 Refractive index of several glasses as a function of wavelength and temperature *J. Opt. Soc. Am.* **59** 774–6

[6] Zaitsev A M 2013 *Optical Properties of Diamond: A Data Handbook* (Berlin: Springer)

[7] Hale G M and Querry M R 1973 Optical constants of water in the 200-nm to 200-μm wavelength region *Appl. Opt.* **12** 555–63

[8] Schott 2018 *Optical Glass Data Sheets* https://www.schott.com/en-gb/products/optical-glass-p1000267/downloads

[9] Edlén B 1966 The refractive index of air *Metrologia* **2** 71

IOP Publishing

Electromagnetic Optics

Gregory J Gbur

Chapter 6

Dispersion and the speed of light

We have now formally characterized the optical properties of matter by introducing the permittivity $\epsilon(\omega)$, permeability $\mu(\omega)$, and refractive index $n(\omega)$; next, we would like to connect these optical properties to the physical properties of matter on an atomic scale. This is a problem that could fill an entire book by itself [1], as matter comes in many varieties and configurations, so we focus primarily on one example of surprising generality: the Lorentz oscillator model of the atom.

The Lorentz model indicates that the refractive index of a material is frequency dependent, and a natural question follows: what effect does this frequency dependence have on the propagation of light? This leads us into a discussion of the speed of light in matter and the phenomenon of dispersion, and we will find a number of surprises along the way.

6.1 Lorentz oscillator model of the atom

In order to relate our physical and auxiliary fields in a quantitative way, we need to construct a model of light–matter interactions that is, at its foundation, a model of interactions between light and individual atoms or molecules. We have not introduced the mathematical or physical tools necessary to create a proper quantum mechanical model, but fortunately, we do not have to. We will employ a simple atomic model first introduced by Hendrik Lorentz in the years before quantum physics was established. The model is quite inaccurate in terms of the structure and physics of the atom, but it is intuitively reasonable and gives remarkably accurate results for many light–matter interactions.

Atoms radiate and absorb light at discrete resonant frequencies. To characterize this interaction, we assume that an electron and its atomic nucleus are bound together by a linear (spring-like) force, as illustrated in figure 6.1. The motion of the electron is taken to be damped, energy being lost to other vibrational and electronic states of the atom. We also assume, for the moment, that the atom is illuminated by a monochromatic field,

doi:10.1088/978-0-7503-6064-7ch6
6-1

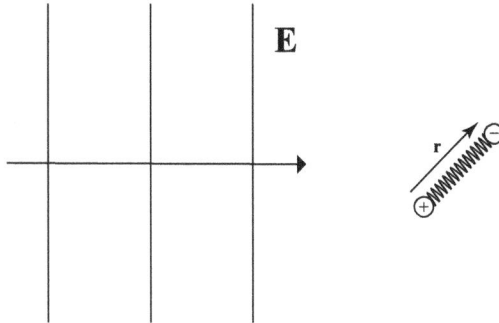

Figure 6.1. Illustration of the Lorentz model of the atom.

$$\mathbf{E}(t) = \mathbf{E}_0 e^{-i\omega t}. \tag{6.1}$$

You will notice that we have not included any spatial variation in the electric field. This is essentially a consequence of the third of three assumptions that we make in this model:

1. The atomic dynamics can be represented by harmonic oscillator forces.
2. The magnetic field has a negligible effect in the calculation.
3. The atom is much smaller than the wavelength of light.

This third assumption indicates that the atom effectively 'sees' a uniform electric field that varies harmonically over time. We may now write a differential equation for the motion of the electron, which is a driven damped harmonic oscillator equation:

$$m\frac{\partial^2 \mathbf{r}}{\partial t^2} = -f\mathbf{r} - g\frac{\partial \mathbf{r}}{\partial t} - e\mathbf{E}_0 e^{-i\omega t}, \tag{6.2}$$

where f is the spring constant, g is the damping constant, and e is the magnitude of the electron charge. The minus sign in the equation makes the electron charge negative. (We almost always write a negative charge as $-e$, where e is a positive number, so that we do not get confused by hidden signs.) This equation describes the net force on the electron as coming from three contributions: the external electric field, the atomic restoring force, and a damping force.

Before we solve this equation, we address one significant question: why do we use a harmonic oscillator model? In an atom, an electron may be viewed as a particle trapped in a potential well, and for small perturbations, almost every stable potential can be described approximately as a harmonic oscillator. Mathematically, if we take a potential function with a minimum at a particular location and Taylor expand around that minimum point, the lowest nontrivial term of the series is the quadratic term. This quadratic term is a harmonic oscillator potential. In developing his model, Lorentz was basically making an educated but quite reasonable guess. In fact, because many stable configurations of charges, like molecules, can be viewed in approximately the same way, the Lorentz model has surprisingly broad applicability. For the moment, we keep the paradigm of an atomic model for conceptual simplicity.

The complete solution to equation (6.2) would include both transient oscillations and steady-state oscillations. We ignore the transients for now and focus on the steady-state solution, so that the electron is oscillating at the same frequency as the field:

$$\mathbf{r}(t) = \mathbf{r}_0 e^{-i\omega t}. \tag{6.3}$$

By substituting this into our differential equation, we can eliminate the time dependence and solve for $\mathbf{r}_0$:

$$\mathbf{r}_0 = \frac{-e\mathbf{E}_0}{m(\omega_f^2 - \omega^2) - i\omega g}, \tag{6.4}$$

where we have introduced

$$\omega_f \equiv \sqrt{\frac{f}{m}} \tag{6.5}$$

as the natural frequency of oscillation of the electron.

We now introduce the dipole moment of our atom as

$$\mathbf{p}_0 = -e\mathbf{r}_0 = \frac{e^2\mathbf{E}_0}{m(\omega_f^2 - \omega^2) - i\omega g}. \tag{6.6}$$

This dipole moment is often written in the form $\mathbf{p}_0 = \alpha\epsilon_0\mathbf{E}_0$, where α is the *polarizability* of the atom,

$$\alpha(\omega) = \frac{e^2\mathbf{E}_0}{\epsilon_0[m(\omega_f^2 - \omega^2) - i\omega g]}. \tag{6.7}$$

The polarizability is a measure of how easily the atom can be polarized by an electric field at frequency ω.

This result already answers a couple of questions we raised in the previous chapter. It illustrates that the simplest atomic model has a linear relationship between $\mathbf{p}$ and $\mathbf{E}$, and it shows that the material properties are frequency dependent. In short, because atoms have characteristic frequencies of vibration, their response to light waves is also frequency dependent.

This simple model, however, is only for a single atom, and we really want to study the interaction between a bulk material consisting of many atoms and an applied electric field. Here, we run into a challenge: our equations on a microscopic scale and on a macroscopic scale need to be reconciled, as different electric fields come into play!

Let us return to the relationship between $\mathbf{P}$ and $\mathbf{E}$ introduced in the previous chapter, but now with more explicit labeling,

$$\mathbf{P}(\mathbf{r}, \omega) = \epsilon_0\chi_e(\mathbf{r}, \omega)\mathbf{E}_M(\mathbf{r}, \omega). \tag{6.8}$$

The quantity **P** is the polarization *density*, and the electric field on the right-hand side of the expression has been labeled 'M' for macroscopic. Because our detectors are of finite size, this macroscopic field always represents a spatial average of the electric field over a finite region of space. Within a system of atoms on the smallest scale, we expect that the electric field varies rapidly because of the tiny positive electrons and nuclei, and we almost never directly measure this microscopic 'm' electric field.

However, our Lorentz model derivation gave us the equation

$$\mathbf{p} = \alpha\epsilon_0\mathbf{E}_m, \qquad (6.9)$$

where 'm' labels the microscopic field. Individual atoms are sensitive to the local fluctuations of the electric field.

It is tempting to relate the two by writing $\mathbf{P} = N\mathbf{p}$, where N is the number of atoms per unit volume, but we then have different **E**'s on either side of the equation. We need to figure out a way to relate the macroscopic and microscopic fields.

The approach we take is to look at an atom in a small spherical cavity within a bulk dielectric. An individual atom sees empty space in its immediate vicinity, and that is what our cavity represents. Let us imagine a block of dielectric resting between the plates of a charged capacitor, as illustrated in figure 6.2. We may assume that the capacitor produces a uniform electric field **E** and induces a uniform polarization **P**.

What do we expect to happen qualitatively inside the cavity? Because of the way the dipoles orient in response to the capacitor field, they produce an enhanced electric field within the cavity. We therefore expect the field to be larger than the capacitor field.

In the absence of the cavity, we just have the macroscopic field $\mathbf{E}_M$. If the cavity is present, we argue that the macroscopic field is just the combination of the field in the hole (the microscopic field) and the field of the spherical plug of dielectric that has been removed, i.e.

$$\mathbf{E}_M = \mathbf{E}_m + \mathbf{E}_{plug}. \qquad (6.10)$$

We just need to calculate the field of the plug, which we can do using ordinary electrostatics. The plug is uniformly polarized and has a surface charge density σ_P

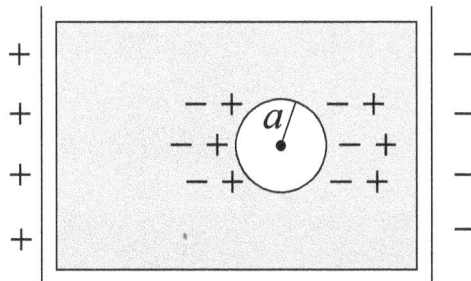

Figure 6.2. An atom in a cavity in a dielectric in a capacitor.

associated with it in accordance with equation (5.16). In spherical coordinates, we have

$$\sigma_P(\mathbf{r}') = -|\mathbf{P}|\cos\theta', \tag{6.11}$$

where θ is the angle from the z-axis, which is assumed to be perpendicular to the capacitor plates. The electric field of the dielectric plug is generally determined using electrostatics theory, specifically by applying the surface charge density of equation (5.15),

$$\mathbf{E}_{plug}(\mathbf{r}) = \frac{1}{4\pi\epsilon_0}\int_S \frac{\sigma_P(\mathbf{r}')\hat{\mathbf{R}}}{R^2}da'. \tag{6.12}$$

What is the electric field in the center of the plug? By symmetry, there can only be a z-component. We may write

$$\hat{\mathbf{z}}\cdot\mathbf{E}_{plug}(0) = \frac{1}{4\pi\epsilon_0}\int_S \frac{\sigma_P(\mathbf{r}')\cos\theta'}{a^2}da', \tag{6.13}$$

where we have used $\hat{\mathbf{z}}\cdot\hat{\mathbf{R}} = \cos\theta'$. We finally use $da' = a^2\sin\theta'd\theta'd\phi'$, perform the trivial integration over ϕ', and obtain

$$\hat{\mathbf{z}}\cdot\mathbf{E}_{plug}(0) = -\frac{2\pi a^2|\mathbf{P}|}{4\pi\epsilon_0 a^2}\int_0^\pi \sin\theta'\cos^2\theta'd\theta' = -\frac{|\mathbf{P}|}{3\epsilon_0}. \tag{6.14}$$

Using this in equation (6.10), we may therefore write

$$\mathbf{E}_m = \mathbf{E}_M + \frac{\mathbf{P}}{3\epsilon_0}. \tag{6.15}$$

This confirms that the microscopic and macroscopic fields are, in general, different from each other.

With this relationship, we can work out an expression for $\mathbf{P}$ from $\mathbf{p}$. First, we have

$$\mathbf{p} = \alpha\epsilon_0\mathbf{E}_m. \tag{6.16}$$

We then note that

$$\mathbf{P} = N\mathbf{p} = \alpha N\epsilon_0\mathbf{E}_m = \alpha N\epsilon_0(\mathbf{E}_M + \mathbf{P}/3\epsilon_0). \tag{6.17}$$

We may substitute from our macroscopic relation, equation (6.8), for $\mathbf{P}$; then everything in the equation is written in terms of $\mathbf{E}_M$, which may be canceled out, leaving

$$\epsilon_0\chi_e = \alpha N\epsilon_0 + \frac{\epsilon_0\chi_e\alpha N}{3}. \tag{6.18}$$

We solve this equation for αN:

$$\alpha N = \frac{\chi_e}{1 + \chi_e/3}. \tag{6.19}$$

This equation is quite remarkable: it connects the microscopic properties of the material (in the form of α and N) to the macroscopic bulk properties (in the form of χ_e). We are used to working with permittivity instead of susceptibility, so we apply

$$\epsilon = \epsilon_0(1 + \chi_e), \tag{6.20}$$

and with a little work, we finally write

$$\frac{\alpha N}{3} = \frac{\epsilon/\epsilon_0 - 1}{\epsilon/\epsilon_0 + 2}. \tag{6.21}$$

This equation is known as the *Clausius–Mossotti equation*, or alternatively as the *Lorentz–Lorenz equation*. The latter name is usually used when ϵ/ϵ_0 is replaced by n^2. The polarizability α is given by equation (6.7). We have derived this equation using electrostatics, but in most cases, it gives accurate results for electromagnetic waves as well. Like the Lorentz model of the atom, it has broader applicability than its simple formulation would suggest.

As written, the Clausius–Mossotti equation has a complicated dependence on ϵ. We can readily solve this equation for ϵ, but as a simple first approach, let us consider the case of a tenuous medium, like a gas, for which $\epsilon \approx \epsilon_0$. We may then approximate the denominator as $\epsilon/\epsilon_0 + 2 \approx 3$ and write

$$\frac{\epsilon(\omega)}{\epsilon_0} - 1 = \frac{Ne^2\mathbf{E}_0}{\epsilon_0[m(\omega_f^2 - \omega^2) - i\omega g]}. \tag{6.22}$$

It should be noted that we have explicitly written ϵ as depending on the frequency of light ω and that $\epsilon(\omega)$ is, in general, complex. This arises from the damping term of our atomic force equation. In short, the permittivity has at least a small imaginary part if atomic absorption is present, i.e. always.

It is convenient to write this expression in terms of the refractive index $n(\omega)$, introduced in section 5.4, which characterizes a number of wave propagation properties. We assume that the medium is nonmagnetic, so $\mu = \mu_0$; we then have

$$n^2(\omega) = \epsilon(\omega)\mu_0 c^2 = \epsilon(\omega)/\epsilon_0, \tag{6.23}$$

where, of course, $c^2 = 1/\epsilon_0\mu_0$. If we use the tenuous approximation for the Lorentz–Lorenz equation, we may write

$$n(\omega) = \sqrt{1 + \alpha N}. \tag{6.24}$$

Since $\alpha N \ll 1$ if the medium is tenuous, we may approximate

$$\sqrt{1 + \alpha N} \approx 1 + \frac{\alpha N}{2}. \tag{6.25}$$

We then have

$$n(\omega) \approx 1 + \frac{N^2 e/2\epsilon_0}{m(\omega_f^2 - \omega^2) - i\omega g}. \tag{6.26}$$

We make one more small change that puts this equation in standard form. We pull out the factor of m from the denominator and write

$$n(\omega) \approx 1 + \frac{\omega_p^2}{2} \frac{1}{(\omega_f^2 - \omega^2) - i\Gamma\omega}, \tag{6.27}$$

where we have introduced

$$\omega_p \equiv \sqrt{\frac{Ne^2}{m\epsilon_0}} \tag{6.28}$$

as the *plasma frequency*, and $\Gamma \equiv g/m$ is the damping rate. The plasma frequency will be of particular importance when we talk about light interaction with conductors in chapter 11. We can readily separate this expression into its real and imaginary parts:

$$n_R(\omega) - 1 = \frac{\omega_p^2}{2} \frac{\omega_f^2 - \omega^2}{(\omega_f^2 - \omega^2)^2 + (\Gamma\omega)^2}, \tag{6.29}$$

$$n_I(\omega) = \frac{\omega_p^2}{2} \frac{\Gamma\omega}{(\omega_f^2 - \omega^2)^2 + (\Gamma\omega)^2}.$$

The general shape of the resulting refractive index around the special frequency ω_f is shown in figure 6.3. We note a few significant features of this curve. At the resonant frequency, the imaginary part of the refractive index becomes large. This represents the absorption of light, which is strongest when the frequency of light matches the frequency ω_f of the atom. The real part of the refractive index has a

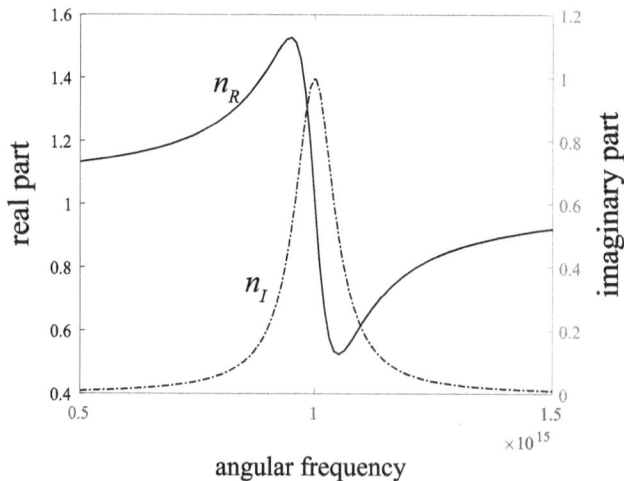

Figure 6.3. The real and imaginary parts of the refractive index arising from a single Lorentz resonance, for $\omega_f = 10^{15}$ s^{-1}, $\omega_p^2 = 10^{29}$ s^{-2}, and $\Gamma = 10^{14}$ s^{-1}.

positive slope everywhere except in a region close to the resonant frequency, where the slope is negative. The region of negative slope is known as the region of *anomalous dispersion*, while the positive slope region is the region of *regular dispersion*. These terms have a historical origin: before the discovery of resonant features, it had been noted that the real part of the refractive index decreased with wavelength for all materials observed and thus became a 'regular' feature of dispersion.

Finally, we note that in the anomalous dispersion region and beyond, there are frequencies where the real part of the refractive index is less than unity. For those who are used to thinking of the refractive index as a measure of how much the speed of light is slowed in matter, this will probably be troubling, as it would seem to imply light traveling faster than the vacuum speed of light. We will resolve this concern in section 6.4.

6.2 The Lorentz model for multiple oscillators

You may have wondered how I chose the parameters for figure 6.3. In fact, I made them up: it is rather difficult to find an example of a Lorentz oscillator that exactly fits the single resonant model of equation (6.27). For instance, one of the first examples of anomalous dispersion studied in the early 1900s by Robert Williams Wood was the sodium D lines [2], which show a real refractive index profile comparable to figure 6.3; however, as the name implies, the D lines are, in fact, two resonances spaced closely together at 589.0 nm and 589.6 nm!

To account for the complexity of real matter, researchers use a multiple oscillator model based on the Lorentz form for the permittivity,

$$\epsilon(\omega) = \epsilon_\infty + \sum_j \frac{\omega_{pj}^2}{(\omega_j^2 - \omega^2) - i\Gamma_j\omega}, \tag{6.30}$$

where the sum over j is over all the relevant resonances being considered, ω_{pj}^2 represents the relative strength of the jth oscillator, and Γ_j represents the damping rate of the jth oscillator. The quantity ϵ_∞ represents the dielectric constant created by all resonances at a higher frequency than those included explicitly. We therefore have two mechanisms for including multiple resonances: the direct inclusion of Lorentz-like terms and an average over higher-frequency resonances.

The Lorentz model is still discussed today because it is remarkably successful in characterizing the frequency behavior of many materials, even in cases where it arguably should not work! In figure 6.4, the real and imaginary parts of the refractive index of water in the infrared frequency range are plotted; the experimental data is taken from Rusk, Williams, and Querry [3]. Water is a polar molecule with an inherent dipole moment, which is very different from the neutral atom model we have been studying. Nevertheless, two Lorentz-like resonances can be seen in the data at approximately 49 THz and 102 THz, or 6.10 μm and 2.94 μm, respectively. These resonances are due to the vibrational modes of the water molecule.

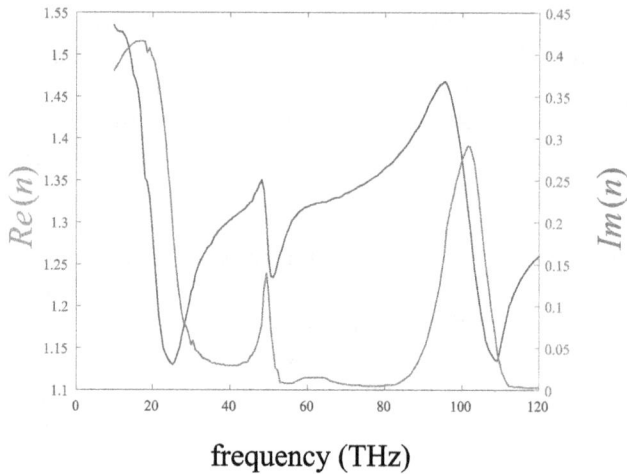

Figure 6.4. The real part and imaginary parts of the refractive index of water, adapted from data of Rusk, Williams, and Querry [3].

At about 16.2 THz, or 18.5 μm, we see what appears to be another broad Lorentz-like resonance; this arises from interactions between molecules: intermolecular vibrations.

None of these cases fit our original picture of an electron vibrating in a neutral molecule, but they all involve charged particles vibrating at discrete frequencies with relatively small damping, meaning they at least qualitatively agree with our Lorentz-type refractive index profile. To me, this is the inadvertent strength of Lorentz's approach: because he chose the simplest model that worked, it ended up being applicable beyond the case he envisioned.

There is, of course, much more that could be said about the features of water, and we refer the reader to Bohren and Huffman [4] for a broad discussion of the topic, as well as the next section.

The Lorentz model, and other models of matter that describe the optical properties as a function of frequency, also have a very practical purpose: They allow large tables of experimental data for a material to be condensed into a single formula with a small number of parameters, provided a good experimental fit to the model can be found. With this in mind, it is worth noting the earliest formula of this type, the *Sellmeier equation*, introduced by Wolfgang Sellmeier in 1872 [5], which is still used to this day to characterize the frequency dependence of the real part of the refractive index. The original form of the Sellmeier equation is given by

$$n^2(\lambda) = 1 + \sum_i \frac{B_i \lambda^2}{\lambda^2 - C_i}, \tag{6.31}$$

where the sum over i is over all absorption resonances of the material in question, each of which is characterized by a strength B_i and a wavelength $\sqrt{C_i}$, and λ represents the free-space wavelength. The Sellmeier equation explicitly demonstrates what we have implied over the previous sections: the refractive index of a material is determined by its resonances.

This formula can be compared to equation (6.30). If we write ω in terms of λ and note that the imaginary part of the Lorentz formula is negligible far from resonance, we can readily see that the Sellmeier formula agrees with the Lorentz model, with ω_{pj}^2 directly proportional to B_j and ω_j^2 directly proportional to C_j.

For transparent materials, typically only a few resonances need to be included for the Sellmeier formula to accurately characterize the frequency dependence of the refractive index; the discrepancy between the formula and experimental data is, in many cases, a fraction of a percent. Modern data tables of optical glasses often include the Sellmeier coefficients.

6.3 The Debye model

Before continuing on to a discussion of how dispersion affects the definition of the speed of light in matter, it is worth discussing another light–matter interaction that is distinct from the Lorentz model and is particular to polar molecules like water. Water molecules are asymmetric, with the hydrogen atoms forming an angle of $105°$ with respect to the oxygen atom between them; this means that they have a strong permanent dipole moment. It is possible to deflect a falling stream of water with a static electric charge, a process that is usually attributed to this strong moment.

In the absence of an applied electric field, the water molecules are randomly oriented and produce no macroscopic polarization on average. When a field is applied, however, the molecules align with the field, and a polarization density is induced. This process is much slower than the excitation of electronic or vibrational modes in the molecules, which are effectively instantaneous in comparison. The model for the response of such a medium is significantly different from the Lorentz model and is often called the Debye model [6]. Because of the relative slowness of the process, it manifests at lower frequencies than Lorentz resonances, specifically in the microwave part of the spectrum.

We roughly follow the derivation of the Debye model as discussed by Bohren and Huffman [4], which in turn follows from Gevers [7]. Let us first imagine that we have applied a constant electric field to our system, allowed it to reach a steady state, and then suddenly switched it off at $t = 0$. Through electric interactions and random collisions, the dipoles tend to rotate until they are, on average, randomly oriented. As a reasonable guess, we assume that the rate of change is proportional to the polarization density, i.e.

$$\frac{dP_d(t)}{dt} = -P_d(t)/\tau, \qquad (6.32)$$

where τ is the characteristic time of relaxation and we have written P_d to signify the polarization density due to the polar molecules. In the absence of an applied field, the force between electric dipoles tends to align them antiparallel to each other, so it is natural to expect the rate of decay to be proportional to $P_d(t)$, which is a measure of dipole alignment.

Equation (6.32), of course, is simply the differential equation of exponential decay, which has a solution

$$P_d(t) = P_d(0)e^{-t/\tau}. \tag{6.33}$$

Whereas the Lorentz model acts like an underdamped harmonic oscillator, we can think of the Debye model as an overdamped oscillator—any oscillations of the polar molecules around their equilibrium positions in the electric field are washed out by the random thermal motion of the liquid.

We also expect that there is a polarization density due to much higher-frequency vibrational and electronic excitations, and we label that contribution as P_v. On the slow timescale of interest for the Debye model, this polarization density is, for all intents and purposes, excited instantaneously. We will include it later in the derivation, but it is assumed that its evolution is independent of the evolution of the permanent dipoles.

Now let us imagine that we abruptly turn on a constant electric field E_0 at time t_0. This causes an increase in the value of $P_d(t)$ that eventually settles at the value $P_d(\infty) = \epsilon_0 \chi_{0d} E_0$, where χ_{0d} is the static susceptibility of the permanent dipoles. It is natural to assume an exponential relaxation to this final state, which suggests an equation

$$P_d(t) = P_d(\infty_0) - [P_d(\infty_0) - P_d(t_0)]e^{-(t-t_0)/\tau}. \tag{6.34}$$

We have used the curious notation ∞_0 to indicate the asymptotic state when the applied field is E_0. Suppose now at a time $t_1 > t_0$, the field is changed to E_1. We then have the new evolution, i.e. for $t > t_1$, that

$$P_d(t) = P_d(\infty_1) - [P_d(\infty_1) - P_d(t_1)]e^{-(t-t_1)/\tau}. \tag{6.35}$$

We may now substitute from equation (6.34) into equation (6.35) and simplify by noting that $P(t_0) = 0$. With a little effort, we end up with the expression

$$P_d(t) = P_d(\infty_1)[1 - e^{-(t-t_1)/\tau}] + P_d(\infty_0)[e^{-(t-t_1)/\tau} - e^{-(t-t_0)/\tau}]. \tag{6.36}$$

We now use $P_d(\infty_0) = \epsilon_0 \chi_{0d} E_0$ and $P_d(\infty_1) = \epsilon_0 \chi_{0d} E_1$ to write

$$P_d(t) = \epsilon_0 \chi_{0d} \{E_1[1 - e^{-(t-t_1)/\tau}] + E_0[e^{-(t-t_1)/\tau} - e^{-(t-t_0)/\tau}]\}. \tag{6.37}$$

The equation in the brackets, however, can be written as an integral, so that we have

$$P_d(t) = \epsilon_0 \chi_{0d} \int_{t_0}^{t} E(t') \frac{d}{dt'}[e^{-(t-t')/\tau}]dt', \tag{6.38}$$

as can be confirmed by performing the integration directly for our two-step electric field; $E(t')$ represents the applied electric field at time t'. This expression tells us that the current polarization state depends on the history of its evolution, which is not surprising in light of the observation that the dipole polarization takes some time to evolve to a new state.

We make one final significant leap and argue that equation (6.38) applies to a situation where the electric field progresses through an arbitrary number of steps.

If a large number of very closely spaced steps are used, we can approximate any continuous electric field $E(t')$ as well. Equation (6.38) represents the dipole polarization density in Green's function form, where the time derivative of the exponential is the Green's function.

Let us now add the polarization due to vibrational and electronic modes, assumed to evolve at a frequency much faster than the frequency of the applied field. If we use χ_{0v} to represent the 'static' susceptibility for these higher-frequency modes, we may write the total polarization density as

$$P(t) = \epsilon_0 \chi_{0v} E(t) + \epsilon_0 \chi_{0d} \int_{t_0}^{t} E(t') \frac{d}{dt'} [e^{-(t-t')/\tau}] dt'. \tag{6.39}$$

Let us assume that we have a monochromatic field $E(t) = E_0 \exp[-i\omega t]$ exciting our system. We then have

$$P(t) = \epsilon_0 \chi_{0v} E_0 e^{-i\omega t} + \epsilon_0 \chi_{0d} \int_{-\infty}^{t} E_0 e^{-i\omega t'} \frac{d}{dt'} [e^{-(t-t')/\tau}] dt', \tag{6.40}$$

where we have pushed t_0 back to $t_0 = -\infty$. The integral can be evaluated explicitly by parts, with the result

$$P(t) = \epsilon_0 \chi_{0v} E_0 e^{-i\omega t} + \frac{\epsilon_0 \chi_{0d} E_0 e^{-i\omega t}}{1 - i\omega\tau}. \tag{6.41}$$

It is standard to write the result in terms of the overall static susceptibility of the material, $\chi_{0s} = \chi_{0d} + \chi_{0v}$. Factoring out the electric field, the susceptibility at frequency ω is therefore

$$\chi(\omega) = \chi_{0v} + \frac{\Delta}{1 - i\omega\tau}, \tag{6.42}$$

where $\Delta \equiv \chi_{0s} - \chi_{0v}$. From the relation of the permittivity to the susceptibility, equations (5.77), we may write

$$\epsilon(\omega) = \epsilon_{0v} + \frac{\epsilon_0 \Delta}{1 - i\omega\tau}, \tag{6.43}$$

where $\epsilon_{0v} = \epsilon_0 \chi_{0v}$.

We have therefore derived the permittivity of a liquid of polar molecules in a low-frequency limit, and this is the main result of the Debye model. The real and imaginary parts of the permittivity are

$$\epsilon_R = \epsilon_{0v} + \frac{\epsilon_0 \Delta}{1 + (\omega\tau)^2}, \tag{6.44}$$

$$\epsilon_I = \frac{\omega\tau\epsilon_0 \Delta}{1 + (\omega\tau)^2}. \tag{6.45}$$

We illustrate these results in figure 6.5 for water at 20 °C. We can see excellent agreement between the experimental data, compiled from a variety of sources, and a

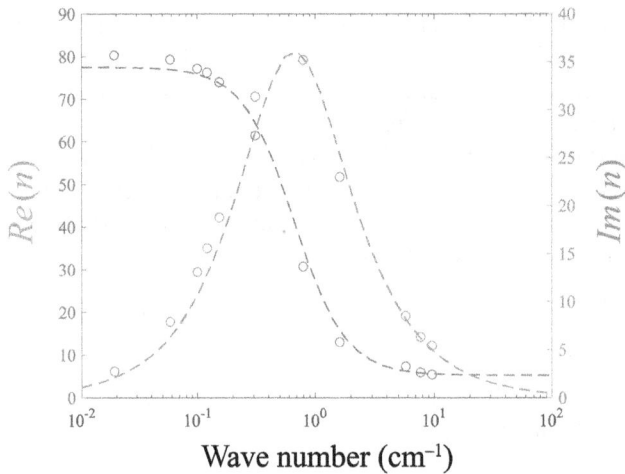

Figure 6.5. The real part and imaginary parts of the refractive index of water in the microwave region at 20 C. The data has been fitted to $\tau = 0.8 \times 10^{-11}$ s, $\epsilon_{0s} = 77.5\epsilon_0$ and $\epsilon_{0v} = 5.27\epsilon_0$ after Bohren and Huffman [4]; the data comes from Cook [9], Lane and Saxton [10]; Grant, Buchanan, and Cook [11]; and Kaatze and Uhlendorf [8].

fit to the Debye model. It should be noted that the Debye model is an extremely simple model, and researchers have found notable discrepancies between the model and the data at certain frequencies [8].

How can we physically interpret the shape of the Debye curves? By taking the derivative of equation (6.45), we can readily find that the maximum absorption occurs at $\omega = 1/\tau$. Apparently, when the applied frequency matches the characteristic relaxation frequency of the dipoles, we transfer a maximal amount of energy to them, roughly analogous to the resonance frequency in the Lorentz model. As ω increases, the real part of ϵ transitions from the static value ϵ_{0s} to the value that neglects vibrational frequencies ϵ_{0v}. When the frequency is below the critical value $1/\tau$, the dipoles relax much faster than the field excites them; to the dipoles, the applied field is effectively static. When the frequency is above the critical value, the dipoles cannot respond fast enough to keep up with the field and are themselves static. In this case, the optical properties are determined by the higher-frequency vibrational and electronic modes of the system alone.

It should also be noted that the permittivity at microwave frequencies is temperature dependent because the relaxation time of the polarization density is determined by the random thermal motion of the dipoles. This is distinct from the electronic transitions covered by the Lorentz model, which are largely temperature independent.

6.4 The speed of light in matter

In most of this book, we work with monochromatic waves, i.e. waves with a single frequency ω_0; this case is sufficient to describe optics used with steady-state light sources that are sufficiently narrowband in frequency, such as continuous-wave

(CW) lasers. For monochromatic waves, the refractive index has a fixed value $n(\omega_0)$ that appears naturally as an effect of the speed of waves in the medium.

However, when we are looking at broadband electromagnetic waves, such as light pulses, the frequency dependence of the refractive index of the medium can change the temporal evolution of the light pulse in a nontrivial way. In this section, we discuss how our understanding of the speed of light must change in such circumstances.

Let us begin with the monochromatic case. We imagine a monochromatic wave propagating in a medium with permittivity $\epsilon(\omega)$ and permeability $\mu(\omega)$ and at a single frequency ω. Maxwell's equations are then of the form

$$\epsilon(\omega)\nabla \cdot \mathbf{E}(\mathbf{r}, \omega) = 0, \tag{6.46}$$

$$\mu(\omega)\nabla \cdot \mathbf{H}(\mathbf{r}, \omega) = 0, \tag{6.47}$$

$$\nabla \times \mathbf{E}(\mathbf{r}, \omega) = i\omega\mu(\omega)\mathbf{H}(\mathbf{r}, \omega), \tag{6.48}$$

$$\nabla \times \mathbf{H}(\mathbf{r}, \omega) = -i\omega\epsilon(\omega)\mathbf{E}(\mathbf{r}, \omega). \tag{6.49}$$

We have assumed that ϵ and μ do not vary with position, i.e. we are dealing with a bulk material. This, combined with the assumption of a monochromatic wave, makes these quantities constants in Maxwell's equations, allowing us to bring them inside or outside of derivatives as needed. For convenience, we write all equations in terms of $\mathbf{E}$ and $\mathbf{H}$; this is a common convention.

For the record, we have simplified Maxwell's equations above from their general form by making three approximations: no sources, ϵ and μ independent of position, and monochromatic fields.

We may combine Maxwell's equations analogously to the way we derived the free-space wave equation, i.e. by taking the curl of Faraday's law. We then have

$$\nabla \times [\nabla \times \mathbf{E}] = i\omega\mu\nabla \times \mathbf{H} = \omega^2\epsilon\mu\mathbf{E}. \tag{6.50}$$

For brevity, we have left off the dependence on $\mathbf{r}$ and ω from the expressions. The left side may be expanded using the usual double curl formula, and with $\nabla \cdot \mathbf{E} = 0$, we get

$$(\nabla^2 + \omega^2\epsilon\mu)\mathbf{E} = 0. \tag{6.51}$$

This is a vector Helmholtz equation. We will not solve this generally, but look for a plane wave solution of the form

$$\mathbf{E}(z, \omega) = \hat{\mathbf{x}}\tilde{E}_0(\omega)e^{ikz}, \tag{6.52}$$

where k is to be determined. The tilde represents a Fourier transform, which we will return to momentarily. If we plug this into our vector Helmholtz equation, we quickly find

$$k^2(\omega) = \omega^2\epsilon(\omega)\mu(\omega). \tag{6.53}$$

Let us assume that ϵ and μ are real-valued for a moment. If we look at the full time-dependent solution for the electric field, we have

$$\mathbf{E}(z, t) = \hat{\mathbf{x}}\tilde{E}_0(\omega)e^{i(kz-\omega t)}. \tag{6.54}$$

This represents a wave traveling in the positive z-direction at speed v, given by

$$v = \frac{1}{\sqrt{\epsilon(\omega)\mu(\omega)}}. \tag{6.55}$$

The fraction by which the speed of light is reduced from its speed in vacuum is thus

$$\frac{c}{v} = \sqrt{\frac{\epsilon(\omega)\mu(\omega)}{\epsilon_0\mu_0}} \equiv n(\omega). \tag{6.56}$$

This is our traditional definition of the refractive index: it is the fraction by which the speed of a monochromatic wave is reduced from the vacuum speed of light. We write our plane wave solution in matter as

$$\mathbf{E}(z, \omega) = \hat{\mathbf{x}}\tilde{E}_0(\omega)e^{ik_0 n(\omega)z}, \tag{6.57}$$

where $k_0 = \omega/c$ is the vacuum wavenumber and $k_0 n$ is the wavenumber in matter. Since $k = 2\pi/\lambda$, this implies that the wavelength of light in matter is reduced from its free-space value λ_0 by

$$\lambda = \frac{\lambda_0}{n}. \tag{6.58}$$

For $n > 1$, the waves therefore become compressed as they enter the medium. This is readily understood by analogy: as a group of cars enters a region of road with a lower speed limit, they end up coming closer together, as the faster-moving cars in the back catch up to the slower-moving cars up front. Similarly, the wavefronts come closer together as they enter a region with a lower wave speed.

Up until now, we have assumed that ϵ and μ, and therefore the refractive index, are all real-valued; however, we have seen that $\epsilon(\omega)$ is generally a complex quantity. We therefore want to study how a complex refractive index changes the behavior of a propagating wave, and then look at what happens if we send a pulse through our medium.

Equation (6.57) was derived for a monochromatic wave, but we can also use it to derive the evolution of a pulse in matter. A pulse has a finite bandwidth frequency spectrum, i.e. $\tilde{E}_0(\omega)$ is a function with a finite width in frequency. We determine the spatiotemporal evolution by taking the inverse temporal Fourier transform of $\mathbf{E}(z, \omega)$,

$$\mathbf{E}(z, t) = \hat{\mathbf{x}}\int_{-\infty}^{\infty} \tilde{E}_0(\omega)e^{ik_0 n(\omega)z}e^{-i\omega t}d\omega. \tag{6.59}$$

If $n(\omega) = 1$, i.e. we are in vacuum, then this inverse transform has the simple form

$$\mathbf{E}(z, t) = \hat{\mathbf{x}}E_0(t - z/c), \tag{6.60}$$

which represents a wave propagating in the positive z-direction with speed c. If $n(\omega)$ is nontrivial, however, it affects the speed and the shape of the wave as it propagates in the medium.

Sticking to a monochromatic wave at first, let us explicitly write $n = n_R + in_I$; we then have

$$\mathbf{E}(z, \omega) = \hat{\mathbf{x}}\tilde{E}_0(\omega)e^{-n_I(\omega)\omega z/c}e^{i\omega[n_R(\omega)z/c-t]}. \tag{6.61}$$

The imaginary part n_I causes the wave to decay exponentially as it propagates; this represents the effect of absorption. A positive imaginary part of the refractive index is always associated with absorption. The real part n_R results in a wave that propagates in the positive z-direction at speed

$$v_p(\omega) = \frac{c}{n_R(\omega)}. \tag{6.62}$$

This is what we call the *phase velocity* of the wave, and it is our first definition of the speed of light in matter. If we are dealing with a narrowband pulse centered on frequency ω_s, then we might view the speed of light as being given by $v_p(\omega_s)$.

There is a problem with this interpretation, as we can readily see by returning to figure 6.3. Around the region of anomalous dispersion, we can have $n_R(\omega) < 1$, which implies that the wave is traveling faster than the vacuum speed of light! This would appear to be a violation of Einstein's special theory of relativity, so we suspect that the phase velocity is not, in general, a practical measure of wave speed.

We can understand this issue by thinking about the physics of a monochromatic wave. A monochromatic wave is infinite and unchanging: it has the same behavior over all time and space. Because it is unchanging, it is impossible to encode information onto such a signal; any encoding would be a distortion of the monochromatic wave, which would make it no longer monochromatic! We conclude that the phase velocity does tell us the speed of waves at a given frequency ω but that speed does not tell us anything useful about how a pulse, consisting of multiple frequencies, travels in matter.

Let us now assume that our pulse is narrowband, i.e. $\tilde{\mathbf{E}}_0(\omega)$ is a narrow function of frequency, and that the central frequency ω_s is far from any resonance of the material. Roughly speaking, we are therefore looking at a pulse that is very broad in time, as indicated by Fourier theory. The basic assumptions are illustrated in figure 6.6.

Because we are far from resonance, we may take $n_I \approx 0$ for all relevant frequencies; this means that $k(\omega) \approx k_R(\omega)$. For notational convenience, we write

$$\tilde{E}_0(\omega) = \tilde{A}(\omega - \omega_s), \tag{6.63}$$

where $A(\omega)$ is a function centered on $\omega = 0$. This allows us to explicitly include the central frequency of our pulse in our derivation. We now may write

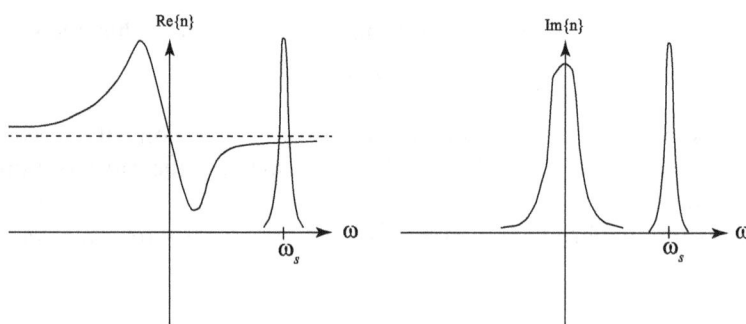

Figure 6.6. Rough illustration of the narrowband approximation and how it relates to the real and imaginary parts of the refractive index.

$$\mathbf{E}(z, t) = \hat{\mathbf{x}} \int_{-\infty}^{\infty} \tilde{A}(\omega - \omega_s) e^{ik_R(\omega)z} e^{-i\omega t} d\omega, \tag{6.64}$$

where $k_R(\omega) \equiv n_R(\omega)\omega/c$. We now make the change of variable to $\omega' = \omega - \omega_s$ to get

$$\mathbf{E}(z, t) = \hat{\mathbf{x}} e^{-i\omega_s t} \int_{-\infty}^{\infty} \tilde{A}(\omega') e^{i[k_R(\omega' + \omega_s)z - \omega' t]} d\omega'. \tag{6.65}$$

We now implement our approximation that $A(\omega')$ is a very narrow function of ω' and centered on $\omega' = 0$. If $k_R(\omega' + \omega_s)$ is slowly varying over all the relevant nonzero frequencies of $A(\omega')$, we may use a Taylor series approximation of this function about $\omega' = 0$,

$$k_R(\omega' + \omega_s) \approx k_R(\omega_s) + \left.\frac{dk_R}{d\omega}\right|_{\omega=\omega_s} \omega'. \tag{6.66}$$

If we substitute this approximation into equation (6.65), we get

$$\mathbf{E}(z, t) = \hat{\mathbf{x}} e^{i[k_R(\omega_s)z - \omega_s t]} \int_{-\infty}^{\infty} \tilde{A}(\omega') e^{-i\omega'[t - z/v_g]} d\omega', \tag{6.67}$$

where we have now defined

$$v_g \equiv \frac{1}{dk_R/d\omega|_{\omega_s}} \tag{6.68}$$

as the *group velocity* of our field. The meaning of this group velocity now becomes apparent, as we may further use the inverse Fourier transform to write

$$\mathbf{E}(z, t) = \hat{\mathbf{x}} e^{i[k_R(\omega_s)z - \omega_s t]} A(t - z/v_g). \tag{6.69}$$

In our approximate solution, we now have two relevant velocities. We see that the phase of the field, represented by the complex exponent, propagates at a speed $\omega_s/k_R(\omega_s)$, which is just the phase velocity we discussed earlier. The overall envelope of the pulse, represented by the function $A(t)$, travels at a speed v_g through the medium. That envelope represents a 'group' of monochromatic waves propagating together, and v_g is therefore named the group velocity.

This group velocity behaves somewhat better than the phase velocity as a description of the speed of light. If we explicitly write $k_R(\omega) = n_R(\omega)\omega/c$, we may readily find that

$$v_g = \frac{c}{n_R(\omega_s) + \omega_s \left. \dfrac{dn_R}{d\omega}\right|_{\omega_s}}. \tag{6.70}$$

Recall that in a region of normal dispersion, as defined earlier, $dn_R/d\omega > 0$; this means that the group velocity is slower than the phase velocity in a region of normal dispersion; a good sign! However, in a region of anomalous dispersion, the opposite is true and $dn_R/d\omega < 0$, meaning that the group velocity is even faster than the phase velocity and can also be faster than the vacuum speed of light!

To make sense of this, let us make an analogy. Imagine that we have a bus filled with passengers and that at every stop, passengers get off the back of the bus; however, no other passengers change their seats. We end up with fewer passengers at the back of the bus, and the center of mass of the passengers ends up moving toward the front of the bus. In fact, this center of mass is moving faster than the bus itself, even though no passengers are otherwise moving!

We can imagine something analogous happens in a region of anomalous dispersion. Light is absorbed by the medium at the rear of the pulse, which effectively shifts the peak of the pulse forward. Even though nothing is traveling faster than the leading edge of the pulse, the peak of the overall pulse envelope is moving forward, which manifests as a 'superluminal' group velocity.

A similar thing can happen in a gain medium, in which atoms of the material add to the energy of the light wave. A gain medium can be roughly approximated using a *negative* imaginary part of the refractive index (though a proper treatment of gain requires nonlinear optics). In such a case, the front of the pulse is amplified, shifting the peak forward and making it move faster than the leading edge of the pulse. The analogous situation on a bus would be passengers getting on at the front of the bus and refusing to redistribute toward the back.

In all these cases, the group velocity approximation—remember, we made an approximation to get that formula—is misleading because it measures velocity as a sort of 'center of mass' of the pulse. But when energy is being added or subtracted from the pulse, that center of mass does not necessarily tell us how long the pulse takes to transit through the medium. In any real situation, the pulse has a finite duration and a definite start time; we can then instead consider the speed at which the front edge of the pulse moves through the medium, called the *front velocity*. For electromagnetic waves in matter, this front velocity is expected to be exactly equal to the speed of light in vacuum, which was argued qualitatively by Sommerfeld and Brillouin [12] as follows. Because the electrons in a medium have a nonzero mass and finite inertia, it always takes them a finite amount of time to respond to an incident electromagnetic wave. By the time they respond, however, the front of the wave will already have passed them, leaving the very front of the pulse experiencing effectively no medium at all. However, Sommerfeld and Brillouin also note that the front of the pulse is typically of very low amplitude and often undetectable, making the front velocity also an impractical measure of the speed of light in matter.

We may introduce a final measure of the speed of light in matter known as the *signal velocity*, defined as the speed at which we can actually transmit information through a medium. The definition here is much more qualitative than the previous ones; Sommerfeld said that the signal velocity 'yields the arrival of the main signal, with intensities of the order of magnitude of the input signal' [12]. In other words, we transmit our data with a localized pulse of significant intensity, and the signal velocity is determined by how fast the bulk of that energy arrives at the detector, often in highly distorted form. Sommerfeld and Brillouin were able to quantify this statement mathematically, but the derivation is rather complicated, and we refer to their text for more information [12]. In regions of normal dispersion, we may say with some confidence that the signal velocity is equivalent to the group velocity.

This lengthy discussion makes an important point about the speed of light: there is no single number that properly characterizes it. Light is 'squishy,' and a pulse can change shape dramatically during propagation. Aside from the leading edge of the pulse, we cannot label a particular part of a pulse and use that as an unambiguous measure of wave speed—and we normally cannot measure the exact leading edge of the pulse anyway.

This confusion about the speed of light was highlighted in 2000, when researchers published a paper with the provocative title, 'Gain-assisted superluminal light propagation' [13]. As reported in this paper, the researchers experimentally studied a material where the group velocity greatly exceeded the speed of light, to the degree that the peak of the pulse would exit the material before it had actually entered it.

To say that the optics community lost their minds over this would be an understatement. There was much confusion about the results, including accusations that it was simply incorrect. The result, however, was perfectly consistent with relativity and the optics we have discussed in this section. The light pulses were propagated in a cesium gas that had been optically pumped to be a gain medium; the pulse frequency was chosen to lie between two of the spectral gain lines in the medium.

We may use the bus analogy again to explain the physics. We now imagine a bus where people get off the back of the bus and an equal number of people get on at the front of the bus. Again, the center of mass of the passengers has shifted forward, though the total number of passengers remains unchanged. In the optical pulse, energy is removed from the back of the pulse and added to the front, causing the peak to shift forward and give the illusion of superluminal propagation. The result was surprising at first because there was no obvious gain or loss of light to account for the superluminal group velocity.

Though several of our defined velocities can exceed the vacuum speed of light c, it is widely assumed that nothing meaningful, be it photons or information, travels faster than c. You will note I said 'assumed.' All of our theoretical and experimental investigations have strongly suggested that the speed of light is relativistically limited, but it is not known for certain. In fact, high-precision experiments have been done to see whether single photons somehow subtly break relativity [14] using the phenomenon of precursors, which we briefly discuss in section 6.7.

Let us make one final observation about group velocity: if we have an expression providing frequency ω in terms of the wavenumber k in a dispersion relation, we may also calculate the group velocity by

$$v_g = \frac{d\omega}{dk}. \tag{6.71}$$

Quite often, this is the case.

6.5 Optical dispersion

The term 'dispersion' is broadly used to refer to the frequency dependence of the refractive index of a medium. A familiar illustration of dispersion is the dispersion of the colors of white light by a prism, caused by each color experiencing a different refractive index and therefore propagating in a different direction. Group velocity may be considered an effect of dispersion, but we note that the term is sometimes reserved only for frequency-dependent effects that cause a change in pulse shape.

For a narrowband pulse, we have seen that the main effect of dispersion is a change in velocity, in the form of group velocity. For a broadband pulse, which might overlap several resonant frequencies, we expect the shape of the pulse to be highly distorted and to broaden upon propagation.

In general, we cannot predict what will happen to a pulse when it is strongly dispersed, but we can look at how pulse evolution changes in a region of normal dispersion when we go beyond the group velocity approximation. We return to equation (6.65) and now consider an additional term in the Taylor expansion of $k_R(\omega)$,

$$k_R(\omega' + \omega_s) \approx k_R(\omega_s) + \frac{dk_R}{d\omega}\bigg|_{\omega=\omega_s} \omega' + \frac{1}{2}\frac{d^2 k_R}{d\omega^2}\bigg|_{\omega=\omega_s} \omega'^2. \tag{6.72}$$

Equation (6.65) now becomes

$$\mathbf{E}(z,\,t) = \hat{\mathbf{x}}e^{ik_s z}e^{-i\omega_s t}\int_{-\infty}^{\infty} \tilde{A}(\omega')e^{i\left[k_s' \omega' z - \omega t + \frac{k_s''}{2}\omega'^2 z\right]}d\omega', \tag{6.73}$$

where we have written for brevity:

$$k_s = k_R(\omega_s), \quad k_s' = \frac{dk_R}{d\omega}\bigg|_{\omega=\omega_s}, \quad k_s'' = \frac{d^2 k_R}{d\omega^2}\bigg|_{\omega=\omega_s}. \tag{6.74}$$

We can only evaluate this integral analytically for certain special choices of $\tilde{A}(\omega')$. Let us consider a Gaussian spectrum,

$$\tilde{A}(\omega') = \frac{E_0 \sigma}{\sqrt{2\pi}}e^{-\sigma^2(\omega-\omega_s)^2/2}. \tag{6.75}$$

This spectrum corresponds to a Gaussian pulse with a central frequency ω_s and width σ,

$$A(t) = E_0 e^{-i\omega_s t} e^{-t^2/2\sigma^2}. \tag{6.76}$$

We may evaluate the integral for this choice of spectrum; the messy result is

$$E(z, t) = \frac{E_0}{\sqrt{1 - \dfrac{i k_s'' z}{\sigma^2}}} e^{i(k_s z - \omega_s t)} e^{-i k_s'' z (k_s' z - t)^2 / 2\beta^4} e^{-(k_s' z - t)^2 \sigma^2 / 2\beta^4}, \tag{6.77}$$

where

$$\beta^4 = \sigma^4 + k_s''^2 z^2. \tag{6.78}$$

The first exponential is again the phase velocity term, while the second term represents a complex phase distortion of the field due to the different frequencies propagating at different speeds. The third term is again a pulse propagating at the group velocity, but with an effective width of β^2/σ. This width increases as z increases, which means that the pulse broadens upon propagation.

A dispersive medium generally causes a pulse to broaden, but it is also possible to design pulse shapes that are compressed over a finite propagation distance in the medium. The propagation of a pulse in a medium under the approximation of equation (6.72) is mathematically analogous to the propagation of a paraxial monochromatic beam in space. Just as a monochromatic beam can be made to focus with an appropriate quadratic phase imposed on the wavefront via a lens, a pulse can be made to compress if an appropriate quadratic spectral phase (a 'chirp') is imposed upon it.

6.6 Kramers–Kronig relations

We have seen that a number of equations describing the response of a material to an electromagnetic wave may be written in a multiplicative form in the frequency domain; for example, in equation (5.73) we saw that the relation between the induced polarization density and the electric field is of the form

$$\tilde{\mathbf{P}}(\omega) = \epsilon_0 \tilde{\chi}_e(\omega) \tilde{\mathbf{E}}(\omega), \tag{6.79}$$

where we have added a tilde to each of the quantities to indicate that they represent the Fourier transforms of their time-domain counterparts. If we look at this expression in the time domain and make a very reasonable and broad physical assumption, we find remarkable relationships between the real and imaginary parts of the material properties; these are known as the *Kramers–Kronig relations*, after Ralph Kronig [15] and Hans Kramers [16], who independently discovered them. The derivation requires significant knowledge of complex analysis, but the result is well worth the effort, as we will see.

Let us for the moment focus on the relationship of equation (6.79), and take the inverse Fourier transform to determine the response of the system in the time domain. This inverse transform may be defined as

$$\mathbf{P}(t) = \int_{-\infty}^{\infty} \tilde{\mathbf{P}}(\omega)e^{-i\omega t}d\omega, \tag{6.80}$$

which then takes on the form

$$\mathbf{P}(t) = \int_{-\infty}^{\infty} \tilde{\chi}_e(\omega)\tilde{\mathbf{E}}(\omega)e^{-i\omega t}d\omega. \tag{6.81}$$

We then use the definition of the forward transform,

$$\tilde{\mathbf{E}}(\omega) = \frac{1}{2\pi}\int_{-\infty}^{\infty} E(t')e^{i\omega t'}dt', \tag{6.82}$$

with a similar definition for $\tilde{\chi}_e(\omega)$, to write

$$\mathbf{P}(t) = \frac{1}{(2\pi)^2}\int_{-\infty}^{\infty}\int_{-\infty}^{\infty}\int_{-\infty}^{\infty} \mathbf{E}(t')\chi_e(t'')e^{-i\omega(t-t'-t'')}d\omega dt'dt''. \tag{6.83}$$

The integral over ω represents the integral form of the Dirac delta function, i.e.

$$\delta(\tau) = \frac{1}{2\pi}\int_{-\infty}^{\infty} e^{i\omega\tau}d\omega, \tag{6.84}$$

which means we may write the polarization density in time as

$$\mathbf{P}(t) = \frac{1}{2\pi}\int_{-\infty}^{\infty}\int_{-\infty}^{\infty} \mathbf{E}(t')\chi_e(t'')\delta(t - t' - t'')dt'dt''. \tag{6.85}$$

Evaluating the Dirac delta integral over t'', we then get the result

$$\mathbf{P}(t) = \frac{1}{2\pi}\int_{-\infty}^{\infty} \mathbf{E}(t')\chi_e(t - t')dt'. \tag{6.86}$$

This expression represents the polarization response to an electric field in the form of a linear, time-shift-invariant system. It is linear because the polarization depends upon the electric field through a linear relation, and it is time-shift invariant because the response, characterized by $\chi_e(t - t')$, only depends upon the time difference $t - t'$.

We could also have evaluated equation (6.85) by integrating over t' to get the equivalent form

$$\mathbf{P}(t) = \frac{1}{2\pi}\int_{-\infty}^{\infty} \mathbf{E}(t - t'')\chi_e(t'')dt''. \tag{6.87}$$

Both equations (6.86) and (6.87) are in the mathematical form of convolutions from the theory of Fourier transforms.

Equation (6.87) is quite general, and we have put no restrictions yet upon χ_e; we now ask what broad behaviors we should expect from it. We note that the integral is over all times t' from $-\infty$ to ∞; however, we expect that the polarization can only be affected by electric fields at times earlier than t. This is the definition of causality, or 'cause precedes effect.' In our integral, causality is clearly represented by the requirement that

$$\chi_e(\tau) = 0, \quad \tau < 0. \tag{6.88}$$

The Fourier transform of $\chi_e(t)$ may therefore be written as

$$\tilde{\chi}_e(\omega) = \frac{1}{2\pi} \int_0^\infty \chi_e(t) e^{i\omega t} dt, \tag{6.89}$$

where causality is the imposition of zero as the lower limit of integration.

We now apply some insights from complex analysis. If we replace the real-valued ω by a complex form, $\tilde{\omega} = \omega_R + i\omega_I$, with ω_R and ω_I real-valued, we see that, for any $t > 0$, $\exp[i\tilde{\omega}t]$ is an analytic function in the upper half-space of the complex plane, i.e. $\omega_I > 0$. This is self-evident because the exponential has the explicit form

$$e^{i\tilde{\omega}t} = e^{i\omega_R t} e^{-\omega_I t}, \tag{6.90}$$

and the function exponentially decays as $\omega_I \to \infty$.

If the function $\chi_e(t)$ is assumed to be square integrable, then it decays rapidly with increasing t, and one can show that the complex function $\tilde{\chi}_e(\tilde{\omega})$ must also be analytic in the upper half-space.

It is well known that the real and imaginary parts of an analytic function are related to each other. We can make this explicit by considering the integral

$$I = \int_{-\infty}^\infty \frac{\tilde{\chi}_e(\omega')}{\omega' - \omega} d\omega'. \tag{6.91}$$

This can be evaluated using the Cauchy residue theorem, using a closed contour that is an infinite semicircular path that follows the real axis and circles the upper half-space, as illustrated in figure 6.7. The integral over the semicircular part vanishes as $R \to \infty$.

The only singularity present is on the contour at $\omega' = \omega$; according to the residue theorem, the integral has the value

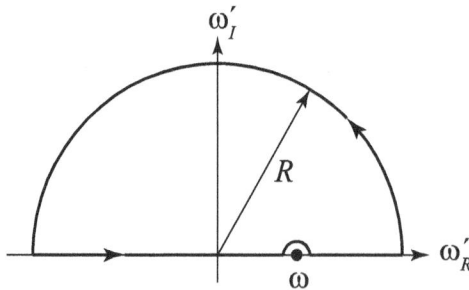

Figure 6.7. The contour of integration for the Kramers–Kronig relations, taken over the upper half-space of the complex plane as $R \to \infty$.

$$I = \int_{-\infty}^{\infty} \frac{\tilde{\chi}_e(\omega')}{\omega' - \omega} d\omega' = \pi i \tilde{\chi}_e(\omega). \qquad (6.92)$$

The only complex part of the integrand is $\tilde{\chi}_e$; if we take the real and imaginary parts of the integral and its residue value, we have

$$\tilde{\chi}_I(\omega) = -\frac{1}{\pi} \int_{-\infty}^{\infty} \frac{\tilde{\chi}_R(\omega')}{\omega' - \omega} d\omega', \qquad (6.93)$$

$$\tilde{\chi}_R(\omega) = \frac{1}{\pi} \int_{-\infty}^{\infty} \frac{\tilde{\chi}_I(\omega')}{\omega' - \omega} d\omega'. \qquad (6.94)$$

These are the Kramers–Kronig relations, which allow us to derive the real part of $\tilde{\chi}(\omega)$ from the imaginary part and vice versa.

The Kramers–Kronig relations are of surprising generality. We have derived them for the complex susceptibility, but they apply to any linear, time-invariant, causal response function, including, with a slight modification, the refractive index of a material. Equation (6.59), introduced when we discussed the propagation of a plane wave in matter, implies that the electric field $\tilde{\mathbf{E}}(\omega)$ of a plane wave after propagating a distance z in matter is related to the input electric field $\tilde{\mathbf{E}}_0(\omega)$ by the expression

$$\tilde{\mathbf{E}}(\omega) = \tilde{\mathbf{E}}_0(\omega) e^{ik_0 n(\omega) z}, \qquad (6.95)$$

where the complex exponential has a role analogous to $\tilde{\chi}_e(\omega)$ in equation (6.79). We expect that the propagation of light in a medium is causal as well, which implies that the exponential in equation (6.95) is analytic in the upper half-space of $\tilde{\omega}$. This, in turn, indicates that $n(\omega)$ must be analytic in this upper half-space.

This suggests that $n(\omega)$ must satisfy the Kramers–Kronig relations, but there is one catch: the relations only apply to functions that are square integrable, and the refractive index $n(\omega)$ approaches unity, the vacuum value, as $\omega \to \infty$. There is a simple fix, however, and we simply apply the conditions to $n(\omega) - 1$ to get

$$n_I(\omega) = -\frac{1}{\pi} \int_{-\infty}^{\infty} \frac{n_R(\omega') - 1}{\omega' - \omega} d\omega', \qquad (6.96)$$

$$n_R(\omega) = 1 + \frac{1}{\pi} \int_{-\infty}^{\infty} \frac{n_I(\omega')}{\omega' - \omega} d\omega'. \qquad (6.97)$$

A similar argument may be made for the permittivity, so that $\epsilon(\omega) - \epsilon_0$ must satisfy the Kramers–Kronig relations.

It should not be surprising at this point to note that the refractive index derived in the Lorentz model, equation (6.27), and the permittivity derived in the Debye model, equation (6.43), satisfy Kramers–Kronig. This is quite remarkable, however, considering we did not explicitly introduce or consider causality in our derivations. A demonstration that the relations hold is left as an exercise.

The Kramers–Kronig relations have a very common practical use: if the real part of the quantity can be measured over a significantly broad spectral range, the imaginary part can be estimated using the Kramers–Kronig relations. For example, the intensity of light reflected from a sample can be measured, and the complex reflectivity can then be determined; from this, the complex optical constants of the material can be found. An illustrative example of this is the work of Philipp and Tait [17], who determined the optical constants of germanium.

Due to their very general derivation based on causality and linear systems, the Kramers–Kronig relations are applicable and useful for fields outside optics, from electron scattering to particle physics to seismic wave physics. They provide a stunning demonstration of the link between causality and analytic functions [18, 19].

6.7 Optical precursors

Throughout this chapter, we have applied various simplifying assumptions in order to analyze the propagation of light in a random medium—in particular, we have assumed a narrowband signal and one that is far from resonance. When these assumptions are not satisfied, the problem becomes difficult to analyze, and the pulse dynamics take on a nontrivial form.

There are some surprisingly general phenomena that arise in the propagation of wideband pulses through dispersive and absorptive media, however. When a short pulse propagates over long distances (the meaning of 'long' in this case depends on the specific circumstances) in a resonant medium, it can evolve two distinct waveforms that typically arrive before the main body of the pulse. These waveforms were first introduced by Sommerfeld and Brillouin in 1914 and were referred to by them as 'forerunners,' though today they are typically known as *precursors*. The first to arrive is a fast, high-frequency wave known as the Sommerfeld precursor, and the second is a slower, low-frequency wave known as the Brillouin precursor.

The mathematics of precursors is dauntingly complicated. The analysis relies on the asymptotic method of steepest descents and requires significant abstract manipulation of integrals in the complex plane. The original work of Sommerfeld and Brillouin was compiled into a book by Brillouin [12]. A refined and improved analysis was presented many years later in an article by Oughstun and Sherman [20], and its length—33 pages!—gives an idea of the difficulty of the problem.

In this section, we briefly note some of the main features of precursors and attempt to give a very rough explanation of their origin; for those who are interested in delving deeper, we refer to the book by Oughstun and Sherman [21].

Precursors arise on propagation for pulses that have very broadband spectra. Sommerfeld and Brillouin's original analysis considered a monochromatic signal that was switched on instantaneously at $t = 0$; others have considered 'top-hat' pulses modulated by a carrier frequency. Many researchers have approximated an ultranarrow temporal pulse by a delta function signal.

The analysis is typically done for pulses in the neighborhood of a single Lorentz-type resonance; however, double Lorentz resonance analyses have been performed [22].

We may summarize the results as reported by Brillouin in his book. The Sommerfeld precursor travels fastest and arrives first, at the vacuum speed of light c. It starts with a small amplitude and a small period of oscillation that increases continuously until it matches the period of the electrons in the medium. The Brillouin precursor arrives next, traveling at a speed $c/n(0)$, i.e. at a speed dictated by the zero-frequency refractive index. It starts with a large period that decreases until it matches the signal. The signal travels at, of course, the signal velocity. These speeds are presented with the usual caveats about the imprecise nature of defining a speed of light in matter.

One of the most striking features of precursors, and one that has caused them to receive significant attention, is that their maximum amplitude decays algebraically as $1/\sqrt{z}$, where z is the propagation distance, instead of exponentially. This has led many researchers over the years to investigate whether these precursors can be used to probe deeper into otherwise opaque media.

Even the simplest explanations of the origins of precursors typically involve a lot of math; see, for example, Sherman and Oughstun [23] and Jakobsen and Mansuripur [24]. Here, I attempt to give a very simple and qualitative explanation of precursors and their features, though it should be noted that my explanation is definitely an oversimplification of the problem.

The manifestation of precursors appears strongly connected to the use of broadband pulses; in particular, pulses with a sharp temporal edge. It is well known from Fourier analysis that a discontinuous function in the time domain has a very broad frequency spectrum that extends down to zero frequency and up to arbitrarily high frequencies (see, for example, section 11.4.1 of [25]). Precursors evidently arise from these high-frequency and low-frequency tails of the pulse spectrum. At extremely high frequencies, electrons cannot respond to the field at all, and the field effectively experiences vacuum propagation; this results in the Sommerfeld precursor, which travels at the speed of light c. At extremely low frequencies, the electrons effectively respond instantaneously to the slow oscillations of the field, responding as in electrostatics; this results in the Brillouin precursor, which travels at the speed $c/n(0)$. These interpretations are bolstered by the observation that the Sommerfeld precursor starts with a high frequency, and the Brillouin precursor starts with a low frequency.

In both the extreme high-frequency and low-frequency limits, there is negligible transfer of energy to the electrons of the medium and hence little absorption. This appears to result in the low absorption of the precursors as a whole.

This explanation captures the most prominent features of precursors, at least qualitatively. It is intended to build the reader's intuition only; any practical study of precursors requires a full dive into the asymptotic behavior of pulse propagation.

It took many years for precursors to be observed experimentally, due in large part to the fact that their amplitude is extremely low in many cases. The first experiments were conducted in 1969 by Pleshko and Palócz [26] in the microwave regime. They studied the propagation of microwaves in a variety of metallic waveguides, whose modal dispersion characteristics are similar to Lorentzian media (these waveguides will be discussed in chapter 13). Both the Sommerfeld and Brillouin precursors were observed. The first precursors at optical frequencies were reported by Aaviksoo,

Kuhl, and Ploog in 1991 [27]. They propagated an optical pulse with a sharp exponential rise through a thin GaAs layer and measured the temporal shape of the pulse at the output. They found that a sharp signal increase arrived at the detector independent of the detuning (frequency shift from resonance) of the input pulse—this was interpreted as the Sommerfeld precursor. The main pulse followed afterward with a delay dependent on detuning.

In 2004, Choi and Österberg observed precursor effects in water [28], using a chirped femtosecond pulse to provide enough bandwidth to stimulate precursor formation. They confirmed their precursors exhibited nonexponential decay with distance.

The Sommerfeld precursor, with its predicted speed of c, has found an interesting physical use: testing whether individual photons can violate relativity. We have noted that the Kramers–Kronig relations for refractive index imply that pulses must be causal and relativistically limited, and experimental measurements have indicated that the Kramers–Kronig relations are satisfied, but we cannot measure that index for arbitrarily high frequencies, where there is the possibility of some sort of subtle violation of Kramers–Kronig. The Sommerfeld precursor, traveling at c and generated by the high-frequency components of a pulse spectrum, provides an intriguing way to hunt for relativity-breaking behavior. Such an experiment was performed in 2011 by Zhang *et al* [14] using single photons propagating through a magneto-optical trap of cold rubidium atoms. Perhaps disappointingly, they did not find any causality-breaking for a single photon, but their approach shows that precursors can find surprising applications.

6.8 Exercises

1. We have seen that Gaussian pulses with second-order dispersion (non-negligible k'') generally become wider as they propagate; however, this is not necessarily the case! Consider a pulse with the following temporal profile at $z = 0$:

$$A(t) = E_0 e^{-i\omega_s t} e^{-t^2/2\sigma^2} e^{-it^2/2\alpha^2},$$

where α is a positive real number. Such a pulse is *chirped*: it has instantaneously higher/lower frequencies at times earlier/later in the pulse, respectively. By repeating the propagation calculation for a Gaussian pulse, show that there are values of k'' and α for which the pulse has a minimum width at $z > 0$.

   Explain how this can happen! (Hint: remember that the phase velocity of light varies by frequency. For a chirped pulse, different regions of the pulse effectively have different frequencies.)

2. In a hollow metal circular waveguide of radius a, the lowest-order ($m = 0$, $n = 1$) mode of the waveguide satisfies the dispersion relation

$$k = \sqrt{\left(\frac{\omega}{c}\right)^2 - \frac{(2.4048)^2}{a^2}}.$$

Calculate the phase velocity and group velocity for this waveguide. If $a = 1$ cm, what range of frequencies ω cannot propagate in this waveguide?

3. For a surface wave on a metal (known as a surface plasmon—to be discussed in chapter 11) at a single interface, the wave satisfies the dispersion relation

$$k = \frac{\omega}{c} \sqrt{\frac{\epsilon_1 \epsilon_2(\omega)}{\epsilon_0(\epsilon_1 + \epsilon_2(\omega))}},$$

where ϵ_1 is taken to be vacuum and ϵ_2 is a metal satisfying the relation

$$\epsilon_2(\omega) = \epsilon_0 \left[1 - \frac{\omega_p^2}{\omega^2} \right],$$

where ω_p is known as the plasma frequency. For $\omega < \omega_p/\sqrt{2}$, calculate the phase and group velocities of the surface plasmon. Are either of these velocities superluminal?

4. A plasma oscillation is a volumetric charge density wave that can propagate in metals. Such oscillations satisfy the dispersion relation

$$\omega^2 = \omega_p^2 + \frac{3k_B T}{m} k^2,$$

where T is the temperature of the metal in Kelvin, k_B is the Boltzmann constant, and ω_p is the plasma frequency. Calculate the group velocity and phase velocity of a plasma oscillation. Assuming $T = 293$ K and $m = 9.11 \times 10^{-31}$ kg, calculate the values of these velocities for $\omega = \sqrt{2}\,\omega_p$.

5. Using the residue theorem, show that the real part of the refractive index can be derived from the imaginary part using the Kramers–Kronig relations in the Lorentz model,

$$n(\omega) = 1 + \frac{\omega_p^2}{2} \frac{1}{(\omega_f^2 - \omega^2) - i\Gamma\omega},$$

where $\Gamma > 0$.

6. Using the residue theorem, show that the real part of the permittivity can be derived from the imaginary part using the Kramers–Kronig relations in the Debye model,

$$\epsilon(\omega) = \epsilon_{0v} + \frac{\epsilon_0 \Delta}{1 - i\omega\tau},$$

where $\Delta > 0$, $\tau > 0$, and ϵ_{0v} is real-valued.

7. In a metal, there are a large number of free electrons, approximately forming a plasma (the Drude model) with the permittivity

$$\epsilon(\omega) = \epsilon_0 \left[1 - \frac{Ne^2}{\epsilon_0(m\omega^2 - i\omega g)} \right],$$

where m, N, and g are all positive-valued. Attempt to use the Kramers–Kronig relations to derive the imaginary part of the permittivity from the real part, and show that the result does *not* satisfy Kramers–Kronig. From the given $\epsilon(\omega)$, can you see what assumption in Kramers–Kronig has been broken?

8. We have seen that the Lorentz–Lorenz condition amounts to

$$\frac{\epsilon/\epsilon_0 - 1}{\epsilon/\epsilon_0 + 2} = \alpha N/3.$$

Instead of making the 'tenuous approximation,' solve the above equation directly for ϵ. Then show that this equation approximately reduces to

$$\epsilon/\epsilon_0 - 1 = N\alpha$$

when $N\alpha \ll 1$. (Hint: in the second part, take the Taylor expansion of the denominator and throw away anything not linear in αN.)

9. The Earth's ionosphere may be approximated by an ideal plasma, with a permittivity

$$\epsilon(\omega) = \epsilon_0 \left[1 - \frac{\omega_p^2}{\omega^2} \right],$$

where the plasma frequency $\omega_p \sim 6 \times 10^7$ s^{-1}. Consider the propagation of a narrowband light pulse of vacuum wavelength $\lambda = 500$ nm and a radio pulse of vacuum wavelength $\lambda = 12.00$ m through 100 km of the ionosphere. How long does it take for each signal to travel this distance? Which signal arrives first, and by how much time? Repeat the calculation for $\lambda = 24$ m.

10. Phonons are collective oscillations in a periodic arrangement of atoms. For a one-dimensional chain of atoms with spacing a bound by an effective spring constant f and with mass m, the dispersion relation of phonons is given by

$$\omega = 2\sqrt{\frac{f}{m}} \left| \sin\left(\frac{ka}{2}\right) \right|,$$

with k restricted to lie within the range $-\pi/a \leqslant k < \pi/a$. Calculate the group velocity for $0 \leqslant k < \pi/a$ by calculating $d\omega/dk$. Invert this dispersion relation to get a formula for $k(\omega)$ and show that the same group velocity results from

$$v_g = \frac{1}{dk/d\omega}.$$

11. In metallic rectangular microwave waveguides with side widths a and b, the different modes of the waveguide satisfy a dispersion relation

$$k = \sqrt{\left(\frac{\omega}{c}\right)^2 - \omega_{mn}^2},$$

where

$$\omega_{mn} = c\pi\sqrt{\left(\frac{m}{a}\right)^2 + \left(\frac{n}{b}\right)^2}.$$

Here, $a > b$ by convention, and m, n are nonnegative integers. Write an expression for the group velocity of the modes. For a waveguide of dimensions $a = 8$ mm and $b = 4$ mm, determine the time delay between the lowest-order transverse electric (TE) mode with $m = 1$, $n = 0$ and the lowest-order transverse magnetic (TM) mode with $m = 1$, $n = 1$ after propagating through a waveguide of length $d = 10$ m, if the waveguide operates at a frequency $\nu = 88.045$ GHz. How does this time delay compare to the period of oscillation?

References

[1] Fox M 2010 *Optical Properties of Solids* 2nd edn (Oxford: Oxford University Press)

[2] Wood R W 1902 The anomalous dispersion of sodium vapour *Lond. Edinb. Dublin Philos. Mag. J. Sci.* **3** 128–44

[3] Rusk A N, Williams D and Querry M R 1971 Optical constants of water in the infrared *J. Opt. Soc. Am.* **61** 895–903

[4] Bohren C F and Huffman D R 1983 *Absorption and Scattering of Light by Small Particles* (New York: Wiley)

[5] Sellmeier W 1872 Ueber die durch die Aetherschwingungen erregten Mitschwingungen der Körpertheilchen und deren Rückwirkung auf die ersteren, besonders zur Erklärung der Dispersion und ihrer Anomalien *Ann. Phys., Lpz.* **223** 386–403

[6] Debye P 1929 *Polar Molecules* (New York: The Chemical Catalogue Company)

[7] Gevers M 1946 The relation between the power factor and the temperature coefficient of the dielectric constant of solid dielectrics *Philips Research Reports 1945/46* **1** (Eindhoven: Philips' Research Laboratory) 197–224

[8] Kaatze U and Uhlendorf V 1981 The dielectric properties of water at microwave frequencies *Z. Phys. Chem.* **126** 151–65

[9] Cook H F 1952 A comparison of the dielectric behaviour of pure water and human blood at microwave frequencies *Br. J. Appl. Phys.* **3** 249

[10] Lane J A and Saxton J A 1952 Dielectric dispersion in pure polar liquids at very high radio-frequencies I. Measurements on water, methyl and ethyl alcohols *Proc. R. Soc.* A **213** 400–8

[11] Grant E H, Buchanan T J and Cook H F 1957 Dielectric behavior of water at microwave frequencies *J. Chem. Phys.* **26** 156–61

[12] Brillouin L 1960 *Wave Propagation and Group Velocity* (New York and London: Academic)

[13] Wang L, Kuzmich A and Dogariu A 2000 Gain-assisted superluminal light propagation *Nature* **406** 277–9

[14] Zhang S, Chen J F, Liu C, Loy M M T, Wong G K L and Du S 2011 Optical precursor of a single photon *Phys. Rev. Lett.* **106** 243602

[15] de L and Kronig R 1926 On the theory of dispersion of x-rays *J. Opt. Soc. Am.* **12** 547–57

[16] Kramers H A 1927 La diffusion de la lumière par les atomes *Atti Cong Intern Fisici (Transactions of Volta Centenary Congress)* **2** 545–57

[17] Philipp H R and Taft E A 1959 Optical constants of germanium in the region 1 to 10 eV *Phys. Rev.* **113** 1002–5

[18] Wolf E 2001 Analyticity, Causality, and Dispersion Relations *Selected Works of Emil Wolf: with commentary* (Singapore: World Scientific) 577–84

[19] Nussenzveig H M 1972 Causality and Dispersion Relations *Mathematics in Science and Engineering* (New York and London: Academic) 3–53

[20] Oughstun K E and Sherman G C 1988 Propagation of electromagnetic pulses in a linear dispersive medium with absorption (the Lorentz medium) *J. Opt. Soc. Am.* B **5** 817–49

[21] Oughstun K E and Sherman G C 1994 *Electromagnetic Pulse Propagation in Casual Dielectrics* (Berlin, Heidelberg: Springer)

[22] Oughstun K E and Xiao H 2001 Influence of precursor fields on ultrashort pulse autocorrelation measurements and pulse width evolution *Opt. Express* **8** 481–91

[23] Sherman G C and Oughstun K E 1981 Description of pulse dynamics in Lorentz media in terms of the energy velocity and attenuation of time-harmonic waves *Phys. Rev. Lett.* **47** 1451–4

[24] Jakobsen P K and Mansuripur M 2020 On the nature of the Sommerfeld–Brillouin forerunners (or precursors) *Quantum Stud. Math. Found.* **7** 315–9

[25] Gbur G J 2011 *Mathematical Methods for Optical Physics and Engineering* (Cambridge: Cambridge University Press)

[26] Pleshko P and Palócz I 1969 Experimental observation of Sommerfeld and Brillouin precursors in the microwave domain *Phys. Rev. Lett.* **22** 1201–4

[27] Aaviksoo J, Kuhl J and Ploog K 1991 Observation of optical precursors at pulse propagation in GaAs *Phys. Rev.* A **44** R5353–6

[28] Choi S H and Österberg U 2004 Observation of optical precursors in water *Phys. Rev. Lett.* **92** 193903

IOP Publishing

Electromagnetic Optics

Gregory J Gbur

Chapter 7

Conservation laws

Electromagnetic waves carry energy and momentum. The fact that they carry energy is obvious from such mundane phenomena as feeling the interior of your car heat up on a sunny day. Visible light passes easily through the car windshield and is absorbed by the material of the car's interior. Some of it is reradiated as infrared light, which cannot escape the interior of the car, making it effectively a greenhouse.

The momentum of light is less obvious, as a photon has a very small 'kick' to it, but it follows implicitly from the Lorentz force law. If a light wave is exerting a force on a charged particle, then it is transferring momentum to it. If momentum is conserved, the light wave itself must carry momentum. James Clerk Maxwell himself estimated the radiation pressure exerted by sunlight on the surface of the Earth (see paragraph 793 of [1]), deriving a value of 8.82×10^{-8} pounds per square foot.

The momentum of light, though negligible on an everyday scale, can be significant in two circumstances, and applications have followed from this:

- *On big objects in a frictionless environment over a long time.* Under these conditions, small forces can build up to a significant effect. The Yarkovsky–O'Keefe–Radzievskii–Paddack (YORP) effect is a verified phenomenon in which sunlight can alter the spin of asteroids [2]. As a practical application of the momentum of light, researchers are developing solar sails that will be propelled by the Sun's radiation pressure [3].
- *On small objects in intense light.* Microscopic particles can experience significant forces that move or trap them, and this has formed the basis of a practical technology called 'optical tweezing' [4].

Both energy and momentum are conserved, and this means that we should be able to formulate conservation laws for these quantities. In doing so, we will arrive at some important definitions of electromagnetic energy density and power flux that cannot be derived in any other way, along with their momentum analogues.

doi:10.1088/978-0-7503-6064-7ch7   7-1   © IOP Publishing Ltd 2025. All rights,

To conclude the chapter, we look at some applications of electromagnetic momentum that can be analyzed directly from Maxwell's equations.

7.1 Conservation of energy

The derivation of an energy conservation law for electromagnetic waves begins with the Lorentz force law,

$$\mathbf{F}(\mathbf{r},\, t) = q[\mathbf{E}(\mathbf{r},\, t) + \mathbf{v} \times \mathbf{B}(\mathbf{r},\, t)]. \tag{7.1}$$

Let us use this law to determine the amount of work done by an electromagnetic field on a charged particle. The infinitesimal work dW done by an electric field on a charge q upon moving it an infinitesimal distance $d\mathbf{r}$ is 'force times distance,' or

$$dW = q\mathbf{E} \cdot d\mathbf{r}. \tag{7.2}$$

Because the magnetic force is always perpendicular to the path of motion of a charged particle, it does not play a direct role in energy transfer or energy conservation.

Let us divide dW by an infinitesimal time dt and note that the velocity of the particle $\mathbf{v} = d\mathbf{r}/dt$; we then find the mechanical power dE_{mech}/dt given to the charge is

$$\frac{dE_{mech}}{dt} = q\mathbf{v} \cdot \mathbf{E}. \tag{7.3}$$

We can now generalize this expression to an extended current distribution using $q\mathbf{v} = \mathbf{J}d^3r$, where d^3r is an infinitesimal volume element and $\mathbf{J}$ is again the current density. The rate of energy transfer to a system of currents can then be written as

$$\frac{dE_{mech}}{dt} = \int_V \mathbf{J}(\mathbf{r},\, t) \cdot \mathbf{E}(\mathbf{r},\, t)d^3r, \tag{7.4}$$

where it is assumed that all charges and currents are contained within a volume V, as illustrated in figure 7.1.

Though equation (7.4) is a perfectly simple and elegant expression, it tells us nothing about the energy contained in the electromagnetic field itself. We reformulate the expression by eliminating the current density through the use of the Ampère–Maxwell law,

$$\nabla \times \mathbf{H}(\mathbf{r},\, t) = \mathbf{J}(\mathbf{r},\, t) + \frac{\partial \mathbf{D}(\mathbf{r},\, t)}{\partial t}. \tag{7.5}$$

Solving this for $\mathbf{J}$, we then substitute the result into equation (7.4), which yields

$$\frac{dE_{mech}}{dt} = \int_V \left[\mathbf{E}(\mathbf{r},\, t) \cdot [\nabla \times \mathbf{H}(\mathbf{r},\, t)] - \mathbf{E}(\mathbf{r},\, t) \cdot \frac{\partial \mathbf{D}(\mathbf{r},\, t)}{\partial t} \right]d^3r. \tag{7.6}$$

Our expression now describes the change of mechanical energy entirely in terms of the electric and magnetic fields. Next, we apply the standard vector calculus identity,

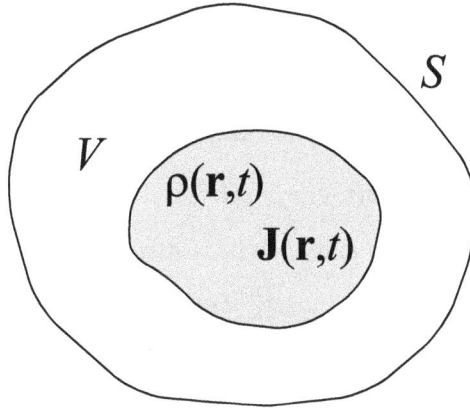

Figure 7.1. Illustration of the notation used in deriving the momentum conservation law. The shaded area indicates the location of sources, and these sources are assumed to remain completely within V.

$$\nabla \cdot (\mathbf{E} \times \mathbf{H}) = \mathbf{H} \cdot (\nabla \times \mathbf{E}) - \mathbf{E} \cdot (\nabla \times \mathbf{H}). \tag{7.7}$$

We can use this expression, along with Faraday's law,

$$\nabla \times \mathbf{E}(\mathbf{r},\, t) = -\frac{\partial \mathbf{B}(\mathbf{r},\, t)}{\partial t}, \tag{7.8}$$

to write the change in mechanical energy as

$$\begin{aligned}
\frac{dE_{mech}}{dt} = &-\int_V \nabla \cdot [\mathbf{E}(\mathbf{r},\, t) \times \mathbf{H}(\mathbf{r},\, t)] d^3 r \\
&- \int_V \left[\mathbf{H}(\mathbf{r},\, t) \cdot \frac{\partial \mathbf{B}(\mathbf{r},\, t)}{\partial t} + \mathbf{E}(\mathbf{r},\, t) \cdot \frac{\partial \mathbf{D}(\mathbf{r},\, t)}{\partial t} \right] d^3 r.
\end{aligned} \tag{7.9}$$

So far, our calculation has been completely general. To gain further insight, we must make some seemingly reasonable physical interpretations of the quantities in equation (7.9). These interpretations, however, can lead to seemingly unphysical results in some cases; we discuss this issue in section 7.2.

The first integral in equation (7.9) can be written as a surface integral using Gauss's theorem,

$$\int_V \nabla \cdot [\mathbf{E}(\mathbf{r},\, t) \times \mathbf{H}(\mathbf{r},\, t)] d^3 r = \oint_S [\mathbf{E}(\mathbf{r},\, t) \times \mathbf{H}(\mathbf{r},\, t)] \cdot d\mathbf{a}, \tag{7.10}$$

where S is the surface bounding the volume V and $d\mathbf{a}$ is the infinitesimal vector area element. The vector that appears in the surface integral is now known as the *Poynting vector* $\mathbf{S}$, named after physicist John Henry Poynting, who first formulated the electromagnetic energy conservation law in 1884 [5]. The Poynting vector is defined as

$$\mathbf{S}(\mathbf{r},\, t) \equiv \mathbf{E}(\mathbf{r},\, t) \times \mathbf{H}(\mathbf{r},\, t). \tag{7.11}$$

To evaluate the second integral in equation (7.9), let us restrict ourselves for the moment to vacuum propagation. In this case, the constitutive relations take on the simple form

$$\mathbf{D}(\mathbf{r}, t) = \epsilon_0 \mathbf{E}(\mathbf{r}, t), \tag{7.12}$$

$$\mathbf{B}(\mathbf{r}, t) = \mu_0 \mathbf{H}(\mathbf{r}, t). \tag{7.13}$$

The quantities in the integrand may then be written as total derivatives in time, e.g.

$$\mathbf{E}(\mathbf{r}, t) \cdot \frac{\partial \mathbf{D}(\mathbf{r}, t)}{\partial t} = \epsilon_0 \mathbf{E}(\mathbf{r}, t) \cdot \frac{\partial \mathbf{E}(\mathbf{r}, t)}{\partial t} = \frac{1}{2} \frac{\partial}{\partial t} \mathbf{E}(\mathbf{r}, t) \cdot \mathbf{D}(\mathbf{r}, t), \tag{7.14}$$

and the second integral of equation (7.9) becomes

$$\int_V \left[\mathbf{H}(\mathbf{r}, t) \cdot \frac{\partial \mathbf{B}(\mathbf{r}, t)}{\partial t} + \mathbf{E}(\mathbf{r}, t) \cdot \frac{\partial \mathbf{D}(\mathbf{r}, t)}{\partial t} \right] d^3 r$$
$$= \frac{1}{2} \frac{\partial}{\partial t} \int_V \left[\mathbf{E}(\mathbf{r}, t) \cdot \mathbf{D}(\mathbf{r}, t) + \mathbf{H}(\mathbf{r}, t) \cdot \mathbf{B}(\mathbf{r}, t) \right] d^3 r. \tag{7.15}$$

The integral as a whole has units of energy, which suggests that we may associate an *electromagnetic energy density* $U(\mathbf{r}, t)$ with the expression

$$U(\mathbf{r}, t) \equiv \frac{1}{2} [\mathbf{H}(\mathbf{r}, t) \cdot \mathbf{B}(\mathbf{r}, t) + \mathbf{E}(\mathbf{r}, t) \cdot \mathbf{D}(\mathbf{r}, t)]. \tag{7.16}$$

It should be noted that the total field energy is simply the sum of the electric field energy U_e and the magnetic field energy U_m.

The total rate of change of electromagnetic energy in our volume can then be defined as

$$\frac{dE_{field}}{dt} = \frac{\partial}{\partial t} \int_V U(\mathbf{r}, t) d^3 r. \tag{7.17}$$

The total energy E_{tot} in our volume is the sum of the mechanical energy and the electromagnetic energy, i.e. $E_{tot} = E_{mech} + E_{field}$. With this definition of E_{tot}, we can write equation (7.9) in the concise form

$$\frac{dE_{tot}}{dt} = -\oint_S \mathbf{S}(\mathbf{r}, t) \cdot d\mathbf{a}. \tag{7.18}$$

Equation (7.18) is the law of energy conservation in electromagnetic systems, which is known as Poynting's theorem. In this equation, the Poynting vector evidently may be interpreted as a flux of energy per unit area, with the surface integral indicating the total power flow through the surface. Poynting's theorem is a conservation law because it indicates that the total energy in our volume can only change when electromagnetic energy flows through the surface. Energy cannot spontaneously appear or disappear; it can only change form and/or move from one region to another. The minus sign on the right of equation (7.18) indicates that a net outward flow results in a decrease in total energy within the system.

This conservation law can also be written in a differential form. Let us consider a region of space free of charge, for which $E_{mech} = 0$ and $E_{tot} = E_{field}$. We use Gauss's theorem to write the right-hand side of equation (7.18) as a volume integral; then, requiring the integrands to be equal, we find that

$$\frac{\partial U(\mathbf{r}, t)}{\partial t} = -\nabla \cdot \mathbf{S}(\mathbf{r}, t). \tag{7.19}$$

Let us take a moment to compare this with the continuity equation, equation (2.57), which represents conservation of charge:

$$\frac{\partial \rho(\mathbf{r}, t)}{\partial t} = -\nabla \cdot \mathbf{J}(\mathbf{r}, t). \tag{7.20}$$

We can see that we have a structurally similar conservation law for charge and electromagnetic energy; in both cases, the law states: 'If (conserved something) is flowing through the surface, then the amount of (conserved something) in the volume must be changing over time.'

So far, we have studied conservation of energy in the time domain because it allowed us to readily interpret the physical meaning of quantities, such as the Poynting vector, that appeared in the derivation. The tradeoff, however, is that we were restricted to studying energy conservation in vacuum. The motivation for this restriction is clear: as we have seen, material properties are, in general, frequency dependent, i.e. dispersive, and in the time domain, the constitutive relations take on a nontrivial integral form (recall section 6.6). It is more natural to study energy conservation in the frequency domain, which we now consider.

We assume our usual time dependence $\exp[-i\omega t]$ and consider the time-averaged power flow into a system of charges. We follow equation (4.64) and write the time average of the product of two vectors as

$$\langle \mathbf{A} \cdot \mathbf{B} \rangle = \frac{1}{2}\text{Re}\{\mathbf{A}^* \cdot \mathbf{B}\}, \tag{7.21}$$

where $\mathbf{A}$ and $\mathbf{B}$ on the right represent the complex monochromatic forms of the real fields. We may then write the time-averaged monochromatic version of equation (7.4) as

$$\left\langle \frac{dE_{mech}}{dt} \right\rangle = \frac{1}{2} \int_V \mathbf{J}^*(\mathbf{r}, \omega) \cdot \mathbf{E}(\mathbf{r}, \omega) d^3r, \tag{7.22}$$

where it is implied that the physical result is found by taking the real part of the equation. In the frequency domain, Maxwell's curl equations take on the form

$$\nabla \times \mathbf{E}(\mathbf{r}, \omega) = i\omega \mathbf{B}(\mathbf{r}, \omega) \tag{7.23}$$

$$\nabla \times \mathbf{H}(\mathbf{r}, \omega) = \mathbf{J}(\mathbf{r}, \omega) - i\omega \mathbf{D}(\mathbf{r}, \omega), \tag{7.24}$$

and we use these to rewrite equation (7.22) entirely in terms of electric and magnetic fields. By following steps analogous to those taken in the time domain, we derive the expression

$$\left\langle \frac{dE_{mech}}{dt} \right\rangle = -\frac{1}{2} \oint_S \mathbf{E}(\mathbf{r}, \omega) \times \mathbf{H}^*(\mathbf{r}, \omega) \cdot d\mathbf{a}$$

$$- \frac{i\omega}{2} \int_V \left[\mathbf{D}^*(\mathbf{r}, \omega) \cdot \mathbf{E}(\mathbf{r}, \omega) - \mathbf{B}(\mathbf{r}, \omega) \cdot \mathbf{H}^*(\mathbf{r}, \omega) \right] d^3r. \tag{7.25}$$

We now define the cycle-averaged Poynting vector, electric energy density, and magnetic energy density as follows:

$$\mathbf{S}(\mathbf{r}, \omega) \equiv \frac{1}{2} \mathbf{E}(\mathbf{r}, \omega) \times \mathbf{H}^*(\mathbf{r}, \omega), \tag{7.26a}$$

$$U_e(\mathbf{r}, \omega) \equiv \frac{1}{4} \mathbf{D}^*(\mathbf{r}, \omega) \cdot \mathbf{E}(\mathbf{r}, \omega), \tag{7.26b}$$

$$U_m(\mathbf{r}, \omega) \equiv \frac{1}{4} \mathbf{H}^*(\mathbf{r}, \omega) \cdot \mathbf{B}(\mathbf{r}, \omega). \tag{7.26c}$$

With these definitions, our energy conservation law for monochromatic fields has the form

$$\left\langle \frac{dE_{mech}}{dt} \right\rangle = -2i\omega \int_V [U_e(\mathbf{r}, \omega) - U_m(\mathbf{r}, \omega)] d^3r - \oint_S \mathbf{S}(\mathbf{r}, \omega) \cdot d\mathbf{a}. \tag{7.27}$$

We now assume the usual linear constitutive relations for the electric and magnetic fields, i.e. $\mathbf{D}(\mathbf{r}, \omega) = \epsilon(\mathbf{r}, \omega)\mathbf{E}(\mathbf{r}, \omega)$ and $\mathbf{B}(\mathbf{r}, \omega) = \mu(\mathbf{r}, \omega)\mathbf{H}(\mathbf{r}, \omega)$. We have already seen in chapter 6 that $\epsilon(\mathbf{r}, \omega)$ and $\mu(\mathbf{r}, \omega)$ are complex, with $\epsilon = \epsilon_R + i\epsilon_I$; the imaginary part is positive-valued and represents absorption in the medium. With this substitution, we have

$$\left\langle \frac{dE_{mech}}{dt} \right\rangle = -\frac{i}{2}\omega \int_V \left[(\epsilon_R - i\epsilon_I)|\mathbf{E}|^2 - (\mu_R + i\mu_I)|\mathbf{H}|^2 \right] d^3r - \oint_S \mathbf{S}(\mathbf{r}, \omega) \cdot d\mathbf{a}. \tag{7.28}$$

We have noted that we must take the real part of any equations to get the physical conservation law; it is straightforward to see that (7.28) can be reduced to

$$\left\langle \frac{dE_{mech}}{dt} \right\rangle + \frac{\omega}{2} \int_V \left[\epsilon_I|\mathbf{E}|^2 + \mu_I|\mathbf{H}|^2 \right] d^3r = -\oint_S \mathbf{S}(\mathbf{r}, \omega) \cdot d\mathbf{a}. \tag{7.29}$$

The first term of this equation is the rate of change of the mechanical energy in the volume. The second term must represent the rate of change of the electromagnetic energy in the volume; because it depends on the imaginary parts of ϵ and μ, it, in particular, represents the loss of energy through absorption. The two terms on the left of the equation therefore represent the net change of energy in the volume, which must be balanced by a flux of power through the surface on the right-hand side of the equation.

For materials which are nonabsorbing, at least to a good approximation, ϵ and μ are real-valued, $\epsilon_I = 0$, and $\mu_I = 0$. In this case, the second term on the left of equation (7.29) vanishes. With no absorption present, the energy density plays no role in our energy conservation law. This may at first seem somewhat paradoxical, but recall that we are looking at a monochromatic system for which the cycle average of all quantities is time-independent. This is effectively a steady-state system, and therefore the energy density is, on average, constant.

To understand the meaning of the terms with $\epsilon_I \neq 0$ and $\mu_I \neq 0$, let us consider a system in the absence of free charges. Our monochromatic Poynting's theorem may then be written in differential form as

$$\nabla \cdot \mathbf{S}(\mathbf{r}, \omega) = -\frac{1}{2}[\epsilon_I |\mathbf{E}|^2 + \mu_I |\mathbf{H}|^2]. \tag{7.30}$$

A negative divergence of a vector field represents the location of a 'sink' of the field; this indicates that there is a net flow of power into those regions with absorption, which is a physically reasonable result.

Even with $\epsilon_I = 0$ and $\mu_I = 0$, equation (7.29) suggests that there can generally be a nonzero net flow of energy into or out of the volume. This, at first glance, appears inconsistent, considering that we are studying a time-harmonic system that is completely independent of time over a cycle average. In this case, the energy couples out of the system by some other mechanism not accounted for in our derivation; the reasoning is similar to that discussed in section 5.4 in the context of Ohm's law. Furthermore, our solution implicitly assumes that the system is in a steady state, where the energy flowing into or out of the volume is balanced by energy loss or gain through physical processes that we have not included in our model.

7.1.1 Energy of a plane wave

When calculated for simple electromagnetic waves, the Poynting vector produces physically reasonable and intuitive results. Let us consider an electromagnetic plane wave propagating through free space in the $\hat{\mathbf{z}}$-direction and take the polarization along $\hat{\mathbf{x}}$. The fields may be written in the familiar form

$$\mathbf{E}(\mathbf{r}, \omega) = \hat{\mathbf{x}} E_0 e^{ikz}, \tag{7.31}$$

$$\mathbf{H}(\mathbf{r}, \omega) = \hat{\mathbf{y}} \frac{E_0}{\mu_0 c} e^{ikz}. \tag{7.32}$$

Using these expressions in the definition for the Poynting vector, equation (7.26), we find

$$\mathbf{S}(\mathbf{r}, \omega) = \frac{|E_0|^2}{2\mu_0 c} \hat{\mathbf{z}}. \tag{7.33}$$

The Poynting vector points in the $\hat{\mathbf{z}}$-direction, which is also the direction of wave front propagation. We may also consider the electric and magnetic energy densities, i.e.

$$U_e = \frac{1}{4}\mathbf{D}^*(\mathbf{r}, \omega) \cdot \mathbf{E}(\mathbf{r}, \omega) = \frac{1}{4}\epsilon_0|E_0|^2, \tag{7.34}$$

$$U_m = \frac{1}{4}\mathbf{H}^*(\mathbf{r}, \omega) \cdot \mathbf{B}(\mathbf{r}, \omega) = \frac{1}{4}\frac{1}{\mu_0 c^2}|E_0|^2 = U_e. \tag{7.35}$$

Thus, the electric and magnetic energy densities are equal in a plane wave. By comparing these expressions to equation (7.33), we also find that

$$|\mathbf{S}| = c(U_e + U_m). \tag{7.36}$$

This result is readily understandable: the power flux can be found by multiplying the density of energy by the speed at which it is moving. Such a simple relationship does not necessarily exist for more complicated waves, however.

7.2 Paradoxical behavior of the Poynting vector

The Poynting vector is an important quantity that allows us to define the amount of power carried in an electromagnetic wave. It is important to note, however, that the Poynting vector can lead to seemingly paradoxical situations, at least at first glance.

Let us begin by noting the root of the problem. In our derivation of Poynting's theorem, the Poynting vector only appears either within an integral over a closed surface or, in differential form, in a divergence, $\nabla \cdot \mathbf{S}$. We may therefore say that the only quantity that has physical meaning is the divergence of the Poynting vector, not the Poynting vector itself. To see this, let us define a well-behaved vector field $\mathbf{X}(\mathbf{r}, \omega)$ and a vector field $\mathbf{Y}(\mathbf{r}, \omega) \equiv \nabla \times \mathbf{X}(\mathbf{r}, \omega)$. The divergence of $\mathbf{Y}(\mathbf{r}, \omega)$ vanishes identically, which means that we may add it to the Poynting vector $\mathbf{S}(\mathbf{r}, \omega)$ without changing the form of our energy conservation law at all. The Poynting vector, as we have defined it, is therefore not a unique and straightforward definition of power flux, and we can anticipate that trying to treat it as such will lead to trouble.

An excellent example of this trouble was highlighted in 1963 by the German physicist Fritz Bopp, who published a paper with the translated title, 'Is the Poynting vector observable?' [6]. Bopp introduced the following puzzling scenario: let us consider a pair of monochromatic plane waves propagating in the xz-plane, one with transverse electric (TE) polarization and one with transverse magnetic (TM) polarization, as illustrated in figure 7.2.

The electric fields are taken to have equal amplitudes and propagate at angles θ_1 and θ_2 with respect to the z-axis; they have the form

$$\mathbf{E}_1(\mathbf{r}) = E_0[\hat{\mathbf{x}} \cos \theta_1 - \hat{\mathbf{z}} \sin \theta_1]e^{ik(\sin \theta_1 x + \cos \theta_1 z)}, \tag{7.37}$$

$$\mathbf{E}_2(\mathbf{r}) = E_0\hat{\mathbf{y}}e^{ik(\sin \theta_2 x + \cos \theta_2 z)}. \tag{7.38}$$

The corresponding magnetic fields can be found through Faraday's law,

$$\mathbf{H}_1(\mathbf{r}) = \frac{E_0}{\mu_0 c}\hat{\mathbf{y}}e^{ik(\sin \theta_1 x + \cos \theta_1 z)}, \tag{7.39}$$

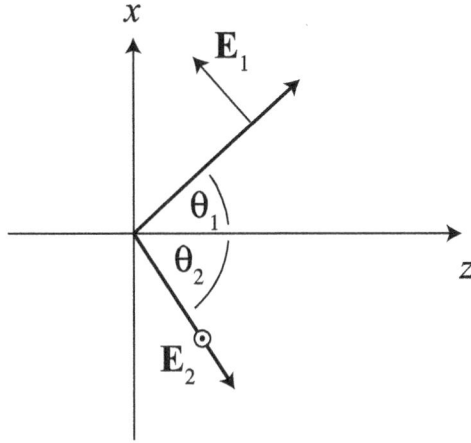

Figure 7.2. The propagation direction and field polarizations for Bopp's Poynting vector example. Reprinted from [7] by permission of the publisher (Taylor & Francis Ltd, http://www.tandfonline.com).

$$\mathbf{H}_2(\mathbf{r}) = \frac{E_0}{\mu_0 c}[-\hat{\mathbf{x}}\cos\theta_2 + \hat{\mathbf{z}}\sin\theta_2]e^{ik(\sin\theta_2 x + \cos\theta_2 z)}. \tag{7.40}$$

Each plane wave, taken in isolation, has a Poynting vector that propagates in the xz-plane, as can readily be determined. Let us consider the Poynting vector of the total field, with $\mathbf{E}_{tot} = \mathbf{E}_1 + \mathbf{E}_2$ and an analogous expression for $\mathbf{H}_{tot}$. The total Poynting vector $\mathbf{S}_{tot}$ now possesses cross terms that result in a component that is out of the plane in the y-direction, given by

$$S_{tot}\big|_y = -\frac{|E_0|^2}{\mu_0 c}\big[\cos\theta_1\sin\theta_2 - \sin\theta_1\cos\theta_2\big]\cos[k(\sin\theta_1 + \sin\theta_2)x$$

$$+ k(\cos\theta_1 + \cos\theta_2)z]. \tag{7.41}$$

It should be noted that this out-of-plane component takes the form of a standing wave, which suggests that there is no net energy flow in that direction. It appears instead that the combined electromagnetic wave oscillates up and down around the xz-plane as it propagates. This is nevertheless quite counterintuitive, as it indicates that the combination of two waves can flow in a direction that neither flows in individually.

Even stranger behavior is possible. In 1967, Boivin, Dow, and Wolf calculated the Poynting vector of an electromagnetic wave focused by a hard aperture, concentrating on the region in the neighborhood of geometric focus [8]. They found that the Poynting vector actually goes in circles in areas of the focal plane, even circling back toward the aperture itself. These circulations of the Poynting vector are now known to be common whenever electromagnetic waves create interference patterns, such as when light passes through a small hole in a metal plate [9].

These latter examples hint at an explanation of the problem with the Poynting vector. We have been implicitly interpreting it as a measure of the geometric flow of electromagnetic power, even though light is inherently a wave phenomenon. When the wave properties of light are important, such as when multiple waves interfere, the Poynting vector no longer gives us a simple picture of power flow. A little reflection makes this clear: when multiple waves cross each other, propagating in different directions, why should we expect a single vector to characterize the complexity of the net power flow? Because of this, we typically only interpret the Poynting vector as a measure of geometric energy flow in the simplest cases, such as plane waves and spherical waves propagating far from any sources.

It is worth noting that the Poynting vector even causes trouble when no waves are present at all. Let us consider a bar magnet sitting next to a static electric charge, with both of them fixed in place. The charge produces an electrostatic field $\mathbf{E}$ and the magnet produces a magnetostatic field $\mathbf{H}$. Our derivation of the energy conservation law still applies in this case, so we expect there to be a nonzero Poynting vector representing electromagnetic power flow around static objects!

This result, popularized by Feynman (see section 27-5 of [10]), is particularly strange because there are no electromagnetic waves to associate with the Poynting vector. What power is then flowing, if any? One way to explain away this seemingly paradoxical result is to note that the divergence of the Poynting vector is identically zero outside the source region, using the vector identity

$$\nabla \times [\mathbf{E}(\mathbf{r}) \times \mathbf{H}(\mathbf{r})] = [\nabla \times \mathbf{E}(\mathbf{r})] \cdot \mathbf{H}(\mathbf{r}) - [\nabla \times \mathbf{H}(\mathbf{r})] \cdot \mathbf{E}(\mathbf{r}). \qquad (7.42)$$

The divergence always vanishes because both curls vanish in the static, source-free case. The vanishing curl implies that the net energy flow into or out of any region of space is zero and that the static Poynting energy, even if it exists, cannot be accessed or observed. This can be made more evident by noting that the net Poynting flux through a closed surface surrounding either the charge or the magnet—or any other static object placed in their vicinity—is always zero. This explanation is still probably not fully satisfying but appears to be the best we can do in the context of classical electromagnetism.

7.3 Conservation of momentum

By considering the conservation of energy in electromagnetic waves, we were led to the definition of important physical quantities, namely the electromagnetic energy density and the power flux, i.e. the Poynting vector. We expect that we will gain similar insights by developing a law of conservation of momentum in electromagnetic waves.

We return to the Lorentz force $\mathbf{F}(\mathbf{r}, t)$ acting on a point particle of charge q and velocity $\mathbf{v}$,

$$\mathbf{F}(\mathbf{r}, t) = q[\mathbf{E}(\mathbf{r}, t) + \mathbf{v} \times \mathbf{B}(\mathbf{r}, t)]. \qquad (7.43)$$

Let us note right away that we will restrict ourselves to the vacuum form of Maxwell's equations in this derivation, working with the fields $\mathbf{E}$ and $\mathbf{B}$ and not with the macroscopic **D** and **H**. There are difficult and ongoing debates about how to properly define the momentum of light in matter; we will say more about this in section 7.4.

It is straightforward to generalize the Lorentz force law to a continuous distribution of sources by replacing the charge q with the quantity $\rho(\mathbf{r}', t)d^3r'$ and by replacing the charge times velocity $q\mathbf{v}$ with the quantity $\mathbf{J}(\mathbf{r}, t)d^3r'$. The sources are again assumed to lie within a volume V bounded by a surface S. Integrating our generalized force law over the volume, we find that the net mechanical force $\mathbf{F}_{mech}$ on the system of charges and currents is given by

$$\mathbf{F}_{mech}(t) = \int_V [\rho(\mathbf{r}', t)\mathbf{E}(\mathbf{r}', t) + \mathbf{J}(\mathbf{r}', t) \times \mathbf{B}(\mathbf{r}', t)]d^3r'. \tag{7.44}$$

As in the conservation of energy case, we will eliminate the source terms from our expression through the use of the vacuum form of Maxwell's equations,

$$\nabla \cdot \mathbf{E}(\mathbf{r}, t) = \rho(\mathbf{r}, t)/\epsilon_0, \tag{7.45a}$$

$$\nabla \cdot \mathbf{B}(\mathbf{r}, t) = 0, \tag{7.45b}$$

$$\nabla \times \mathbf{E}(\mathbf{r}, t) = -\frac{\partial \mathbf{B}(\mathbf{r}, t)}{\partial t}, \tag{7.45c}$$

$$\nabla \times \mathbf{B}(\mathbf{r}, t) = \mu_0 \mathbf{J}(\mathbf{r}, t) + \mu_0 \epsilon_0 \frac{\partial \mathbf{E}(\mathbf{r}, t)}{\partial t}. \tag{7.45d}$$

We eliminate $\rho(\mathbf{r}, t)$ by the use of equation (7.45a) and eliminate $\mathbf{J}(\mathbf{r}, t)$ by the use of equation (7.45d), with the result

$$\mathbf{F}_{mech} = \int_V \left[\epsilon_0 \mathbf{E}(\nabla \cdot \mathbf{E}) - \frac{1}{\mu_0}\mathbf{B} \times (\nabla \times \mathbf{B}) - \epsilon_0 \frac{\partial \mathbf{E}}{\partial t} \times \mathbf{B} \right]d^3r'. \tag{7.46}$$

Going forward, we suppress the arguments of the functions in the integrand for the sake of brevity.

If we now apply the vector product rule formula for derivatives,

$$\frac{\partial}{\partial t}[\mathbf{E} \times \mathbf{B}] = \mathbf{E} \times \frac{\partial \mathbf{B}}{\partial t} + \frac{\partial \mathbf{E}}{\partial t} \times \mathbf{B}, \tag{7.47}$$

we can alter the last term of equation (7.46), resulting in the more suggestive expression

$$\mathbf{F}_{mech} = \int_V \left[\epsilon_0 \mathbf{E}(\nabla \cdot \mathbf{E}) - \frac{1}{\mu_0}\mathbf{B} \times (\nabla \times \mathbf{B}) - \epsilon_0 \frac{\partial}{\partial t}(\mathbf{E} \times \mathbf{B}) + \epsilon_0 \mathbf{E} \times \frac{\partial \mathbf{B}}{\partial t} \right]d^3r'. \tag{7.48}$$

The last term of this new equation can also be rewritten using Faraday's law, equation (7.45), and we then have

$$\mathbf{F}_{mech} = \int_V \left[\mathbf{Q} - \epsilon_0 \frac{\partial}{\partial t}(\mathbf{E} \times \mathbf{B}) \right] d^3 r', \tag{7.49}$$

where we have defined a new vector $\mathbf{Q}$ as

$$\mathbf{Q} \equiv \epsilon_0 \mathbf{E}(\nabla \cdot \mathbf{E}) - \epsilon_0 \mathbf{E} \times (\nabla \times \mathbf{E}) + \frac{1}{\mu_0} \mathbf{B}(\nabla \cdot \mathbf{B}) - \frac{1}{\mu_0} \mathbf{B} \times (\nabla \times \mathbf{B}). \tag{7.50}$$

We have sneakily added a new term to this equation: $\mathbf{B}(\nabla \cdot \mathbf{B})$. It is identically equal to zero due to the 'no magnetic monopoles' law, equation (7.45), but including it makes the expression for $\mathbf{Q}$ symmetric with respect to $\mathbf{E}$ and $\mathbf{B}$.

We are now in a position to begin to physically interpret equation (7.49). First, we note that the mechanical force on the charges may be expressed as the rate of change of the mechanical momentum $\mathbf{P}_{mech}$, i.e.

$$\mathbf{F}_{mech} = \frac{\partial \mathbf{P}_{mech}}{\partial t}. \tag{7.51}$$

It follows that the quantity within the integral on the right-hand side of equation (7.49) may be viewed as the time derivative of the *momentum density of the electromagnetic field*, and the total field momentum in the volume may consequently be written as

$$\mathbf{P}_{field} = \int_V \epsilon_0 \mathbf{E} \times \mathbf{B} d^3 r'. \tag{7.52}$$

Equation (7.49) may then be rewritten in the enlightening form

$$\frac{\partial \mathbf{P}_{tot}}{\partial t} = \frac{\partial}{\partial t}[\mathbf{P}_{mech} + \mathbf{P}_{field}] = \int_V \mathbf{Q} d^3 r', \tag{7.53}$$

where $\mathbf{P}_{tot}$ is the total momentum of fields and charges within the volume.

In order to properly interpret this equation in its entirety, we need to write the integral over $\mathbf{Q}$ in a more intuitive form. Let us switch momentarily to tensor notation for vectors and derivatives, where E_j is the jth Cartesian component of $\mathbf{E}$ and ∂_j is the partial derivative with respect to the jth component. In tensor notation, we may write

$$[\mathbf{E}(\nabla \cdot \mathbf{E}) - \mathbf{E} \times (\nabla \times \mathbf{E})]_i = \partial_j [E_i E_j - \frac{1}{2} \delta_{ij} E_k E_k], \tag{7.54}$$

where δ_{ij} is the Kronecker delta; we apply Einstein's summation convention of summing over repeated indices.

Let us introduce a tensor known as the *Maxwell stress tensor*,

$$T_{ij} = -\left[\epsilon_0 E_i E_j + \frac{1}{\mu_0} B_i B_j - \frac{1}{2} \delta_{ij} \left(\epsilon_0 E_k E_k + \frac{1}{\mu_0} B_k B_k \right) \right]. \tag{7.55}$$

With the stress tensor, equation (7.53) may be written as

$$\left[\frac{\partial \mathbf{P}_{tot}}{\partial t}\right]_i = -\int_V \partial_j T_{ij} d^3 r'. \tag{7.56}$$

Let us consider a single component i of this equation. For this single component, the integrand on the right takes the form of a divergence, $\partial_j T_{ij}$. For each component i, then, the right-hand side of equation (7.56) satisfies Gauss's theorem, and we may rewrite it as a surface integral over the bounding surface S,

$$\left[\frac{\partial \mathbf{P}_{tot}}{\partial t}\right]_i = -\oint_S T_{ij} da_j, \tag{7.57}$$

where da_j is the jth component of the infinitesimal area element $d\mathbf{a}$.

Equation (7.57) is the momentum conservation law for electromagnetic waves. It shows that a change of momentum within the volume is associated with a flux of the stress tensor through the surface. It follows that the Maxwell stress tensor may also be considered a *momentum flux density* of the field. Furthermore, as already noted, we may define the *momentum density* $\mathbf{p}_{field}$ of the field as

$$\mathbf{p}_{field}(\mathbf{r}, t) = \epsilon_0 \mathbf{E}(\mathbf{r}, t) \times \mathbf{B}(\mathbf{r}, t). \tag{7.58}$$

It should be noted that our definition of the Maxwell stress tensor, equation (7.55), is defined with a minus sign different from the convention used in many books such as Jackson [11]. I have done this so that the momentum conservation law has the same form and signs as the energy conservation law introduced earlier. To see this, we note that we can convert equation (7.57) into a differential form using Gauss's law:

$$\left[\frac{\partial \mathbf{p}_{tot}}{\partial t}\right]_i = -\partial_j T_{ij}. \tag{7.59}$$

For comparison, the energy conservation law was of the form,

$$\frac{\partial U(\mathbf{r}, t)}{\partial t} = -\nabla \cdot \mathbf{S}(\mathbf{r}, t). \tag{7.60}$$

As in the case of the energy conservation law, our momentum conservation law and the quantities introduced in it must be interpreted with some care, as strange behaviors analogous to those in section 7.2 for energy can arise.

We need a tensor to define the momentum flux density because momentum is a vector quantity. The second index of T_{ij} represents the direction of flux, or normal to the surface it is flowing across, while the first index represents the direction of the transferred momentum.

In the derivations of both our energy and momentum conservation laws, we implicitly assumed that no charge is transferred across the surface S. A transfer of charge would imply a change in mechanical energy and momentum within the volume, but characterizing this change would require a knowledge of the charge/

mass ratios of the charge carriers, which takes us beyond the scope of Maxwell's equations and into solid-state physics.

7.3.1 Momentum of a plane wave

It is worthwhile to apply our results to the simple plane wave case, which allows us to estimate the order of magnitude of quantities as well as develop some intuition about the stress tensor. We again assume a plane wave propagating in the $\hat{\mathbf{z}}$-direction with polarization along $\hat{\mathbf{x}}$, for which the fields may be written as

$$\mathbf{E}(\mathbf{r},\,\omega) = \hat{\mathbf{x}}E_0 e^{ikz}, \tag{7.61}$$

$$\mathbf{B}(\mathbf{r},\,\omega) = \hat{\mathbf{y}}\frac{E_0}{c}e^{ikz}. \tag{7.62}$$

Let us suppose we wish to calculate the amount of pressure that this plane wave applies when at normal incidence to a surface. The dimensions of the stress tensor are 'momentum/(area-time),' or force per unit area, which is already a pressure. If we consider the T_{zz}-component of the stress tensor, the first subscript indicates we are interested in the z-component of the momentum, while the second subscript indicates we are interested in how it flows across a surface normal to $\hat{\mathbf{z}}$. We have

$$T_{zz} = -\frac{1}{2}\left[\epsilon_0 E_z^* E_z + \frac{1}{\mu_0}B_z^* B_z - \frac{1}{2}\left(\epsilon_0 |\mathbf{E}|^2 + \frac{1}{\mu_0}|\mathbf{B}|^2\right)\delta_{zz}\right]. \tag{7.63}$$

Because the field is purely transverse, we find that the first two terms in the brackets vanish, and we are left with

$$T_{zz} = \frac{1}{4}\left(\epsilon_0 |E_0|^2 + \frac{1}{\mu_0 c^2}|E_0|^2\right) = U_e + U_m. \tag{7.64}$$

The pressure may be written simply as the total energy density at the surface. Referring back to equation (7.36), we may also note that

$$T_{zz} = \frac{|\mathbf{S}|}{c}. \tag{7.65}$$

This final expression allows us to easily determine the pressure exerted by a highly directional beam of light, based on its total power and its size. For example, a standard laser pointer can have a total power no greater than 5 mW and typically has a beam diameter on the order of 1 mm. The Poynting vector can be found from power/area; its magnitude is 6.4×10^3 W m^{-2}, a rather large number and a good indication of why one should not stare directly into even a weak laser. This corresponds to a pressure $P = 2.1 \times 10^{-5}$ N m^{-2}. For comparison, standard atmospheric pressure is 101 325 pascals, or 1.01×10^5 N m^{-2}. The pressure of a gentle breeze is orders of magnitude larger than the pressure due to a laser pointer.

7.4 Momentum in matter and the Abraham–Minkowski controversy

In deriving our momentum conservation law and related quantities, we worked entirely with free charges and currents in vacuum and avoided the use of macroscopic properties such as permittivity and permeability. The reason for this is that it is surprisingly difficult to determine the momentum of light in matter, and the proper definition has been the topic of an ongoing debate for over 100 years. The original attempts to derive formulas for the momentum of light in matter were made by Hermann Minkowski [12] in 1908 and Max Abraham [13] in 1909, and because of this, the problem is referred to as the *Abraham–Minkowski controversy.*

At first glance, it is surprising that there is any controversy at all, since it would appear that the results should follow in a systematic manner directly from Maxwell's equations. Let us recall, however, that we were required in our derivations of energy and momentum conservation to provide physical interpretations for quantities like the Poynting vector and the Maxwell stress tensor. When attempting to derive a momentum conservation law in matter, an overabundance of terms appears in the integrals, and one must choose which to associate with the momentum of light and which to associate with the momentum of the medium.

Let us consider just the field momentum density for simplicity. Minkowski argued that the momentum density should be given by the expression

$$\mathbf{p}_M = \mathbf{D} \times \mathbf{B}, \tag{7.66}$$

while Abraham argued it should be given by the expression

$$\mathbf{p}_A = \frac{1}{c^2}\mathbf{E} \times \mathbf{H}. \tag{7.67}$$

We have used subscripts M and A to represent Minkowski and Abraham, respectively.

We can use a monochromatic plane wave to show what the implications are for the momentum of a single photon in each case. We again let

$$\mathbf{E} = \hat{\mathbf{x}}E_0 e^{i(kz-\omega t)}, \tag{7.68}$$

where now $k = nk_0 = n\omega/c$. Using Faraday's law, we have

$$\mathbf{B} = \hat{\mathbf{y}}\frac{n\omega}{c}E_0 e^{i(kz-\omega t)}. \tag{7.69}$$

The time-averaged form of the Minkowski momentum density is

$$|\mathbf{p}_M| = \frac{1}{2}|\mathbf{D} \times \mathbf{B}^*| = \frac{1}{2}\frac{\epsilon n}{c}|E_0|^2, \tag{7.70}$$

and the time-averaged form of the Abraham momentum density is

$$|\mathbf{p}_A| = \frac{1}{2c^2}|\mathbf{E} \times \mathbf{H}^*| = \frac{1}{2c^2}\frac{n\omega}{\mu c}|E_0|^2. \tag{7.71}$$

To determine the average momentum per photon, we must divide the momentum density by the photon density. The photon density can be found by dividing the energy density of a plane wave by the quantum energy of a photon, $\hbar\omega$. We have already seen that the energy density U of a monochromatic electromagnetic wave is

$$U = \frac{1}{4}(\mathbf{D}^* \cdot \mathbf{E} + \mathbf{H}^* \cdot \mathbf{B}) = \frac{1}{2}\epsilon |E_0|^2. \tag{7.72}$$

We may then write the density N of photons as

$$N = \frac{U}{\hbar\omega} = \frac{1}{2\hbar\omega}\epsilon |E_0|^2. \tag{7.73}$$

We may finally get the Minkowski and Abraham momenta per photon, p_M and p_A, by dividing each momentum density by the photon density:

$$p_M = n\hbar k_0, \tag{7.74}$$

$$p_A = \frac{\hbar k_0}{n},$$

where $k_0 = \omega/c$.

The difference between these two results is extreme. If, for example, we assume that the photon is propagating through a medium like glass or water, Minkowski predicts that the momentum of the photon *increases* in the medium, while Abraham predicts that the momentum *decreases* in the medium. This is, in principle, a measurable result; because momentum is conserved, one would expect that the medium gets pulled backward (against the direction of propagation) if Minkowski is correct, and pulled forward (along the direction of propagation) if Abraham is correct.

We illustrate this with an experimental setup that was used in 2008 by She, Yu, and Feng [14] to test the Abraham–Minkowski controversy. They fired a short optical pulse downward out of a vertically hanging optical fiber and looked at the response of the fiber as the pulse emerged. According to the Minkowski prediction, the pulse momentum should decrease as it leaves the fiber, indicating that the fiber is pulled downward. In principle, this should produce no observable effect, but in practice, imperfections at the output of the fiber cause it to be kicked slightly sideways. According to the Abraham prediction, the pulse momentum should increase as it leaves the fiber, and the fiber should be pushed upward. These scenarios are illustrated in figure 7.3. In the experiment, the fiber was kicked upward, apparently in agreement with Abraham's theory.

However, a variety of experiments have been conducted over the years, and results have been found that support both the Abraham and the Minkowski views. One of the earliest experiments was undertaken by Jones and Richards in 1954 [15]. The researchers mounted a reflective metal vane on a torsional pendulum in a container and illuminated the vane with light to measure the applied pressure. By filling the container with liquids of varying refractive indices, they could determine

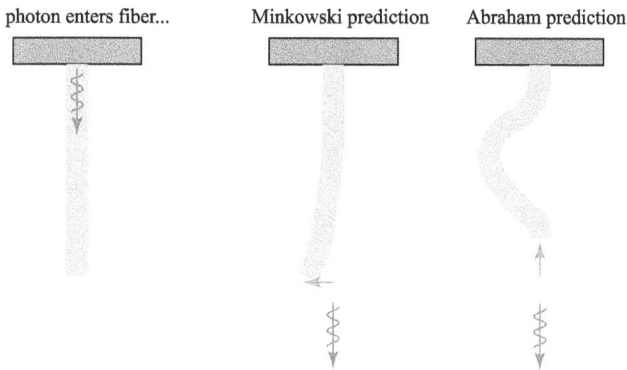

Figure 7.3. Illustration of the Minkowski and Abraham predictions for a pulse exiting a free-hanging optical fiber.

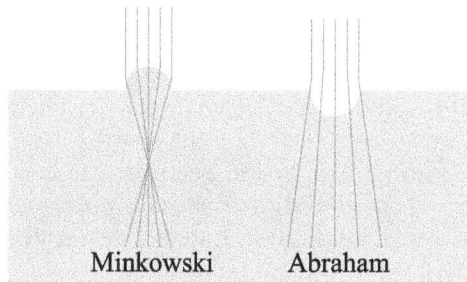

Figure 7.4. Simple idea behind the Ashkin and Dziedzic experiment, showing the different predictions made by Abraham and Minkowski. Reprinted from [7] by permission of the publisher (Taylor & Francis Ltd, http://www.tandfonline.com).

whether the pressure increased or decreased, in accordance with Minkowski or Abraham, respectively. Their data matched the Minkowski expression.

In 1973, Ashkin and Dziedzic performed an ingenious experiment of a very different nature [16]. A transparent liquid was illuminated from above by a pulsed and focused beam of light. According to the Minkowski interpretation, the liquid should bulge outward; according to the Abraham interpretation, the liquid should bulge inward. This will make the curved surface of the fluid either a natural convex or concave lens, as illustrated in figure 7.4, and the change in focus of the light beam was used as an indicator of which theory is correct. The Ashkin and Dziedzic experiment appeared to be consistent with the Minkowski result; however, a recent version of the experiment conducted by Zhang, She, Peng, and Leonhardt [17] was instead consistent with Abraham's prediction.

We have already noted that our expressions for energy and momentum are valid even in the static or near-static limit, where dielectric constants can be extremely large. In 1975, Walker and Lahoz looked at the oscillations of a disk of barium titanate, with a normalized dielectric constant of nearly 4000, suspended on a torsional pendulum between the poles of a static electromagnet and driven by a low-frequency sinusoidal voltage [18]. Their result agreed with Abraham's.

In 1980, Gibson *et al* investigated the controversy using the photon drag effect [19]. Photon drag is the creation of induced currents in a semiconductor due to a transfer of momentum from radiation to the electrons in the material. This experiment had the benefit of testing photon momentum using a different mechanism than those previously studied; furthermore, the high refractive index of semiconductors such as germanium ($n = 4$ in the infrared) results in large predicted differences between the Abraham and Minkowski results. In this experiment, Gibson *et al* found results consistent with the Minkowski prediction.

So where does this leave us, after over 100 years of conflicting experimental and theoretical results? There has been a flurry of activity in the last twenty years that seems to indicate that the answer depends upon other properties of the system, often outside the scope of Maxwell's equations. We note that derivations of the momentum conservation law typically use bulk permittivity and permeability, and these quantities do not take into account the mechanical properties of the medium at all, which can include effects like the aforementioned photon drag, electrostriction and magnetostriction, lattice vibrations, and more. Even materials with superficially identical refractive indices can have different mechanical responses to an electromagnetic wave. In 2014, Leonhardt [20] analyzed the effects of electromagnetic momentum on fluids and concluded that the appropriate formula depends on whether the fluid is set into motion by the electromagnetic wave. In 2010, Barnett [21] argued that both Abraham and Minkowski's formulas are correct, but represent different forms of momenta that appear in Lagrangian and Hamiltonian mechanics. He specifically argues that the Abraham momentum represents the kinetic momentum and that the Minkowski momentum represents the canonical momentum.

A number of years earlier, Pfeifer, Nieminen, Heckenberg, and Rubinsztein-Dunlop [22] came to similar conclusions, suggesting that any differences in experimental outcomes are the result of neglecting the material contributions to the momentum. This view is strengthened by the observation that there are many other formulations for the momentum of light in matter beyond Abraham and Minkowski, most notably one introduced by Einstein and Laub [23].

In concluding this discussion, it is worth noting that the Abraham–Minkowski controversy may be viewed physically as a debate over whether a photon behaves like a wave or a particle as far as its momentum is concerned. We have seen in chapter 6 that the phase velocity of light is inversely proportional to the refractive index n in matter, i.e. $v_p = c/n$. If we view light as a particle, we should predict that its momentum is proportional to its velocity and that it consequently decreases when it enters matter, i.e.

$$p_A = p_0/n, \tag{7.75}$$

where p_0 is the momentum in vacuum. On comparison with equation (7.74), we can see that this is the Abraham prediction for the momentum of light in matter.

However, we also know that the wavelength of light λ decreases by a factor n in matter, which implies that the wavenumber k is scaled by a multiplicative factor n. If we view light as a wave, the momentum is proportional to the wavenumber, so it should increase in matter according to

$$p_M = np_0. \tag{7.76}$$

On comparison with equation (7.74), we see that this is the Minkowski result for the momentum of light in matter.

Overall, the Abraham–Minkowski controversy seems to be on a much more stable foundation than it has been for the last century. I would not rule out additional surprises in the future, as that same century has seen multiple papers that prematurely declared a 'resolution' of the issue.

7.5 Optical trapping

The momentum conservation law of section 7.3 is somewhat cumbersome to use in practice due to its complicated tensor form. In some special cases, however, we can circumvent that formalism and directly calculate the forces that an electromagnetic wave exerts on matter. The most important such case is the use of a single tightly focused light beam to trap a microscopic particle at the geometric focus; this technique is now known as optical tweezing, and it is a special case of the more general technique of optical trapping. The fundamentals of optical trapping can be directly derived by calculating the electromagnetic forces on a small particle using Maxwell's equations, and that is where we begin our investigation.

Let us suppose that we have a small dipole with positive and negative charges $\pm q$ separated by a vector distance $\mathbf{d}$, as shown in figure 7.5(a). It readily follows from the Lorentz force law that the electric force acting on the dipole is given by

$$\mathbf{F} = q\mathbf{E}(\mathbf{r} + \mathbf{d}) - q\mathbf{E}(\mathbf{r}). \tag{7.77}$$

Let us rewrite this in the form

$$\mathbf{F} = qd \left\{ \frac{\mathbf{E}(\mathbf{r} + \mathbf{d}) - \mathbf{E}(\mathbf{r})}{d} \right\}. \tag{7.78}$$

In the limit $d \to 0$, the bracketed term approaches the directional derivative in the $\hat{\mathbf{d}}$-direction. We may then approximate this term by $\hat{\mathbf{d}} \cdot \nabla \mathbf{E}(\mathbf{r})$; if we further define

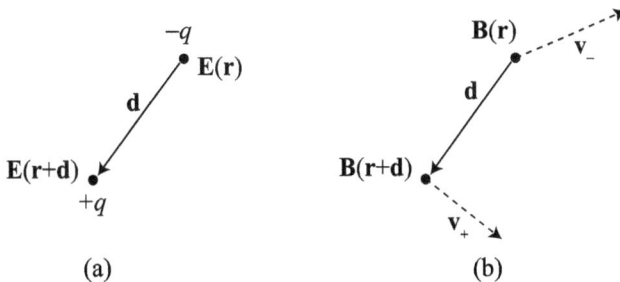

(a)   (b)

Figure 7.5. Illustration of the (a) electric force and (b) the magnetic force acting on an electric dipole. Reprinted from [7] by permission of the publisher (Taylor & Francis Ltd, http://www.tandfonline.com).

the dipole moment as $\mathbf{p} \equiv q\mathbf{d}$, we find that the electric force on a point-like dipole has the form

$$\mathbf{F} = (\mathbf{p} \cdot \nabla)\mathbf{E}(\mathbf{r}). \tag{7.79}$$

What about the magnetic force? We assume that each charge has its own velocity $\mathbf{v}_{\pm}$, as illustrated in figure 7.5(b). The Lorentz force law then readily gives us

$$\mathbf{F} = q\mathbf{v}_+ \times \mathbf{B}(\mathbf{r} + \mathbf{d}) - q\mathbf{v}_- \times \mathbf{B}(\mathbf{r}), \tag{7.80}$$

where

$$\mathbf{v}_+ = \frac{d(\mathbf{r} + \mathbf{d})}{dt}, \quad \mathbf{v}_- = \frac{d\mathbf{r}}{dt}. \tag{7.81}$$

Let us assume that the magnetic field is effectively constant over the region of the dipole, so that $\mathbf{B}(\mathbf{r} + \mathbf{d}) \approx \mathbf{B}(\mathbf{r})$. The magnetic force then has the form

$$\mathbf{F} = \frac{d(q\mathbf{d})}{dt} \times \mathbf{B}(\mathbf{r}) = \frac{d\mathbf{p}}{dt} \times \mathbf{B}(\mathbf{r}),$$

and the total Lorentz force on our point dipole may be written as

$$\mathbf{F} = (\mathbf{p} \cdot \nabla)\mathbf{E}(\mathbf{r}) + \frac{d\mathbf{p}}{dt} \times \mathbf{B}(\mathbf{r}). \tag{7.82}$$

As always, we are primarily concerned with monochromatic electromagnetic waves. We can convert equation (7.82) to a cycle-averaged form by conjugating one of the two vectors in each product and introducing a factor of one half:

$$\mathbf{F} = \text{Re}\left\{ \frac{1}{2}(\mathbf{p} \cdot \nabla)\mathbf{E}^*(\mathbf{r}) - \frac{1}{2}i\omega\mathbf{p} \times \mathbf{B}^*(\mathbf{r}) \right\}. \tag{7.83}$$

For the particular case of optical tweezing, we are interested in the dipole moment *induced* in a small dielectric particle by the illuminating electromagnetic wave in the Rayleigh scattering limit (which we discuss later in section 15.5). Following reasoning similar to that used in section 6.1, we assume a linear relationship,

$$\mathbf{p} = \alpha\mathbf{E}, \tag{7.84}$$

where α is the complex polarizability of the particle. If we substitute this dipole moment into equation (7.83), we get

$$\mathbf{F} = \text{Re}\left\{ \frac{\alpha}{2}(\mathbf{E} \cdot \nabla)\mathbf{E}^* - \frac{i\omega}{2}\alpha\mathbf{E} \times \mathbf{B}^* \right\}. \tag{7.85}$$

We now have to do some rearranging to extract some physical meaning from this expression. We first use Faraday's law on the second term to get

$$\mathbf{F} = \text{Re}\left\{ \frac{\alpha}{2}[(\mathbf{E} \cdot \nabla)\mathbf{E}^* + \mathbf{E} \times (\nabla \times \mathbf{E}^*)] \right\}. \tag{7.86}$$

Because α itself is complex, we may expand this expression into the form

$$\mathbf{F} = \frac{\alpha_R}{2}\text{Re}\{(\mathbf{E} \cdot \nabla)\mathbf{E}^* + \mathbf{E} \times (\nabla \times \mathbf{E}^*)\}$$
$$- \frac{\alpha_I}{2}\text{Im}\{(\mathbf{E} \cdot \nabla)\mathbf{E}^* + \mathbf{E} \times (\nabla \times \mathbf{E}^*)\}, \qquad (7.87)$$

where $\alpha = \alpha_R + i\alpha_I$. If we use the following vector calculus identity:

$$\nabla(\mathbf{A} \cdot \mathbf{B}) = (\mathbf{A} \cdot \nabla)\mathbf{B} + (\mathbf{B} \cdot \nabla)\mathbf{A} + \mathbf{A} \times (\nabla \times \mathbf{B}) + \mathbf{B} \times (\nabla \times BA), \qquad (7.88)$$

we can readily write the real part of $\mathbf{F}$ as

$$\mathbf{F}_R = \frac{\alpha_R}{2}\nabla|\mathbf{E}|^2. \qquad (7.89)$$

The imaginary part of $\mathbf{F}$ can be written, using Faraday's law again, as

$$\mathbf{F}_I = \frac{\alpha_I\omega}{2}\text{Re}\{\mathbf{E} \times \mathbf{B}^*\} - \frac{\alpha_I}{2}\text{Im}\{(\mathbf{E} \cdot \nabla)\mathbf{E}^*\}. \qquad (7.90)$$

Finally, we may write the total force on our dipole as

$$\mathbf{F} = \frac{1}{4}\alpha_R\nabla|\mathbf{E}|^2 + \frac{1}{2}\alpha_I\omega\text{Re}\{\mathbf{E} \times \mathbf{B}^*\} - \frac{1}{2}\alpha_I\text{Im}\{(\mathbf{E} \cdot \nabla)\mathbf{E}^*\}. \qquad (7.91)$$

We now consider each term of this expression in order. The first term is a force proportional to the gradient of the electric field squared, or intensity, and is thus referred to as the *gradient force*. This force pulls particles toward regions of high intensity, and it is the foundation of optical trapping and tweezing.

What is the origin of the gradient force? We can understand it using a geometrical optics model, as illustrated in figure 7.6(a). When a ray of light is incident upon a

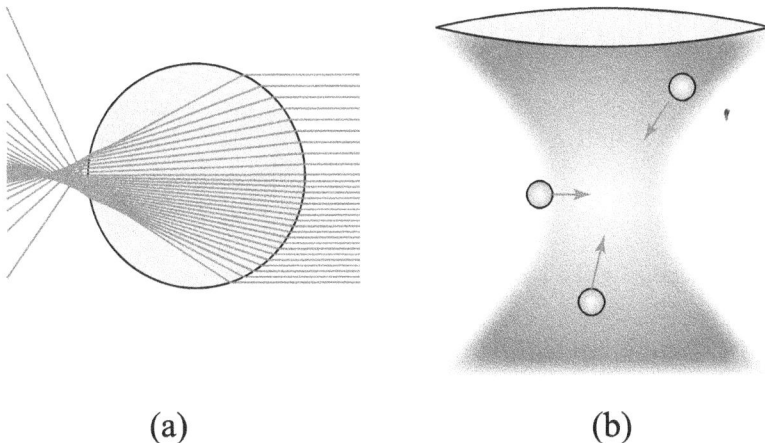

(a)   (b)

Figure 7.6. (a) A geometrical model of the gradient force. More rays, and hence more light momentum, are refracted upward than downward, resulting in a net downward push on the particle. (b) Illustration of an optical tweezers setup, showing the direction in which particles are pulled.

transparent particle, it is partially refracted in the transverse direction. This change of transverse momentum of the light ray must be accompanied by a change in the transverse momentum of the particle. If the particle is illuminated by a beam of uniform intensity, these transverse momentum components cancel out due to symmetry. For a beam of nonuniform intensity, however, the particle experiences a net push in the direction of higher intensity. Calculations of the forces on particles were first done by Ashkin in the geometrical optics approximation [24]; however, it turns out that analogous forces appear for small particles in the Rayleigh scattering regime, with refraction replaced by scattering.

The second term in equation (7.91) is proportional to the momentum density, equation (7.58), and is known as the *radiation pressure force* or *scattering force*. Because it depends on the imaginary part of the polarizability, it represents the momentum transferred to the particle through light absorption. The derivation of the gradient force and the radiation pressure force on Rayleigh particles was first done by Gordon [25] and elaborated upon some time later by Chaumet and Nieto-Vesperinas [26].

A conceptual illustration of an optical tweezers system is shown in figure 7.6(b). The gradient force pulls a particle toward the geometric focus of the light regardless of direction; however, the radiation pressure force always tends to push a particle downstream. To ensure that the gradient force dominates the radiation pressure force, a high-numerical-aperture lens is used. In this way, a particle can be trapped in three-dimensional space by a single focused beam of light. This technique of optical tweezing was first introduced by Ashkin, Dziedzic, Bjorkholm, and Chu in 1986 [27]. Ashkin, however, laid the foundations for this discovery with research on optical trapping performed in the 1970s [28]. This early work focused on the acceleration of micron-sized particles by a single laser beam and the trapping of particles between counterpropagating beams. Over the course of the next few years, Ashkin developed the technique of laser trapping and cooling of single atoms [29]; this is now a standard technique for studying fundamental quantum physics [30].

But what about the third term of equation (7.91), which we will refer to as the *spin force*? Historically, this term has been considered negligible compared to the other two; in 2009, however, researchers noted that this component of the force can be significant if the illuminating field is circularly or elliptically polarized [31], and we briefly consider this case.

Let us expand the imaginary operation in the expression for the spin force,

$$\mathbf{F}_{spin} = -\frac{1}{4i}\alpha_I [\mathbf{E} \cdot \nabla \mathbf{E}^* - \mathbf{E}^* \cdot \nabla \mathbf{E}]. \qquad (7.92)$$

This looks suspiciously like the form of a vector triple product, and in fact, we may simplify it using the identity

$$\nabla \times (\mathbf{E} \times \mathbf{E}^*) = (\mathbf{E}^* \cdot \nabla)\mathbf{E} - \mathbf{E}^*(\nabla \cdot \mathbf{E}) + \mathbf{E}(\nabla \cdot \mathbf{E}^*) - (\mathbf{E} \cdot \nabla)\mathbf{E}^* \qquad (7.93)$$

along with Gauss's law, $\nabla \cdot \mathbf{E} = 0$, with no sources present. The spin force therefore simplifies to

$$\mathbf{F}_{spin} = \frac{\alpha_I}{4i}\nabla \times (\mathbf{E} \times \mathbf{E}^*). \qquad (7.94)$$

Let us further introduce a new vector $\mathbf{L}_S$, defined as

$$\mathbf{L}_S \equiv \frac{\alpha_I}{4i}\mathbf{E} \times \mathbf{E}^*. \tag{7.95}$$

The spin force is then simply the curl of $\mathbf{L}_S$,

$$\mathbf{F}_{spin} = \nabla \times \mathbf{L}_S. \tag{7.96}$$

The quantity $\mathbf{L}_S$ is interpreted as the time-averaged spin density of the electromagnetic field. Because it is the cross product of the electric field with its complex conjugate, it can only be nonzero if there is a complex relationship between the polarization components. If we consider linear polarization, for example, with $\mathbf{E}(\mathbf{r}) = \hat{\mathbf{x}}E(\mathbf{r})$, we can readily see that $\mathbf{L}_S = 0$. If instead we take $\mathbf{E}(\mathbf{r}) = (\hat{\mathbf{x}} + i\hat{\mathbf{y}})E(\mathbf{r})$, we find that $\mathbf{L}_S = -2i\hat{\mathbf{z}}|E(\mathbf{r})|^2$. Therefore, only fields with circulation can have a spin force, justifying the name.

To further elucidate the effect of the spin force, we can consider a region in which

$$E(\mathbf{r}) = E_0[1 - x/x_0], \tag{7.97}$$

and for this field amplitude, the spin force takes the form

$$\mathbf{F}_{spin} = -\frac{\alpha_I}{2}\hat{\mathbf{y}}\frac{|E_0|^2}{x_0}[1 - x/x_0]. \tag{7.98}$$

The spin force may be thought of as analogous to *shear forces* in fluids. It is well known that a nonuniform fluid flow results in a net fluid vorticity. In our electromagnetic case, we have the reverse: a nonuniform vorticity results in the equivalent of a shear force.

It is worth noting that the analogy between light forces acting on particles and fluid forces can be extended even further. In 2011, Swartzlander *et al* [32] demonstrated that asymmetrically shaped refractive particles can be effectively 'flown' on a beam of light. An optical analogue of aerodynamic lift is generated by the uneven refraction of light in different directions. It has been suggested that a large array of such particles could allow solar sails to tack against the Sun's radiation pressure, much as sailboats tack against the wind.

7.6 Conservation of angular momentum

Before concluding this chapter, it is worth extending our calculations one step further to consider the conservation of angular momentum in electromagnetic fields. Fortunately, the initial calculations follow in a straightforward way from our discussion of linear momentum. Instead of considering the force on a charged particle, we investigate the torque $\boldsymbol{\tau}$, which is given by

$$\boldsymbol{\tau} = \mathbf{r} \times \mathbf{F} = \mathbf{r} \times [q\mathbf{E} + \mathbf{v} \times \mathbf{B}]. \tag{7.99}$$

We can follow the same approach that we used in the derivation of the momentum conservation law. We first introduce the net torque on a system of charges and currents, which comes from integrating over a charge and current density:

$$\tau_{mech}(t) = \int_D \mathbf{r}' \times [\rho(\mathbf{r}', t)\mathbf{E}(\mathbf{r}', t) + \mathbf{J}(\mathbf{r}', t) \times \mathbf{B}(\mathbf{r}', t)]d^3r'. \tag{7.100}$$

We now use Maxwell's equations to eliminate the charge and current density from the integrand. It can readily be seen that the result should follow from the momentum result as

$$\tau_{mech}(t) = \int_D \mathbf{r}' \times \left[\mathbf{Q} - \epsilon_0 \frac{\partial}{\partial t}(\mathbf{E} \times \mathbf{B})\right]d^3r', \tag{7.101}$$

where we use τ_{mech} to specify the mechanical torque on the charges in the system. The quantity $\mathbf{Q}$ is still defined by equation (7.50). We define an angular momentum density $\mathbf{l}_{field}$ of the field as

$$\mathbf{l}_{field}(\mathbf{r}', t) = \epsilon_0 \mathbf{r}' \times [\mathbf{E}(\mathbf{r}', t) \times \mathbf{B}(\mathbf{r}', t)], \tag{7.102}$$

and if we introduce $\mathbf{L}$ as the angular momentum, we may write the total change of angular momentum as

$$\frac{\partial \mathbf{L}_{tot}}{\partial t} = \frac{\partial}{\partial t}[\mathbf{L}_{mech} + \mathbf{L}_{field}] = \int_D \mathbf{r}' \times \mathbf{Q}d^3r'. \tag{7.103}$$

The only difficult part of the calculation is to convert the integral on the right into one that is related to the flux through a surface. In tensor notation, we may write

$$[\mathbf{r} \times \mathbf{Q}]_i = -\epsilon_{ijk}r_j\partial_l T_{kl}, \tag{7.104}$$

where ϵ_{ijk} is the Levi-Civita tensor (see section 1.3.3 of [33]), r_j is the jth component of the position vector, and T_{kl} is a component of the Maxwell stress tensor. Due to the antisymmetry of ϵ_{ijk} and the symmetry of T_{kl}, the following equation holds:

$$\epsilon_{ijk}r_j\partial_l T_{kl} = \partial_l[\epsilon_{ijk}r_j T_{kl}]. \tag{7.105}$$

This allows us to finally write

$$\frac{\partial \mathbf{L}_{tot}}{\partial t} = -\int_D \nabla' \cdot (\mathbf{r}' \times \mathbf{T})d^3r' = -\oint_S (\mathbf{r}' \times \mathbf{T}) \cdot d\mathbf{a}', \tag{7.106}$$

where, of course, we have used Gauss's theorem in the final step. This becomes our angular momentum conservation law: any change in the total angular momentum within the volume must be the result of an electromagnetic angular momentum flux density flowing through the surface, of the form

$$\mathbf{M}(\mathbf{r}', t) \equiv \mathbf{r}' \times \mathbf{T}(\mathbf{r}', t). \tag{7.107}$$

How does angular momentum manifest in an electromagnetic wave, or specifically an optical beam? In 1909, Poynting himself hypothesized [34] that circularly polarized light, with its inherent handedness, must possess angular momentum. He also recognized that this angular momentum could be transferred to physical objects to induce rotation but was pessimistic about ever detecting such a weak force. Nevertheless, in 1936, Beth [35] detected and measured the torque induced in a wave plate by circularly polarized light. The mechanism by which this works is quite simple to understand: if a quarter wave plate is used to convert circularly polarized light to linear polarization, the light itself loses its angular momentum. Because angular momentum is conserved, it must be transferred to the wave plate itself.

The angular momentum of light associated with circular polarization is now known as 'spin' angular momentum, to distinguish it from angular momentum that can be encoded on the phase structure of a light wave, which is now known as orbital angular momentum (OAM). The Laguerre–Gaussian beams of section 3.5, which possess an azimuthal phase twist, are the simplest class of beams possessing orbital angular momentum, as was first recognized and demonstrated by Allen *et al* [36] in 1992. Beams with OAM have become another tool for optical trapping, as it is possible to use them to trap and spin particles [37] or even create light-powered micromachines such as fluid pumps [38].

For a highly directional beam, it is possible to write explicit expressions for the spin and orbital angular momentum components; for more information, we refer the reader to the author's book on singular optics [7].

7.7 Exercises

1. Let us consider two counterpropagating monochromatic plane waves in vacuum, with electric fields of the form

   $$\mathbf{E}_+(z) = E_+\hat{\mathbf{x}}e^{ikz}, \quad \mathbf{E}_-(z) = E_-\hat{\mathbf{x}}e^{-ikz},$$

   where $k = \omega/c$. E_+ and E_- are complex constants. Calculate the Poynting vector of the total field. How does the result depend on the relative amplitudes E_+ and E_-? Does the relative phase between the amplitudes affect the results at all?

2. Complete the derivation of the monochromatic version of Poynting's theorem, from equation (7.22) to equation (7.25).

3. A transverse plane wave in vacuum is normally incident on a perfectly reflective flat screen.

   (a) From the law of conservation of linear momentum, show that the pressure exerted on the screen is equal to twice the field energy per unit volume in the wave.

   (b) In the neighborhood of the Earth, the flux of electromagnetic energy from the Sun is approximately 0.14 W cm^{-2}. If an interplanetary 'sailplane' had a sail of mass density 10^{-4} g cm^{-2}, and negligible other weight, what would be its maximum acceleration in centimeters per second squared due to the solar radiation pressure?

4. Using the observation that the flux of sunlight on Earth is 1373 W m^{-2}, calculate the force on 1 m^2 of the Earth's surface due to sunlight. Treating the Earth as a flat circular disk facing the Sun with a radius of 6400 km, calculate the total force of the sunlight acting on the Earth.

5. Making an easy problem hard! Consider two equal point charges q, separated by a distance $2a$. Construct the plane equidistant from the two

charges. By integrating Maxwell's stress tensor over this plane, determine the force of one charge on the other. Do the same for charges that are opposite in sign.

6. The observed dust tail of a comet is the result of radiation pressure from sunlight. Consider a dust grain with a mass of 10^{-11} g and a radius of 1 μm [39] orbiting at the radius of the Earth, where the solar flux is 1400 W m^{-2}. Calculate the force of radiation on the particle and compare it to the force of gravity. Would you say that the solar radiation has a significant influence on the motion of the grain?

7. Suppose

$$\mathbf{E}(r, \theta, \phi, t) = A\frac{\sin\theta}{r}[\cos(kr - \omega t) - (1/kr)\sin(kr - \omega t)]\hat{\phi},$$

which represents a simple electromagnetic spherical wave.

(a) Show that $\mathbf{E}$ obeys all four of Maxwell's equations in vacuum, which requires finding the associated magnetic field.

(b) Calculate the Poynting vector. Average $\mathbf{S}$ over a full cycle to get the intensity vector $\mathbf{I}$. Does it point in the expected direction?

(c) Integrate $\mathbf{I} \cdot d\mathbf{a}$ over a spherical surface to determine the total power radiated.

8. A surgical CO_2 laser operates at 10.6 μm and produces 40 watts of power with a beam spot 1 mm in diameter. Treating the beam as a plane wave, calculate the pressure it exerts when normally incident on (a) a perfectly reflecting surface, and (b) a perfectly absorbent surface.

9. Let us consider the charging of an air-filled capacitor, as illustrated in figure 7.7. The capacitor is charging with a constant current I and the circular plates of radius a are separated by a distance d. The capacitor is uncharged at $t = 0$.

(a) Assuming that the plates are large enough and close enough together that the field between them can be approximated by the field between

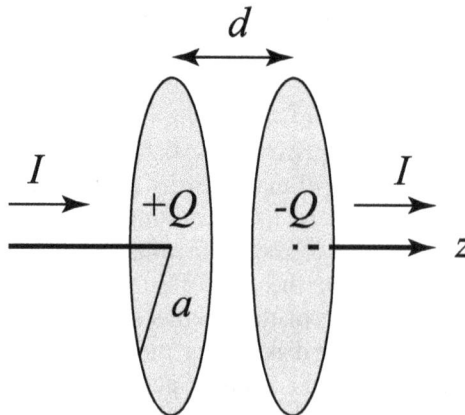

Figure 7.7. Illustration of the problem of a charging capacitor.

two infinite plates, calculate the electric field $\mathbf{E}(\mathbf{r}, t)$ between the plates.

(b) From this electric field, calculate the magnetic field $\mathbf{B}(\mathbf{r}, t)$ between the plates.

(c) Calculate the Poynting vector of the fields between the plates. Is the Poynting vector pointing in the direction you expect?

(d) From the Poynting vector, calculate the total power flux entering the capacitor. Calculate the electric and magnetic energy densities in the capacitor, and show that the rate of change of total energy is equal to the total power flux.

References

[1] Maxwell J C 1892 *A Treatise on Electricity and Magnetism* 3rd edn (Oxford: Clarendon)

[2] Lowry S C *et al* 2007 Direct detection of the asteroidal YORP effect *Science* **316** 272–4

[3] Gong S and Macdonald M 2019 Review on solar sail technology *Astrodynamics* **3** 93–125

[4] Polimeno P *et al* 2018 Optical tweezers and their applications *J. Quant. Spectrosc. Radiat. Transfer* **218** 131–50

[5] Poynting J H 1884 On the transfer of energy in the electromagnetic field *Philos. Trans.* **175** 343–61

[6] Bopp F 1963 Ist der Poynting-Vektor beobachtbar? *Ann. Physik.* **466** 35–51

[7] Gbur G J 2016 *Singular Optics* (Boca Raton, FL: CRC Press)

[8] Boivin A, Dow J and Wolf E 1967 Energy flow in the neighborhood of the focus of a coherent beam *J. Opt. Soc. Am.* **57** 1171–5

[9] Schouten H F, Visser T D, Lenstra D and Blok H 2003 Light transmission through a subwavelength slit: waveguiding and optical vortices *Phys. Rev.* E **67** 036608

[10] Feynman R, Leighton R B and Sands M L 1964 *The Feynman Lectures on Physics* **vol 2** (Reading, MA: Addison-Wesley)

[11] Jackson J D 1975 *Classical Electrodynamics* 2nd edn (New York: Wiley)

[12] Minkowski H 1908 Die Grundgleichungen für die elektromagnetischen Vorgänge in bewegten Körpern *Nachr. von Ges. Wiss. Gött.* **1908** 53–111

[13] Abraham M 1909 Zur Elektrodynamik bewegter Körper *Rend. Circ. Matem. Palermo.* **28** 1–28

[14] She W, Yu J and Feng R 2008 Observation of a push force on the end face of a nanometer silica filament exerted by outgoing light *Phys. Rev. Lett.* **101** 243601

[15] Jones R V and Richards J C S 1954 The pressure of radiation in a refracting medium *Proc. R. Soc.* 221 480–98

[16] Ashkin A and Dziedzic J M 1973 Radiation pressure on a free liquid surface *Phys. Rev. Lett.* **30** 139–42

[17] Zhang L, She W, Peng N and Leonhardt U 2015 Experimental evidence for Abraham pressure of light *New J. Phys.* **17** 053035

[18] Walker G B and Lahoz D G 1975 Experimental observation of Abraham force in a dielectric *Nature* **253** 339–40

[19] Gibson A F, Kimmitt M F, Koohian A O, Evans D E and Levy G F D 1980 A study of radiation pressure in a refractive medium by the photon drag effect *Proc. R. Soc.* A **370** 303–11

[20] Leonhardt U 2014 Abraham and Minkowski momenta in the optically induced motion of fluids *Phys. Rev.* A **90** 033801

[21] Barnett S M 2010 Resolution of the Abraham-Minkowski dilemma *Phys. Rev. Lett.* **104** 070401

[22] Pfeifer R N C, Nieminen T A, Heckenberg N R and Rubinsztein-Dunlop H 2007 Colloquium: momentum of an electromagnetic wave in dielectric media *Rev. Mod. Phys.* **79** 1197–216

[23] Einstein A and Laub J 1908 Über die im elektromagnetischen Felde auf ruhende Körper ausgeübten ponderomotorischen Kräfte *Ann. Phys. (Leipzig)* **26** 541–50

[24] Ashkin A 1992 Forces of a single-beam gradient laser trap on a dielectric sphere in the ray optics regime *Biophys. J.* **61** 569–82

[25] Gordon J P 1973 Radiation forces and momenta in dielectric media *Phys. Rev.* A **8** 14–21

[26] Chaumet P C and Nieto-Vesperinas M 2000 Time-averaged total force on a dipolar sphere in an electromagnetic field *Opt. Lett.* **25** 1065–7

[27] Ashkin A, Dziedzic J M, Bjorkholm J E and Chu S 1986 Observation of a single-beam gradient force optical trap for dielectric particles *Opt. Lett.* **11** 288–90

[28] Ashkin A 1970 Acceleration and trapping of particles by radiation pressure *Phys. Rev. Lett.* **24** 156–9

[29] Ashkin A 1978 Trapping of atoms by resonance radiation pressure *Phys. Rev. Lett.* **40** 729–32

[30] Metcalf H J and van der Straten P 1999 *Laser Cooling and Trapping* (New York: Springer)

[31] Albaladejo S, Marqués M I, Laroche M and Sáenz J J 2009 Scattering forces from the curl of the spin angular momentum of a light field *Phys. Rev. Lett.* **102** 113602

[32] Swartzlander G A, Peterson T J, Artusio-Glimpse A B and Raisanen A D 2011 Stable optical lift *Nat. Photon.* **5** 48–51

[33] Gbur G J 2011 *Mathematical Methods for Optical Physics and Engineering* (Cambridge: Cambridge University Press)

[34] Poynting J H 1909 The wave motion of a revolving shaft, and a suggestion as to the angular momentum in a beam of circularly polarized light *Proc. R. Soc.* A **82** 560–7

[35] Beth R A 1936 Mechanical detection and measurement of the angular momentum of light *Phys. Rev.* **50** 115–25

[36] Allen L, Beijersbergen M W, Spreeuw R J C and Woerdman J P 1992 Orbital angular momentum of light and the transformation of Laguerre-Gaussian laser modes *Phys. Rev.* A **45** 8185–9

[37] Simpson N B, Dholakia K, Allen L and Padgett M J 1997 Mechanical equivalence of spin and orbital angular momentum of light: an optical spanner *Opt. Lett.* **22** 52–4

[38] Ladavac K and Grier D G 2004 Microoptomechanical pumps assembled and driven by holographic optical vortex arrays *Opt. Exp.* **12** 1144

[39] Gruen E, Massonne L and Schwehm G 1987 New properties of cometary dust *Diversity and Similarity of Comets* vol 278 ed E J Rolfe, B Battrick, M Ackerman, M Scherer and R Reinhard (Paris: ESA Special Publication) 305–14

IOP Publishing

Electromagnetic Optics

Gregory J Gbur

Chapter 8

Anisotropic media

So far, our efforts have been focused on isotropic media, in which the permittivity and permeability are scalar quantities independent of direction. This implies that the **D**-field and **E**-field are parallel, as well as the **H**-field and **B**-field, i.e.

$$\mathbf{D} = \epsilon \mathbf{E}, \quad \mathbf{B} = \mu \mathbf{H}. \tag{8.1}$$

But this is not necessarily true in general. Some media are *anisotropic*, in which there is a tensor relationship between the vacuum fields and the auxiliary fields,

$$\mathbf{D} = \epsilon \cdot \mathbf{E}, \quad \mathbf{B} = \mu \cdot \mathbf{H}, \tag{8.2}$$

where ϵ and μ represent matrices or tensors. Anisotropy literally means 'unequal turn' and refers to the fact that the propagation of light in such a material now depends specifically on the direction of propagation and the state of polarization.

The classic example of an anisotropic material is optical calcite, also known as Iceland spar. When an object is viewed through calcite with unpolarized light, there are two images, as demonstrated in figure 8.1. We will see that the origin of this double refraction (or birefringence) is that unpolarized light may be decomposed into a pair of orthogonal polarization states, each of which has a different speed—and angle of refraction—in the anisotropic crystal.

In this chapter, we consider the propagation of electromagnetic waves in a bulk anisotropic material. The details of double refraction, which depends on the behavior of light at an interface, will be deferred to chapter 9.

The mathematics of anisotropy is often quite involved. Though we will endeavor to prove the most significant results, many physically intuitive results require a lot of unenlightening math to prove rigorously. In these cases, we will skip the proof and refer the reader to chapter 15 of the classic book by Born and Wolf [1].

doi:10.1088/978-0-7503-6064-7ch8

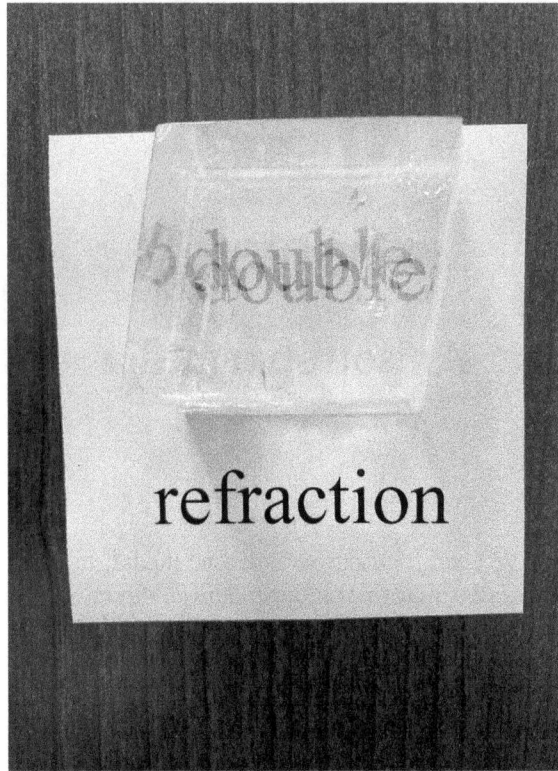

Figure 8.1. Illustration of double refraction in optical calcite.

8.1 Basic concepts of anisotropy

What is the origin of anisotropy? Calcite and other anisotropic crystals have a periodic molecular structure that is strongly dependent on direction. If we think back to our simple Lorentz model of the atom, this suggests that electrons may experience different restoring forces depending on which way they are forced to oscillate, making the restoring force a tensor relation as well:

$$\mathbf{F} = -\mathbf{Q} \cdot \mathbf{r}, \tag{8.3}$$

where $\mathbf{Q}$ is a tensor. This translates to a direction-dependent permittivity, which can be written explicitly as

$$\epsilon = \begin{bmatrix} \epsilon_{xx} & \epsilon_{xy} & \epsilon_{xz} \\ \epsilon_{yx} & \epsilon_{yy} & \epsilon_{yz} \\ \epsilon_{zx} & \epsilon_{zy} & \epsilon_{zz} \end{bmatrix}. \tag{8.4}$$

It is also possible to have a direction-dependent permeability of the form

$$\boldsymbol{\mu} = \begin{bmatrix} \mu_{xx} & \mu_{xy} & \mu_{xz} \\ \mu_{yx} & \mu_{yy} & \mu_{yz} \\ \mu_{zx} & \mu_{zy} & \mu_{zz} \end{bmatrix}. \tag{8.5}$$

An anisotropic medium adds new complexity and mathematical challenges to the wave propagation problem. For simplicity, let us make two big assumptions. The first is that $\mu = \mu_0$, i.e. that there is no anisotropy in the permeability and that the material is nonmagnetic (as we have said, this is usually true at optical frequencies). The second assumption is that the permittivity matrix is a real symmetric matrix. We will show momentarily that a real symmetric matrix represents a material that is nonabsorbing. Furthermore, such a matrix can always be diagonalized by a rotation of the coordinate axes; if we work in a coordinate system where the matrix is diagonal, we may write

$$\boldsymbol{\epsilon} = \begin{bmatrix} \tilde{\epsilon}_x & 0 & 0 \\ 0 & \tilde{\epsilon}_y & 0 \\ 0 & 0 & \tilde{\epsilon}_z \end{bmatrix}. \tag{8.6}$$

The axes that diagonalize the permittivity matrix are called the *principal axes*, and the values $\tilde{\epsilon}_x$, $\tilde{\epsilon}_y$, and $\tilde{\epsilon}_z$ are called the *principal dielectric constants*. We may then refer to three distinct cases:

- $\tilde{\epsilon}_x = \tilde{\epsilon}_y = \tilde{\epsilon}_z$. This case, where all three principal constants are equal, is simply the isotropic case for which ϵ can be treated as a scalar.
- $\tilde{\epsilon}_x = \tilde{\epsilon}_y \neq \tilde{\epsilon}_z$. This case, where one constant is different from the others, represents a *uniaxial* crystal.
- $\tilde{\epsilon}_x \neq \tilde{\epsilon}_y \neq \tilde{\epsilon}_z$. This case, where all constants are different, represents a *biaxial* crystal.

The terms 'uniaxial' and 'biaxial' literally mean 'one (special) axis' and 'two (special) axes,' respectively. The meanings of these terms will be explained later; they refer to the observation that there are particular directions of propagation where the behavior of light in a crystal is special. For brevity, we will simply refer to an anisotropic medium as a 'crystal' going forward, though isotropic crystals, such as diamond, also exist.

The anisotropy of the crystal is determined by the underlying crystal structure and can be characterized by the shape of its primitive cell, the smallest building block from which a crystal can be constructed. In three dimensions, primitive cells are parallelepipeds, and the type of crystal—and its anisotropy—depend on the relationships between the lengths of the edges of the parallelepipeds and the angles between edges. In this way, all crystals can be classified as one of seven *crystal systems*. In order of generally decreasing symmetry, the systems are: cubic, tetragonal, trigonal, hexagonal, orthorhombic, monoclinic, and triclinic; the corresponding primitive cells are illustrated in figure 8.2. Cubic crystals are isotropic.

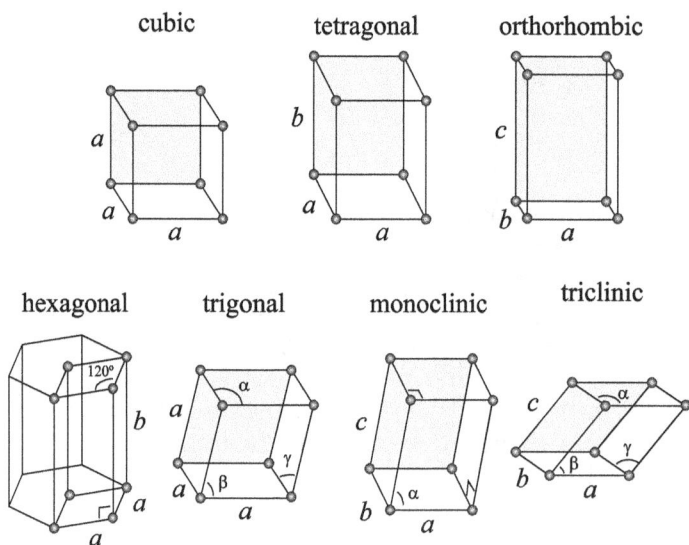

Figure 8.2. Illustration of the primitive cells of the seven crystal systems. Edges with different labels are of different length, and angles with different labels are different.

Table 8.1. Principal refractive indices of selected crystals. Many of these early measurements were made using the D line of sodium (589 nm).

Material	Wavelength (nm)	$\tilde{n}_x$	$\tilde{n}_y$	$\tilde{n}_z$	[Ref]	System
Calcite	589.29	1.486	1.658	1.658	[2]	Trigonal
Aragonite	589.29	1.530	1.681	1.686	[2]	Orthorhombic
Zinc sulfide	500.0	2.425	2.421	2.421	[3]	Hexagonal
Cadmium sulfide	530.0	2.654	2.649	2.649	[3]	Hexagonal
Rutile	546.1	2.958	2.652	2.652	[4]	Tetragonal
Common mica	589	1.5632	1.5971	1.6021	[5]	Monoclinic
Naphthalene	546.1	1.525	1.722	1.945	[6]	Monoclinic
Ulexite	589.3	1.493	1.505	1.529	[7]	Triclinic
Quartz	589	1.553	1.544	1.544	[8]	Trigonal

Hexagonal, trigonal, and tetragonal crystals, which possess a strong axis of symmetry, are uniaxial, while orthorhombic, monoclinic, and triclinic crystals are biaxial.

We may also define principal refractive indices $\tilde{n}_x$, $\tilde{n}_y$, $\tilde{n}_z$ for each principal dielectric constant. Some examples of these indices are given in table 8.1. In uniaxial crystals, the index that appears twice is usually called the ordinary index, while the third index value is called the extraordinary index; we will discuss these terms in detail soon.

Quartz is a crystal commonly used for making wave plates, and rutile is often used for making polarizing prisms. The triclinic mineral ulexite deserves special mention as a fibrous rock that acts as a natural fiber optic system. It is sometimes referred to and sold as 'TV rock,' as a specimen can transfer a flat image placed beneath it to the top of the crystal surface.

As in the isotropic case, anisotropic materials exhibit dispersion and temperature dependence, and in the crystals of the lowest symmetries—orthorhombic, monoclinic, and triclinic—the principal axes themselves can vary with frequency, a phenomenon known as dispersion of the axes. We will not consider any of these effects here: the mathematics is complicated enough already.

For uniaxial crystals, the birefringence is often quantified as the difference $\Delta n = n_e - n_o$. For calcite, for example, the birefringence is $\Delta n = -0.172$, and it is a negative uniaxial crystal.

8.2 Plane waves in crystals

We develop our understanding of crystal optics in the now familiar way: we look for plane-wave solutions to Maxwell's equations. We assume a monochromatic solution of the form

$$\mathbf{D}(\mathbf{r}) = \mathbf{D}_0 e^{ik_0 n \mathbf{s} \cdot \mathbf{r}}, \tag{8.7}$$

$$\mathbf{E}(\mathbf{r}) = \mathbf{E}_0 e^{ik_0 n \mathbf{s} \cdot \mathbf{r}}, \tag{8.8}$$

where $k_0 = \omega/c$ is the free-space wavenumber and the refractive index n is to be determined. It should be noted that it is not necessarily obvious that there is a plane-wave solution in this case, so it will be part of our job to determine the circumstances in which such a solution exists. We will find that there are only two distinct plane-wave solutions for each direction of propagation, each with specific polarization properties; this is quite different from the isotropic case, where a plane wave can be polarized in any transverse direction for a given direction of propagation.

It should also be noted that we are considering $\mathbf{D}$ and $\mathbf{E}$ separately, as the relation between them is no longer as simple as in the isotropic case. The two fields are not parallel in general, as can readily be seen in the principal axes system,

$$\mathbf{D}_0 = \hat{\mathbf{x}} D_x + \hat{\mathbf{y}} D_y + \hat{\mathbf{z}} D_z = \hat{\mathbf{x}} \tilde{\epsilon}_x E_x + \hat{\mathbf{y}} \tilde{\epsilon}_y E_y + \hat{\mathbf{z}} \tilde{\epsilon}_z E_z. \tag{8.9}$$

The two fields are proportional only if: all three $\tilde{\epsilon}$ are equal, the $\mathbf{E}$-field points directly along a principal axis, or the $\mathbf{E}$-field lies in a plane where two permittivities are equal in a uniaxial crystal.

We turn to Maxwell's equations to explore plane-wave solutions in a crystal. In this case, the equations are:

$$\nabla \cdot \mathbf{D} = 0, \tag{8.10}$$

$$\nabla \cdot \mathbf{H} = 0, \tag{8.11}$$

$$\nabla \times \mathbf{E} = i\omega\mu_0 \mathbf{H}, \tag{8.12}$$

$$\nabla \times \mathbf{H} = -i\omega \mathbf{D}, \tag{8.13}$$

where we have used $\mathbf{B} = \mu_0 \mathbf{H}$ and have assumed no sources are present.

From Gauss's law, we can say that the **D**-field is transverse to the direction of propagation **s**,

$$\mathbf{s} \cdot \mathbf{D}_0 = 0. \tag{8.14}$$

From the 'no magnetic monopoles law,' we also know that the **H**-field is transverse, i.e.

$$\mathbf{s} \cdot \mathbf{H}_0 = 0. \tag{8.15}$$

Faraday's law and the Ampère–Maxwell law for plane waves have the forms

$$k_0 n \mathbf{s} \times \mathbf{E}_0 = \omega \mu_0 \mathbf{H}_0. \tag{8.16}$$

$$k_0 n \mathbf{s} \times \mathbf{H}_0 = -\omega \mathbf{D}_0. \tag{8.17}$$

The Ampère–Maxwell law implies that $(\mathbf{D}_0, \mathbf{H}_0, \mathbf{s})$ form a right-handed triplet of vectors, with the cross product of any two in cyclic order being parallel to the third. But because $\mathbf{D}_0$ is not parallel to $\mathbf{E}_0$, we have, in general,

$$\mathbf{s} \cdot \mathbf{E}_0 \neq 0, \tag{8.18}$$

i.e. that the electric field is *not* transverse in a crystal, unless **s** is parallel to a principal axis. We can see this latter case from equation (8.9); if **s** lies along z, then $\mathbf{D}_0$ lies in the xy-plane, and $\mathbf{E}_0$ must necessarily lie in the same plane.

Unlike the isotropic case, for a crystal there are, in general, only two distinct plane-wave solutions for a given direction of propagation. We can demonstrate this by looking at the consistency of Maxwell's equations. We take the curl of equation (8.16) and then substitute from equation (8.17) to find

$$-k_0^2 n^2 \mathbf{s} \times (\mathbf{s} \times \mathbf{E}_0) = \omega^2 \mu_0 \epsilon \mathbf{E}_0. \tag{8.19}$$

We need to rewrite this equation in a form that allows us to determine what values of n and $\mathbf{E}_0$ are allowed. We do so by first expanding the vector triple product,

$$k_0^2 n^2 [\mathbf{E}_0 - \mathbf{s}(\mathbf{s} \cdot \mathbf{E}_0)] = \omega^2 \mu_0 \epsilon \mathbf{E}_0. \tag{8.20}$$

We can reinterpret this as a vector-matrix equation. We have

$$-\mathbf{s} \times (\mathbf{s} \times \mathbf{E}_0) = [\mathbf{I} - \mathbf{s} \otimes \mathbf{s}]\mathbf{E}_0 \equiv \mathbf{T}\mathbf{E}_0, \tag{8.21}$$

where **I** is the identity matrix and $\otimes$ represents the outer product of vectors, i.e. the column matrix of one vector multiplies the row matrix of the second. (This quantity is known as a 'dyad.' We will explore dyads in detail in chapter 14.) The matrix **T** on the right is consequently the identity matrix with the component of the matrix along the **s**-direction removed. If the direction of propagation were the z-direction, for example, the matrix **T** would be

$$\mathbf{T} = \begin{bmatrix} 1 & 0 & 0 \\ 0 & 1 & 0 \\ 0 & 0 & 0 \end{bmatrix}. \tag{8.22}$$

We refer to it as 'T,' as it filters and keeps only the transverse (T) component of the field. Our consistency equation (8.20) then is of the form

$$k_0^2 n^2 \mathbf{T} \mathbf{E}_0 = \omega^2 \mu_0 \epsilon \mathbf{E}_0. \tag{8.23}$$

The matrix ϵ is a real symmetric matrix with nonzero eigenvalues, which are the principal dielectric constants. This means that the matrix can be inverted, and we may write

$$\epsilon_0 \epsilon^{-1} \mathbf{T} \mathbf{E}_0 = \frac{1}{n^2} \mathbf{E}_0, \tag{8.24}$$

where ϵ^{-1} is the inverse of ϵ. What we now have is an eigenvalue equation, where $1/n^2$ is the eigenvalue and $\mathbf{E}_0$ is the eigenvector. At first, we might assume that it has three nontrivial solutions, but $\mathbf{T}$ reduces this number to two. We may make this more explicit by multiplying both sides by $\mathbf{T}$ and noting that $\mathbf{T} = \mathbf{T}^2$. We then have

$$\epsilon_0 \mathbf{T} \epsilon^{-1} \mathbf{T} \mathbf{T} \mathbf{E}_0 = \frac{1}{n^2} \mathbf{T} \mathbf{E}_0. \tag{8.25}$$

Let us define $\mathbf{T} \mathbf{E}_0 \equiv \mathbf{E}_T$ as the transverse component of the electric field vector. We then write

$$\epsilon_0 \mathbf{T} \epsilon^{-1} \mathbf{T} \mathbf{E}_T = \frac{1}{n^2} \mathbf{E}_T. \tag{8.26}$$

The matrix $\mathbf{T} \epsilon^{-1} \mathbf{T}$, in diagonal form, has zeros for the $\mathbf{s}$ component. This means that the eigenvalue equation in the $\mathbf{s}$-direction has a trivial zero solution and that we effectively have a 2×2 matrix eigenvalue problem. For a given direction of propagation $\mathbf{s}$, there are two possible refractive index values n and two distinct electric field vector solutions $\mathbf{E}_T$. We could, in principle, solve this equation for $\mathbf{E}_T$ and n and then determine the component of the electric field along $\mathbf{s}$ by $\nabla \cdot \mathbf{D} = 0$, though there are more elegant strategies, as we will see.

It is important to note that the eigenvalues and eigenvectors of a real symmetric matrix, such as the one in equation (8.26), are real-valued, which means that the electric field vectors $\mathbf{E}_0$ are always real-valued, i.e. linear polarization. Moving forward, we will always maintain the complex form of general equations involving $\mathbf{E}_0$ and $\mathbf{H}_0$, but those vectors will always be, at their most complicated, a real vector multiplied by a complex quantity.

What happens if we send a light wave with an arbitrary polarization state into a crystal at normal incidence along a principal axis? We decompose the electric field vector into its two distinct linear polarization components and propagate each of them in the normal direction with the appropriate refractive index n. This is the approach we will take in studying wave plates later in this chapter.

We have not yet explicitly determined the allowed values of n and $\mathbf{E}_0$ for a given direction of propagation $\mathbf{s}$. That will come soon, but for the moment, we assume we are working with one of the allowed plane-wave solutions.

8.3 Energy flow in crystals

The nontransversality of the electric field leads to unusual properties related to the flow of energy. Let us first consider the electric and magnetic energy densities, U_e and U_m. The formulas (7.26b) and (7.26c) may be applied directly, as they do not depend on the specific definitions of **D** and **H**. For a crystal, this leads to the expressions

$$U_e = \frac{1}{4}\mathbf{D}^*\cdot\mathbf{E} = \frac{1}{4}\mathbf{E}^*\cdot\epsilon^*\cdot\mathbf{E}, \tag{8.27}$$

$$U_m = \frac{1}{4}\mathbf{B}^*\cdot\mathbf{H} = \frac{1}{4}\mu_0\mathbf{H}^*\cdot\mathbf{H}, \tag{8.28}$$

where, for the moment, we allow the possibility of a complex ϵ. For a plane wave, these two quantities are still equal to each other; this can be demonstrated using equations (8.16) and (8.17), which allows us to write

$$U_e = -\frac{k_0 n}{4\omega}(\mathbf{s} \times \mathbf{H}_0^*) \cdot \mathbf{E}_0, \tag{8.29}$$

$$U_m = \frac{k_0 n}{4\omega}(\mathbf{s} \times \mathbf{E}_0) \cdot \mathbf{H}_0^*. \tag{8.30}$$

These can be shown to be equal using the cyclic permutation of the scalar triple product.

We can use equation (8.27) to show that our real symmetric matrix represents a non-absorbing material. Referring back to our monochromatic energy conservation formula, equation (7.27), we expect an absence of absorption because U_e is a purely real-valued quantity. Therefore, the imaginary part must be zero, which we can write in component form as

$$U_e - U_e^* = \frac{1}{4}\Big[E_i^* \epsilon_{ij}^* E_j - E_i \epsilon_{ij} E_j^* \Big] = 0, \tag{8.31}$$

where summation over repeated indices is assumed. We may reverse the labeling of the dummy indices in one of the two terms in this formula, giving us

$$U_e - U_e^* = \frac{1}{4}\Big[E_i(\epsilon_{ji}^* - \epsilon_{ij})E_j^* \Big] = 0. \tag{8.32}$$

Because **E** is an arbitrary electric field vector, this can only be satisfied by a Hermitian matrix, $\epsilon_{ji}^* = \epsilon_{ij}$. If we further require that the matrix can be diagonalized by a real-valued rotation matrix, then the matrix must be a real symmetric matrix. Is it possible to have a material represented by a more general Hermitian matrix? We discuss this in section 8.7.

For future reference, we write $U \equiv U_e + U_m$ as the total energy density, which may also be written in the form

$$U = \frac{k_0 n}{2\omega}(\mathbf{E}_0 \times \mathbf{H}_0^*) \cdot \mathbf{s} = \frac{n}{c}\mathbf{S} \cdot \mathbf{s}, \qquad (8.33)$$

where we have applied equations (8.29) and (8.30).

We may also use equation (8.16) to write the Poynting vector of a plane wave as

$$\mathbf{S} = \frac{k_0 n}{2\omega\mu_0}\mathbf{E}_0 \times (\mathbf{s} \times \mathbf{E}_0^*) = \frac{n}{2\mu_0 c}\Big[\mathbf{s}(\mathbf{E}_0 \cdot \mathbf{E}_0^*) - \mathbf{E}_0^*(\mathbf{s} \cdot \mathbf{E}_0)\Big]. \qquad (8.34)$$

The first term on the right is parallel to $\mathbf{s}$, but the second term—which is nonzero according to equation (8.18)—is not. We therefore find that the Poynting vector $\mathbf{S}$ is nonparallel to the direction of wave propagation $\mathbf{s}$. In other words, the wave's direction of propagation is different from the direction in which energy is actually flowing. Furthermore, they have different speeds in the crystal. We note that the phase velocity is $v_p = c/n$, as usual; the speed of energy flow, the *ray velocity* v_r, may be defined as

$$v_r = \frac{|\mathbf{S}|}{U} = \frac{c}{n}\frac{1}{\cos\alpha}, \qquad (8.35)$$

where we have applied equation (8.33) and defined α as the angle between the direction of propagation $\mathbf{s}$ and the Poynting vector $\mathbf{S}$. We can then relate the ray and phase velocities by

$$v_p = v_r \cos\alpha. \qquad (8.36)$$

The phase velocity is always smaller than the ray velocity and may be thought of as the projection of the ray velocity onto the direction of wave propagation $\mathbf{s}$. For future reference, we refer to the unit vector in the direction of the ray velocity as $\mathbf{t}$; then, we may write

$$v_p = v_r \mathbf{s} \cdot \mathbf{t}. \qquad (8.37)$$

We can show that $(\mathbf{E}_0, \mathbf{H}_0, \mathbf{t})$ form a right-handed triplet of vectors, which we leave as an exercise. This triplet may be considered complementary to the one formed by $(\mathbf{D}_0, \mathbf{H}_0, \mathbf{s})$. The two triplets share $\mathbf{H}_0$, and there is an angle α between $\mathbf{s}$ and $\mathbf{t}$; this, in turn, implies that the angle between $\mathbf{E}_0$ and $\mathbf{D}_0$ is also α. We therefore have that $\mathbf{E}_0$ is transverse to one velocity in a crystal—the ray velocity—and $\mathbf{D}_0$ is transverse to another velocity in a crystal—the phase velocity.

To get a feel for the magnitude of this angle α, let us calculate it for light propagating in optical calcite, a material with significant anisotropy for which the ordinary index is 1.658 and the extraordinary index is 1.486, and we take the extraordinary axis to lie along z. As an example, let us take $\mathbf{t}$ to lie in the xz-plane at an angle $\theta = 45°$ and the transverse $\mathbf{H}_0$-field to lie in the y-direction. Using the triplet relation $(\mathbf{E}_0, \mathbf{H}_0, \mathbf{t})$, we may determine that

$$\mathbf{t} = \hat{\mathbf{x}}\sin(\pi/4) + \hat{\mathbf{z}}\cos(\pi/4), \qquad (8.38)$$

$$\mathbf{E}_0 = E_0[\hat{\mathbf{x}}\cos(\pi/4) - \hat{\mathbf{z}}\sin(\pi/4)]. \qquad (8.39)$$

We now use equation (8.9) to calculate the direction of the **D**-field,

$$\mathbf{D}_0 = \epsilon \cdot \mathbf{E}_0 = \hat{\mathbf{x}} E_0 \tilde{\epsilon}_{xx} \cos(\pi/4) - \hat{\mathbf{z}} E_0 \tilde{\epsilon}_{zz} \sin(\pi/4). \tag{8.40}$$

Next, we apply the triplet relation ($\mathbf{D}_0$, $\mathbf{H}_0$, **s**) to determine that **s** must have the form

$$\mathbf{s} = \frac{\hat{\mathbf{x}}\tilde{\epsilon}_{xx} \sin(\pi/4) + \hat{\mathbf{z}}\tilde{\epsilon}_{zz} \cos(\pi/4)}{\sqrt{[\tilde{\epsilon}_{xx} \sin(\pi/4)]^2 + [\tilde{\epsilon}_{zz} \cos(\pi/4)]^2}}. \tag{8.41}$$

Finally, we find that

$$\cos\alpha = \mathbf{s} \cdot \mathbf{t} = \frac{\tilde{\epsilon}_{xx} \sin^2(\pi/4) + \tilde{\epsilon}_{zz} \cos^2(\pi/4)}{\sqrt{[\tilde{\epsilon}_{xx} \sin(\pi/4)]^2 + [\tilde{\epsilon}_{zz} \cos(\pi/4)]^2}}. \tag{8.42}$$

Plugging in the numbers, we find that $\alpha = 3.13°$, a relatively small number.

The recognition that there are two velocities in different directions again raises the question: which velocity do we use as the 'physical' velocity? There is an intuitive appeal to viewing the velocity of energy transport as more physical, but we must keep in mind the concerns we introduced about the Poynting vector in section 7.2. We will see at least one demonstration that shows the physical appeal of the energy transport velocity at the end of the chapter.

Why do the ray velocity and the phase velocity point in different directions? Evidently, this is a consequence of the nontrivial interaction between the anisotropic material and the light wave. We will see an even more extreme difference when we discuss negative-refractive-index materials in chapter 12.

8.4 The Fresnel equation of wave normals

We can derive a general relationship between $\mathbf{D}_0$ and $\mathbf{E}_0$ by substituting from equation (8.16) into equation (8.17):

$$\mathbf{D}_0 = [\mathbf{E}_0 - \mathbf{s}(\mathbf{s} \cdot \mathbf{E}_0)]\epsilon_0 n^2. \tag{8.43}$$

This equation shows again that $\mathbf{E}_0$ possesses a longitudinal component because $\mathbf{D}_0$ is transverse, and we must subtract the component of $\mathbf{E}_0$ along **s** to get $\mathbf{D}_0$. It also allows us to do something much more significant: derive an explicit equation for the refractive index n.

To derive such an equation, we now switch to an index notation for vectors. If we combine equation (8.9) and equation (8.43), we may write

$$D_j = \tilde{\epsilon}_j E_j = \epsilon_0 n^2 E_j - \epsilon_0 n^2 s_j \left(\sum_k s_k E_k \right). \tag{8.44}$$

It should be noted that we are not using the Einstein summation convention, which involves an automatic sum over repeated indices; the leftmost product in the above equation is a straight product of $\tilde{\epsilon}_j$ and E_j. To avoid confusion, we will explicitly write any summation symbol in this section.

We have an E_j on either side of this equation, and we may solve for it, giving

$$E_j = \frac{s_j(\mathbf{s} \cdot \mathbf{E})\epsilon_0 n^2}{\epsilon_0 n^2 - \tilde{\epsilon}_j}. \tag{8.45}$$

We may now use a clever trick: we multiply both sides of this expression by s_j and then sum over j. Since $\sum_j s_j E_j = \mathbf{s} \cdot \mathbf{E}$, we get

$$\mathbf{s} \cdot \mathbf{E} = \mathbf{s} \cdot \mathbf{E} \epsilon_0 n^2 \left[\frac{s_x^2}{\epsilon_0 n^2 - \tilde{\epsilon}_x} + \frac{s_y^2}{\epsilon_0 n^2 - \tilde{\epsilon}_y} + \frac{s_z^2}{\epsilon_0 n^2 - \tilde{\epsilon}_z} \right]. \tag{8.46}$$

We may cancel the $\mathbf{s} \cdot \mathbf{E}$ on each side and reorganize to write

$$\frac{1}{n^2} = \left[\frac{s_x^2}{n^2 - \tilde{\epsilon}_x/\epsilon_0} + \frac{s_y^2}{n^2 - \tilde{\epsilon}_y/\epsilon_0} + \frac{s_z^2}{n^2 - \tilde{\epsilon}_z/\epsilon_0} \right]. \tag{8.47}$$

This is one form of the *Fresnel equation of wave normals*, and it is an equation that can be solved for the refractive index n for a given direction of propagation.

This is a very awkward equation, but it already provides some interesting insights. If our wave propagates along a principal axis, e.g. $s_x = 1$, $s_y = s_z = 0$, then the equation is inconsistent unless $n = \sqrt{\tilde{\epsilon}_y/\epsilon_0}$ or $n = \sqrt{\tilde{\epsilon}_z/\epsilon_0}$, which results in an indeterminate 0/0 situation. For propagation along a principal axis, the allowed indices are easy to determine.

The Fresnel equation can be written in a more useful form by first multiplying by n^2 and then replacing the 1 on the left of the resulting equation with $s_x^2 + s_y^2 + s_z^2 = 1$. Grouping together the terms by components of $\mathbf{s}$, we end up with

$$\frac{s_x^2}{v_p^2 - v_x^2} + \frac{s_y^2}{v_p^2 - v_y^2} + \frac{s_z^2}{v_p^2 - v_z^2} = 0, \tag{8.48}$$

where $v_p = c/n$ is the phase velocity, $v_x = c\sqrt{\epsilon_0/\tilde{\epsilon}_x}$, and so on. This is another form of the Fresnel equation. We note that v_x is the phase velocity for light that is polarized in the x-direction, with similar interpretations for v_y and v_z.

We can use equation (8.48) to determine the phase velocities of the two orthogonal polarization states. We rewrite it as

$$s_x^2(v_p^2 - v_y^2)(v_p^2 - v_z^2) + s_y^2(v_p^2 - v_x^2)(v_p^2 - v_z^2) + s_z^2(v_p^2 - v_x^2)(v_p^2 - v_y^2) = 0. \tag{8.49}$$

This is a quadratic equation in v_p^2, which implies that there are two values of v_p^2 and consequently two refractive indices for a given direction of propagation. These two values correspond to two orthogonal polarizations, though polarization does not appear in the formula. We note that the unit vector $\mathbf{s}$ may be written in terms of angles of propagation θ_p and ϕ_p; equation (8.49) is therefore an equation for the vector phase velocity in spherical coordinates (v_p, θ_p, ϕ_p). With three degrees of freedom and one constraint, we find that the solutions are surfaces. Because it is a quadratic equation, there are two surfaces. These two surfaces correspond to the two phase velocities and two refractive indices for every direction of propagation.

In Cartesian coordinates, these surfaces exist in v_{px}, v_{py}, v_{pz} space. In general, they intersect (i.e. they have the same phase velocity and refractive index) on two lines passing through the origin. These lines are called the *optic axes*. Now we finally have an explanation for the terms 'uniaxial' and 'biaxial': a uniaxial crystal has one optic axis, and a biaxial crystal has two.

Let us focus on the uniaxial case. We take $v_x = v_y = v_o$, where o denotes 'ordinary,' and $v_z = v_e$, where e denotes 'extraordinary.' For a uniaxial crystal, the optic axis is simply the axis of v_e. In this case, we can simplify equation (8.49) to the form

$$(v_p^2 - v_o^2)\left[(s_x^2 + s_y^2)(v_p^2 - v_e^2) + s_z^2(v_p^2 - v_o^2)\right] = 0. \tag{8.50}$$

Let us define θ as the angle with respect to the extraordinary axis. We then have $s_x^2 + s_y^2 = \sin^2 \theta$ and $s_z^2 = \cos^2 \theta$. We then write

$$(v_p^2 - v_o^2)\left[\sin^2 \theta(v_p^2 - v_e^2) + \cos^2 \theta(v_p^2 - v_o^2)\right] = 0. \tag{8.51}$$

We have two solutions for v_p^2. From the parenthesis, we have $v_p = v_o$, and from the square brackets, we have

$$v_p^2 = v_o^2 \cos^2 \theta + v_e^2 \sin^2 \theta. \tag{8.52}$$

We therefore find that one phase velocity (and refractive index) in a uniaxial crystal is always constant. The other depends upon the relative angle between the direction of propagation and the optic axis. If a piece of optical calcite is rotated, one image seen through the calcite therefore remains fixed while the second changes direction—this is why the latter is called the 'extraordinary' image. The surfaces in this case are a sphere and an ovaloid that is rotationally symmetric around the optic axis.

These surfaces are a good way to visualize the phase velocities of light in a crystal and how they depend on the direction of propagation. Uniaxial crystals may be classified as *negative uniaxial* if $v_o < v_e$ ($n_o > n_e$) or *positive uniaxial* if $v_o > v_e$ ($n_o < n_e$). These two cases are illustrated in figure 8.3. At the wavelengths given in

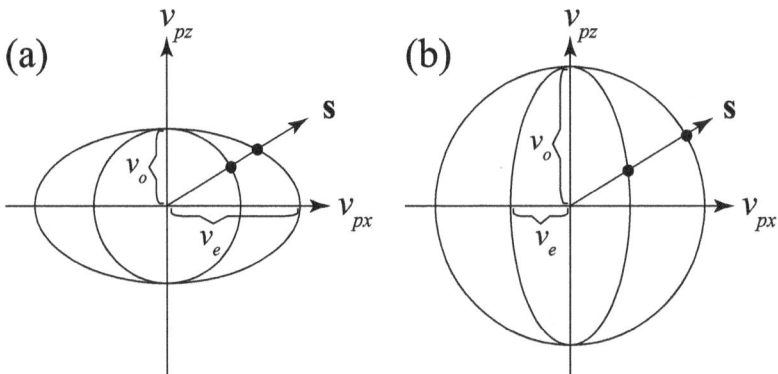

Figure 8.3. Illustration of (a) a negative uniaxial crystal and (b) a positive uniaxial crystal. The dots indicate the solutions for v_p.

table 8.1, for example, optical calcite is a negative uniaxial crystal, while rutile is a positive uniaxial crystal.

One thing that is difficult to do using this geometric approach is evaluate the effect of a crystal on the state of polarization of an arbitrary field. In section 8.6, we look at a special case where the mathematics is much clearer.

8.5 Ellipsoid of wave normals

We may also introduce a different geometrical construction to visualize and determine the allowed directions of the **D**-field for a given direction of propagation. We again note that we are only considering linear polarization states, for which the phases between the field components at most differ by a sign. We now write the energy density in terms of the **D**-field in the principal axis system:

$$U = \frac{1}{2}\mathbf{D}^* \cdot \epsilon^{-1} \cdot \mathbf{D} = \frac{D_x^2}{2\tilde{\epsilon}_x} + \frac{D_y^2}{2\tilde{\epsilon}_y} + \frac{D_z^2}{2\tilde{\epsilon}_z}. \tag{8.53}$$

Let us divide both sides of this expression by U, and introduce new variables x, y, z, with $x \equiv D_x/\sqrt{2U}$, and so forth. Our equation simplifies to the form

$$\frac{x^2}{\tilde{\epsilon}_x} + \frac{y^2}{\tilde{\epsilon}_y} + \frac{z^2}{\tilde{\epsilon}_z} = 1. \tag{8.54}$$

This is an equation for an ellipsoid, and the semiaxes of the ellipsoid are equal to the square roots of the principal dielectric constants; this surface is referred to as the *ellipsoid of wave normals*. It can be used to determine both the directions of the **D**-field components and their corresponding phase velocities for a given direction of propagation **s**.

To do so, we intersect the ellipsoid with a plane that passes through the origin and whose normal is the propagation direction **s**. The intersection is, in general, an ellipse. Because the **D**-field must lie in a plane perpendicular to **s**, the field components lie within this plane, and the major and minor axes of the ellipse are inversely proportional to the allowed phase velocities and give the polarization directions associated with those indices. In the intersection of a plane with an ellipsoid, there are, in general, two directions where the intersection is a circle; these represent the optic axes.

A similar construction can be used to construct the *ray ellipsoid*, which determines the allowed directions of the **E**-field for a given ray direction **t**. We write the energy density instead as

$$U = \frac{1}{2}\mathbf{E} \cdot \epsilon \cdot \mathbf{E} = \frac{\tilde{\epsilon}_x E_x^2}{2} + \frac{\tilde{\epsilon}_y E_y^2}{2} + \frac{\tilde{\epsilon}_z E_z^2}{2}. \tag{8.55}$$

We introduce new variables $\tilde{x} \equiv E_x/\sqrt{2U}$, $\tilde{y}$, and $\tilde{z}$, which presents us with the ellipsoid equation

$$\tilde{\epsilon}_x \tilde{x}^2 + \tilde{\epsilon}_y \tilde{y}^2 + \tilde{\epsilon}_z \tilde{z}^2 = 1. \tag{8.56}$$

We can intersect this ellipsoid with a plane that passes through the origin and whose normal is **t**. The major and minor axes of the resulting ellipse are proportional to the two allowed ray velocities v_r and provide the two associated electric field directions. Because we again have the intersection of a plane with an ellipsoid, there are generally two directions where the intersection is a circle; these are known as the *ray axes*.

If we consider a uniaxial crystal, we can readily see that the single optic axis and the single ray axis lie on the same line, due to the symmetry of the respective ellipse equations. For a biaxial crystal, however, the optic axes and ray axes generally lie in different directions; this leads to the fascinating and counterintuitive phenomenon of conical refraction, which we discuss in section 8.9.

8.6 Anisotropy and wave plates

One of the significant applications of anisotropic crystals is in the creation of wave plates, which change the relative phase between electric field components by a fixed amount. We now take a look at how the design of such a device can be accomplished.

We begin again with Maxwell's equations in the absence of sources,

$$\nabla \times \mathbf{E} = i\omega\mu\mathbf{H}, \tag{8.57}$$

$$\nabla \times \mathbf{H} = -i\omega\mathbf{D}, \tag{8.58}$$

$$\nabla \cdot \mathbf{D} = 0, \tag{8.59}$$

$$\nabla \cdot \mathbf{B} = 0. \tag{8.60}$$

We take the curl of equation (8.57) to develop a wave equation, as we have done in the past. Using the formula for a double curl, we readily find

$$\nabla(\nabla \cdot \mathbf{E}) - \nabla^2\mathbf{E} = \omega^2\mu\mathbf{D}, \tag{8.61}$$

where we have used equation (8.58) as well. The problem here is that we cannot, in general, reduce this to a simple wave equation, because $\nabla \cdot \mathbf{E} \neq 0$. Let us consider the special case of a plane wave propagating in the direction of a principal axis, which we take to be the z-axis. The two distinct plane-wave solutions in this direction have polarizations along the other two principal axes, and in these cases, $\nabla \cdot \mathbf{E} = 0$. We then use $\mathbf{D} = \epsilon\mathbf{E}$, obtaining

$$\nabla^2\mathbf{E} + \omega^2\mu\epsilon\mathbf{E} = 0. \tag{8.62}$$

Since we are assuming a plane wave propagating in the z-direction, all derivatives except the z derivative are zero. We finally have

$$\frac{\partial^2\mathbf{E}}{\partial z^2} + \omega^2\mu\epsilon\mathbf{E} = 0. \tag{8.63}$$

We note that $\mathbf{E}$ lies in the transverse xy-plane. Working in the principal axes coordinate system, equation (8.63) can be decomposed into two equations:

$$\frac{\partial^2 E_x}{\partial z^2} + \omega^2 \mu \epsilon_o E_x = 0, \tag{8.64}$$

$$\frac{\partial^2 E_y}{\partial z^2} + \omega^2 \mu \epsilon_e E_y = 0, \tag{8.65}$$

where we have taken $\tilde{\epsilon}_x = \tilde{\epsilon}_z = \epsilon_o$ and $\tilde{\epsilon}_y = \epsilon_e$. We then have the following solutions:

$$E_x(z) = E_{x0} e^{ik_o z}, \tag{8.66}$$

$$E_y(z) = E_{y0} e^{ik_e z}, \tag{8.67}$$

where we have introduced

$$k_o = \frac{\omega}{c} \sqrt{\frac{\epsilon_o}{\epsilon_0}}, \tag{8.68}$$

$$k_e = \frac{\omega}{c} \sqrt{\frac{\epsilon_e}{\epsilon_0}}. \tag{8.69}$$

As light propagates through a crystal over any finite distance, a phase difference accumulates between the x- and y-components of the field. The axis with the smallest n is called the 'fast axis' because it has the fastest phase velocity. The axis with the largest n is called the 'slow axis.' It should be noted that the relationship between the fast and slow axes and the ordinary and extraordinary directions depends on whether the crystal is a positive or negative uniaxial crystal.

It is quite straightforward to show how to create a wave plate from a uniaxial crystal. The relevant geometry is illustrated in figure 8.4.

Let us write the field incident upon the plate at the plane $z = 0$ as

$$\mathbf{E}_{in} = E_{x0} \hat{\mathbf{x}} + E_{y0} \hat{\mathbf{y}}. \tag{8.70}$$

Assuming that the reflection at the input and output faces is negligible, the output field at $z = d$ can be found from equations (8.66) and (8.67) to be

$$\mathbf{E}_{out} = e^{ik_o d} [E_{x0} \hat{\mathbf{x}} + E_{y0} \hat{\mathbf{y}} e^{i(k_e - k_o)d}], \tag{8.71}$$

indicating that the y-component of the field has acquired a phase $\phi = (k_e - k_o)d$ relative to the x-component.

Let us suppose that this phase difference is $\phi = \pm\pi/2$, i.e. that $d = \pi/2|k_e - k_o|$. A field sent in with $E_{0x} = E_{0y} = E_0$ now has an output state of

$$\mathbf{E}_{out} \propto E_0(\hat{\mathbf{x}} \pm i\hat{\mathbf{y}}), \tag{8.72}$$

which is circular polarization. This device is a quarter-wave plate, named as such because it imparts a $\pi/2$ phase shift to the field, i.e. one quarter of a full 2π wave

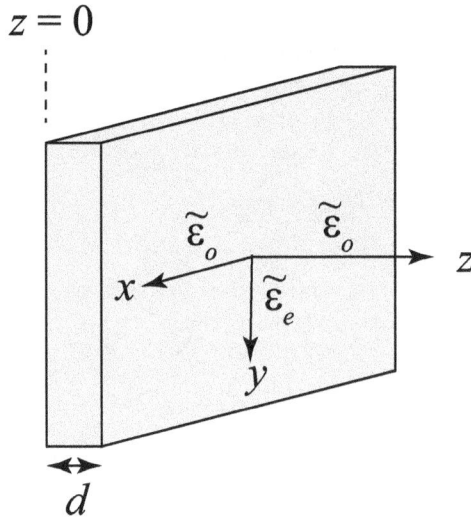

Figure 8.4. Illustration of the geometry of a wave plate.

cycle. A half-wave plate is similarly constructed with a thickness such that $d = \pi/|k_e - k_o|$.

The design of a wave plate, given a crystal with known principal dielectric constants, therefore requires two steps: (i) choosing an axis to be the direction of propagation and (ii) choosing a thickness that provides the appropriate phase difference between components.

The formulas above result in the thinnest wave plates possible for a given material; however, 'thinnest' is not always ideal, as the thinnest plates can be on the order of a few wavelengths thick, making them exceedingly delicate. However, since phase is periodic, we can design a quarter-wave plate, for example, with

$$(k_e - k_o)d = \pm\pi/2 + 2\pi m, \tag{8.73}$$

where m is an integer referred to as the *order* of the plate. The disadvantage of the multiple-order wave plate is that, with the thickness increased, the device becomes more sensitive to temperature and wavelength variations.

An alternative to a multiple-order wave plate is a zero-order wave plate. Such plates are constructed from two multiple-order wave plates stacked together so that the fast axis of one plate is aligned with the slow axis of the other. This results in a partial cancellation of the phase retardance between the plates and can be designed so that the total phase retardance is a true $m = 0$ effect. For example, if two quartz wave plates are stacked together, one which produces a $11\lambda/2$ retardance and the second aligned to produce a $-10\lambda/2$ retardance, the combination will be a $\lambda/2$ wave plate.

It should be noted that we can also discuss the actions of polarizers using a similar formalism, where one of the principal dielectric constants is taken to be complex.

This, in turn, indicates that the associated refractive index is complex, resulting in the absorption of light polarized along that axis.

8.7 Optical rotation

We have so far considered anisotropy where the matrix can be diagonalized in a linear polarization basis. It is also possible to produce anisotropic media that are diagonal in a circular polarization basis, and these media have their own unusual properties.

We begin a discussion of such materials by introducing the most famous historical example. In 1845, at the age of 54, Michael Faraday was continuing his research into electricity, magnetism, and light [9]. Heartened by his discovery of Faraday induction, he further wondered if there might be a way for magnetism to affect a beam of light. His key experiment is illustrated in figure 8.5. Collimated light from a lamp was reflected off a glass surface at the Brewster angle (to be discussed later), which resulted in a horizontally polarized beam. This beam passed through the field of an electromagnet, propagating parallel to the field lines. The beam's polarization state was then analyzed by a Nicol prism, an early type of polarizer.

The experiment was performed as follows. With the magnet off, light was passed through the system, and the eyepiece oriented to allow the maximum light transmission. The magnet was then turned on, and any change in the direction of polarization resulted in a decrease in light transmission through the eyepiece. By rotating the eyepiece, Faraday could determine how much the polarization had rotated, if at all.

A key component of the experiment, not originally included, was a sample of material, often a liquid, placed in the path of the light and within the effect of the magnetic field. With the sample present, Faraday found that the direction of polarization rotated through an angle θ according to the very simple formula

$$\theta = \mathcal{V}Bd, \tag{8.74}$$

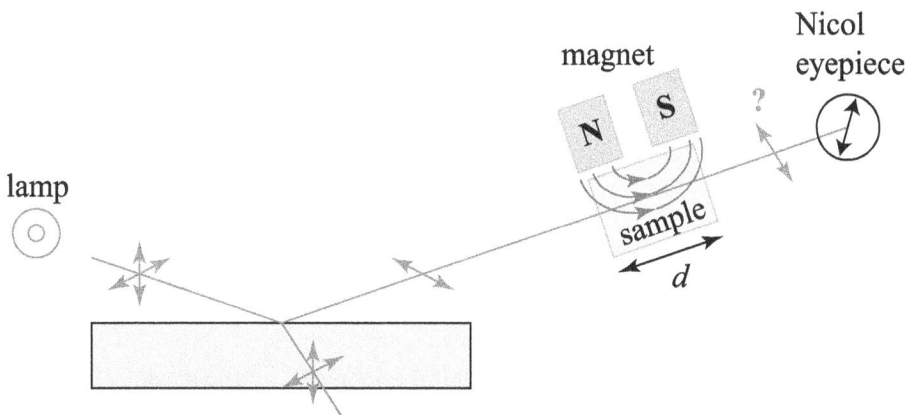

Figure 8.5. Illustration of Faraday's rotation experiment.

Table 8.2. List of Verdet constants for selected materials. All Verdet constants are measured in arcminutes per gauss-centimeter. Jena glass O.451 is a commercial flint glass.

Material	Wavelength (nm)	Verdet constant	Temperature (C)	[ref]
Air	589.3	6.83×10^{-6}	20	[10]
Water	495.8	1.903×10^{-2}	0	[10]
Water	589.3	1.311×10^{-2}	0	[10]
Acetone	589.3	1.109×10^{-2}	15.1	[10]
Diamond	589.3	1.28×10^{-2}	16	[10]
Amber	589.3	-9.60×10^{-3}	19	[10]
$TiCl_2$	589.3	-1.521×10^{-2}	16	[10]
Jena glass O.451	589.3	3.17×10^{-2}	18	[10]
BK7 glass	589.3	1.8×10^{-2}	20	[11]

where d is the thickness of the sample, B is the strength of the field, and $\mathcal{V}$ is a constant called the Verdet constant, which depends on the material and is in essence a measure of the strength of the material's response. A list of some values of the Verdet constant is given in table 8.2. Faraday, a thorough experimentalist, tested almost every liquid he possessed in his laboratory to show that this rotation was a universal effect, with only the Verdet constant differing between samples. A positive angle represents a counterclockwise rotation with respect to the observer; it should be noted that some materials can exhibit a negative Verdet constant and clockwise rotation.

Faraday's discovery, which became known as Faraday rotation, demonstrated a previously unproven link between magnetism and light, further paving the way for Maxwell's discovery of electromagnetic waves. It is one example of what is known as *optical rotation*, where the linear polarization of light is rotated on propagation through a material. We can explain this rotation as arising from an anisotropy in the optical properties of materials with respect to left- and right-handed circular polarization. This phenomenon is occasionally called *gyrotropy*, though it appears that this term is sometimes used specifically to describe Faraday rotation and sometimes used to describe any material that exhibits optical rotation; we will see that there are different types of optical rotation with very different physical origins.

To explain Faraday rotation, we first present a simple, crude (and quite inaccurate) picture that highlights the physics, and then we provide a more quantitative analysis based on our old friend the Lorentz oscillator.

Let us imagine a material that consists of a number of atoms with electrons all orbiting in the xy-plane. In the absence of a magnetic field, we might expect roughly half of these atoms to orbit clockwise and half to orbit counterclockwise, as illustrated in figure 8.6(a).

If we illuminate this material with circularly polarized light, we can imagine that the electrons that are orbiting in the same sense as the polarization then experience a stronger optical interaction than those orbiting in the opposite sense. However,

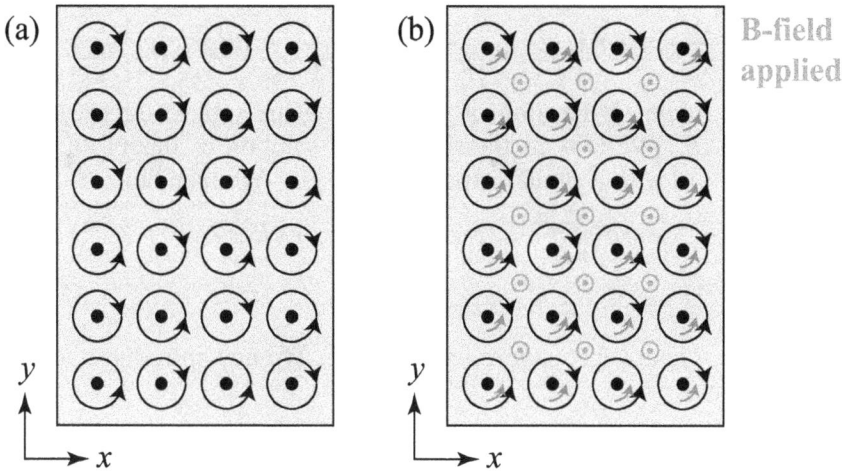

Figure 8.6. Crude explanation of Faraday rotation: (a) the material in the absence of a magnetic field, and (b) the material in the presence of the magnetic field. The z-direction is taken to be out of the page, i.e. toward the reader.

because there are equal numbers of clockwise and counterclockwise electrons, both circular polarization states experience the same average optical response in the system.

Now let us imagine that we apply a constant magnetic field along the z-axis. The effect of this magnetic field on the electrons is that the left-handed (counter-clockwise) electrons speed up, and the right-handed electrons slow down, as sketched in figure 8.6(b). There is now a fundamental asymmetry in the behavior of clockwise and counterclockwise electrons, and this means that left-handed and right-handed circular polarization states experience different responses in the material, which naturally translates into a difference in refractive index for the two circular polarization states.

This picture of Faraday rotation is quite inaccurate, as we do not expect that a material has one set of atoms rotating clockwise and another set rotating counter-clockwise. In reality, we have a collection of essentially identical atoms that experience an asymmetric response to circular polarization states in the presence of a magnetic field. In other words, the magnetic field introduces a left–right asymmetry in the resonance states of an individual atom.

We can make this a bit more quantitative by applying the Lorentz oscillator model of section 6.1, following the approach of Sommerfeld [12]. Let us consider a Lorentz electron oscillating in a constant magnetic field pointing in the z-direction. Using the Lorentz force, equation (2.10), we may write equations for the motion of the x- and y-components of the electron motion as

$$m\frac{\partial^2 x(t)}{\partial t^2} = -fx(t) - g\frac{\partial x(t)}{\partial t} - eB_0\frac{\partial y(t)}{\partial t} - eE_x e^{-i\omega t}, \qquad (8.75a)$$

$$m\frac{\partial^2 y(t)}{\partial t^2} = -fy(t) - g\frac{\partial y(t)}{\partial t} + eB_0\frac{\partial x(t)}{\partial t} - eE_y e^{-i\omega t}, \qquad (8.75b)$$

where we have assumed the atom is experiencing an applied electric field with components E_x, E_y. We look at the steady-state solution of this problem, where $x(t) = x_0 \exp[-i\omega t]$ with a similar expression for $y(t)$. We then have

$$(-m\omega^2 + f - i\omega g)x_0 = i\omega B_0 y_0 - eE_x, \qquad (8.76a)$$

$$(-m\omega^2 + f - i\omega g)y_0 = -i\omega B_0 x_0 - eE_y. \qquad (8.76b)$$

We have a pair of coupled equations for x_0 and y_0. We now introduce the sum and difference variables

$$z_\pm = x_0 \pm iy_0, \quad E_\pm = E_x \pm iE_y, \qquad (8.77)$$

and if we take the appropriate combinations, we end up with a pair of separated equations

$$(-m\omega^2 + f - i\omega g)z_\pm = \pm\omega B_0 z_\pm - eE_\pm. \qquad (8.78)$$

The quantities $z_\pm$ represent the components of the position vector in a circular polarization basis $\hat{\mathbf{e}}_\pm$ (recall equation (4.78)), and similarly, $E_\pm$ represent the circular polarization components of the field. These results suggest a circular-polarization-dependent polarizability of the atom of the form

$$\alpha_\pm(\omega) = \frac{e^2}{\epsilon_0[m(\omega_f^2 - \omega^2) - i\omega g \mp \omega B_0]}. \qquad (8.79)$$

The polarizability of the atom therefore depends upon the circular polarization state driving it, and consequently, so does the refractive index. Equation (8.79) furthermore suggests that a single optical resonance splits into two in the presence of a static magnetic field. This is a classical derivation of the so-called 'normal' Zeeman effect, and the Lorentz model was the prevailing model for the effect prior to the introduction of quantum mechanics. There is also an anomalous Zeeman effect, which is the splitting of spectral lines due to a nonzero net electron spin in an atom, and this effect can only be explained quantum mechanically.

Let us turn to the practical effect of Faraday rotation on a light wave. From the preceding discussion, it appears that we may mathematically model a medium exhibiting optical rotation by assuming that the medium possesses a permittivity that depends on the handedness of light; that is, assuming propagation in the $+z$-direction,

$$D_+ = \epsilon_+ E_+, \quad D_- = \epsilon_- E_-, \qquad (8.80)$$

where E_+ and E_- represent the amplitudes of the left- and right-circular polarizations,

$$\mathbf{E}_0 = E_+(\hat{\mathbf{x}} + i\hat{\mathbf{y}})/\sqrt{2} + E_-(\hat{\mathbf{x}} - i\hat{\mathbf{y}})/\sqrt{2}. \qquad (8.81)$$

We can write this electric field in a linear basis simply by rearranging the terms:

$$\mathbf{E}_0 == \frac{E_+ + E_-}{\sqrt{2}}\hat{\mathbf{x}} + i\frac{E_+ - E_-}{\sqrt{2}}\hat{\mathbf{y}}. \tag{8.82}$$

In matrix form, we may write

$$\begin{bmatrix} E_x \\ E_y \end{bmatrix} = \begin{bmatrix} 1/\sqrt{2} & 1/\sqrt{2} \\ i/\sqrt{2} & -i\sqrt{2} \end{bmatrix}\begin{bmatrix} E_+ \\ E_- \end{bmatrix}, \tag{8.83}$$

or

$$\begin{bmatrix} E_+ \\ E_- \end{bmatrix} = \begin{bmatrix} 1/\sqrt{2} - i/\sqrt{2} \\ 1/\sqrt{2} \quad i\sqrt{2} \end{bmatrix}\begin{bmatrix} E_x \\ E_y \end{bmatrix}. \tag{8.84}$$

We may determine the form of the permittivity in the xy-basis by transforming the relation

$$\begin{bmatrix} D_+ \\ D_- \end{bmatrix} = \begin{bmatrix} \epsilon_+ & 0 \\ 0 & \epsilon_- \end{bmatrix}\begin{bmatrix} E_+ \\ E_- \end{bmatrix}, \tag{8.85}$$

and the result is

$$\begin{bmatrix} D_x \\ D_y \end{bmatrix} = \begin{bmatrix} \epsilon_+ + \epsilon_- & i(\epsilon_- - \epsilon_+) \\ -i(\epsilon_- - \epsilon_+) & \epsilon_+ + \epsilon_- \end{bmatrix}\begin{bmatrix} E_x \\ E_y \end{bmatrix}. \tag{8.86}$$

This is a 2×2 Hermitian matrix. We may write a full three-dimensional Cartesian dielectric tensor for such a medium in the form

$$\epsilon = \begin{bmatrix} \kappa & i\kappa' & 0 \\ -i\kappa' & \kappa & 0 \\ 0 & 0 & \kappa_0 \end{bmatrix}, \tag{8.87}$$

where κ, κ', and κ_0 all have units of permittivity. It should be noted that this matrix is complex, though there is no absorption in the system. It is Hermitian, and, as discussed in section 8.3, is therefore nonabsorbing. A complex permittivity in an anisotropic medium does not automatically imply absorption.

What is the effect of such a medium on linearly polarized light? We follow the same approach that we used in section 8.6 for designing wave plates: we look at a normally incident wave propagating through the medium. We assume an incident field propagating in the z-direction,

$$\mathbf{E}_0 = E_0\hat{\mathbf{x}} = E_0(\hat{\mathbf{e}}_+ + \hat{\mathbf{e}}_-)/\sqrt{2}, \tag{8.88}$$

where $\hat{\mathbf{e}}_\pm = (\hat{\mathbf{x}} \pm i\hat{\mathbf{y}})/\sqrt{2}$, and note that after a distance z, it has the form,

$$\mathbf{E}(z) = E_0(\hat{\mathbf{e}}_+ e^{ik_0 n_+ z} + \hat{\mathbf{e}}_- e^{ik_0 n_- z})/\sqrt{2}. \tag{8.89}$$

If we write this in a Cartesian basis, we have

$$\mathbf{E}(z) = E_0[\hat{\mathbf{x}}(e^{ik_0 n_+ z} + e^{ik_0 n_- z}) + i\hat{\mathbf{y}}(e^{ik_0 n_+ z} - e^{ik_0 n_- z})]/2. \tag{8.90}$$

On pulling out an appropriate complex exponential, we finally have

$$\mathbf{E}(z) = E_0 e^{ik_0 z(n_+ + n_-)/2}[\hat{\mathbf{x}}\cos(k\Delta z) + \hat{\mathbf{y}}\sin(k\Delta z)], \tag{8.91}$$

where $\Delta \equiv (n_- - n_+)/2$. We see that our field remains linearly polarized but rotates through an angle $k\Delta z$, in agreement with Faraday's observation that the angle depends on the thickness of the sample.

Faraday rotation is not the only way to create optical rotation. Many types of molecules have *chirality*: they have a left-handed or right-handed nature, and when light passes through such a material, it also experiences an asymmetric response based on handedness. These materials are labeled based on the direction in which they rotate polarization with respect to the observer:

- Dextrorotary (clockwise): examples include sucrose and camphor; rotation is defined as positive.
- Levorotary (counterclockwise): examples include cholesterol; rotation is defined as negative.

A chiral material that rotates the plane of polarization is said to exhibit *optical activity*.

To characterize optical activity, a quantity known as the *specific rotation* is used. For a given measurement, we use the labels T for the temperature in Celsius, λ for the wavelength of light used, c for the concentration of the solution in g/ml, and d for the length of the cell holding the solution in decimeters. The specific rotation $[\alpha]_\lambda^T$ is then defined in this case as

$$[\alpha]_\lambda^T = \frac{\alpha}{cd}, \tag{8.92}$$

where α is the measured rotation in degrees. In older texts, the convention was to measure concentration in g/100 ml, in which case the formula above was multiplied by 100. $[\alpha]_\lambda^T$ has units of degree $\cdot$ ml/(g $\cdot$ dm). The subscript λ used to label the wavelength of operation was often a 'D' in early work, which stood for the sodium D line.

The specific rotation is usually measured for a given concentration and solvent, and it can depend on both of these quantities in a nontrivial way. The specific rotation is thus often written with the concentration and solvent in parentheses after the measured value. For example, L-lysine has a specific rotation given by [13]

$$\text{Specific rotation of L$-$lysine: } [\alpha]_D^{20} + 14.6 \quad (w, c = 6), \tag{8.93}$$

where the solvent is water and the concentration is in g/100 ml.

The formula describing the optical activity of liquids is roughly analogous to that for Faraday rotation, with the concentration of the solution replacing the strength of

Table 8.3. List of specific rotation values for selected materials. Specific rotation is given in units of degree · ml/(g · dm). '(us)' indicates unspecified.

Material	λ (nm)	T (C)	Solvent	c (g/100 ml)	$[\alpha]_\lambda^T$	Reference
D-fructose	589	20	Water	2	−92	[13]
D-glucose	589	20	Water	(us)	+52.7	[13]
Sucrose	589	20	Water	(us)	+66.37	[13]
Camphor	589	20	(us)	(us)	+44.26	[13]
Cholesterol	589	20	Ethanol	2	−31.5	[13]

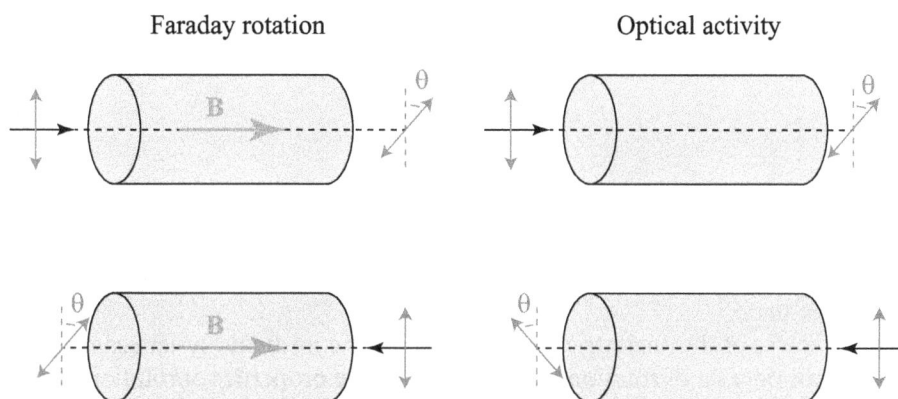

Figure 8.7. Illustration showing how the angle of rotation depends on propagation direction for Faraday rotation and optical activity.

the magnetic field; the caveat is that the rotation often depends in a nonlinear way upon the concentration. Several values of specific rotation are given in table 8.3.

Solids may also exhibit optical activity, and in such cases, the activity is measured in degrees per unit length.

There is one practical difference between Faraday rotation and optical activity, namely, how the rotation depends on the direction of propagation through the material; this is illustrated in figure 8.7. In optical activity, light rotates in the same direction with respect to the observer regardless of its direction of travel in the medium; the rotation is determined by the direction of light propagation. For Faraday rotation, light rotates in the opposite direction with respect to the observer from opposite ends of the sample. This is because Faraday rotation depends specifically on the direction of the **B**-field passing through the material. Faraday rotation is an example of a system that exhibits *nonreciprocal behavior*: that is, the optical response of the medium is different for counterpropagating light beams. The result for Faraday rotation, as seen in the figure, is that the polarization rotates in the same absolute direction regardless of the direction of propagation. This nonreciprocal behavior allows for the construction of devices such as optical isolators, i.e. 'light diodes,' that only permit light to propagate through them in one direction.

8.8 Anisotropic media with absorption

We have assumed in our derivations up to this point that the anisotropic medium is nonabsorbing, but this excluded the case of some of the most important anisotropic materials, i.e. those used to construct polarizers. Such materials are strongly absorbing along one principal axis and nonabsorbing (or at least significantly less absorbing) along the others.

To better understand the properties of absorbing anisotropic materials, let us first assume that we have a material for which we can still find a principal axis coordinate system, so that it has a dielectric tensor

$$\epsilon = \begin{bmatrix} \tilde{\epsilon}_x & 0 & 0 \\ 0 & \tilde{\epsilon}_y & 0 \\ 0 & 0 & \tilde{\epsilon}_z \end{bmatrix}, \tag{8.94}$$

where now the $\tilde{\epsilon}_j$ are complex-valued. The construction of a polarizer from such a material follows along similar lines to section 8.6, where now the thickness of the wave plate is chosen so that the electric field is sufficiently attenuated on propagation through the plate.

The structure of this dielectric tensor in any other coordinate system can be found by an appropriate set of rotations. It follows from the properties of rotations that the tensor is, in general, a *complex symmetric matrix*, i.e. that $\epsilon_{ij} = \epsilon_{ji}$, with no complex conjugation. This is distinct from the complex matrix of equation (8.87) for optical activity, which was a Hermitian matrix. The difference is that the optical activity matrix is energy conserving, whereas the complex symmetric matrix is not.

It is worth noting that not every complex symmetric matrix can be diagonalized by a rotation [14]. This suggests that there are absorbing anisotropic materials that cannot have their permittivity written in a single principal axis coordinate system. However, the real and imaginary parts of the matrix are individually symmetric and can be diagonalized separately by different rotations, with the principal axes generally pointing in different directions for each part. This is typically the case for absorbing triclinic and monoclinic crystals.

Materials with anisotropic absorption are often referred to as *dichroic*. This term arose from observations of minerals such as tourmaline, which exhibit different colors when observed from different directions. Changing the direction of observation changes the allowed axes of polarization, which may then have more or less absorption at certain wavelengths.

8.9 Conical refraction

We have noted the peculiar property of crystals in which the wave direction **s** and the ray direction **t** are generally different. In biaxial crystals, this difference leads to a remarkable phenomenon known as *conical refraction*, which we will briefly summarize.

We consider the process known as *internal conical refraction* first. Let us imagine that we have a biaxial crystal cut so that one of the optic axes is normal to the input face and output face, and we illuminate the crystal with a normally incident unpolarized plane wave. In air, the plane wave may be considered a combination of waves with every possible transverse direction of the **D**-field and an orthogonal **H**-field. Upon entering the crystal, the wave continues in the normal direction, and the **D**-field and **H**-field remain transverse.

Because we are propagating along an optic axis, all possible **D** directions are allowed; with unpolarized light, all these directions are present when the plane wave enters the crystal. We have seen that **E** is not parallel to **D**, which means that every **D** polarization state in the transverse plane has an **E**-field propagating in a distinct direction. Each of these polarization states has a distinct ray propagation direction **t**; the complete set of these ray propagation directions forms a cone. Upon exiting the crystal, this cone of light becomes a hollow circular beam.

We may roughly illustrate this process using some basic matrix manipulations. Our strategy is to construct a permittivity matrix that has an optic axis aligned in the $\hat{z}$-direction and then calculate the propagation of light in that direction.

We begin with a crystal in the principal axis system, given by equation (8.6), and rotate around the y-axis by an angle α using the rotation matrix

$$\mathbf{R} = \begin{bmatrix} \cos\alpha & 0 & -\sin\alpha \\ 0 & 1 & 0 \\ \sin\alpha & 0 & \cos\alpha \end{bmatrix}. \tag{8.95}$$

The new form of the permittivity matrix ϵ' in this coordinate system is

$$\epsilon' = \mathbf{R}\epsilon\mathbf{R}^T = \begin{bmatrix} \tilde{\epsilon}_x\cos^2\alpha + \tilde{\epsilon}_z\sin^2\alpha & 0 & (\tilde{\epsilon}_z - \tilde{\epsilon}_x)\cos\alpha\sin\alpha \\ 0 & \tilde{\epsilon}_y & 0 \\ (\tilde{\epsilon}_z - \tilde{\epsilon}_x)\cos\alpha\sin\alpha & 0 & \tilde{\epsilon}_x\sin^2\alpha + \tilde{\epsilon}_z\cos^2\alpha \end{bmatrix}$$

$$\equiv \begin{bmatrix} \epsilon_{xx} & 0 & \epsilon_{xz} \\ 0 & \epsilon_{yy} & 0 \\ \epsilon_{xz} & 0 & \epsilon_{zz} \end{bmatrix}. \tag{8.96}$$

Under what condition does the new z-axis represent an optic axis of the crystal? We have seen from equation (8.54), the ellipsoid of wave normals, that light propagating along the z-axis is an optic axis if the elements of the matrix ϵ^{-1} perpendicular to that axis are equal. We can readily find that the inverse matrix is of the form

$$\epsilon^{-1} = \frac{1}{\epsilon_D^2} \begin{bmatrix} \epsilon_{zz} & 0 & -\epsilon_{xz} \\ 0 & \epsilon_D^2/\epsilon_{yy} & 0 \\ -\epsilon_{xz} & 0 & \epsilon_{xx} \end{bmatrix}, \tag{8.97}$$

where $\epsilon_D^2 = \epsilon_{xx}\epsilon_{zz} - \epsilon_{xz}^2$. For ϵ^{-1} to be isotropic with respect to the x- and y-directions, we thus must have $\epsilon_D^2/\epsilon_{yy} = \epsilon_{zz}$. A matrix with an optic axis aligned in the z-direction is therefore

$$\epsilon^{-1} = \frac{1}{\epsilon_D^2}\begin{bmatrix} \epsilon_{zz} & 0 & -\epsilon_{xz} \\ 0 & \epsilon_{zz} & 0 \\ -\epsilon_{xz} & 0 & \epsilon_{xx} \end{bmatrix}, \tag{8.98}$$

Let us now consider a plane wave propagating in the $\hat{\mathbf{z}}$-direction, with $\mathbf{D}$-field and $\mathbf{H}$-field of the form

$$\mathbf{D} = D_0[\cos\phi\hat{\mathbf{x}} + \sin\phi\hat{\mathbf{y}}], \tag{8.99}$$

$$\mathbf{H} = \beta D_0[-\sin\phi\hat{\mathbf{x}} + \cos\phi\hat{\mathbf{y}}], \tag{8.100}$$

where β is a real-valued constant whose specific value is of no concern to us. Since $\mathbf{E} = \epsilon^{-1}\mathbf{D}$, we find that the electric field in the crystal may be written as

$$\mathbf{E} = \frac{D_0}{\epsilon_D^2}[\epsilon_{zz}\cos\phi\hat{\mathbf{x}} + \epsilon_{zz}\sin\phi\hat{\mathbf{y}} - \epsilon_{xz}\cos\phi\hat{\mathbf{z}}]. \tag{8.101}$$

Finally, considering that the $\mathbf{H}$-field is the same in the crystal as outside, we find that the Poynting vector has the form

$$\mathbf{S} = \frac{\beta D_0^2}{2\epsilon_D^2}\{\epsilon_{zz}\hat{\mathbf{z}} + \epsilon_{xz}\cos\phi\sin\phi\hat{\mathbf{y}} + \epsilon_{xz}\cos^2\phi\hat{\mathbf{x}}\}. \tag{8.102}$$

We may use trigonometric identities to simplify this to

$$\mathbf{S} = \frac{\beta D_0^2}{4\epsilon_D^2}\left\{2\epsilon_{zz}\hat{\mathbf{z}} + \frac{1}{2}\epsilon_{xz}[\sin(2\phi)\hat{\mathbf{y}} + (1 + \cos(2\phi))]\hat{\mathbf{x}}\right\}. \tag{8.103}$$

For unpolarized light, $0 \leqslant \phi < 2\pi$. The appearance of 2ϕ as the argument of the sine and cosine functions reflects the fact that a linear polarization state and elliptical polarization states are invariant with respect to a π rotation about the propagation axis. We can see that the Poynting vector has a constant $\hat{\mathbf{z}}$ component and forms a circle in an xy-plane displaced from the z-axis. The unpolarized input beam therefore fans out into a ray cone displaced from the axis, as illustrated in figure 8.8 (a). Upon leaving the crystal, each ray refracts to the $\hat{\mathbf{z}}$-direction, resulting in a hollow cylinder of light. In practice, the plane wave is replaced by a collimated beam, which is collimated by the use of a small aperture at the input face of the crystal.

Internal conical refraction involves a single wave direction $\mathbf{s}$ fanning out inside the crystal into a cone of ray directions $\mathbf{t}$. We may also produce *external conical refraction*, which is, in essence, the reverse process: a single ray direction $\mathbf{t}$ within a crystal fans out into a cone of wave directions upon leaving the crystal. In this case, the unpolarized plane wave must have its ray direction aligned with a ray

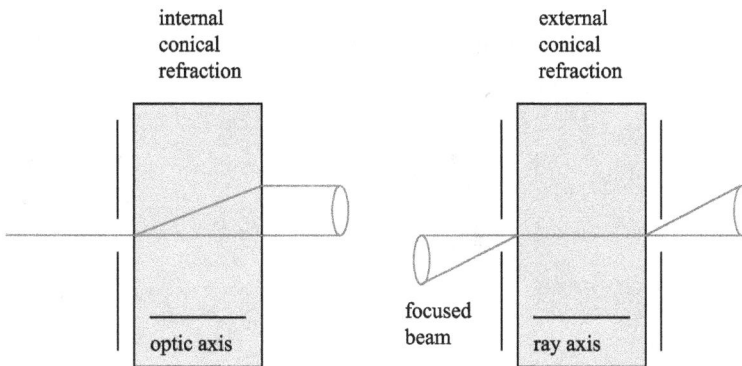

Figure 8.8. Illustrations of (a) internal and (b) external conical refraction.

axis of the crystal. Along this axis, all transverse **E**-field states are allowed; upon leaving the crystal, each of these converts into a distinct **D**-direction, forming a cone.

It is much more difficult to get a collection of plane waves of different polarizations to all propagate in the same ray direction than it is to get them to propagate in the same wave direction. In practice, one focuses an unpolarized plane wave onto a narrow aperture adjacent to a crystal cut with a ray axis normal to the surface. The focused wave represents a collection of unpolarized plane waves all propagating at different angles, and some of these couple to the ray axis direction. An aperture on the opposite crystal face is used to filter out all rays that are not aligned with the ray axis. This arrangement is illustrated in figure 8.8(b).

It is also not easy to construct a matrix technique to analyze the exterior conical refraction case. For interior conical refraction, the input field has both **E** and **D** transverse to the surface normal; in exterior conical refraction, the **D**-field impinges upon the surface at an oblique angle, suggesting that we must carefully consider the effects of the interface and both the reflected and refracted light. We will simply be satisfied with the qualitative description above and leave a crude matrix calculation of external conical refraction as an exercise.

In a uniaxial crystal, the optic axis coincides with the extraordinary axis of the crystal. The **E**-field remains proportional to the **D**-field, and both remain transverse while propagating along this axis. Under these conditions, no conical refraction occurs.

Conical refraction was first predicted by William Rowan Hamilton in 1832 [15] and observed experimentally one year later by Lloyd [16], using aragonite crystal. This demonstration was a dramatic confirmation of Fresnel's wave theory of light and was one of the major factors that led to its broad acceptance.

Conical refraction is inherently a wave phenomenon and manifests features that are not accounted for in our simple matrix analysis above. The most significant of these is the appearance of two rings of light instead of one; this is an effect connected to the fact that no beam of light is perfectly collimated. These dual rings were first

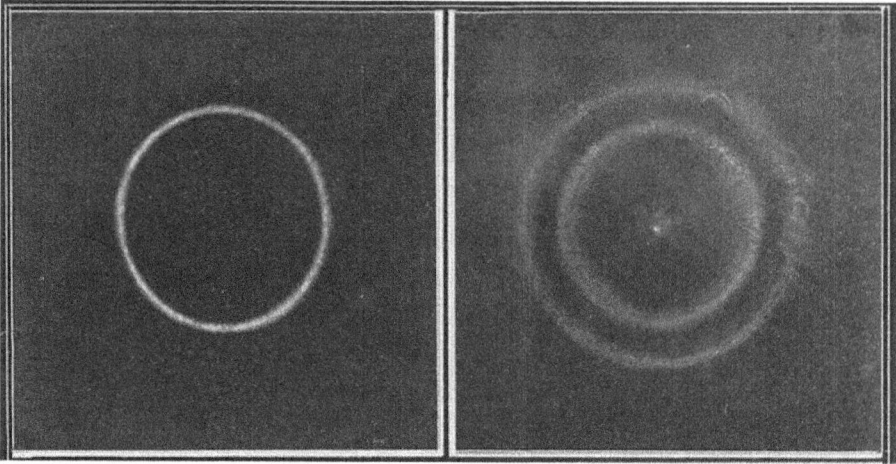

Figure 8.9. Photographs of (a) external and (b) internal conical refraction. Reproduced from [6], with permission from Springer Nature.

observed in 1839 by Poggendorff [17]; an excellent demonstration was performed by Raman, Rajagopalan, and Nedungadi in 1941 [6] using the strongly birefringent material naphthalene; their photographs of external and internal conical refraction are shown in figure 8.9. For a discussion of the finer details of the theory of conical refraction, we refer the reader to the paper by Berry [18].

The beams produced by conical refraction have complex polarization structures, making them, in effect, a natural form of polarization-structured light (discussed in section 4.8). The use of conical refraction to generate structured light was investigated by Turpin *et al* [19]. Recognizing the advantages of using polarization to multiplex data in free-space optical communication, the same researchers looked at using conical refraction as a means of generating robust communication beams [20]. Other applications involving structured light have found uses for conical refraction: in 2012, O'Dwyer *et al* [21] investigated cascaded conical refraction—sending light through multiple biaxial crystals in series—as a means to create tunable optical traps. Conical refraction, once a very obscure phenomenon, is attracting increasing attention for optical applications.

8.10 Exercises

1. Suppose we have an anisotropic crystal with a permittivity tensor at wavelength $\lambda = 500$, nm of the form

$$\epsilon = \epsilon_0 \begin{bmatrix} 1.75 & 0 & 0.25 \\ 0 & 2.00 & 0 \\ 0.25 & 0 & 1.75 \end{bmatrix}.$$

(a) Determine the principal dielectric constants of the medium.
(b) Is this a uniaxial or biaxial crystal?

   (c) Suppose we want to construct a quarter-wave plate from this crystal; i.e. we want light propagating through the crystal at orthogonal polarizations to have a $\pi/2$ phase difference between polarizations when the light leaves the crystal. Describe how you would align the principal axes of the crystal with respect to the propagation direction to achieve this, and calculate the smallest thickness of the crystal that results in a quarter-wave plate.

2. Let us consider an anisotropic material. At a wavelength of 532 nm, its permittivity tensor has the form

$$\epsilon = \epsilon_0 \begin{bmatrix} 1.5 & 0 & 0 \\ 0 & 1.4 & 0 \\ 0 & 0 & 1.5 + 0.5i \end{bmatrix}.$$

   (a) Imagine that you want to construct a quarter-wave plate from this material. Describe how the principal axes should be oriented with respect to the propagation direction and determine the thickness of the thinnest plate possible.

   (b) Imagine that you want to construct a polarizer from this material. Again, describe how the principal axes should be oriented with respect to the propagation direction, and determine the thickness of the polarizer needed so that the intensity of one polarization is attenuated by e^{-4}.

3. Let us imagine that we have light propagating at an angle of $\theta = 30°$ with respect to the z-axis in a uniaxial crystal with *normalized* principal dielectric constants $\tilde{\epsilon}_x = \tilde{\epsilon}_y = 1.80$ and $\tilde{\epsilon}_z = 1.30$. Using one form of the Fresnel equation, calculate the two phase velocities of light propagating in this direction, and specify which is the 'ordinary' wave (velocity independent of direction) and which is the 'extraordinary' wave (velocity dependent on direction). Determine the effective refractive indices of the two waves. (Give the index values to three significant figures.)

4. Show that $(\mathbf{E}_0, \mathbf{H}_0, \mathbf{t})$ form a right-handed triplet of vectors in a crystal, where the cross product of any two in cyclic order is proportional to the third.

5. A quarter-wave plate made out of quartz is 1.212 mm thick and designed to operate at a wavelength of 1064 nm. Assuming the normal to the surface is oriented along an ordinary principal axis, what is the order of the wave plate? (Use the refractive indices of quartz given in the chapter.)

6. Let us construct a zero-order quarter-wave plate out of two plates of quartz, designed to operate at 589 nm. If we desire the first wave plate to have a retardance of $13\lambda/4$, how thick should the plate be? How thick should the second plate be if the result is to be a zero-order quarter-wave plate?

7. Let us attempt a crude illustration of exterior conical refraction. We assume that we have a biaxial crystal rotated so that the z-axis is the ray axis, as we

did for interior conical refraction; determine what condition the permittivity must satisfy for this to be the case. Assume that we now have a collection of plane waves propagating along the ray axis with

$$\mathbf{E} = E_0[\cos\phi\hat{\mathbf{x}} + \sin\phi\hat{\mathbf{y}}],$$
$$\mathbf{H} = \beta E_0[-\sin\phi\hat{\mathbf{x}} + \cos\phi\hat{\mathbf{y}}],$$

and $0 \leqslant \phi < 2\pi$. Determine the range of **D**-vectors within the crystal. Finally, assume that the **D**-field is unchanged upon exiting the crystal and calculate the Poynting vector of the transmitted rays.

References

[1] Born M and Wolf E 1999 *Principles of Optics* 7th edn (Cambridge: Cambridge University Press)

[2] Bragg W L 1924 The refractive indices of calcite and aragonite *Proc. R. Soc.* A **105** 370–86

[3] Bieniewski T M and Czyzak S J 1963 Refractive indexes of single hexagonal ZnS and CdS crystals *J. Opt. Soc. Am.* **53** 496–7

[4] DeVore J R 1951 Refractive indices of rutile and sphalerite *J. Opt. Soc. Am.* **41** 416–9

[5] Record F and Jones W D 1932 Determination of the refractive indices of a material such as muscovite mica *J. Sci. Instrum.* **9** 24

[6] Raman C V, Rajagopalan V S and Nedungadi T M K 1941 Conical refraction in napthalene crystals *Proc. Indian Acad. Sci. (Math. Sci.)* **14** 221–7

[7] Fleischer M, Wilcox R E and Matzko J J 1984 *Microscopic Determination of the Non-Opaque Materials* 3rd edn (Washington, DC: U.S. Government Printing Office)

[8] Gifford J W 1984 The refractive indices of fluorite, quartz, and calcite *Proc. R. Soc.* **70** 329–40

[9] Faraday M 1846 Experimental researches in electricity: nineteenth series *Phil. Trans. R. Soc.* **136** 1–20

[10] Washburn W D (ed) 1929 *International Critical tables of Numerical Data* **vol 6** 1st edn (New York: McGraw-Hill)

[11] Weber M J (ed) 1985 *CRC Handbook of Laser Science and Technology* **vol 4** (Boca Raton, FL: CRC Press)

[12] Sommerfeld A 1964 *Optics* (New York and London: Academic)

[13] Weast R C 1974 CRC Handbook of Chemistry and Physics: 1974-1975 *CRC Handbook of Chemistry and Physics: A Ready-Reference Book of Chemical and Physical Data* 55th edn (Boca Raton, FL: CRC Press)

[14] Craven B D 1969 Complex symmetric matrices *J. Aust. Math. Soc.* **10** 341–54

[15] Hamilton W R 1831 Third supplement to an essay on the theory of systems of rays *Trans. R. Irish Acad.* **17** pp v–x 1–144

[16] Lloyd H 1833 XXI. On the phaenomena presented by light in its passage along the axes of biaxal crystals *London, Edinburgh Dublin Phil. Mag. J. Sci.* **2** 112–20

[17] Poggendorff 1839 Ueber die konische Refraction *Ann. Phys., Lpz.* **124** 461–2

[18] Berry M V 2004 Conical diffraction asymptotics: fine structure of Poggendorff rings and axial spike *J. Opt. A: Pure Appl. Opt.* **6** 289

[19] Turpin A, Loiko Y V, Peinado A, Lizana A, Kalkandjiev T K, Campos J and Mompart J 2015 Polarization tailored novel vector beams based on conical refraction *Opt. Express* **23** 5704–15

[20] Turpin A, Loiko Y, Kalkandjiev T K and Mompart J 2012 Free-space optical polarization demultiplexing and multiplexing by means of conical refraction *Opt. Lett.* **37** 4197–9

[21] O'Dwyer D P, Ballantine K E, Phelan C F, Lunney J G and Donegan J F 2012 Optical trapping using cascade conical refraction of light *Opt. Express* **20** 21119–25

IOP Publishing

Electromagnetic Optics

Gregory J Gbur

Chapter 9

Interface effects

So far, we have been concerned with light propagating in unbounded space, but, of course, much of optics is concerned with the behavior of light at the interface between two media. The simplest case to consider is the propagation of light as it goes from one dielectric medium to another across a planar interface, a situation where we expect the laws of reflection and refraction to apply. Part of our goal will be to show that these laws follow directly from Maxwell's equations. In this chapter, we consider the propagation of light across a single planar interface and discuss the phenomena associated with it. We further discuss the propagation of light across an interface into anisotropic media, magnetic media, and absorptive media.

To begin, however, we consider a fundamental experiment in the history of optics, which shows that even the simplest interface effects can lead to nontrivial and important insights.

9.1 Wiener's experiment

In his foundational work in 1864, James Clerk Maxwell showed theoretically that electromagnetic waves should exist and speculated that light itself is an electromagnetic wave. Though his results were elegant and intuitive, explicit experimental confirmation took some time to follow. The first step was taken by Heinrich Hertz, who performed experiments over the period from 1886 to 1889 to show that electromagnetic waves exist. He used what we would now call a dipole antenna to produce radio waves with a wavelength of approximately 4 m. He used a metal plate to reflect these waves, producing standing waves. By measuring the positions of the nodes and antinodes of these waves with a spark gap receiver, he could determine their wavelength. Since the frequency was known, he could then confirm that the speed of the electromagnetic waves was equal to the vacuum speed of light, using $f = c/\lambda$.

Hertz's work demonstrated that electromagnetic waves exist, but this did not specifically prove that light is an electromagnetic wave. In 1890, the physicist Otto

doi:10.1088/978-0-7503-6064-7ch9

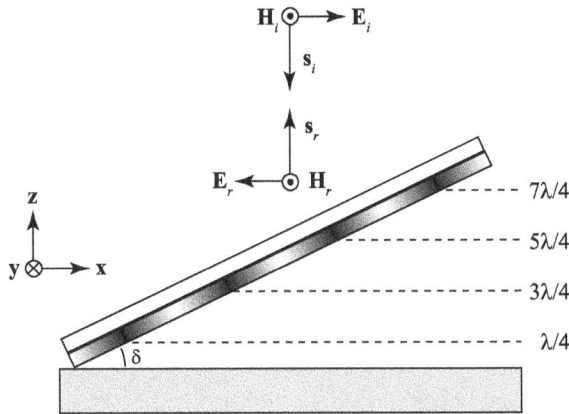

Figure 9.1. Illustration of Wiener's 1890 experiment.

Wiener performed an experiment similar to Hertz's to specifically show that light possesses the same properties as other electromagnetic waves [1].

Wiener's experiment is illustrated in figure 9.1. A plane wave is incident from above and reflects off a metal mirror, with the incident and reflected waves forming standing waves. Wiener's experiment had an additional difficulty compared to Hertz's because the wavelength of light is on the order of a billionth of a meter, making the standing waves much too small to observe directly. In an ingenious move, Wiener deposited an extremely thin photographic film on the surface of a glass plate and oriented the plate at a small angle to the mirrored surface. This effectively 'stretched out' the interference pattern over the surface of the plate, making it large enough to measure.

We take the incident field to be of the form

$$\mathbf{E}_i(z) = \hat{\mathbf{x}} E_0 e^{-ikz}, \tag{9.1}$$

$$\mathbf{H}_i(z) = -\hat{\mathbf{y}} H_0 e^{-ikz}, \tag{9.2}$$

where $k = \omega/c$ is the wavenumber, as usual.

Let us assume that the mirror is a perfect conductor, which means that the free electrons within it move rapidly to cancel out any electric fields parallel to the surface, or

$$\mathbf{E}_{tot}(z = 0) = 0, \tag{9.3}$$

where $\mathbf{E}_{tot} = \mathbf{E}_r + \mathbf{E}_i$. This implies that the reflected field $\mathbf{E}_r$ must satisfy

$$\mathbf{E}_r(z = 0) = -\mathbf{E}_i(z = 0). \tag{9.4}$$

But since our reflected wave is propagating in the positive z-direction, and our wave satisfies the right-hand rule, we must have

$$\mathbf{E}_r(z) = -\hat{\mathbf{x}} E_0 e^{ikz}, \tag{9.5}$$

$$\mathbf{H}_r(z) = -\hat{\mathbf{y}}H_0 e^{ikz}. \tag{9.6}$$

The total fields in the vicinity of the mirror must have the form

$$\mathbf{E}_{tot}(z) = 2iE_0\hat{\mathbf{x}}\sin(kz) = 2iE_0\hat{\mathbf{x}}\sin(2\pi z/\lambda), \tag{9.7}$$

$$\mathbf{H}_{tot}(z) = -2H_0\hat{\mathbf{y}}\cos(kz) = -2H_0\hat{\mathbf{y}}\cos(2\pi z/\lambda), \tag{9.8}$$

where we have used $k = 2\pi/\lambda$.

The striking thing about this result is that the electric and magnetic fields have maxima and minima in different locations. The electric field has antinodes at the positions

$$z_{max}^E = (2m + 1)\lambda/4, \quad m = 1, 2, 3, \dots, \tag{9.9}$$

while the magnetic field has antinodes at positions

$$z_{max}^H = m\lambda/2, \quad m = 1, 2, 3, \dots. \tag{9.10}$$

This led to an interesting and important conclusion by Wiener. He found that the photographic film he used was exposed at the locations where the antinodes of the electric field were present. This suggested to him that the electric field is the 'active ingredient' in light–matter interactions, not the magnetic field. Or, as he described it,

> In the nodes of the electric forces a minimum takes place, in the antinodes of the same a maximum of the chemical effect; or: the chemical effect of the light wave is attached to the presence of the oscillations of the electric and not the magnetic forces.

Wiener's observation explains in part why we have chosen to use the electric field as our descriptor of the polarization of light: because, in most light–matter interactions, it is the electric field that is dominant. (This is not always the case, as we will note when we discuss metamaterials.)

Wiener's experiment also shows that the interaction of electromagnetic waves at an interface can produce interesting and nontrivial phenomena, even in the simplest cases imaginable. We now turn to a general discussion of light at an interface.

9.2 Boundary conditions

As a first step in investigating interface effects, we must consider the behavior of electromagnetic fields crossing a boundary between media. We return to Maxwell's equations in their general form,

$$\nabla \cdot \mathbf{D}(\mathbf{r}, t) = \rho_f(\mathbf{r}, t), \tag{9.11}$$

$$\nabla \cdot \mathbf{B}(\mathbf{r}, t) = 0, \tag{9.12}$$

$$\nabla \times \mathbf{E}(\mathbf{r}, t) = -\frac{\partial \mathbf{B}(\mathbf{r}, t)}{\partial t}, \tag{9.13}$$

$$\nabla \times \mathbf{H}(\mathbf{r}, t) = \mathbf{J}_f(\mathbf{r}, t) + \frac{\partial \mathbf{D}(\mathbf{r}, t)}{\partial t}. \tag{9.14}$$

We will consider isotropic materials for the moment, so ϵ is a scalar.

Let us begin by examining Gauss's law in its integral form, which may be written using Gauss's theorem as

$$\oint \mathbf{D} \cdot d\mathbf{a} = Q_{f,en}, \tag{9.15}$$

where $Q_{f,en}$ is again the free charge enclosed by the surface. We consider a smooth interface without sharp edges or corners; if the interface includes sharp edges, the physics is very different [2]. Locally, any smooth interface looks like a flat surface if we zoom in far enough, just as the Earth looks flat locally to us even though it is curved.

We then consider the integration surface illustrated in figure 9.2, often referred to as a 'Gaussian pillbox.' The pillbox intersects the interface between two media with permittivities ϵ_1 and ϵ_2. This interface possesses a free surface charge density σ. The normal to the surface points from medium 2 to medium 1 and is labeled $\mathbf{n}$. The pillbox itself extends to a height h away from the interface in both directions and has a circular endcap of area A.

If we assume that the pillbox is sufficiently small, then the **D**-field may be assumed to be approximately constant over the surface. Furthermore, we will take the limit where $h \to 0$; then, assuming that the field is finite on the pillbox, we expect the contribution of the tube of the cylinder to vanish. We then have only the integral over the two endcaps of the cylinder, and the integral form of Gauss's law becomes

$$A[\mathbf{D}_1 \cdot \mathbf{n} - \mathbf{D}_2 \cdot \mathbf{n}] = \sigma A, \tag{9.16}$$

where $\mathbf{D}_1$ and $\mathbf{D}_2$ are the fields immediately on either side of the interface. The right-hand side of this equation represents the free charge enclosed, which is the surface charge density times A. Dividing out by A, we obtain

$$[\mathbf{D}_1 - \mathbf{D}_2] \cdot \mathbf{n} = \sigma. \tag{9.17}$$

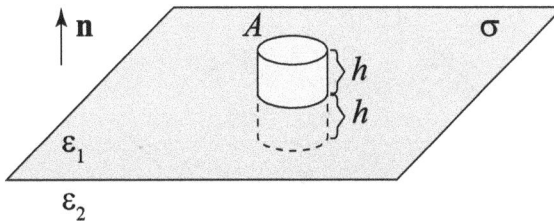

Figure 9.2. The surface used to determine boundary conditions in Gauss's law.

We have therefore found that the normal component of the **D**-field is discontinuous across the surface by an amount equal to the surface charge density σ.

In dielectrics, we do not have a free surface charge density that is oscillating at the same frequency as the field, so $\sigma = 0$. In this case, although the normal component of the **D**-field is continuous, we can readily see that the normal component of the **E**-field is generally discontinuous, since $\mathbf{D}_1 = \epsilon_1 \mathbf{E}_1$, etc. so that

$$[\epsilon_1 \mathbf{E}_1 - \epsilon_2 \mathbf{E}_2] \cdot \mathbf{n} = 0. \tag{9.18}$$

Since $\epsilon_1 \neq \epsilon_2$, this implies that the normal component of **E** is, in general, discontinuous across the interface.

We may do an analogous calculation for the 'no magnetic monopoles' law and find

$$[\mathbf{B}_1 - \mathbf{B}_2] \cdot \mathbf{n} = 0. \tag{9.19}$$

This indicates that the normal component of the **B**-field is always continuous; however, this implies that the normal component of **H** is generally discontinuous, since $\mathbf{B} = \mu \mathbf{H}$:

$$[\mu_1 \mathbf{H}_1 - \mu_2 \mathbf{H}_2] \cdot \mathbf{n} = 0. \tag{9.20}$$

Since we usually work with nonmagnetic materials, where $\mu_1 = \mu_2 = \mu_0$, we can typically say that the normal component of **H** is continuous.

We now turn to Faraday's law and the boundary condition associated with it. The integral form of Faraday's law is

$$\oint_C \mathbf{E} \cdot d\mathbf{l} = -\frac{\partial}{\partial t} \int_S \mathbf{B} \cdot d\mathbf{a}, \tag{9.21}$$

where the closed path C and open surface S are illustrated in figure 9.3, analogous to the Ampère loop used in the integral form of Ampère's law. The loop is of length l parallel to the surface and extends to a height δ above and below the surface. The unit vector **t** indicates a direction parallel to the in-plane part of the loop, and **n** is again normal to the surface.

We assume that both l and δ are small, so that the field is effectively constant throughout the area of the loop, and we look at the limit $\delta \to 0$. In this limit, the area of the loop vanishes, and we expect that

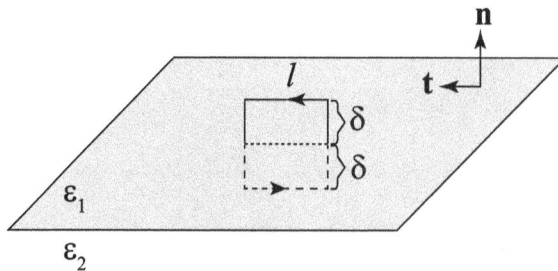

Figure 9.3. The surface used to determine boundary conditions in Faraday's law.

$$\int_S \mathbf{B} \cdot d\mathbf{a} = 0. \tag{9.22}$$

The integral over the vertical parts of the path is negligible, and we take the field to be constant over the horizontal parts of the path. This leads to

$$\oint \mathbf{E} \cdot d\mathbf{l} = [\mathbf{E}_1 \cdot \mathbf{t} - \mathbf{E}_2 \cdot \mathbf{t}] = 0. \tag{9.23}$$

This indicates that the tangential component of the electric field is always continuous across the interface.

It is somewhat unsatisfying to have the boundary condition written in terms of an arbitrary unit vector $\mathbf{t}$, which is unspecified—it can refer to any direction in the plane of the interface. We may rewrite our result by noting first that the field may be written as $\mathbf{E} = E_n \mathbf{n} + E_t \mathbf{t}$, where E_n and E_t are the normal and tangential components of the field. If we take the cross product of this field with $\mathbf{n}$, we have

$$\mathbf{n} \times \mathbf{E} = (\mathbf{n} \times \mathbf{t})E_t. \tag{9.24}$$

Since $\mathbf{n} \times \mathbf{t} \neq 0$, we have an expression for the tangential component of the field that does not depend on a specific choice of $\mathbf{t}$. We may rewrite equation (9.23) as

$$\mathbf{n} \times (\mathbf{E}_1 - \mathbf{E}_2) = 0. \tag{9.25}$$

Finally, we turn to the Ampère–Maxwell law. We again use a loop across an interface, as illustrated in figure 9.4. We have now introduced two explicit tangential unit vectors $\mathbf{t}_1$ and $\mathbf{t}_2$, with the relation $\mathbf{t}_1 = \mathbf{t}_2 \times \mathbf{n}$.

The Ampère–Maxwell law in integral form becomes

$$\oint_C \mathbf{H} \cdot d\mathbf{l} = \int_S \mathbf{J}_f \cdot d\mathbf{a} + \frac{\partial}{\partial t} \int_S \mathbf{D} \cdot d\mathbf{a}. \tag{9.26}$$

We will now follow the same process that we used for Faraday's law. As we let $\delta \to 0$, we find that the integral over $\mathbf{D}$ vanishes, again assuming the fields are nonsingular. The integral over $\mathbf{J}_f$ only includes any currents right at the surface, i.e. a free surface current $\mathbf{K}_f$; we have

$$\int \mathbf{J}_f \cdot d\mathbf{a} = l\mathbf{K}_f \cdot \mathbf{t}_2. \tag{9.27}$$

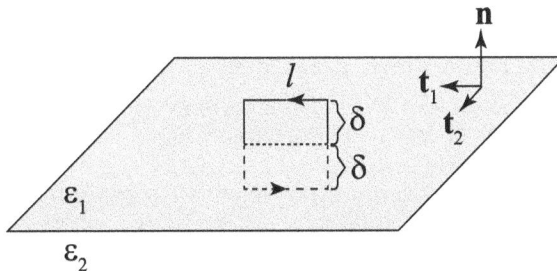

Figure 9.4. The surface used to determine boundary conditions in the Ampère–Maxwell law.

Assuming l is sufficiently small, the **H**-field is constant over the path, and we have

$$\oint_C \mathbf{H} \cdot d\mathbf{l} = [\mathbf{H}_1 - \mathbf{H}_2] \cdot \mathbf{t}_1 l. \tag{9.28}$$

Putting all these pieces into the integral form of the Ampère–Maxwell law and dividing out an arbitrary l, we get

$$[\mathbf{H}_1 - \mathbf{H}_2] \cdot (\mathbf{t}_2 \times \mathbf{n}) = \mathbf{K}_f \cdot \mathbf{t}_2. \tag{9.29}$$

It should be noted that the left-hand side is of the form of a scalar triple product. We may permute the terms to get

$$\mathbf{t}_2 \cdot [\mathbf{n} \times (\mathbf{H}_1 - \mathbf{H}_2)] = \mathbf{K}_f \cdot \mathbf{t}_2. \tag{9.30}$$

This expression must be true regardless of the choice of $\mathbf{t}_2$, so we have

$$\mathbf{n} \times (\mathbf{H}_1 - \mathbf{H}_2) = \mathbf{K}_f. \tag{9.31}$$

In other words, the tangential component of **H** is discontinuous in the presence of a free surface current.

We therefore have a large number of boundary conditions, which we summarize below:

$$[\mathbf{D}_1 - \mathbf{D}_2] \cdot \mathbf{n} = \sigma_f, \tag{9.32}$$

$$[\epsilon_1 \mathbf{E}_1 - \epsilon_2 \mathbf{E}_2] \cdot \mathbf{n} = \sigma_f, \tag{9.33}$$

$$[\mathbf{B}_1 - \mathbf{B}_2] \cdot \mathbf{n} = 0, \tag{9.34}$$

$$[\mu_1 \mathbf{H}_1 - \mu_2 \mathbf{H}_2] \cdot \mathbf{n} = 0, \tag{9.35}$$

$$\mathbf{n} \times (\mathbf{E}_1 - \mathbf{E}_2) = 0, \tag{9.36}$$

$$\mathbf{n} \times (\mathbf{H}_1 - \mathbf{H}_2) = \mathbf{K}_f. \tag{9.37}$$

We may make a number of observations related to these boundary conditions. First, it should be noted that, in the absence of sources, all fields are continuous except the normal components of **E** and **H**. For nonmagnetic materials, the normal component of **H** is continuous as well. The **H**-field is often used in place of the **B**-field because it is completely continuous for nonmagnetic materials, and it is generally easier to work with continuous boundary conditions.

These formulas are linear, which means that they not only hold in general but also frequency by frequency. For optical fields, the source terms are usually absent because any induced free charges and currents can be incorporated into a complex permittivity, as discussed in section 5.4.

Again, we note that these conditions, though we derived them for a flat surface, should apply to any locally smooth surface. In looking at electromagnetic fields interacting with a sphere, for example, we can use the same conditions. We cannot use these conditions when studying the fields near a sharp edge, however, like the fields near the tip of a pointed metal lightning rod. (The sharp tip is what makes lightning rods so interesting.) See, for instance, Van Bladel [2, 3].

These boundary conditions will form the foundation of our investigations of light-wave interactions at a variety of interfaces.

9.3 Reflection and refraction at an interface

The foundations of geometrical optics are the law of reflection,

$$\theta_i = \theta_r, \tag{9.38}$$

where θ_i and θ_r are the angles of the incident and reflected rays with respect to the normal to the surface, and the law of refraction,

$$n_1 \sin \theta_i = n_2 \sin \theta_t, \tag{9.39}$$

where θ_t is the angle of the transmitted wave in the second medium. These laws were empirically determined in ancient times, but they can be explicitly derived from Maxwell's equations with plane waves, as we now demonstrate. Maxwell's equations can further determine the *amount* of light reflected and refracted, something that is not covered by the laws of reflection and refraction.

We consider the situation shown in figure 9.5. We assume a plane wave of wave vector $\mathbf{k}_i$ is traveling from a medium of index n_1 into a medium of index n_2, and that the wave vector lies in the yz-plane. For the moment, we assume a real-valued refractive index, i.e. no absorption, and we discuss the complex case in section 9.6.

At the interface, part of the wave is reflected with wave vector $\mathbf{k}_r$ and part is transmitted with wave vector $\mathbf{k}_t$. We assume that all three wave vectors lie in the same yz-plane; on a smooth interface with constant refractive indices, there is nothing to break the symmetry of the system and allow out-of-plane effects. It is worth noting, however, that it is possible to create metasurfaces with a structure that causes out-of-plane reflection and refraction [4, 5].

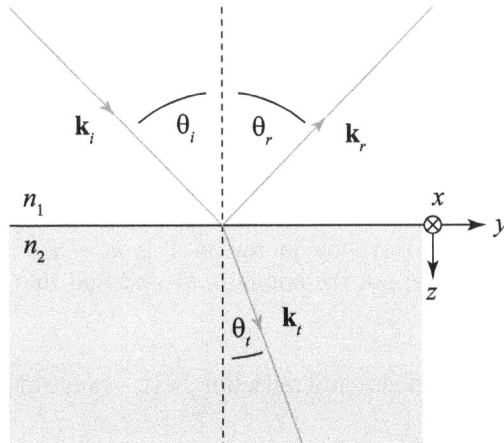

Figure 9.5. The basic geometry used to derive the Fresnel formulas and the laws of reflection and refraction.

We assume that we have monochromatic plane waves for all three contributions, i.e.

$$E_i(\mathbf{r}) = E_{0i} e^{i\mathbf{k}_i \cdot \mathbf{r}}, \tag{9.40}$$

$$E_r(\mathbf{r}) = E_{0r} e^{i\mathbf{k}_r \cdot \mathbf{r}}, \tag{9.41}$$

$$E_t(\mathbf{r}) = E_{0t} e^{i\mathbf{k}_t \cdot \mathbf{r}}, \tag{9.42}$$

where all three waves have the same time dependence $\exp[-i\omega t]$. The three wavenumbers can be written explicitly as

$$\mathbf{k}_i = k_i(\hat{\mathbf{y}} \sin \theta_i + \hat{\mathbf{z}} \cos \theta_i), \tag{9.43}$$

$$\mathbf{k}_r = k_r(\hat{\mathbf{y}} \sin \theta_r - \hat{\mathbf{z}} \cos \theta_r), \tag{9.44}$$

$$\mathbf{k}_t = k_t(\hat{\mathbf{y}} \sin \theta_t + \hat{\mathbf{z}} \cos \theta_t). \tag{9.45}$$

It should be noted that we have made no assumptions about the values of θ_r and θ_t other than the aforementioned assumption that all wave vectors lie within the same plane.

We now ask what our boundary conditions tell us about the behavior of the fields at the interface. Because the tangential component of the fields must be continuous, we have

$$(\mathbf{E}_i + \mathbf{E}_r) \cdot \mathbf{t}|_{z=0} = \mathbf{E}_t \cdot \mathbf{t}|_{z=0}. \tag{9.46}$$

In terms of the explicit form of the plane waves, we have

$$\mathbf{E}_{0i} \cdot \mathbf{t} e^{i\mathbf{k}_i \cdot \mathbf{r}}|_{z=0} + \mathbf{E}_{0r} \cdot \mathbf{t} e^{i\mathbf{k}_r \cdot \mathbf{r}}|_{z=0} = \mathbf{E}_{0t} \cdot \mathbf{t} e^{i\mathbf{k}_t \cdot \mathbf{r}}|_{z=0}. \tag{9.47}$$

This equation has to be true for every point within the boundary. The fact that the vectors $\mathbf{E}_{0i}$, etc. are constant implies that the equation can only be true if the exponentials are all the same, i.e.

$$\mathbf{k}_i \cdot \mathbf{r}|_{z=0} = \mathbf{k}_r \cdot \mathbf{r}|_{z=0} = \mathbf{k}_t \cdot \mathbf{r}|_{z=0}. \tag{9.48}$$

Writing this explicitly in terms of angles, we have

$$k_i \sin \theta_i = k_r \sin \theta_r = k_t \sin \theta_t, \tag{9.49}$$

where we have divided out a common y dependence.

Let us consider the first two terms of the above equation. The magnitude of the wavenumber of any wave traveling in region 1 is $k_1 = n_1 \omega / c$, so we have that $k_i = k_r = k_1$. We may cancel out the common k_1 and find that

$$\sin \theta_i = \sin \theta_r. \tag{9.50}$$

Since both the angles of incidence and reflection have a range of zero to $\pi/2$, we may write

$$\theta_i = \theta_r, \tag{9.51}$$

which is the *law of reflection.*

Considering the latter two terms of our equation above, we have

$$k_i \sin \theta_i = k_t \sin \theta_t. \qquad (9.52)$$

But we may also write $k_i = n_1 \omega / c$, $k_t = n_2 \omega / c$, and the above expression simplifies to

$$n_1 \sin \theta_i = n_2 \sin \theta_t, \qquad (9.53)$$

which is the *law of refraction.* We have therefore found that both the law of reflection and the law of refraction can be derived from the boundary conditions of electromagnetic waves.

What is the physical interpretation of such a result? Equation (9.49) says that the component of the **k**-vector parallel to the surface must be the same for the incident, reflected, and refracted waves. The momentum of a light wave, or a photon, is proportional to this **k**-vector. The laws of reflection and refraction say that the momentum of the wave parallel to the surface must be continuous. This makes sense for a smooth surface, as there is no mechanism by which the momentum can change. We will see that this way of thinking is important when we later discuss surface plasmons.

9.4 Fresnel equations

We now turn to calculating the amount of light that is reflected and refracted at an interface; the formulas that result are known as the *Fresnel equations.* They were first deduced by Augustin-Jean Fresnel based purely on the assumption that light is a transverse wave, even before the recognition of light as an electromagnetic wave.

To find the Fresnel equations, we will apply the boundary conditions derived in previous sections for the electric and magnetic fields. Because the polarization of light can be described using two orthogonal states of linear polarization, we will solve the equations for two distinct polarization cases: one in which the electric field vector points out of the plane of incidence (traditionally called s-polarization), and one in which the electric field vector lies within the plane of incidence (traditionally called p-polarization). These are illustrated in figure 9.6. The s-polarization state is often referred to as transverse electric (TE), as the polarization is transverse to the plane of incidence, and the p-polarization state is called transverse magnetic (TM).

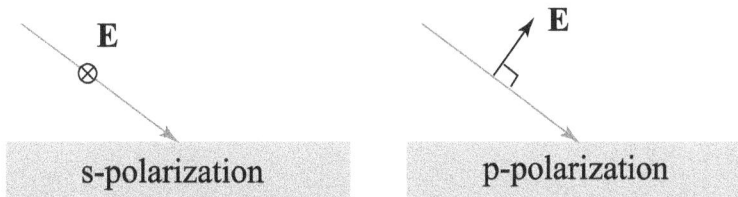

Figure 9.6. Illustration of the difference between s- and p-polarization. The 'X' represents an arrow pointing into the page.

However, these terms are also used in a very different manner when talking about guided waves, as we will see later.

We now consider each case in turn.

9.4.1 The s-polarization case

The s-polarization geometry is illustrated in figure 9.7. The electric fields may be written as

$$\mathbf{E}_i(\mathbf{r}) = \hat{\mathbf{x}} E_{0i} e^{i\mathbf{k}_i \cdot \mathbf{r}}, \tag{9.54}$$

$$\mathbf{E}_r(\mathbf{r}) = \hat{\mathbf{x}} E_{0r} e^{i\mathbf{k}_r \cdot \mathbf{r}}, \tag{9.55}$$

$$\mathbf{E}_t(\mathbf{r}) = \hat{\mathbf{x}} E_{0t} e^{i\mathbf{k}_t \cdot \mathbf{r}}. \tag{9.56}$$

We now begin to impose boundary conditions in order to determine the values of E_{0r} and E_{0t} relative to E_{0i}. The first condition we impose is our tangential condition for the electric field,

$$\mathbf{E}_i(\mathbf{r}) \cdot \hat{\mathbf{x}} + \mathbf{E}_r(\mathbf{r}) \cdot \hat{\mathbf{x}} = \mathbf{E}_t(\mathbf{r}) \cdot \hat{\mathbf{x}}, \tag{9.57}$$

which may be written simply as

$$E_{0i} + E_{0r} = E_{0t}. \tag{9.58}$$

We need one more equation to get a unique solution for the electric fields. For this, we calculate the magnetic fields via Faraday's law, using equation (3.38),

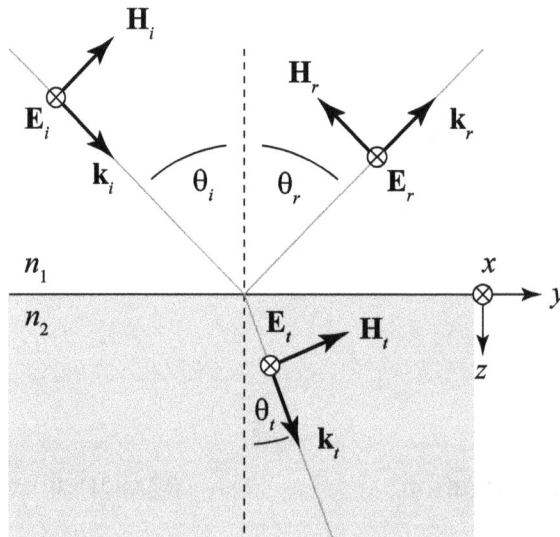

Figure 9.7. The geometry for s-polarization, showing the direction of the fields.

$$\mathbf{H}_0 = \frac{1}{\mu_0 \omega} \mathbf{k} \times \mathbf{E}_0. \tag{9.59}$$

This leads to three equations for the **H**-fields,

$$\mathbf{H}_{0i} = \frac{k_i E_{0i}}{\mu_0 \omega} (\cos \theta_i \hat{\mathbf{y}} - \sin \theta_i \hat{\mathbf{z}}), \tag{9.60}$$

$$\mathbf{H}_{0r} = \frac{k_r E_{0r}}{\mu_0 \omega} (-\cos \theta_r \hat{\mathbf{y}} - \sin \theta_r \hat{\mathbf{z}}), \tag{9.61}$$

$$\mathbf{H}_{0t} = \frac{k_t E_{0t}}{\mu_0 \omega} (\cos \theta_t \hat{\mathbf{y}} - \sin \theta_t \hat{\mathbf{z}}). \tag{9.62}$$

It should be remembered that $\theta_i = \theta_r$ and $k_i = k_r$. We now impose our condition that the tangential component of the **H**-field must be continuous,

$$\hat{\mathbf{z}} \times (\mathbf{H}_1 - \mathbf{H}_2) = 0, \tag{9.63}$$

which from the above set of equations results in

$$k_i \cos \theta_i (E_{0i} - E_{0r}) = k_t \cos \theta_t E_{0t}. \tag{9.64}$$

We finally normalize equations (9.58) and (9.64) to the incident field using the definitions

$$r_\perp \equiv \frac{E_{0r}}{E_{0i}}, \quad t_\perp \equiv \frac{E_{0t}}{E_{0i}}, \tag{9.65}$$

and we also note that $k_i = k_r = n_1 \omega / c$ and $k_t = n_2 \omega / c$. We then have the pair of conditions

$$1 + r_\perp = t_\perp, \tag{9.66}$$

$$n_1 \cos \theta_i [1 - r_\perp] = n_2 \cos \theta_t t_\perp. \tag{9.67}$$

We may readily solve this pair of equations for $r_\perp$ and $t_\perp$ to find

$$r_\perp = \frac{n_1 \cos \theta_i - n_2 \cos \theta_t}{n_1 \cos \theta_i + n_2 \cos \theta_t}, \tag{9.68}$$

$$t_\perp = \frac{2 n_1 \cos \theta_i}{n_1 \cos \theta_i + n_2 \cos \theta_t}. \tag{9.69}$$

These are the Fresnel equations for the case where the electric field is perpendicular to the plane of incidence. They give us the relative amplitudes of the reflected and transmitted fields.

9.4.2 The p-polarization case

We immediately continue on to the p-polarization case. The relevant system geometry is shown in figure 9.8. With this arrangement, we may write the electric fields as

$$\mathbf{E}_{0i} = E_{0i}(-\sin\theta_i\hat{\mathbf{z}} + \cos\theta_i\hat{\mathbf{y}}), \tag{9.70}$$

$$\mathbf{E}_{0r} = E_{0r}(-\sin\theta_i\hat{\mathbf{z}} - \cos\theta_i\hat{\mathbf{y}}), \tag{9.71}$$

$$\mathbf{E}_{0t} = E_{0t}(-\sin\theta_t\hat{\mathbf{z}} + \cos\theta_t\hat{\mathbf{y}}). \tag{9.72}$$

We know that the tangential components of the **E**-field must be continuous; from this, we immediately get one equation,

$$E_{0i}\cos\theta_i - E_{0r}\cos\theta_i = E_{0t}\cos\theta_t. \tag{9.73}$$

For the second equation, we again use Faraday's law. We readily find that the magnetic field vectors are of the simple forms

$$\mathbf{H}_{0i} = -\frac{k_i E_{0i}}{\mu_0\omega}\hat{\mathbf{x}}, \tag{9.74}$$

$$\mathbf{H}_{0r} = -\frac{k_r E_{0r}}{\mu_0\omega}\hat{\mathbf{x}}, \tag{9.75}$$

$$\mathbf{H}_{0t} = -\frac{k_t E_{0t}}{\mu_0\omega}\hat{\mathbf{x}}. \tag{9.76}$$

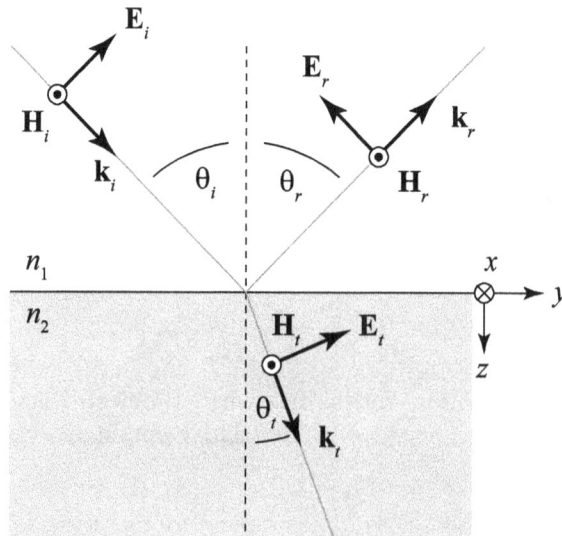

Figure 9.8. The geometry for p-polarization, showing the direction of the fields.

We require the tangential components of the **H**-field to be continuous, which gives us the equation

$$n_1 E_{0i} + n_1 E_{0r} = n_2 E_{0t}. \tag{9.77}$$

Putting together equations (9.73) and (9.77) and defining $r_\parallel$ and $t_\parallel$ in a similar manner as in the s-polarization case, we may write

$$r_\parallel = \frac{n_2 \cos\theta_i - n_1 \cos\theta_t}{n_2 \cos\theta_i + n_1 \cos\theta_t}, \tag{9.78}$$

$$t_\parallel = \frac{2n_1 \cos\theta_i}{n_2 \cos\theta_i + n_1 \cos\theta_t}, \tag{9.79}$$

which are the Fresnel equations for the case where the electric field is parallel to the plane of incidence, i.e. lies within it.

9.4.3 Fresnel equations: observations

The Fresnel equations encapsulate a surprising amount of physics, and we will take some time to discuss them.

First, we note that in both the s- and p-polarization cases, we only used two of the four possible boundary conditions, namely the conditions that the tangential components of **E** and **H** must be continuous. What happens if we evaluate the remaining two conditions for the normal components? We find that one condition gives redundant information and the other gives trivial information (i.e. 'zero equals zero').

We have solved the problem for two orthogonal polarization states that are a natural choice for this system. To determine the reflection and refraction of a light wave with an arbitrary polarization state, one should decompose the incident field into its s- and p-polarized components, find the relative amplitudes of each of the reflected and transmitted components, and then recombine.

In the limit that $\theta_i = 0$, the light wave is normally incident, and then there is no difference between the s- and p-polarization states. However, if we look at the formulas, we find something troubling:

$$r_\perp(\theta_i = 0) = \frac{n_1 - n_2}{n_1 + n_2}, \tag{9.80}$$

$$r_\parallel(\theta_i = 0) = \frac{n_2 - n_1}{n_1 + n_2}. \tag{9.81}$$

The formulas seem to differ by a sign, even though they should be the same! To resolve this, look again at figures 9.7 and 9.8. It should be noted that, *by definition*, we have assumed that the s-polarized reflected electric field points in the same direction as the incident electric field, while the p-polarized reflected electric field points in the opposite direction relative to the incident field. In essence, in our definition of the p-polarization case, we introduced a π phase change upon reflection into the definition. When we take this into account, we see the two results are the same. For normal incidence, it is usually good to use the s-polarization definition, where the phase change is explicitly present in the formula.

A little historical note: I was working with Emil Wolf when the 7th edition of *Principles of Optics* was coming out; Wolf noticed the discrepancy in s- and p-polarization for normal incidence, while noting that other books, such as Hecht's *Optics*, do not have this discrepancy. Wolf gathered together his students to figure out what was going on, and this is how we noticed that the signs depend in this subtle way on the definition of the direction of the reflected **E**-field. Hecht uses the opposite convention and thus does not have the phase change. The convention we are using is sometimes referred to as the *Verdet convention*, while the convention used by Hecht is referred to as the *Fresnel convention*. The lesson here: pay careful attention to the way vector directions are defined in problems like these.

Examples of the reflected amplitudes for both s- and p-polarization are shown in figure 9.9, with light going from air ($n_1 = 1$) to water ($n_2 = 1.33$). We note several features of the results. First, it should be noted that the magnitude of the reflected amplitude approaches unity as $\theta_i = \pi/2$. In essence, a field is perfectly reflected when at a grazing incidence angle, regardless of polarization. Curiously, something similar happens even for surfaces that are not perfectly smooth and have a degree of roughness [6]. As the angle of incidence increases, the component of the wave vector normal to the surface gets smaller, and at $\theta_i = \pi/2$, the effective wavelength normal to the surface is infinite. In this limit, the surface appears locally smooth as far as the light is concerned—provided that the spatial variations of the roughness parallel to

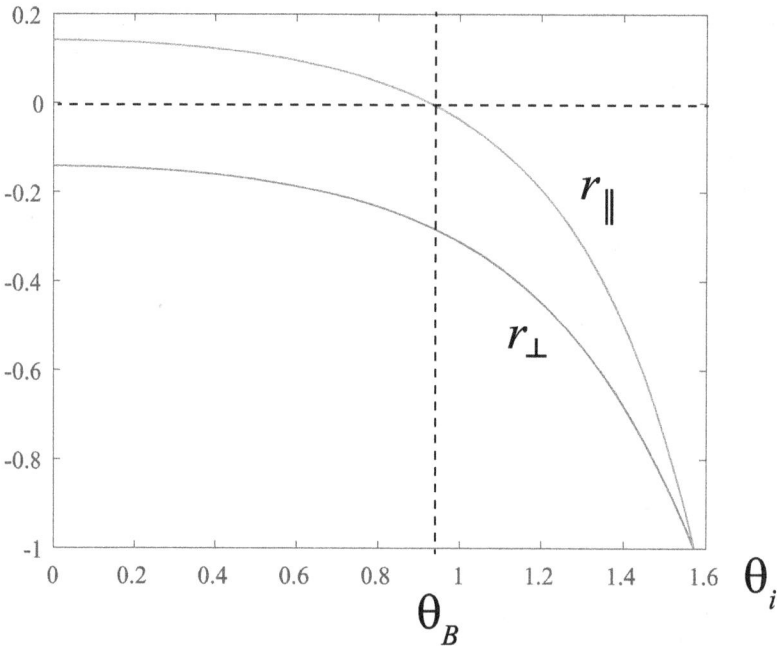

Figure 9.9. The reflected field amplitudes for both s- and p-polarization for light traveling from air $n_1 = 1$ to water $n_2 = 1.33$.

the surface are also smaller than the wavelength. You can verify this high reflectivity result yourself with a sheet of paper. A sheet of paper is hardly reflective, but if you look at the image of a light source such as a lamp at grazing incidence, you will see a faint image of the lamp itself.

The most striking feature of figure 9.9 is that there is an angle, θ_B, at which $r_\parallel = 0$. There is no reflected light of that polarization at that angle, which is known as the *Brewster angle*. Reflection at the Brewster angle became a standard method for polarizing a beam of light before the advent of good polarizing optical elements. When unpolarized light is reflected by a dielectric at the Brewster angle, p-polarized light is perfectly transmitted, leaving only s-polarized light reflected. When Malus discovered the polarization of light by looking through Iceland spar at the reflection of sunlight on palace windows, he was looking at reflection near the Brewster angle.

The Brewster angle depends on the properties of the media at the interface and is given by the formula

$$\theta_B = \tan^{-1}(n_2/n_1). \tag{9.82}$$

When light travels from air ($n = 1$) into water ($n = 1.33$), for example, the Brewster angle is $\theta_B = 0.9261$ in radians and $\theta_B = 53°$ in degrees.

This simple formula is surprisingly tricky to derive, so we spend a few moments doing so. We begin by noting that, at the Brewster angle, $r_\parallel = 0$. Using equation (9.78), this amounts to

$$n_2 \cos \theta_i = n_1 \cos \theta_t. \tag{9.83}$$

We square this equation to get

$$\cos^2 \theta_t = \left(\frac{n_2}{n_1} \cos \theta_i \right)^2. \tag{9.84}$$

We also have Snell's law, of the form

$$n_1 \sin \theta_i = n_2 \sin \theta_t. \tag{9.85}$$

We may square this formula as well, to write

$$\sin^2 \theta_t = \left(\frac{n_1}{n_2} \sin \theta_i \right)^2. \tag{9.86}$$

We add equations (9.84) and (9.86) to get

$$1 = \left(\frac{n_2}{n_1} \right)^2 \cos^2 \theta_i + \left(\frac{n_1}{n_2} \right)^2 \sin^2 \theta_i = \sin^2 \theta_i + \cos^2 \theta_i. \tag{9.87}$$

We now rearrange the rightmost two parts of this expression to get

$$\left[\left(\frac{n_1}{n_2} \right)^2 - 1 \right] \sin^2 \theta_i = - \left[\left(\frac{n_2}{n_1} \right)^2 - 1 \right] \cos^2 \theta_i. \tag{9.88}$$

If we put each side of this expression under a common denominator, we obtain

$$\frac{n_1^2 - n_2^2}{n_2^2} \sin^2 \theta_i = \frac{n_1^2 - n_2^2}{n_1^2} \cos^2 \theta_i. \qquad (9.89)$$

Finally, we cancel terms in the numerator and take the square root; we are left with

$$\frac{\sin \theta_i}{\cos \theta_i} = \tan \theta_i = \frac{n_2}{n_1}, \qquad (9.90)$$

which is effectively our Brewster result.

This is the mathematical derivation of the Brewster angle, but what is the physics? We can elucidate it by deriving another result, beginning by writing our Brewster equation above in the form

$$\sin \theta_i = \frac{n_2}{n_1} \cos \theta_i. \qquad (9.91)$$

We may also return to the condition $r_\parallel = 0$ and write it in the form

$$\frac{n_2}{n_1} \cos \theta_i = \cos \theta_t. \qquad (9.92)$$

Equating these two formulas, we have

$$\sin \theta_i = \cos \theta_t. \qquad (9.93)$$

But $\cos \theta = \sin(\pi/2 - \theta)$, so we may write our Brewster condition as

$$\sin \theta_i = \sin(\pi/2 - \theta_t). \qquad (9.94)$$

Because the angles must be the same, we have $\theta_i = \pi/2 - \theta_t$, or

$$\theta_i + \theta_t = \pi/2. \qquad (9.95)$$

Geometrically, the result is illustrated in figure 9.10. It can be seen that, at the Brewster angle, the electric field vector of the transmitted field is parallel to the direction of the reflected wave.

Physically, we may view reflection and refraction as arising from an incident light wave exciting electric dipoles in the second medium. These dipoles, set into motion, produce not only the transmitted wave but the reflected wave. However, at the Brewster angle, the oscillation of the dipoles, and the electric field generated by them, takes place in the direction of propagation of the reflected wave. But an electromagnetic wave must be transverse; because there is no transverse component to the electric field generated in the second medium, no reflected wave can be produced.

Moving on from the Brewster angle, it is of general interest to study the flow of power across an interface using the Poynting vector. As seen in equation (7.33), the Poynting vector $\mathbf{S}$ of a plane wave may be written as

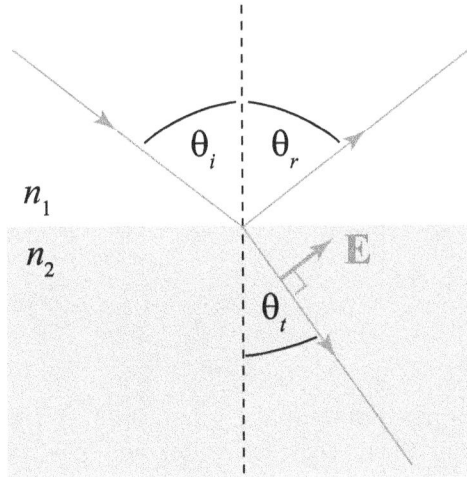

Figure 9.10. Illustration of the fields and wave vectors at the Brewster angle.

$$\mathbf{S} = \frac{n|E_0|^2}{2\mu_0 c}\hat{\mathbf{s}}, \tag{9.96}$$

where $\hat{\mathbf{s}}$ is the unit vector in the direction of wave propagation, and we have incorporated the refractive index n into the calculation. The Poynting vector of our three relevant plane waves may then be written as

$$\mathbf{S}_i = \frac{n_1|E_{0i}|^2}{2\mu_0 c}[\cos\theta_i\hat{\mathbf{z}} + \sin\theta_i\hat{\mathbf{y}}], \tag{9.97}$$

$$\mathbf{S}_r = \frac{n_1|E_{0r}|^2}{2\mu_0 c}[-\cos\theta_i\hat{\mathbf{z}} + \sin\theta_i\hat{\mathbf{y}}], \tag{9.98}$$

$$\mathbf{S}_t = \frac{n_2|E_{0t}|^2}{2\mu_0 c}[\cos\theta_t\hat{\mathbf{z}} + \sin\theta_t\hat{\mathbf{y}}]. \tag{9.99}$$

Let us consider the case when there is no absorption, i.e. the refractive index is real-valued. We expect energy to be conserved in our problem and want to see whether this is reflected in our calculations. We consider the power flux through the surface of the reflected and transmitted waves, defined as

$$R \equiv \frac{|\mathbf{S}_r \cdot \hat{\mathbf{z}}|}{|\mathbf{S}_i \cdot \hat{\mathbf{z}}|}, \tag{9.100}$$

$$T \equiv \frac{|\mathbf{S}_t \cdot \hat{\mathbf{z}}|}{|\mathbf{S}_i \cdot \hat{\mathbf{z}}|}. \tag{9.101}$$

The dot product with $\hat{\mathbf{z}}$ indicates that we are only concerned with the power that flows across the interface; the component of the Poynting vector parallel to the interface does not cross it and is not relevant to energy conservation. The quantities R and T are normalized to the incident field, and if energy conservation holds, we have

$$R + T = 1. \qquad (9.102)$$

Upon direct substitution from our Poynting formulas into the formulas for R and T, we find

$$R = |r|^2, \qquad (9.103)$$

$$T = |t|^2 \frac{n_2 \cos \theta_t}{n_1 \cos \theta_i}. \qquad (9.104)$$

The reader might be surprised to see that T is not simply $|t|^2$. The extra factors in the formula take into account the change of direction and speed of the wave as it passes from medium 1 to medium 2, which changes the amount of power flow in the $\hat{z}$ direction.

By substituting $r_\perp$ and $t_\perp$ or $r_\parallel$ and $t_\parallel$ into the equations for R and T, we can readily show that $R + T = 1$ and that power flow is conserved in reflection and refraction.

9.4.4 Total internal reflection

In the discussions above, we considered examples where light traveled from a rarer medium (lower refractive index) to a denser medium (higher refractive index). The Fresnel formulas apply both ways, however, and we can also consider the physics when a plane wave travels from a dense medium to a rare medium. This introduces a new phenomenon called *total internal reflection* (TIR), which we discuss in some detail.

When $n_1 < n_2$, the Fresnel formulas imply that there is always some transmission of light. But is this true in the case $n_1 > n_2$? We investigate the problem by first considering Snell's law, written in the form

$$\sin \theta_t = \frac{n_1}{n_2} \sin \theta_i. \qquad (9.105)$$

In going from a rare to a dense medium, $n_1/n_2 < 1$, which means that the right-hand side of the formula is always less than unity, and therefore an angle θ_t can be found. But for $n_1/n_2 > 1$, i.e. the rare-to-dense case, there is an angle θ_i beyond which the right-hand side of the formula is greater than one. In such a case, there exists no real-valued θ_t that can satisfy the equation. The *critical angle* θ_c at which this happens is readily seen to be given by

$$\sin \theta_c = \frac{n_2}{n_1}. \qquad (9.106)$$

For the case where $n_1 = 1.33$ (water) and $n_2 = 1.0$ (air), the critical angle is $\theta_c = 48.75°$. Beyond this incident angle, we see that there is no transmission into the rare medium.

We recall that the incident wave vector $\mathbf{k}_i$ may be written as

$$\mathbf{k}_i = k_{iy}\hat{\mathbf{y}} + k_{iz}\hat{\mathbf{z}} = k_i \sin\theta_i \hat{\mathbf{y}} + k_i \cos\theta_i \hat{\mathbf{z}}, \tag{9.107}$$

and that this implies that

$$k_{iy}^2 + k_{iz}^2 = k_i^2. \tag{9.108}$$

A similar expression applies for $\mathbf{k}_t$,

$$k_{ty}^2 + k_{tz}^2 = k_t^2. \tag{9.109}$$

Let us rewrite this equation in terms of k_{tz}. We have

$$k_{tz} = \sqrt{k_t^2 - k_{ty}^2}. \tag{9.110}$$

We note that $k_i = n_1\omega/c$ and $k_t = n_2\omega/c$; furthermore, we recall from our discussion of boundary conditions and refraction that $k_{ty} = k_{iy} = k_i \sin\theta_i$. We may then write

$$k_{tz} = k_i \sqrt{\frac{n_2^2}{n_1^2} - \sin^2\theta_i}. \tag{9.111}$$

The quantity in the square root becomes negative when $\theta_i > \theta_c$, which implies that k_{tz} becomes imaginary beyond the critical angle. For this case, we may write

$$k_{tz} = ik_i \sqrt{\sin^2\theta_i - \frac{n_2^2}{n_1^2}} \equiv i\beta(\theta_i). \tag{9.112}$$

We have introduced the real-valued function $\beta(\theta_i)$ for convenience. The total wave vector in the transmitted region is therefore

$$\mathbf{k}_t = \hat{\mathbf{y}}k_{iy} + \hat{\mathbf{z}}i\beta. \tag{9.113}$$

If we input this into equation (9.56), we find that

$$\mathbf{E}_t(\mathbf{r}) = \mathbf{E}_{0t} e^{ik_{iy}y} e^{-\beta z}. \tag{9.114}$$

The z-dependent exponential shows that the electric field amplitude decays exponentially away from the interface for $z > 0$. The field, therefore, penetrates into the rarer medium, but it is localized to the region near the $z = 0$ interface. A wave of this form is sometimes called an *inhomogeneous plane wave*, as it satisfies our definition of a plane wave but has behavior that is strongly dependent on the z value. It is more commonly referred to as an *evanescent wave*. The term 'evanescent' means 'soon passing out of sight, memory, or existence; quickly fading or disappearing,' referring to the limited range of the wave in the z-direction. We noted in section 3.4 that evanescent waves are natural features of wave problems near planar interfaces.

What happens to the reflected field in the case of total internal reflection? We find that, formally, the expression for $\cos\theta_t$ becomes

$$\cos \theta_t = i\sqrt{\sin^2 \theta_i - \frac{n_2^2}{n_1^2}} = i \sinh \theta_t', \tag{9.115}$$

where we have introduced a new real-valued θ_t' for convenience. In other words, the y-component of the field becomes purely imaginary. We substitute this expression into equation (9.78) and find

$$r_\parallel = \frac{n_2 \cos \theta_i - i n_1 \sinh \theta_t'}{n_2 \cos \theta_i + i n_1 \sinh \theta_t'} \equiv \frac{C}{C^*}, \tag{9.116}$$

where C represents the complex numerator of the expression. We therefore find that $|r_\parallel| = 1$, and therefore $R_\parallel = 1$; all of the power of the incident field is reflected, and none is transmitted. The field is totally reflected in the interior of the denser medium.

We may show the same thing for $r_\perp$,

$$r_\perp = \frac{n_1 \cos \theta_i - i n_2 \sinh \theta_t'}{n_1 \cos \theta_i + i n_2 \sinh \theta_t'} \equiv \frac{D}{D^*}, \tag{9.117}$$

where D represents the complex numerator in this case. Again, we have $|r_\perp| = 1$ and $R_\perp = 1$.

It is natural to assume that the transmitted amplitudes must be zero, but this is, in fact, not the case; we have $|t_\parallel| > 0$ and $|t_\perp| > 0$ because the field is nonzero in the evanescent region. To understand how all the power can be reflected yet the field can be nonzero, we return and do the Poynting vector calculation again, this time for evanescent waves. We stick to just the TE case for convenience and write

$$\mathbf{E}_t = \hat{\mathbf{x}} E_{0t} e^{ik n_1 \sin \theta_i y} e^{-k n_1 \sinh \theta_t' z}, \tag{9.118}$$

where we have taken advantage of the fact that the y-contribution to the plane wave has the same form as the incident wave. Using Faraday's law, we may readily find that the magnetic field is

$$\mathbf{H}_t = E_{0t}\left[\frac{i\hat{\mathbf{y}} n_1}{\mu_0 c} \sinh \theta_t' - \frac{\hat{\mathbf{z}} n_1}{\mu_0 c} \sin \theta_i \right] e^{ik n_1 \sin \theta_i y} e^{-k n_1 \sinh \theta_t' z}. \tag{9.119}$$

When we calculate the Poynting vector in the usual manner, we have

$$\mathbf{S}_t = \frac{|E_{0t}|^2}{2\mu_0 c}\mathrm{Re}\left\{\hat{\mathbf{y}} n_1 \sin \theta_i - i\hat{\mathbf{z}} n_1 \sinh \theta_t'\right\} e^{-2k n_1 \sinh \theta_t' z}. \tag{9.120}$$

In fact, the Poynting vector has a complex form, but the contribution in the z-direction is purely imaginary, so we may finally write the physical (real-valued) Poynting vector as

$$\mathbf{S}_t = \frac{\hat{\mathbf{y}} n_1 |E_{0t}|^2}{2\mu_0 c} \sin \theta_i e^{-2k n_1 \sinh \theta_t' z}. \tag{9.121}$$

We find that there is power flow in the rare medium, but it is entirely parallel to the surface! There is no power flow in the z-direction through the surface, so

$$T = \frac{\mathbf{S}_t \cdot \hat{\mathbf{z}}}{\mathbf{S}_i \cdot \hat{\mathbf{z}}} = 0. \tag{9.122}$$

How far does the evanescent wave extend into the rarer medium? We note that the exponential has a $1/e$ decay length δ of

$$\delta = \frac{1}{2kn_1 \sinh \theta_t'} = \frac{\lambda}{4\pi n_1 \sinh \theta_t'}. \tag{9.123}$$

The evanescent wave may be said to be significant over a distance of only several wavelengths into the rarer medium.

A physically minded reader might wonder how there can be any fields at all in the rare medium if there is no power flow through the boundary, since the existence of fields implies that there is a nonzero energy density in the rare medium. It should be recalled that we are working with monochromatic fields, i.e. a steady-state system, where it is assumed that the fields have been on for a long time. Presumably, when we first switch on our light source and the light first reflects from the interface, there is a temporary flow of energy into the forbidden region, which then becomes a net flow of zero as a steady state is achieved and an equal amount of energy flows back into the dense medium.

The angle θ_t' may be considered a formal extension of the angle of refraction into the complex plane. As θ_i approaches θ_c, θ_t approaches $\pi/2$. At that point, we may imagine that the angle of refraction takes a 'right turn' into the complex plane, so that $\theta_t = \pi/2 - i\theta_t'$. Using trigonometric identities, we can determine $\cos(\pi/2 - i\theta_t') = i\sinh(\theta_t')$, where θ_t' has values ranging from zero to ∞ in principle. For our purposes, there is no particular reason to introduce θ_t' other than to make formulas such as equation (9.116) appear more symmetrical.

The existence of total internal reflection appears to have been recognized for many centuries, but one of its most dramatic illustrations was made by the Swiss physicist Jean-Daniel Colladon in 1842 [7]. He demonstrated that it is possible to guide light within a falling stream of water and make the basin glow at the bottom. An illustration of his experiment is shown in figure 9.11(a), and my recreation of it using a green laser pointer is shown in figure 9.11(b).

An unexpected outcome of my own experiment can be seen by looking closely at figure 9.11(b). Thanks to the use of a little liquid fabric softener to scatter the light and a large stream of water, you can actually see the internal bounces of the laser beam in the column of water as it flows into the basin.

Colladon's experiment may also be used as a simple illustration of how fiber optic cables guide light. A fiber optic cable is basically a glass cylinder with low optical absorption. When light is sent into this tube, it is trapped within by total internal reflection and guided to the other end; a simple illustration of this is shown in figure 9.12. In practice, fiber optic cables usually have a diameter comparable to the wavelength of light, and a simple ray picture of light guiding is inappropriate; we

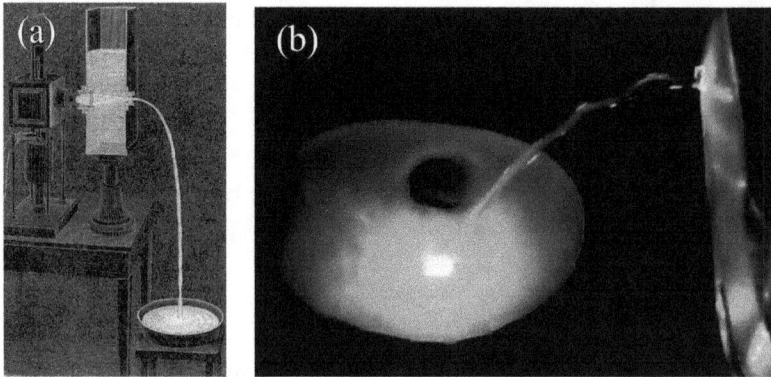

Figure 9.11. (a) Illustration of Colladon's light-guiding experiment, from [8] and (b) my recreation of it.

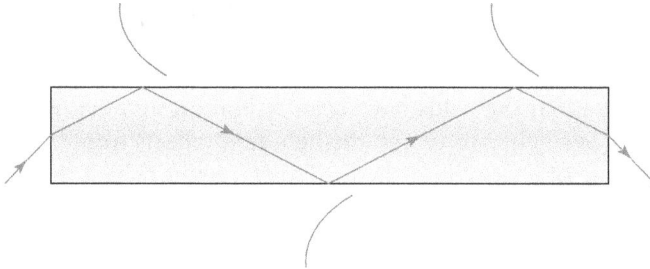

Figure 9.12. Ray picture of light guiding in a fiber optic cable. The curved lines outside the cable are stylistic representations of exponential decay.

will discuss the physics of fiber optics in more detail in chapter 13. However, it is accurate to say that fiber optics is based on the phenomenon of total internal reflection.

A fiber optic cable is usually surrounded by a thick dielectric coating called the cladding, whose purpose is to prevent the evanescent wave surrounding the cable from interacting with—and losing light to—other media. When such an interaction happens, it is referred to as *frustrated total internal reflection.*

Such frustrated total internal reflection can be demonstrated using a pair of prisms. The first prism is used to create TIR by sending a beam of light within it beyond the critical angle. In such a case, all of the light is reflected within the prism and exits through reflection. If a second prism is brought close enough to interact with the evanescent wave of the first, however, one finds that the light gets converted back into a propagating wave within it, with the same angle of incidence θ_i. This is illustrated in figure 9.13.

How do we explain this? We will do the rigorous math of frustrated TIR in chapter 10, after we have introduced the mathematics of multilayer systems. For now, we note that, according to the law of reflection, the y-component of the wave vector is the same in the two prisms and in the gap between them. Though k_z is imaginary in the gap region, it has the same real value in both prisms. When the

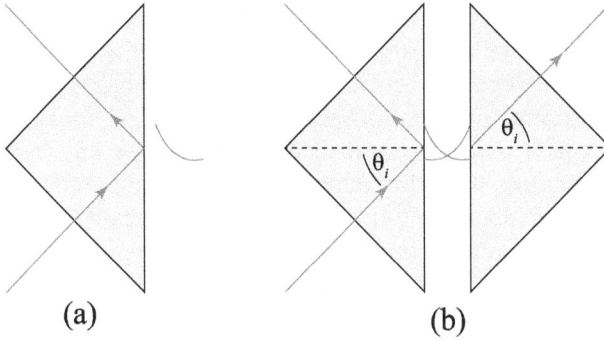

Figure 9.13. (a) Ordinary TIR, (b) frustrated TIR.

evanescent wave penetrates into the second prism, the z-component becomes real-valued again and the evanescent wave is converted back into a propagating wave.

Total internal reflection can also be frustrated by individual particles near the interface. The evanescent wave scatters from the particles, converting into a propagating wave that can be measured. This forms the basis of total internal reflection fluorescence (TIRF) microscopy, a technique which is applied to image individual molecules or cellular processes [9].

There is one more observation to make about evanescent waves. A general plane wave propagating in the xz-plane that satisfies Maxwell's equations has the following form, regardless of whether it is homogeneous or inhomogeneous:

$$\mathbf{E}(\mathbf{r}) = \mathbf{E}_0 e^{i(k_y y + k_z z)}, \tag{9.124}$$

where

$$k_y^2 + k_z^2 = k^2, \tag{9.125}$$

and k is the wavenumber of the medium (recall section 3.4). In terms of k_z, we have

$$k_z = \sqrt{k^2 - k_y^2}. \tag{9.126}$$

Our wave is evanescent if $k_y > k$, so that we then have

$$\mathbf{E}(\mathbf{r}) = \mathbf{E}_0 e^{ik_y y} e^{-k_z' z}, \tag{9.127}$$

with $k_z' = \sqrt{k_y^2 - k^2}$.

Note that there is no limit on the choice of k_y: any value of k_y still satisfies Maxwell's equations, even if $k_y \gg k$. We may define an effective wavelength *in the y-direction* as

$$\lambda_y = 2\pi/k_y. \tag{9.128}$$

Because $k_y > k$, $\lambda_y < \lambda$, where λ is the wavelength of light in the medium. This means that an evanescent wave can be used to create a field with arbitrarily fast oscillations in the y-direction. Since the resolution of imaging systems is typically restricted to imaging objects no smaller than the working wavelength of the system, this suggests that we can use evanescent waves to image objects at a resolution much finer than conventional imaging systems can achieve. This is the basis of the technique of *near-field optics*. In near-field optics, evanescent waves are typically produced by propagating light through a subwavelength-diameter fiber waveguide, and this fiber is brought extremely close to the object to be imaged. Near-field optics has now been a field of research since the 1980s [10], though it is a difficult technique and one that relatively few research groups have mastered.

9.5 The Goos–Hänchen effect

Though an incident plane wave is completely reflected in total internal reflection, it is reflected with a nontrivial phase, i.e. the reflection amplitude r is complex. For a beam-like field, the result of this phase is a displacement of the reflected beam relative to the incident beam. When the input state of polarization is linear, this shift is longitudinal (in the plane of reflection) and is known as the Goos–Hänchen effect; when the input state of polarization is elliptical, the shift is transverse (out of the plane of reflection) and is known as the Imbert–Fedorov effect. It is worthwhile to spend some time deriving basic formulas related to each of these effects.

In total internal reflection, we have seen that there is a nonzero energy density and power flux in the rare medium, with the power flowing along the surface. In a steady state, this implies that light incident upon the interface propagates some distance along it as an evanescent wave before reverting to reflected light, resulting in a longitudinal shift in the position of the reflected wave. This idea was first suggested by Picht [11], and the shift was demonstrated experimentally by Fritz Goos and Hilda Hänchen in 1943 and published in 1947 [12]. As we will see, the shift is typically very small, on the order of a few microns for visible light, so Goos and Hänchen used a simple and clever solution to enhance the effect: they used a long piece of glass in which the beam of light is internally reflected multiple times, making the total shift much greater. An illustration of the method, from their original paper, is shown in figure 9.14.

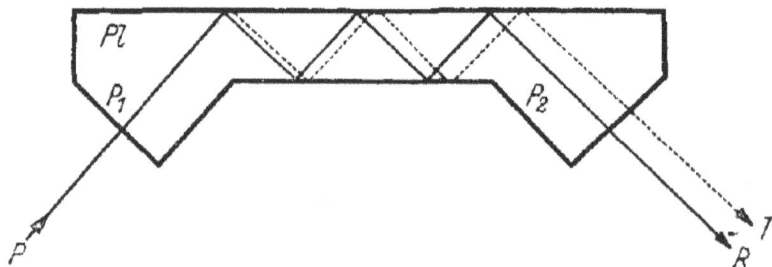

Figure 9.14. Illustration of the experimental approach used to measure the Goos–Hänchen effect. The solid line shows the path the light would have taken without a shift, while the dashed line shows the path after accumulated shifts [12] John Wiley and Sons. Copyright © 1947 WILEY-VCH Verlag GmbH & Co. KGaA, Weinheim.

There are a number of approaches to calculating the Goos–Hänchen shift, and not all of them give the same result. (I interpret this discrepancy as analogous to our trouble with determining the speed of light—in order to determine the 'shift' of the beam, one must define the 'center' of the beam, which is also, in general, not a well-defined quantity.) The most direct approach for calculating the shift is to use an angular spectrum method (such as described in section 3.4) to decompose a beam into a collection of plane waves; this was done by McGuirk and Carniglia [13]. Another approach, undertaken by Wolter [14, 15], uses two plane waves propagating at slightly different angles of incidence to represent the simplest 'beam' possible. The two interfering plane waves create lines of zero intensity along their average angle of incidence, and the reflected fields produce zero lines along the average angle of reflection but shifted along the surface; this shift is interpreted as the Goos–Hänchen shift. Wolter analyzed the Poynting vector of the total fields near the interface and discovered circulating vortices of power flow: the first optical vortices recognized in optics.

We consider a simpler method for deriving the Goos–Hänchen shift that uses energy conservation arguments; this approach is attributed to Kristoffel [16] and Renard [17]. In addition to its simplicity, it can also be applied to determine the Imbert–Fedorov effect.

Let us focus for the moment on the s-polarization case. Equation (9.121) gives the power flux parallel to the surface in the rare medium due to an incident plane wave; let us consider the total power flowing through the surface $y = 0$ by integrating over z from $-\infty$ to 0:

$$P_y = \int_{-\infty}^{0} \mathbf{S}_t \cdot \hat{\mathbf{y}} dz = \frac{n_1 |E_{0t}|^2}{4kn_1\mu_0 c \sinh \theta_t'} \sin \theta_i. \tag{9.129}$$

(Technically, it is total power per unit length in x, but this will not be important in the final result.) Under steady-state conditions, this power must come from the incident field flowing across the interface over a length L_{GH}; this power P_z is written as

$$P_z = \int_{0}^{L_{GH}} \mathbf{S}_i \cdot \hat{\mathbf{z}} dy = \frac{n_1 |E_{0i}|^2}{2\mu_0 c} \cos \theta_i L_{GH}. \tag{9.130}$$

For energy to be conserved, we must have $P_y = P_z$; solving for L_{GH}, we obtain

$$L_{GH} = \frac{|t_\perp|^2 \sin \theta_i}{2kn_1 \cos \theta_i \sinh \theta_t'}. \tag{9.131}$$

This length is interpreted as the Goos–Hänchen shift. Our analysis, in effect, suggests that the evanescent wave in the rare medium is generated by parts of the incident wave extended over a distance L_{GH}; the reflected wave must then include a phase shift for the implied propagation in the evanescent region. We get a similar result for p-polarization, which we derive momentarily.

Let us now turn to the Imbert–Fedorov shift, in which a beam in an elliptical polarization state experiences a shift out of the plane of reflection. This effect was first recognized by Fedor Ivanovich Fedorov in 1955 [18]; however, due to Cold War isolation, it was not seen in the West and was independently discovered by Christian Imbert in 1972 [19].

Our reasoning is similar to that of the Goos–Hänchen shift, except now we consider the net power flowing through the surface $x = 0$, i.e.

$$P_x = \int_{-\infty}^{0} \mathbf{S}_t \cdot \hat{\mathbf{x}} dz. \tag{9.132}$$

Let us consider an incident field in a general polarization state, of the form

$$\mathbf{E}_i(\mathbf{r}) = \{E_\perp \hat{\mathbf{x}} + E_\parallel (\cos \theta_i \hat{\mathbf{y}} - \sin \theta_i \hat{\mathbf{z}})\} e^{i\mathbf{k}_i \cdot \mathbf{r}}, \tag{9.133}$$

$$\mathbf{H}_i(\mathbf{r}) = \frac{n_1}{\mu_0 c} \{E_\perp (-\sin \theta_i \hat{\mathbf{z}} + \cos \theta_i \hat{\mathbf{y}}) - E_\parallel \hat{\mathbf{x}}\} e^{i\mathbf{k}_i \cdot \mathbf{r}}, \tag{9.134}$$

where $E_\perp$ and $E_\parallel$ are the s- and p-polarization components of the incident plane wave. It is straightforward to determine that the incident field's Poynting vector is given by

$$\mathbf{S}_i(\mathbf{r}) = \frac{n_1}{2\mu_0 c} [|E_\perp|^2 + |E_\parallel|^2][\sin \theta_i \hat{\mathbf{y}} + \cos \theta_i \hat{\mathbf{z}}]. \tag{9.135}$$

For the transmitted field, we must take a bit more care, but the expressions are

$$\mathbf{E}_t(\mathbf{r}) = \left\{ t_\perp E_\perp \hat{\mathbf{x}} + t_\parallel E_\parallel \frac{n_1}{n_2} (i \sinh \theta_t' \hat{\mathbf{y}} - \sin \theta_i \hat{\mathbf{z}}) \right\} e^{i\mathbf{k}_t \cdot \mathbf{r}}, \tag{9.136}$$

$$\mathbf{H}_t(\mathbf{r}) = \frac{1}{\mu_0 c} \left\{ n_1 t_\perp E_\perp (-\sin \theta_i \hat{\mathbf{z}} + i \sinh \theta_t' \hat{\mathbf{y}}) - n_2 t_\parallel E_\parallel \hat{\mathbf{x}} \right\} e^{i\mathbf{k}_t \cdot \mathbf{r}}, \tag{9.137}$$

where $t_\perp$ and $t_\parallel$ represent the transmitted s- and p-polarization amplitudes, respectively. The total Poynting vector can be written in the form

$$\mathbf{S}_t(\mathbf{r}) = \frac{n_1}{2\mu_0 c} \left\{ [|t_\perp|^2 |E_\perp|^2 + |t_\parallel|^2 |E_\parallel|^2] \sin \theta_i \hat{\mathbf{y}} + i[|t_\parallel|^2 |E_\parallel|^2 - |t_\perp|^2 |E_\perp|^2] \sinh \theta_t' \hat{\mathbf{z}} \right\} e^{-2kn_1 \sinh \theta_t' z}$$
$$- \frac{in_1^2}{2\mu_0 n_2 c} t_\parallel E_\parallel t_\perp^* E_\perp^* \sin \theta_i \sinh \theta_t' \hat{\mathbf{x}} e^{-2kn_1 \sinh \theta_t' z}. \tag{9.138}$$

Let us now define the Goos–Hänchen shift as

$$L_{GH} = \frac{\int_{-\infty}^{0} \mathbf{S}_t \cdot \hat{\mathbf{y}} dz}{\mathbf{S}_i \cdot \hat{\mathbf{z}}}. \tag{9.139}$$

We then get the formula for arbitrary polarization,

$$L_{GH} = \frac{\sin \theta_i}{2kn_1 \cos \theta_i \sinh \theta_t'} \frac{|t_\perp|^2 |E_\perp|^2 + |t_\parallel|^2 |E_\parallel|^2}{|E_\perp|^2 + |E_\parallel|^2}. \tag{9.140}$$

We note that the Goos–Hänchen shift depends in a nontrivial way upon the state of polarization of the incident field.

For the Imbert–Fedorov shift, we write

$$L_{IF} = \frac{\int_{-\infty}^{0} \mathbf{S}_t \cdot \hat{\mathbf{x}} \, dz}{\mathbf{S}_i \cdot \hat{\mathbf{z}}}, \tag{9.141}$$

where the numerator now represents the total flow of power transverse to the direction of propagation. With some basic manipulations, we get the result

$$L_{IF} = -\frac{i \sin \theta_i}{4kn_2 \cos \theta_i} \frac{t_\parallel t_\perp^* E_\parallel E_\perp^* - t_\parallel^* t_\perp E_\parallel^* E_\perp}{|E_\perp|^2 + |E_\parallel|^2}. \tag{9.142}$$

Because $t_\perp$ and $t_\parallel$ are complex, we always have an Imbert–Fedorov shift as long as the polarization of light is not pure s- or p-polarization.

Let us estimate the magnitude of these shifts for circularly polarized light, i.e. $E_\perp = E_0$ and $E_\parallel = \pm iE_0$. We take $n_1 = 1.52$ (crown glass) and $n_2 = 1.0$ (air). The critical angle is $41.1°$ and we take $\theta_i = 43°$. For a wavelength of 589 nm, we find that $L_{GH} = 160$ nm and $L_{IF} = 19$ nm. Clearly, both of these numbers are much smaller than the width of an optical beam, which is typically significantly larger than a wavelength.

It should be noted that the equations for L_{GH} and L_{IF} are only estimates, as we did not directly calculate the shift of a beam. To be more rigorous, we should use an angular spectrum method, as mentioned earlier; more information can be found in the review by Bliokh and Aiello [20].

We have said that the Goos–Hänchen shift arises from a complex reflection amplitude; it follows that light reflected from an absorbent medium with a complex refractive index should also exhibit such shifts. This was first noted by Wild and Giles [21], who also pointed out that the shift in such cases can be quite large and even negative. Large negative Goos–Hänchen shifts in metals were further explored by Leung et al [22].

As an effect that depends on evanescent waves, the Goos–Hänchen shift has been used to develop sensitive optical detectors. We mention only one of these, a sensor designed by Benam et al [23] that can be used to detect low concentrations of biological molecules.

9.6 Refraction in complex media

All the scenarios considered so far assume a real-valued refractive index for both media at an interface. However, we have also seen that materials that are absorptive due to internal molecular mechanisms or free electrons should be modeled by a complex-valued refractive index. In this section, we consider how the refractive behavior of isotropic materials changes with a complex index.

You might wonder why we did not consider the complex case right from the beginning. In practice, we are usually looking at materials that are either (a) very weakly absorbing, in which case we can safely ignore the imaginary part of the refractive index, or (b) very strongly absorbing, like metals, in which case the field is absorbed after a very small propagation distance in the material, and refraction itself can be largely ignored. Here, we present the complex solutions for those situations that fall between the two extremes mentioned.

Let us start in reverse, with the Fresnel equations. A complex refractive index results in complex wave vectors, which means it is more useful to write the Fresnel equations in terms of wave vector components instead of angles. Fortunately, the derivation of section 9.4 did not really depend on quantities being real-valued, so we can make a straightforward conversion from representing things as angles to representing them as wave vectors, using equation (9.107) and analogous formulas for the reflected and transmitted components. We end up with Fresnel equations in the form

$$r_\perp = \frac{k_{iz} - k_{tz}}{k_{iz} + k_{tz}}, \tag{9.143a}$$

$$t_\perp = \frac{2k_{iz}}{k_{iz} + k_{tz}}, \tag{9.143b}$$

$$r_\parallel = \frac{\dfrac{n_2}{n_1}k_{iz} - \dfrac{n_1}{n_2}k_{tz}}{\dfrac{n_2}{n_1}k_{iz} + \dfrac{n_1}{n_2}k_{tz}}, \tag{9.143c}$$

$$t_\parallel = \frac{2k_{iz}}{\dfrac{n_2}{n_1}k_{iz} + \dfrac{n_1}{n_2}k_{tz}}, \tag{9.143d}$$

where all quantities now have the potential to be complex. In these expressions, we have

$$k_i^2 = k_y^2 + k_{iz}^2, \tag{9.144a}$$

$$k_t^2 = k_y^2 + k_{tz}^2, \tag{9.144b}$$

where $k_y = k_i \sin \theta_i$, $k_i = n_1 \omega/c$, and $k_t = n_2 \omega/c$. Because n_2 is complex, k_t and its components are generally complex as well.

It should be noted that there is a practical use for the formulas for $r_\perp$ and $r_\parallel$ in this case: the ratios of the reflected intensities for various angles of incidence can be used to deduce the complex value of the refractive index of the sample [24].

Now let us turn to Snell's law. Equation (9.49) still applies, in that the y-component of the wave vector is continuous across the interface. However, because n_2 is now complex, we must have a complex angle $\theta_t = \theta_R + i\theta_I$ as well. Using properties of trigonometric functions, we may write

$$\sin(\theta_R + i\theta_I) = \sin \theta_R \cosh \theta_I + i \sinh \theta_I \cos \theta_R. \tag{9.145}$$

We may determine the real and imaginary parts of this sine function by writing $n_2 = n_R + in_I$ and requiring

$$(n_R + in_I)[\sin \theta_R \cosh \theta_I + i \sinh \theta_I \cos \theta_R] = n_1 \sin \theta_i. \tag{9.146}$$

Matching the real and imaginary parts of this expression, we can find that

$$\sinh \theta_I \cos \theta_R = -\frac{n_I n_1 \sin \theta_i}{n_R^2 + n_I^2}, \tag{9.147a}$$

$$\cosh \theta_I \sin \theta_R = \frac{n_R n_1 \sin \theta_i}{n_R^2 + n_I^2}. \tag{9.147b}$$

This approach, while correct, is somewhat unsatisfying because it does not clearly show what happens to the refracted wave. Because n_2 is complex, we know that there is an exponential decay in the z-direction, but because n_2 has a nonzero real part, there is still wave propagation into the second medium, in contrast to the case of total internal reflection. As an alternative approach, we note that we may write

$$k_{tz}^2 = \tilde{\epsilon}_2 k_t^2 - k_y^2 = (\tilde{\epsilon}_R k^2 - k_y^2) + i\tilde{\epsilon}_I k^2, \tag{9.148}$$

where $\tilde{\epsilon} = \epsilon / \epsilon_0$ is the normalized permittivity. We therefore have

$$k_{tz} = \sqrt{(\tilde{\epsilon}_R k^2 - k_y^2) + i\tilde{\epsilon}_I k^2}. \tag{9.149}$$

In the absence of absorption, i.e. $\tilde{\epsilon}_I = 0$, we would have $k_{tz0} = \sqrt{\tilde{\epsilon}_R k^2 - k_y^2}$. The complex k_{tz} can be written as

$$k_{tz} = \sqrt{k_{tz0}^2 + i\tilde{\epsilon}_I k^2}. \tag{9.150}$$

This quantity can be written in terms of its real and imaginary parts using the formula

$$\sqrt{x + iy} = \sqrt{\frac{1}{2}(\sqrt{x^2 + y^2} + x)} + i \, \text{sgn}(y)\sqrt{\frac{1}{2}(\sqrt{x^2 + y^2} - x)}, \tag{9.151}$$

where $\text{sgn}(y)$ is ± 1 according to the sign of y. This allows us to write k_{tz} in the form

$$k_{tz} = \sqrt{\frac{1}{2}\left[\sqrt{k_{tz0}^4 + \tilde{\epsilon}_I^2 k^4} + k_{tz0}^2\right]} + i\sqrt{\frac{1}{2}\left[\sqrt{k_{tz0}^4 + \tilde{\epsilon}_I^2 k^4} - k_{tz0}^2\right]}. \tag{9.152}$$

If we focus on the real part of k_{tz}, which contributes to the direction of wave propagation, we can see that the imaginary part of the permittivity increases the value of k_{tz}. This indicates that the wave vector points more toward the normal to the interface, i.e. we can see that the effect of an imaginary component of ϵ_2 is to decrease the angle of refraction, causing stronger refraction. As we have already indicated, this effect is typically small because we can only observe refraction when the imaginary component of ϵ_2 is small; i.e. the material is weakly absorbing.

9.7 Refraction in anisotropic media

It is worth spending some time discussing refraction into an anisotropic medium. In principle, we can derive a set of Fresnel equations for this case just as we did for the isotropic case, but in practice, this is quite a difficult problem to solve. A little thought illustrates the problem: we simplified our solution of Maxwell's equations in the isotropic case by dividing an incident beam into s-polarization and p-polarization cases, but in a crystal, we now have to contend with the fact that these polarization states do not generally match the polarization of the natural states of the crystal. So, in this section, we talk somewhat generally about what we expect for refraction in anisotropic media; a Fresnel-type analysis can be found in Yeh [25].

One important result that we just take as given is that the wave vectors of the incident, reflected, and transmitted waves still lie within a single plane; i.e. the plane of reflection.

Let us begin by imagining an unpolarized plane wave illuminating our surface. Two orthogonal polarization states of the incident wave match the two natural states of the crystal, each of which has its own refractive index n and thus its own angle of refraction. We expect there to be two rays in the crystal; if these rays exit the crystal at a surface parallel to the incident surface, they are parallel and spatially separated, as illustrated in figure 9.15.

Even with polarized light, we have the potential for two images. A polarized light beam, upon entering the medium, must be decomposed into the natural states of the crystal. If it has appreciable amplitude in both states, there are two images at the output.

We have previously discussed (in section 8.4) that one image in a birefringent crystal is 'ordinary' and independent of the rotation of the crystal, while the other is 'extraordinary' and rotates as the crystal moves. We can roughly illustrate this directly from Maxwell's equations, starting with the ϵ of a uniaxial crystal in its principal coordinate system,

$$\epsilon = \begin{bmatrix} \epsilon_1 & 0 & 0 \\ 0 & \epsilon_1 & 0 \\ 0 & 0 & \epsilon_2 \end{bmatrix}, \tag{9.153}$$

where the z-axis has a different ϵ value. We now imagine rotating this crystal around the y-axis by an angle θ, using the rotation matrix

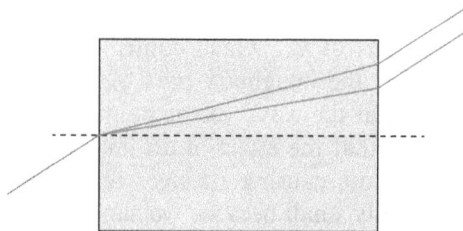

Figure 9.15. The parallel rays that emerge from a birefringent material.

$$\mathbf{R} = \begin{bmatrix} \cos\theta & 0 & -\sin\theta \\ 0 & 1 & 0 \\ \sin\theta & 0 & \cos\theta \end{bmatrix}. \tag{9.154}$$

In the new coordinate system, $\epsilon' = \mathbf{R}\epsilon\mathbf{R}^T$, where T indicates the transpose. We find

$$\epsilon' = \begin{bmatrix} \epsilon_1\cos^2\theta + \epsilon_2\sin^2\theta & 0 & (\epsilon_1 - \epsilon_2)\cos\theta\sin\theta \\ 0 & \epsilon_1 & 0 \\ (\epsilon_1 - \epsilon_2)\cos\theta\sin\theta & 0 & \epsilon_1\sin^2\theta + \epsilon_2\cos^2\theta \end{bmatrix}. \tag{9.155}$$

We return to Maxwell's equations and generate a wave equation with the usual combination of Ampère–Maxwell and Faraday:

$$\nabla^2\mathbf{E} = [-\omega^2\mu_0\epsilon' + \nabla(\nabla\cdot)]\mathbf{E}. \tag{9.156}$$

Let us assume that we have plane waves propagating in the z-direction. Then, all derivatives are z-derivatives, and we simplify our expression to

$$\partial_z^2\mathbf{E} = [-\omega^2\mu_0\epsilon' + \hat{\mathbf{z}}\partial_z^2\hat{\mathbf{z}}\cdot]\mathbf{E}. \tag{9.157}$$

We may write this out component by component to get

$$\partial_z^2 E_x + \omega^2\mu_0\epsilon_{xx} + \omega^2\mu_0\epsilon_{xz}E_z = 0, \tag{9.158}$$

$$\partial_z^2 E_y + \omega^2\mu_0\epsilon_1 E_y = 0, \tag{9.159}$$

$$\omega^2\mu_0\epsilon_{xz}E_x + \omega^2\mu_0\epsilon_{zz}E_z = 0. \tag{9.160}$$

The first equation depends on both E_x and E_z. However, the third equation provides a direct relationship between these components, as all the derivative terms have canceled out, and we may write

$$E_z = -\frac{\epsilon_{xz}}{\epsilon_{zz}}E_x. \tag{9.161}$$

Substituting back in, we get two distinct wave equations,

$$\partial_z^2 E_x + \omega^2\mu_0\left[\epsilon_{xx} - \frac{\epsilon_{xz}^2}{\epsilon_{zz}}\right]E_x = 0, \tag{9.162}$$

$$\partial_z^2 E_y + \omega^2\mu_0\epsilon_1 E_y = 0. \tag{9.163}$$

We therefore have two possible waves, dependent on polarization and with refractive indices given by

$$n^2 = \frac{1}{\epsilon_0}\left[\epsilon_{xx} - \frac{\epsilon_{xz}^2}{\epsilon_{zz}}\right], \tag{9.164}$$

$$n^2 = \frac{\epsilon_1}{\epsilon_0}. \tag{9.165}$$

The quantities ϵ_{xx}, ϵ_{xz}, and ϵ_{zz} depend on the angle θ, whereas ϵ_1 does not. We therefore again find that there is one ordinary solution and one extraordinary solution. The ordinary solution has a purely TE field because it points along one of the principal axes.

A little thought can convince you that there is always one ordinary solution for a uniaxial crystal. Because two principal dielectric constants are the same, there is an entire plane associated with that dielectric value. No matter what direction we choose for propagation, one perpendicular vector can always be taken to lie in this plane.

9.8 Refraction in magnetic materials

The Fresnel formulas can be further generalized to the case of magnetic materials, where the permeability of one medium or both media is different from the vacuum value. Though this does not occur in most natural materials at optical frequencies, it is a key aspect of artificial materials known as metamaterials, to be discussed in chapter 12.

We return to the s-polarization case of figure 9.7 but with material properties ϵ_1, μ_1 for the incident medium and ϵ_2, μ_2 for the transmitted medium. We use the same boundary conditions for the transverse fields, namely

$$\mathbf{E}_i(\mathbf{r}) \cdot \hat{\mathbf{x}} + \mathbf{E}_r(\mathbf{r}) \cdot \hat{\mathbf{x}} = \mathbf{E}_t(\mathbf{r}) \cdot \hat{\mathbf{x}}, \tag{9.166}$$

$$\mathbf{H}_i(\mathbf{r}) \cdot \hat{\mathbf{y}} + \mathbf{H}_r(\mathbf{r}) \cdot \hat{\mathbf{y}} = \mathbf{H}_t(\mathbf{r}) \cdot \hat{\mathbf{y}}, \tag{9.167}$$

but now the **H**-fields are of the form

$$\mathbf{H}_{0i} = \frac{k_i E_{0i}}{\mu_1 \omega}(\cos\theta_i \hat{\mathbf{y}} - \sin\theta_i \hat{\mathbf{z}}), \tag{9.168}$$

$$\mathbf{H}_{0r} = \frac{k_r E_{0r}}{\mu_1 \omega}(-\cos\theta_r \hat{\mathbf{y}} - \sin\theta_r \hat{\mathbf{z}}), \tag{9.169}$$

$$\mathbf{H}_{0t} = \frac{k_t E_{0t}}{\mu_2 \omega}(\cos\theta_t \hat{\mathbf{y}} - \sin\theta_t \hat{\mathbf{z}}). \tag{9.170}$$

We use $k_i = \sqrt{\epsilon_1 \mu_1}\,\omega$, $k_r = \sqrt{\epsilon_1 \mu_1}\,\omega$, and $k_t = \sqrt{\epsilon_2 \mu_2}\,\omega$.

Our two boundary conditions then simplify to

$$1 + r = t, \tag{9.171}$$

$$\frac{\cos\theta_i}{\eta_1}(1 - r) = \frac{\cos\theta_t}{\eta_2}t, \tag{9.172}$$

where we have again used the normalized reflected and transmitted amplitudes r and t, and we have introduced the new quantity

$$\eta_i \equiv \sqrt{\frac{\mu_i}{\epsilon_i}}, \tag{9.173}$$

which is the *electromagnetic impedance*. This term is inspired by the fact that the units of η are ohms, just like those of electrical impedance. It possesses the value $\eta_0 = 376.7 \ \Omega$ in vacuum.

The solution of the equations is straightforward, and we present both the s- and p-polarization solutions together:

$$r_\perp = \frac{\eta_2 \cos \theta_i - \eta_1 \cos \theta_t}{\eta_2 \cos \theta_i + \eta_1 \cos \theta_t}, \tag{9.174}$$

$$t_\perp = \frac{2\eta_2 \cos \theta_i}{\eta_2 \cos \theta_i + \eta_1 \cos \theta_t}, \tag{9.175}$$

$$r_\parallel = \frac{\eta_1 \cos \theta_i - \eta_2 \cos \theta_t}{\eta_1 \cos \theta_i + \eta_2 \cos \theta_t}, \tag{9.176}$$

$$t_\parallel = \frac{2\eta_2 \cos \theta_i}{\eta_1 \cos \theta_i + \eta_2 \cos \theta_t}. \tag{9.177}$$

One important check for any result like this is to see whether it is consistent with previous results. If we set $\mu = \mu_0$, we may multiply the numerator and denominator of all four equations by $\sqrt{\epsilon_0/\mu_0}$ and then note that $\eta_j \sqrt{\epsilon_0/\mu_0} = 1/n_j$. With appropriate multiplications, we can find that the above expressions reduce to the original Fresnel equations.

These equations give us enough context to see that there are two relevant parameters that describe the transmission of light at an interface between isotropic media: (i) the index of refraction n, which dictates how the direction of the wave changes on transmission, and (ii) the impedance η, which dictates how much light is transmitted. (For a nonmagnetic material, we can say that the two quantities are equivalent.)

Let us consider the special case of normal incidence, $\theta_i = 0$. Then, of course, $\theta_t = 0$ and

$$r_\perp = \frac{\eta_2 - \eta_1}{\eta_2 + \eta_1}, \tag{9.178}$$

$$t_\perp = \frac{2\eta_2}{\eta_2 + \eta_1}. \tag{9.179}$$

We now see that it is possible to have perfect transmission, $t = 1$, and no reflection, $r = 0$, if $\eta_1 = \eta_2$. But since η_j depends on the ratios of μ_j and ϵ_j, this can be achieved with different material properties. Two different materials that have the same value of η are said to be impedance matched. We will have more to say about impedance matching in chapter 12.

We may make one final generalization of these formulas, which have been written only for plane waves. If we associate $k_z = \sqrt{\epsilon\mu}\,\omega\cos\theta$ for each region, with some manipulation, we may write

$$r_\perp = \frac{k_{iz}/\eta_1 n_1 - k_{tz}/\eta_2 n_2}{k_{iz}/\eta_1 n_1 + k_{tz}/\eta_2 n_2}, \tag{9.180}$$

$$t_\perp = \frac{2k_{iz}/\eta_1 n_1}{k_{iz}/\eta_1 n_1 + k_{tz}/\eta_2 n_2}, \tag{9.181}$$

$$r_\parallel = \frac{k_{iz}\eta_1/n_1 - k_{tz}\eta_2/n_2}{k_{iz}\eta_1/n_1 + k_{tz}\eta_2/n_2}, \tag{9.182}$$

$$t_\parallel = \frac{2k_{iz}\eta_2/n_1}{k_{iz}\eta_1/n_1 + k_{tz}\eta_2/n_2}. \tag{9.183}$$

These are the most general forms of the Fresnel formulas, at least for isotropic media.

9.9 Exercises

1. A little intuition building: diamond has a high refractive index, $n = 2.417$, and diamonds are said to be sparkly because they have a lot of total internal reflection that makes light bounce around and shine from all directions. Calculate the critical angle for light traveling from diamond to air, and calculate the Brewster angle for light traveling from air into diamond. Repeat these two calculations for water ($n = 1.33$).

2. Some more intuition building: assuming $n_1 = 1.52$ (crown glass) and $n_2 = 1.0$ (air), find the critical angle for total internal reflection. Plot the reflected power, R, as a function of the angle of incidence θ_i for both TE and TM polarizations.

3. Explicitly show that power flow is conserved across an interface for the p-polarization case, i.e. that $R + T = 1$ for $\mathbf{E}_\perp$. Calculate R and T for the $\mathbf{E}_\perp$ and $\mathbf{E}_\parallel$ cases with $n_1 = 1.0$, $n_2 = 1.33$, and $\theta_i = 25°$, and show that $R + T = 1$ for both.

4. When light is normally incident on uncoated fused silica, 3.47% of the power is reflected. From this, determine the refractive index of fused silica. Determine the reflected power R for both s- and p-polarizations when the light is incident at an angle of 30°. (Keep four significant figures in your results.)

5. When light is incident from air upon sapphire at 60°, the reflected s- and p-polarization amplitudes are $r_\perp = -0.3239$ and $r_\parallel = 0.2297$. Using the Fresnel formulas, solve for the refractive index n_2 of sapphire, deriving a general formula that determines the value of n_2 directly from $r_\perp$ and $r_\parallel$. Comment on the dependence of the result on θ_i.

6. Suppose light with a vacuum wavelength of 500 nm is totally internally reflected from an air–water interface at an angle of 60°. Determine the intensity of the evanescent field at a distance $d = 50$ nm relative to the intensity at the surface of the water, i.e. calculate $|\mathbf{E}(d)|^2/|\mathbf{E}(0)|^2$. Repeat this calculation for $\theta_i = 75°$ and $\theta_i = 49°$. (Take $n_1 = 1.33$ as the refractive index of water.)

7. By shining a beam of light on a transparent dielectric surface, you find that the Brewster angle is $\theta_B = 59.5°$. What is the index of refraction of the dielectric? If the light is incident from within the glass, what is the angle of total internal reflection? Plot $R_{\parallel}$ as a function of the incident angle for light illuminating each side of the interface.

8. How sensitive is the polarization effect to deviations from the Brewster angle? Consider light incident from air to glass with a refractive index $n_2 = 1.5$. The Brewster angle is 56.3°. Calculate and plot the ratio

$$\mathcal{R} \equiv \frac{R_{\perp}}{R_{\parallel} + R_{\perp}},$$

the ratio of the s-polarized reflected power to the total reflected power, from the range $\theta_i = 46.3°$ to 66.3°, using at least two-degree increments. When this ratio is high, most of the reflected power is s-polarized. Estimate how collimated a beam must be (i.e. the maximum angular spread) in order for approximately 99% of the reflected power to be s-polarized. (There is no exact answer in this case; you should provide a reasonable estimate.)

9. Suppose a wave is incident from air into a medium with a permittivity ϵ_2 and a permeability μ_2. When the wave is incident at an angle $\theta_i = 30°$, the angle of refraction is $\theta_t = 14.48°$ and the wave is perfectly transmitted (i.e. there is zero reflection) for the perpendicular ($\perp$) polarization. Determine the normalized values ϵ_2/ϵ_0 and μ_2/μ_0 for the second medium.

10. Consider light traveling from air into a material with a permittivity ϵ_2 and a permeability μ_2. Give the formula for the Brewster angle, previously derived only for nonmagnetic materials, for this generalized case, and calculate the Brewster angle for the case $\epsilon_2 = 2$ and $\mu_2 = 1.5$.

11. Let us define r_{12} as the reflection amplitude from medium 1 to medium 2, θ_i as the incident angle in medium 1, and θ_t as the transmission angle in medium 2. Conversely, r_{21} represents the amplitude in the opposite direction (from medium 2 to medium 1); in this case, the incident angle is denoted by θ_t, and the transmission angle is denoted by θ_i. Similarly, the transmitted amplitudes in the respective directions are denoted by t_{12} and t_{21}. Prove the following general relation for s-polarization, which holds regardless of the angle of incidence or whether medium 2 is complex:

$$r_{12}^2 + t_{12}t_{21} = 1.$$

(The result also holds for p-polarization, though we only ask for the s-polarization proof.)

References

[1] Wiener O 1890 Stehende Lichtwellen und die Schwingungsrichtung polarisirten Lichtes *Ann. Phys. Chem.* **38** 203–43

[2] Bladel J V 1991 *Singular Electromagnetic Fields and Sources* (New York: IEEE)

[3] Van Bladel J 1985 Field singularities at metal-dielectric wedges *IEEE Trans. Antennas Propag.* **33** 450–5

[4] Yu N, Genevet P, Kats M A, Aieta F, Tetienne J P, Capasso F and Gaburro Z 2011 Light propagation with phase discontinuities: generalized laws of reflection and refraction *Science* **334** 333–7

[5] Aieta F, Genevet P, Yu N, Kats M A, Gaburro Z and Capasso F 2012 Out-of-Plane Reflection and Refraction of Light by Anisotropic Optical Antenna Metasurfaces with Phase Discontinuities *Nano Lett.* **12** 1702–6

[6] Tavassoli M T, Nahal A and Ebadi Z 2004 Image formation in rough surfaces *Opt. Commun.* **238** 252–60

[7] Colladon J D 1842 Sur les réflexions d'un rayon de lumière àl'intérieur d'une veine liquide parabolique *Comptes Rendus* **6** 800–2

[8] Colladon J D 1884 La fontaine Colladon *La Nature* **12** 525–6

[9] Axelrod D 2001 Total internal reflection fluorescence microscopy in cell biology *Traffic* **2** 764–74

[10] Novotny L and Hecht B 2012 *Principles of Nano-Optics* 2nd edn (Cambridge: Cambridge University Press)

[11] Picht J 1929 Beitrag zur Theorie der Totalreflexion *Ann. Phys., Lpz.* **395** 433–96

[12] Goos F and Hänchen H 1947 Ein neuer und fundamentaler Versuch zur Totalreflexion *Ann. Phys., Lpz.* **436** 333–46

[13] McGuirk M and Carniglia C K 1977 An angular spectrum representation approach to the Goos-Hänchen shift *J. Opt. Soc. Am.* **67** 103–7

[14] Wolter H 1950 Zur Frage des Lichtweges bei Totalreflexion *Z. Naturforsch.* A **5** 276–83

[15] Wolter H 2009 Concerning the path of light upon total reflection *J. Opt.* A **11** 090401

[16] Scient K N 1956 Acta, Tartu State Univ (private communication)

[17] Renard R H 1964 Total reflection: a new evaluation of the Goos-Hänchen shift *J. Opt. Soc. Am.* **54** 1190–7

[18] Fedorov F I 1955 To the theory of total reflection (translated title) *Dokl. Akad. Nauk SSSR* **105** 465

[19] Imbert C 1972 Calculation and experimental proof of the transverse shift induced by total internal reflection of a circularly polarized light beam *Phys. Rev.* D **5** 787–96

[20] Bliokh K Y and Aiello A 2013 Goos-Hänchen and Imbert-Fedorov beam shifts: an overview *J. Opt.* **15** 014001

[21] Wild W J and Giles C L 1982 Goos-Hänchen shifts from absorbing media *Phys. Rev.* A **25** 2099–101

[22] Leung P T, Chen C W and Chiang H P 2007 Large negative Goos-Häanchen shift at metal surfaces *Opt. Commun.* **276** 206–8

[23] Benam E R, Sahrai M and Bonab J P 2020 High sensitive label-free optical sensor based on Goos-Hänchen effect by the single chirped laser pulse *Sci. Rep.* **10** 17176

[24] Avery D G 1952 An improved method for measurements of optical constants by reflection *Proc. Phys. Soc.* B **65** 425

[25] Yeh P 2005 *Optical Waves in Layered Media* (Hoboken, NJ: Wiley)

IOP Publishing

Electromagnetic Optics

Gregory J Gbur

Chapter 10

Light propagation in stratified media

Now that we have studied the propagation of light through a single interface in detail, the next natural step is to consider the propagation of light through multiple planar interfaces. There are many practical and physically interesting situations involving such stratified media, which justifies developing a general formalism to handle such cases. We begin with this general formalism, often referred to as the transfer matrix method, and then consider special cases of optical thin films, periodic multilayers, and frustrated total internal reflection (TIR). We will also find this formalism useful for other cases moving forward.

10.1 General considerations

We consider a situation where the medium possesses a nontrivial permittivity $\epsilon(z)$ and permeability $\mu(z)$. We write them formally as general functions of z but limit ourselves to the case of multiple layers with constant ϵ and μ.

Let us begin with a general observation connected to the monochromatic Maxwell's equations in such a medium. We have the divergence equations,

$$\nabla \cdot \mathbf{E}(\mathbf{r}) = 0, \tag{10.1}$$

$$\nabla \cdot \mathbf{H}(\mathbf{r}) = 0, \tag{10.2}$$

and the curl equations,

$$\nabla \times \mathbf{H}(\mathbf{r}) = -i\epsilon\omega\mathbf{E}(\mathbf{r}), \tag{10.3}$$

$$\nabla \times \mathbf{E}(\mathbf{r}) = i\mu\omega\mathbf{H}(\mathbf{r}). \tag{10.4}$$

At this point, we have dropped the explicit use of ω as an argument when working with monochromatic fields.

Looking at Maxwell's equations, we can see that they remain invariant in form if we make the substitutions

doi:10.1088/978-0-7503-6064-7ch10

$$\mathbf{E} \leftrightarrow \mathbf{H} \text{ and } \epsilon \leftrightarrow -\mu. \tag{10.5}$$

We have already noted that reflection and refraction at a single interface can be separated into two independent cases: transverse electric (TE) and transverse magnetic (TM). This separation extends to an arbitrary number of interfaces, so TE and TM remain separated in a stratified medium. A TE wave has an $\hat{\mathbf{x}}$-polarized electric field and a $\hat{\mathbf{y}}$, $\hat{\mathbf{z}}$-polarized magnetic field, while a TM wave has an $\hat{\mathbf{x}}$-polarized *magnetic* field and an $\hat{\mathbf{y}}$, $\hat{\mathbf{z}}$-polarized *electric* field.

The previous two observations combined suggest that if we find a solution for TE waves propagating through a stratified medium, we can immediately determine the TM solution by making the replacements $\mathbf{E} \leftrightarrow \mathbf{H}$ and $\epsilon \leftrightarrow -\mu$. We therefore focus on calculating the TE case and determine the TM case by this symmetry argument.

10.2 Matrix methods for stratified media

We are interested in calculating the propagation of light through multiple layers. The most straightforward way to do this would be to introduce a pair of fields, left-going and right-going, for each layer of the medium, and match the boundary conditions at each interface. If we have N layers, including the starting layer and the ending layer, we have $N - 2$ interfaces, and because we must match two boundary conditions at each interface, we will end up needing to simultaneously solve a set of $2N - 4$ equations. This quickly becomes a calculation that is difficult to manage, and furthermore, it is a wasteful calculation: we typically only care about the fields at the ends of the stratified medium, namely the reflected and transmitted amplitudes.

Fortunately, there is a more elegant approach, in which each interior layer is represented by a matrix, and the effect of the combined stratified medium is reduced to matrix multiplication. This method is somewhat reminiscent of using Jones vectors to calculate light propagation through polarization-sensitive optical elements, as done in section 4.7, with the added complication that we must consider internal reflections in the stratified medium.

We start by looking at a single layer, or slab, in isolation, as in figure 10.1, of thickness d_f and refractive index n_f. We know that the transverse component of the wave vector of a plane wave remains unchanged as it passes across a planar interface, and we label this component k_y. Inside the layer, the z-component of the wave vector has the magnitude

$$k_z = \sqrt{n_f^2 k^2 - k_y^2}. \tag{10.6}$$

The total wavenumber in the medium is $k_f = n_f k$, with $k_f^2 = k_y^2 + k_z^2$. Writing k_z in the form of equation (10.6) allows us to take into account both plane waves ($k_y > n_f k$) and evanescent waves ($k_y < n_f k$). We consider the possibility of both a right-going wave $\mathbf{k}_1$ and a left-going wave $\mathbf{k}_2$, and we write the total wavenumbers as

$$\mathbf{k}_1 = k_y \hat{\mathbf{y}} + k_z \hat{\mathbf{z}}, \tag{10.7}$$

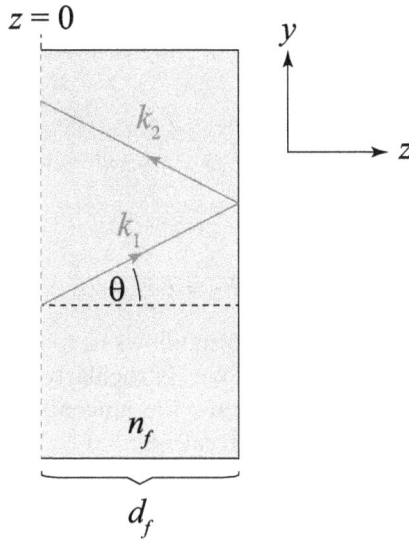

Figure 10.1. Notation related to a single layer of a stratified medium.

$$\mathbf{k}_2 = k_y \hat{\mathbf{y}} - k_z \hat{\mathbf{z}}. \tag{10.8}$$

We are considering a TE wave, so there is only an x-component to the electric field, which we write in the form

$$E_x(\mathbf{r}) = E_+ e^{i\mathbf{k}_1 \cdot \mathbf{r}} + E_- e^{i\mathbf{k}_2 \cdot \mathbf{r}}. \tag{10.9}$$

Using the Ampère–Maxwell law, we may write the components of the magnetic field as

$$H_y(\mathbf{r}) = \frac{E_+}{\eta_f} \frac{k_z}{kn_f} e^{i\mathbf{k}_1 \cdot \mathbf{r}} - \frac{E_-}{\eta_f} \frac{k_z}{kn_f} e^{i\mathbf{k}_2 \cdot \mathbf{r}}, \tag{10.10}$$

$$H_z(\mathbf{r}) = -\frac{E_+}{\eta_f} \frac{k_y}{kn_f} e^{i\mathbf{k}_1 \cdot \mathbf{r}} - \frac{E_-}{\eta_f} \frac{k_y}{kn_f} e^{i\mathbf{k}_2 \cdot \mathbf{r}}, \tag{10.11}$$

where η_f is the impedance of the layer.

Because we have a pair of counterpropagating waves in a single slab, it is more useful to rewrite the above electric and magnetic fields in terms of sine and cosine standing waves. (We first wrote them as propagating waves for conceptual clarity.) We write the z-component of the plane waves using Euler's formula, leaving the y-component as a complex exponential,

$$e^{i\mathbf{k}_1 \cdot \mathbf{r}} = [\cos(k_z z) + i \sin(k_z z)] e^{i k_y y}. \tag{10.12}$$

The fields then take on the form

$$E_x(\mathbf{r}) = [E_1 \cos(k_z z) + E_2 \sin(k_z z)] e^{i k_y y}, \tag{10.13}$$

$$H_y(\mathbf{r}) = \frac{ik_z}{\eta_f k n_f}[E_1 \sin(k_z z) - E_2 \cos(k_z z)]e^{ik_y y}, \tag{10.14}$$

$$H_z(\mathbf{r}) = \frac{k_y}{\eta_f k n_f}[-E_1 \cos(k_z z) - E_2 \sin(k_z z)]e^{ik_y y}, \tag{10.15}$$

where

$$E_1 = E_+ + E_-, \quad E_2 = i(E_+ - E_-). \tag{10.16}$$

Writing the fields in standing-wave form allows us to easily implement boundary conditions on the left of the slab. We particularly focus on the tangential components of the field, because we know the tangential components of the field are continuous across the boundary (recall section 9.2). By setting $z = 0$ in our formulas for the field components, we immediately make the associations

$$E_1 = E_x(0), \quad E_2 = \frac{i\eta_f k n_f}{k_z}H_y(0). \tag{10.17}$$

This naturally leads us to express the transverse fields in matrix form:

$$|\mathbf{Q}\rangle = \mathbf{N}|\mathbf{Q}_0\rangle, \tag{10.18}$$

where

$$|\mathbf{Q}\rangle = \begin{bmatrix} E_x(z) \\ H_y(z) \end{bmatrix}, \quad |\mathbf{Q}_0\rangle = \begin{bmatrix} E_x(0) \\ H_y(0) \end{bmatrix}, \tag{10.19}$$

and

$$\mathbf{N} = \begin{bmatrix} \cos(k_z z) & i\sin(k_z z)/p \\ ip\sin(k_z z) & \cos(k_z z) \end{bmatrix}, \tag{10.20}$$

where

$$p = k_z/\eta_f k n_f. \tag{10.21}$$

The matrix $\mathbf{N}$ transforms the field vector $|\mathbf{Q}_0\rangle$ at $z = 0$ in the layer into the field $|\mathbf{Q}\rangle$ at position z within the layer. Because the matrix is nonsingular, it can be inverted, and it is more convenient to write equation (10.18) in the inverted form

$$|\mathbf{Q}_0\rangle = \mathbf{M}|\mathbf{Q}\rangle, \tag{10.22}$$

where

$$\mathbf{M} = \begin{bmatrix} \cos(k_z z) & -i\sin(k_z z)/p \\ -ip\sin(k_z z) & \cos(k_z z) \end{bmatrix} \tag{10.23}$$

is known as the *characteristic matrix*.

We now make an important observation: the matrix $\mathbf{M}$ depends on k_y, which is constant for a plane wave through every layer of a stratified medium and is independent of the field amplitudes. The effect of multiple layers can then be determined through a simple multiplication of the characteristic matrices of each layer. If there are N layers and layer 1 is the leftmost, the total characteristic matrix $\mathbf{M}$ is simply the product of all the matrices, from left to right,

$$\mathbf{M} = \mathbf{M}_1 \mathbf{M}_2 \mathbf{M}_3 \cdots \mathbf{M}_N. \tag{10.24}$$

We still need to do a bit more to calculate the transmission and reflection of light through a multilayer medium. We now consider a typical transmission geometry, as illustrated in figure 10.2. There are $N - 2$ internal layers and therefore $N - 2$ characteristic matrices that must be multiplied to get the total system characteristic matrix. The input layer is layer 1, with refractive index n_1, and the output layer is layer N, with refractive index n_N. The total thickness of the layers is taken to be z_0.

The incident electric field is taken to have an amplitude A, and the reflected field is taken to have an amplitude R; the transmitted field has an amplitude T. We then have the input fields

$$E_x(0) = A + R, \quad H_y(0) = p_1(A - R), \tag{10.25}$$

and the output fields are

$$E_x(z_0) = T, \quad H_y(z_0) = p_N T. \tag{10.26}$$

The quantities p_1 and p_N can be calculated using equation (10.21). If the incident field is a plane wave, p_1 simplifies to $p_1 = \cos \theta_1 / \eta_1$.

To finish our calculation, we write the total characteristic matrix $\mathbf{M}$ explicitly as

$$\mathbf{M} = \begin{bmatrix} m_{11} & m_{12} \\ m_{21} & m_{22} \end{bmatrix}. \tag{10.27}$$

Now, from equation (10.22), we get a pair of equations,

$$A + R = (m_{11} + m_{12} p_N) T, \tag{10.28}$$

$$p_1(A - R) = (m_{21} + m_{22} p_N) T. \tag{10.29}$$

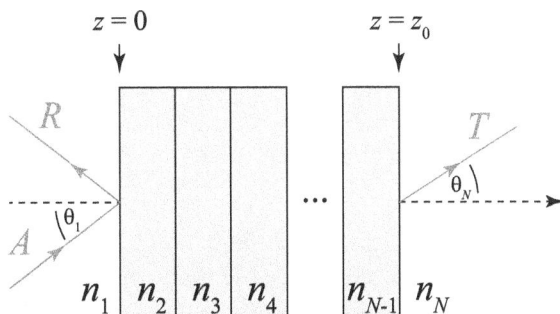

Figure 10.2. Notation related to the reflection and refraction of light through a stratified medium.

Solving these equations for R and T, we end up with a really remarkable result: we can write the reflected and transmitted field amplitudes as a simple pair of equations that apply to any stratified medium. The complexity of the problem is buried in the characteristic matrix. The normalized field amplitudes are given by the expressions

$$r_{TE} = \frac{R}{A} = \frac{(m_{11} + m_{12}p_N)p_1 - (m_{21} + m_{22}p_N)}{(m_{11} + m_{12}p_N)p_1 + (m_{21} + m_{22}p_N)}, \tag{10.30}$$

$$t_{TE} = \frac{T}{A} = \frac{2p_1}{(m_{11} + m_{12}p_N)p_1 + (m_{21} + m_{22}p_N)}, \tag{10.31}$$

where we have labeled these as *TE* to indicate the polarization state we are considering.

We may also determine the TM case by applying the symmetry transformations of equation (10.5). Our calculation then results in the reflected and transmitted amplitudes of the magnetic field H_x, and we replace p in each region with q, defined as

$$q = -k_z \eta_f / k n_f, \tag{10.32}$$

which takes into account the change $\epsilon \leftrightarrow = -\mu$. We then immediately have

$$r_{TM}^H = \frac{R}{A} = \frac{(m_{11} + m_{12}q_N)q_1 - (m_{21} + m_{22}q_N)}{(m_{11} + m_{12}q_N)q_1 + (m_{21} + m_{22}q_N)}, \tag{10.33}$$

$$t_{TM}^H = \frac{T}{A} = \frac{2q_1}{(m_{11} + m_{12}q_N)q_1 + (m_{21} + m_{22}q_N)}. \tag{10.34}$$

It should be noted, however, that these formulas give us the normalized magnetic field amplitude in each region. If we want to directly compare these formulas to the TE case, we must take into account how the relative values of the electric and magnetic field amplitudes change in different media. The reflected amplitude can be used as is, because the reflected wave is in the same medium as the incident wave; for the transmitted amplitude, however, we must correct for the relative impedances of the input and output layers:

$$r_{TM} = \frac{(m_{11} + m_{12}q_N)q_1 - (m_{21} + m_{22}q_N)}{(m_{11} + m_{12}q_N)q_1 + (m_{21} + m_{22}q_N)}, \tag{10.35}$$

$$t_{TM} = \frac{\eta_N}{\eta_1} \frac{2q_1}{(m_{11} + m_{12}q_N)q_1 + (m_{21} + m_{22}q_N)}. \tag{10.36}$$

With these general formulas, we may now look at some illustrative cases.

10.3 Single interface

The simplest case we can consider is a single interface, i.e. the subject of the previous chapter. If we apply our formulas to a single interface, we should reproduce the Fresnel formulas of section 9.8. But what is the characteristic matrix for a single interface? We imagine a 3-layer system, with the incident and transmitted layers being layer 1 and layer 2, and take the limit as we let the thickness d of the intermediate slab go to zero. For the TE case, we have

$$\mathbf{M} = \lim_{d \to 0} \begin{bmatrix} \cos(k_z d) & -i \sin(k_z d)/p \\ -ip \sin(k_z d) & \cos(k_z d) \end{bmatrix} = \begin{bmatrix} 1 & 0 \\ 0 & 1 \end{bmatrix}, \tag{10.37}$$

i.e. the identity matrix! The TE formulas for reflection and transmission from the previous section thus become

$$r_{TE} = \frac{p_1 - p_2}{p_1 + p_2} = \frac{k_{z1}/\eta_1 n_1 - k_{z2}/\eta_2 n_2}{k_{z1}/\eta_1 n_1 + k_{z2}/\eta_2 n_2}, \tag{10.38}$$

$$t_{TE} = \frac{2p_1}{p_1 + p_2} = \frac{2k_{z1}/\eta_1 n_1}{k_{z1}/\eta_1 n_1 + k_{z2}/\eta_2 n_2}. \tag{10.39}$$

If we use $k_{z1} = k n_1 \cos\theta_1$, it can readily be shown with some simple algebra that these formulas correspond to equations (9.175) and (9.176).

For the TM case, we have

$$r_{TM} = \frac{q_1 - q_2}{q_1 + q_2} = \frac{k_{z1}\eta_1/n_1 - k_{z2}\eta_2/n_2}{k_{z1}\eta_1/n_1 + k_{z2}\eta_2/n_2}, \tag{10.40}$$

$$t_{TM} = \frac{\eta_2}{\eta_1} \frac{2q_1}{q_1 + q_2} = \frac{2k_{z1}\eta_2/n_1}{k_{z1}\eta_1/n_1 + k_{z2}\eta_2/n_2}. \tag{10.41}$$

Through simple manipulation, these can be shown to be in agreement with equations (9.177) and (9.178).

At normal incidence, the two reflection formulas should be the same; we find that

$$r_{TE} = \frac{\eta_2 - \eta_1}{\eta_1 + \eta_2}, \tag{10.42}$$

$$r_{TM} = \frac{\eta_1 - \eta_2}{\eta_1 + \eta_2}. \tag{10.43}$$

Referring back to section 9.4.3, we find that our formulas, as written, use the same sign convention as our original Fresnel formulas, which creates the illusion of a difference between the TE and TM cases.

In hindsight, we could have skipped deriving the Fresnel formulas in chapter 9 and simply treated them as a special case of our matrix formulation of stratified media. However, this would have been pedagogically unwise for the same reason the matrix method is computationally advantageous: it hides the interesting physics within a lot of mathematics.

There is one more useful thing we can do with our matrix formulas. Though we have derived the results for general isotropic media, most materials we consider will be nonmagnetic, with $\mu = \mu_0$. Because the p_α and q_α parameters only appear in ratios in the final reflected and transmitted amplitudes, we can ignore the constant μ_0 in these equations and instead use

$$p = k_z/k, \quad q = -k_z/(k\epsilon). \tag{10.44}$$

For nonevanescent waves, these further simplify to

$$p = n\cos\theta, \quad q = -\cos\theta/n. \tag{10.45}$$

Since, in most situations, we are considering plane wave propagation, these latter equations allow us to simplify our transmission and reflection formulas significantly.

10.4 Single thin films

For nonmagnetic materials, the Fresnel equations indicate that there is always some reflection when light is normally incident on an interface between two media; for instance, when light travels from air into a lens. If a system has many lenses or other optical elements, with reflection at every interface, its throughput can be severely reduced. The traditional approach used to mitigate this loss involves adding antireflective coatings to the surface of the lens, the simplest of which is a single thin film. An appropriately designed thin film can, in principle, completely eliminate the reflected light at a chosen wavelength.

This effect is achieved through destructive interference between the waves reflected at the outer and inner surfaces of the film. We illustrate this concept with a simple model before delving into the quantitative calculation. Figure 10.3 shows the basic arrangement, with a thin film of refractive index n_2 bordered by an external medium of index n_1 and an internal medium of index n_3. The optical thickness of the film, i.e. the index multiplied by the thickness d_2, is assumed to be $\lambda/4$. It is assumed that $n_1 < n_2 < n_3$.

There are two effects that can contribute to the changes of phase ϕ_1 and ϕ_2 of the reflected waves. The first is a phase change that arises on reflection, directly from the Fresnel formulas, which we write as δ_1 and δ_2 for the moment. We know that for light traveling from medium A to medium B, the reflected amplitude for normal incidence is given by

$$r_\perp = \frac{n_A - n_B}{n_A + n_B}. \tag{10.46}$$

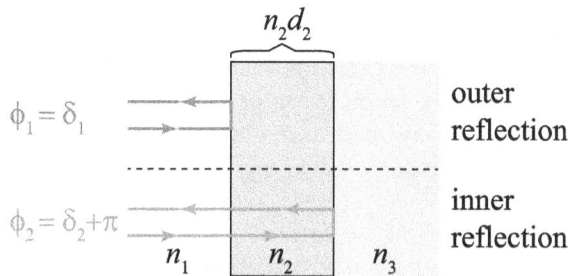

Figure 10.3. Simple model of the working of an antireflective coating.

If $n_A < n_B$, this implies a π phase change on reflection. Since $n_1 < n_2$ and $n_2 < n_3$, we have $\delta_1 = \pi$ and $\delta_2 = \pi$.

The second effect is simply the phase accumulated on propagation. Since the internally reflected wave travels a total round-trip distance of $\lambda/2$, the phase accumulated is π. The phase difference between the internally and externally reflected waves is therefore $\phi_2 - \phi_1 = \pi$, meaning the waves are completely out of phase and destructively interfere, reducing the reflected light.

If we instead assume that $n_1 < n_2$ but $n_2 > n_3$, then we have $\delta_1 = \pi$ and $\delta_2 = 0$, and $\phi_2 - \phi_1 = 0$, resulting in constructive interference. We therefore note that both the thickness of the film as well as the relative values of the refractive indices can affect the interference properties of the film.

We now turn to the quantitative analysis of the coating. Because we have only a single layer, the characteristic matrix takes the form of equation (10.23),

$$\mathbf{M}_2 = \begin{bmatrix} \cos(k_z d_2) & -i\sin(k_z d_2)/p_2 \\ -ip_2 \sin(k_z d_2) & \cos(k_z d_2) \end{bmatrix}. \tag{10.47}$$

We consider only the TE case for simplicity, since, for normal incidence, the TM results are the same. The reflection and transmission can be written as

$$r = \frac{[\cos\beta - ip_3 \sin\beta/p_2]p_1 - [-ip_2 \sin\beta + p_3 \cos\beta]}{[\cos\beta - ip_3 \sin\beta/p_2]p_1 + [-ip_2 \sin\beta + p_3 \cos\beta]}, \tag{10.48}$$

$$t = \frac{2p_1}{[\cos\beta - ip_3 \sin\beta/p_2]p_1 + [-ip_2 \sin\beta + p_3 \cos\beta]}, \tag{10.49}$$

where $\beta = k_z d_2$ and $p_\alpha = n_\alpha \cos\theta_\alpha$, with $\alpha = 1, 2, 3$. This formula is rather complicated and already illustrates the fact that although the matrix method gives us analytic results for transmission and reflection, those results are not easily interpreted.

With this complexity in mind, we turn to an illustrative example. We consider a thin film of refractive index n_2 between air and a glass of refractive index $n_3 = 1.55$, with a plane wave of wavelength λ normally incident upon it. We plot the reflected power $R = |r|^2$ as a function of the optical thickness $n_2 d_2$ of the film, measured in units of wavelength, for two different refractive indices n_2.

We can see that there are periodic oscillations in the reflected power; when the index $n_2 > n_3$, the reflection increases dramatically, and when $n_2 < n_3$, the reflection decreases and even becomes zero for special values of the optical thickness. In the end, we can make the following broad statement:

> For a film of optical thickness $\lambda/4$, $3\lambda/4$, $5\lambda/4$, ... , R reaches a maximum or minimum for $n_2 > n_3$ or $n_2 < n_3$, respectively; for a film of thickness $\lambda/2$, λ, $3\lambda/2$, ..., the opposite is true.

We may further look for a true antireflective film: under what conditions does r vanish for normal incidence? We return to equation (10.48) and let $\beta = \pi/2$, i.e. the optical thickness is a quarter wavelength. Under this condition, we have

$$r_\perp = \frac{-\dfrac{ip_3 p_1}{p_2} + ip_2}{-\dfrac{ip_3 p_1}{p_2} - ip_2}. \tag{10.50}$$

Simplifying, we look for solutions of this equation for which $r_\perp = 0$, or

$$\frac{p_2^2 - p_3 p_1}{p_2^2 + p_3 p_1} = 0. \tag{10.51}$$

We find that we must have $p_2 = \sqrt{p_3 p_1}$, or

$$n_2 = \sqrt{n_1 n_3}. \tag{10.52}$$

For $n_1 = 1$ and $n_3 = 1.55$, we have $n_2 = 1.22$, which motivated the choice in figure 10.4. We therefore have a quantitative pair of conditions, equation (10.52) combined with $\beta = \pi/2$, for designing antireflective coatings for lenses.

It should be noted that a single thin film produces highly selective antireflection at discrete wavelengths. This is illustrated in figure 10.5, where d_2 has been chosen to make the film antireflective at the wavelength $\lambda = 500$ nm; for simplicity, dispersion effects have been neglected. It can be seen that the film is only perfectly nonreflective at the designed wavelength. It still has quite low reflectivity over the entire visible spectrum, but there is a catch: it is not easy to find a thin-film material with a refractive index as low as $n_2 = 1.22$. In practice, thin films with a larger index are used, and these films have much more significant spectral limitations.

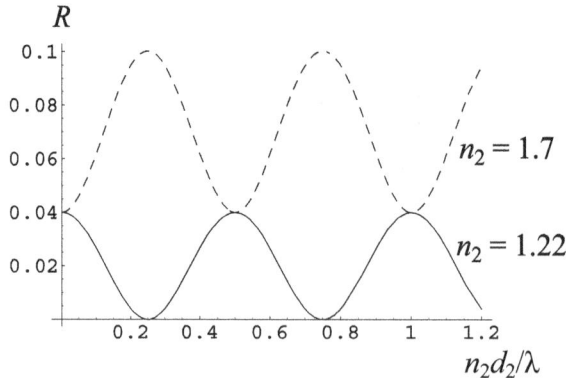

Figure 10.4. Reflected power R as a function of optical thickness $n_2 d_2$, for the values $n_2 = 1.22$ and $n_2 = 1.7$, with $n_1 = 1$ and $n_3 = 1.55$.

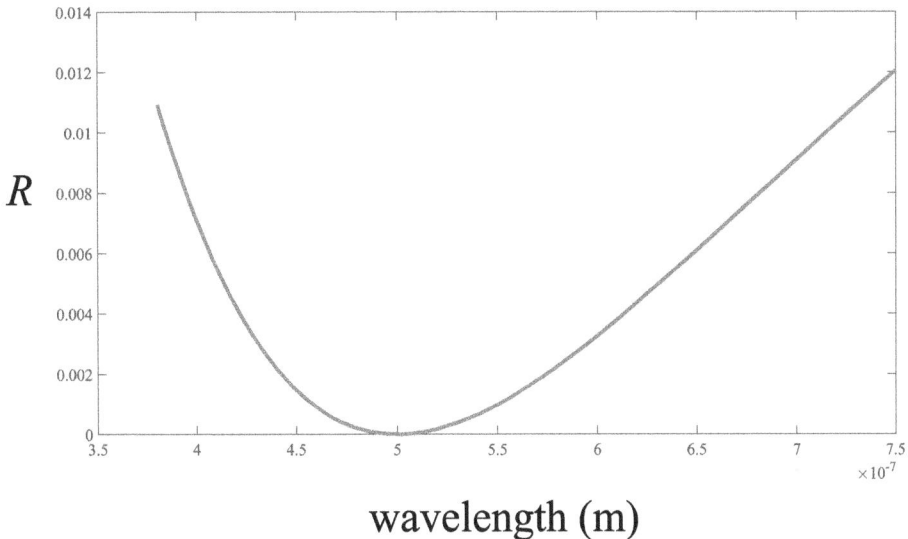

wavelength (m)

Figure 10.5. Reflected power R as a function of wavelength, with $n_1 = 1$, $n_2 = 1.22$, $n_3 = 1.55$, and the thickness is chosen to make the film antireflective at $\lambda = 500$ nm.

One solution for this wavelength selectivity is to use a multilayer thin film, where the additional layers provide the degrees of freedom needed to suppress additional wavelengths and can be achieved with higher index values. For more information on this and other aspects of thin-film optics, we refer the reader to the review article by Heavens [1], the handbook by Willey [2], and the chapter exercises.

10.5 Frustrated total internal reflection

In section 9.4.4, we introduced the phenomenon of TIR, in which a plane wave traveling from a dense medium to a rare medium beyond a critical angle is totally internally reflected. We also noted that this TIR can be frustrated by the introduction of another dense medium parallel to the first. This geometry is ideal for the use of characteristic matrices, and so we briefly analyze frustrated TIR using this approach.

The geometry is illustrated in figure 10.6. A plane wave is incident from a medium with refractive index n_1 into a medium of thickness d of index $n_2 < n_1$ above the critical angle θ_c of equation (9.107). Within the second layer, we expect the waves to be evanescent; a third medium of index n_1 allows for the conversion of these evanescent waves back into plane waves.

We may directly apply our matrix method to this problem, though we note that both p_2 and k_z are purely imaginary. It is therefore convenient to write $p_2 = i|p_2|$ and $k_z = i|k_z|$ for our calculations. If we further use the trigonometric identities $\cos(ix) = \cosh(x)$ and $\sin(ix) = i\sinh(x)$, we may write the characteristic matrix $\mathbf{M}$ of the rare medium as

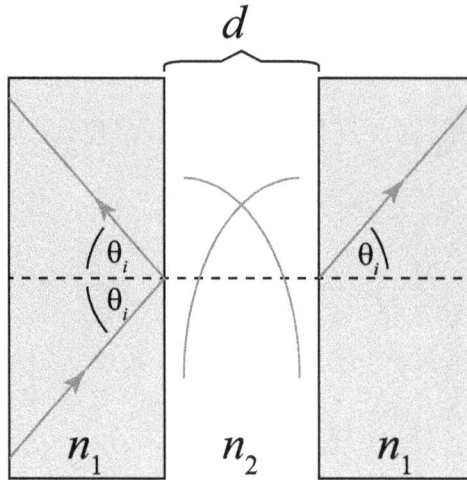

Figure 10.6. Notation related to frustrated TIR.

$$\mathbf{M} = \begin{bmatrix} \cosh(|k_z|d) & -i\sinh(|k_z|d)/|p_2| \\ i|p_2|\sinh(|k_z|d) & \cosh(|k_z|d) \end{bmatrix}. \tag{10.53}$$

The reflected and transmitted amplitudes can be found with a little algebra to be of the form

$$r_{TE} = \frac{-i\left(\dfrac{p_1}{|p_2|} + \dfrac{|p_2|}{p_1}\right)\sinh(|k_z|d)}{2\cosh(|k_z|d) + i\left(\dfrac{|p_2|}{p_1} - \dfrac{p_1}{|p_2|}\right)\sinh(|k_z|d)}, \tag{10.54}$$

$$t_{TE} = \frac{2}{2\cosh(|k_z|d) + i\left(\dfrac{|p_2|}{p_1} - \dfrac{p_1}{|p_2|}\right)\sinh(|k_z|d)}. \tag{10.55}$$

It is readily shown that $|r|^2 + |t|^2 = 1$; because the third medium has the same refractive index as the first, $T = |t|^2$ for this case.

Figure 10.7 shows an example of the transmitted and reflected power in frustrated TIR as a function of film thickness. The bounding media are taken to be glass with index $n_1 = 1.52$, such as two prisms with parallel faces, and the film is taken to be an air gap, $n_2 = 1$. The wavelength is 500 nm. It can be seen that the transmission becomes negligible when the gap thickness becomes greater than a wavelength, confirming our intuition that the evanescent waves in TIR are largely confined to within a wavelength of the surface.

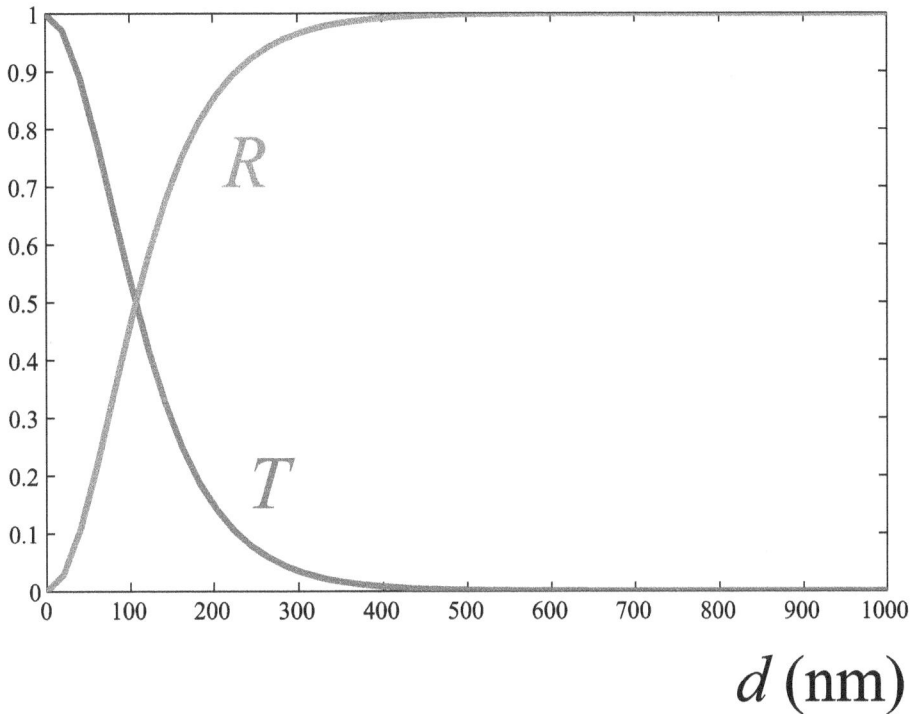

Figure 10.7. Simulation of transmitted and reflected power in frustrated TIR as a function of film thickness. Here, $n_1 = 1.52$, $n_2 = 1$, $\lambda = 500$ nm, and $\theta_i = 50°$. The critical angle is $\theta_c = 33.3°$.

10.6 Dielectric mirrors and photonic bandgaps

So far, we have restricted our examples to single thin films, which represent the simplest application of our matrix method. At the other extreme is a medium that consists of a large number of layers arranged periodically, an example of which is illustrated in figure 10.8. Such periodic multilayers can exhibit extremely useful optical properties, such as a one-dimensional photonic bandgap, and we consider such cases here.

We restrict our attention to a periodic multilayer where each period consists of two layers with permittivities ϵ_1 and ϵ_2 and thicknesses d_1 and d_2, respectively. The total thickness of one period is h, and the total number of layers is taken to be N. If we label the characteristic matrix of a single period as $\mathbf{M}(h)$, the matrix of the entire periodic multilayer is

$$\mathbf{M}(Nh) = [\mathbf{M}(h)]^N. \tag{10.56}$$

We can, of course, directly calculate such a matrix multiplication computationally, but we may also employ a powerful theorem of linear algebra to simplify the calculation analytically [3]. This theorem indicates that the Nth power of the matrix may be written in the form

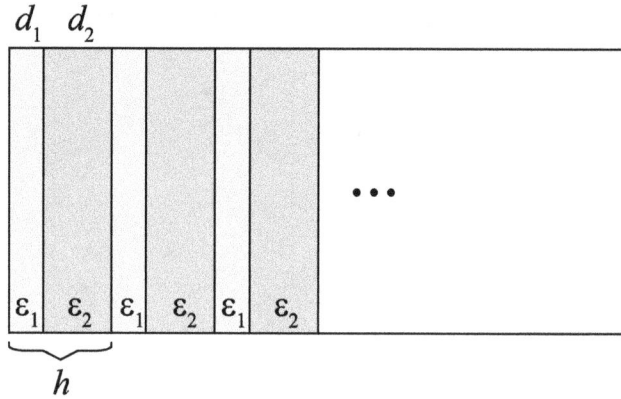

Figure 10.8. Illustration of the notation for a periodic multilayer film. In this case, each period consists of two layers with permittivities ϵ_1 and ϵ_2 with thicknesses d_1 and d_2, respectively. The total thickness of a period is h.

$$[\mathbf{M}(h)]^N = \begin{bmatrix} m_{11}\mathcal{U}_{N-1}(a) - \mathcal{U}_{N-2}(a) & m_{12}\mathcal{U}_{N-1}(a) \\ m_{12}\mathcal{U}_{N-1}(a) & m_{22}\mathcal{U}_{N-1}(a) - \mathcal{U}_{N-2}(a) \end{bmatrix}, \qquad (10.57)$$

where $a \equiv (m_{11} + m_{22})/2$, m_{ij} are the components of the matrix of one period of the multilayer, and $\mathcal{U}_N(x)$ is the Nth-order Chebyshev polynomial of the second kind. These polynomials can be expressed in the compact form [4]

$$\mathcal{U}_n(x) = \frac{(-1)^n(n+1)\pi^{1/2}}{2^{n+1}(n+1/2)!(1-x^2)^{1/2}} \frac{d^n}{dx^n}(1-x^2)^{n+1/2}. \qquad (10.58)$$

If we write $x \equiv \cos\theta$, these polynomials take on the exceedingly simple form

$$\mathcal{U}_n(\cos\theta) = \frac{\sin[(n+1)\theta]}{\sin\theta}. \qquad (10.59)$$

Let us use these formulas to numerically calculate the reflection of a periodic multilayer for different values of N. The system is taken to be alternating layers of zinc sulfide ($n_1 = 2.3$) and cryolite ($n_2 = 1.35$). Each of these layers is taken to have an optical path length of one quarter wavelength at the wavelength $\lambda_0 = 546$ nm; this amounts to $d_1 = 59.3$ nm and $d_2 = 101.1$ nm. Neglecting dispersion, figure 10.9 shows the reflection of the multilayer film as a function of wavelength. The most striking feature of the reflection spectrum is that, even for a relatively small number of layers, the system has a broad range of wavelengths over which the reflectivity is effectively unity. Even though the system is completely dielectric, it acts as a perfect mirror over these wavelengths; the system is thus often referred to as a *dielectric mirror*.

If we consider the same system from the perspective of transmission, we can say that there is a range of wavelengths over which no light can propagate through the multilayer film, at least from normal incidence. This behavior is analogous to the energy bandgaps that electrons experience when propagating through a periodic arrangement of atoms, and today it is said that the dielectric mirror has a *photonic*

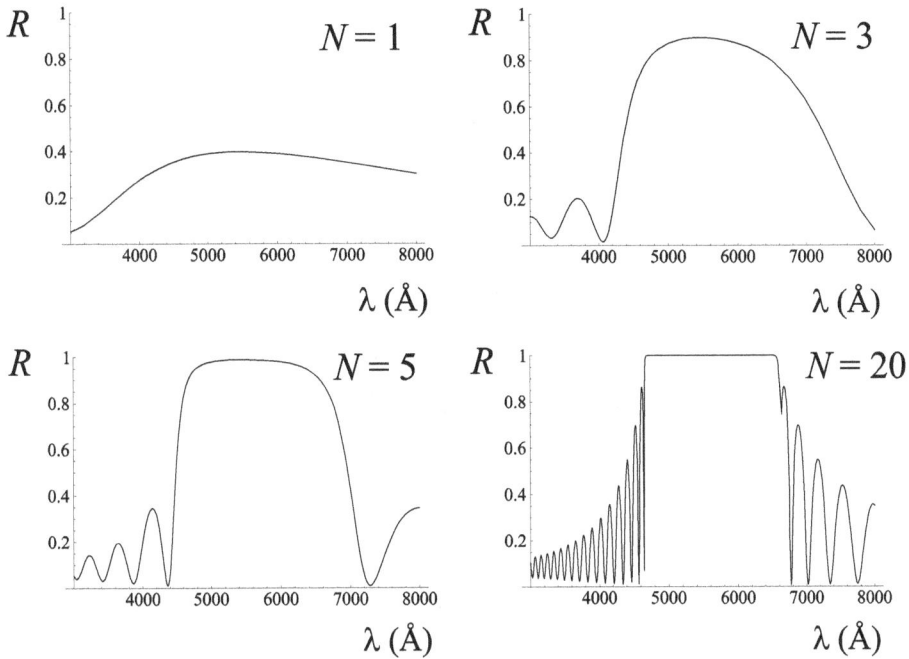

Figure 10.9. Reflection of a periodic multilayer consisting of zinc sulfide and cryolite, as a function of wavelength, for numbers of layers N.

bandgap. In both cases, the bandgaps arise from the cumulative interference that the respective waves experience upon multiple reflections. In recent years, the recognition of photonic bandgaps has led to the construction of materials that have periodic optical properties in one, two, or three dimensions. The materials are known as *photonic crystals*, and they have been used for a number of applications to guide and manipulate light, notably in fiber optics as well as in integrated circuits. A dielectric mirror only has a perfect bandgap for light propagating normally through the layers; a three-dimensional photonic crystal, in principle, has a perfect bandgap for any direction of propagation.

We are unable to calculate the optical properties of two-dimensional and three-dimensional photonic crystals semi-analytically as we have done for dielectric mirrors, and computational methods must be used to analyze the optical properties of such structures in general. We refer the interested reader to the book by Joannopoulos *et al* [5] as a starting point.

10.7 Exercises

1. Let us suppose, for some reason, that we want to make an antireflective coating for diamond ($n = 2.417$) that functions at $\lambda = 600$ nm and is nonreflective at normal incidence. How thick should the film be, and what should its refractive index be?

2. A thin film, freestanding in air, consists of two layers. The first is 500 nm thick and has an index $n = 1.5$, and the second is 1000 nm thick and has an index $n = 2.3$. For the $E_\perp$ case, find the reflection R and transmission T of this film for $\theta_i = 0$ and $\theta_i = 45°$. Plot the reflected power R and the transmitted power T for wavelengths between 500 nm and 700 nm. (You will have to use a computer for this, obviously.)

3. For glass with a refractive index of $n = 1.55$, the ideal antireflective coating would be a material with a refractive index $n_f = 1.22$. However, it is difficult to produce materials with such a low refractive index. Two options are magnesium fluoride, MgF_2, with $n_f = 1.38$, or a fluoropolymer like Teflon, with a refractive index of $n_f = 1.31$. Calculate the amount of reflected power for each of these cases, and compare it to the reflection from uncoated glass.

4. Thin-film interference effects can often survive even in absorbent layers, as shown by Kats *et al* [6], provided the layer is extremely thin. The index of gold at 600 nm is $n_{Au} = 0.248\,73 + i3.0740$, and the index of germanium at the same wavelength is $n_{Ge} = 5.6959 + i1.4059$. Let us consider light incident on a germanium film on a gold substrate. Over the spectral range from 0.4 μm to 1.0 μm, plot the reflected power for the following thicknesses of germanium: $d = 7$ nm, $d = 10$ nm, and $d = 13$ nm. Compare this reflectivity to the reflectivity of uncoated gold. You can neglect dispersion effects for your results. Comment on your results and how they support the explanation of a thin-film interference effect.

5. MgF_2 (index 1.38) is often used as an antireflective coating on crown glass (index 1.52), but the coating index falls short of the ideal value of 1.23, leaving some reflection. To correct for this, a 'V-coat' is often used, in which an additional high-index layer of quarter-wave thickness is inserted between the substrate and the MgF_2. Taking this extra layer to have an index of 1.7, plot the reflected power for normal incidence as a function of wavelength from 400 nm to 700 nm, assuming the coating is designed for the wavelength 500 nm. Compare the result to the glass coated with only the MgF_2 coating. From your result, can you guess why it is called a V-coat? Dispersion effects can be ignored for this problem.

6. We may derive the amplitude reflected from a thin film at normal incidence in a more intuitive way by summing all possible reflected fields in an infinite series. The first few contributions are shown in figure 10.10. In this expression, r_{ij} represents the reflection of light coming from medium 1 at the interface with medium 2, t_{ij} represents the transmission from medium 1 into medium 2, and $\beta = n_2 k_z d_2$, where n_2 and d_2 are the index and thickness of the film. Write an infinite series expression for the total reflection from a thin film in terms of r_{ij} and t_{ij}. Using the properties of a nonabsorbing film and the sum of the geometric series, show that the total reflection has the form

$$r = \frac{r_{12} + r_{23}e^{2i\beta}}{1 + r_{12}r_{23}e^{2i\beta}}.$$

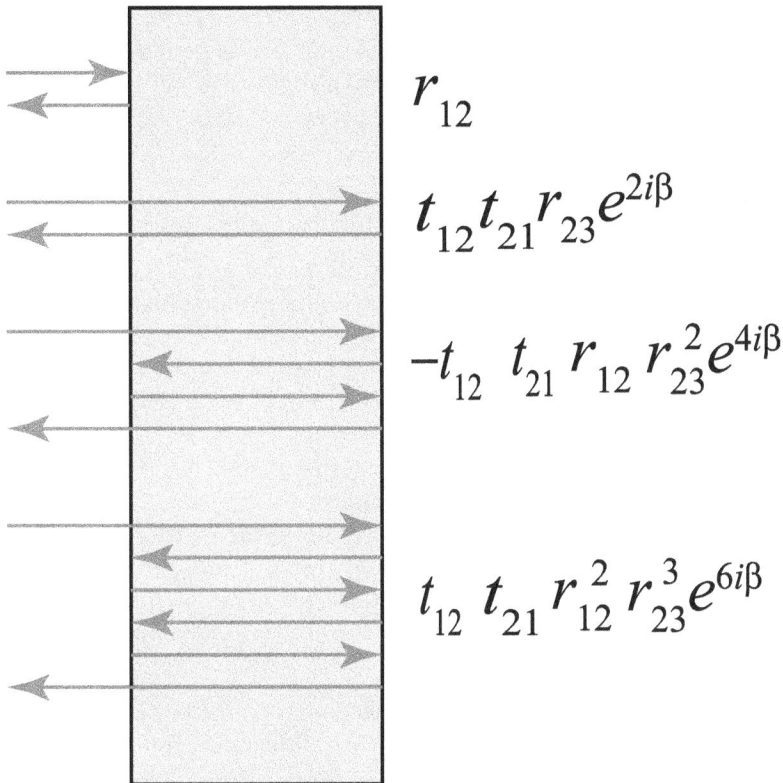

$$r_{12}$$

$$t_{12}t_{21}r_{23}e^{2i\beta}$$

$$-t_{12}\,t_{21}\,r_{12}\,r_{23}^2e^{4i\beta}$$

$$t_{12}\,t_{21}\,r_{12}^2\,r_{23}^3e^{6i\beta}$$

Figure 10.10. The first few contributions to the reflection from a thin film, in order of decreasing importance.

7. Because it is typically difficult to find a sufficiently low-index coating to produce perfect antireflection, a two-layer coating known as a V-coat is often used. For the V-coat, let us take n_0 as the index of air, n_4 as the index of the glass substrate, n_2 as the index of a low-index coating, and n_3 as the index of a high-index coating sandwiched between the low-index coating and the substrate. We assume that thicknesses d_2 and d_3 are taken to be quarter-wave thickness at a designed wavelength and normal incidence. Using the characteristic matrix method, derive an expression for the reflected amplitude at the designed wavelength. Assuming $n_2 = 1.38$ (MgF$_2$) and $n_4 = 1.607$ (flint glass), determine the ideal value of n_3.

8. By making an antireflective coating with multiple layers, a broadband optical coating can be designed which has a low reflectivity over a significant range of the visible spectrum. Consider a multilayer film designed for 500 nm at normal incidence. The substrate is crown glass ($n_5 = 1.52$). The next high-index layer has the thickness of a quarter wave with $n_4 = 1.70$, followed by a layer with the thickness of a half-wave with $n_3 = 2.20$, and then a quarter-wave MgF$_2$ layer with $n_2 = 1.38$. Plot the reflected power for normal incidence over the wavelength range 400 nm to 700 nm, ignoring dispersive effects.

9. In 1956, Kay and Moses [7] demonstrated a remarkable result: it is possible to design a continuously stratified medium that is perfectly nonreflective for any angle of incidence at a given wavelength. The refractive index of one such medium is given by

$$n(z) = \sqrt{1 + 2\frac{\kappa_1^2}{k_0^2}\text{sech}^2(2\kappa_1 z)},$$

where κ_1 is a characteristic inverse width of the medium.

Let us approximate the Kay and Moses result using a finite number of layers N, each of constant refractive index. We take $\lambda = 0.5 \, \mu$m and $\kappa_1 = 40 \, \mu\text{m}^{-1}$. We divide the range $-0.1 \, \mu$m $\leqslant z \leqslant 0.1 \, \mu$m into N constant layers; the index of each layer is determined by the value of $n(z)$ at the position of the layer center.

Calculate the reflected power R as a function of the incident angle for the cases $N = 4$, $N = 20$, $N = 200$, and confirm that the reflection decreases as the number of layers, and the accuracy of our approximation, increases. (Also, plot the refractive index for each case to confirm that we are not simply constructing a medium with a unit refractive index.)

References

[1] Heavens O S 1960 Optical properties of thin films *Rep. Prog. Phys.* **23** 1
[2] Willey R R 2006 *Field Guide to Optical Thin Films* (Bellingham, WA: SPIE Press)
[3] Abelès F 1950 Recherches sur la propagation des ondes électromagnétiques sinusoïdales dans les milieux stratifiés *Ann d Phys.* **12** 706–82
[4] Gbur G J 2011 *Mathematical Methods for Optical Physics and Engineering* (Cambridge: Cambridge University Press)
[5] Joannopoulos J D, Johnson S G, Winn J N and Meade R D 2008 *Photonic Crystals: Molding the Flow of Light* (Princeton, N J and Oxford: Princeton University Press)
[6] Kats M A, Blanchard P C and Romainand Genevet F 2013 Nanometre optical coatings based on strong interference effects in highly absorbing media *Nat. Mater.* **12** 20–4
[7] Kay I and Moses H E 1956 Reflectionless transmission through dielectrics and scattering potentials *J. Appl. Phys.* **27** 1503–8

IOP Publishing

Electromagnetic Optics

Gregory J Gbur

Chapter 11

Surface plasmons

There is one important class of interface effects whose behavior is qualitatively different from those considered in the previous chapters. Under the right conditions, certain metals can support electron charge-density waves that propagate along their surface. These waves are highly localized to the surface of the metal and possess a high intensity; they can also propagate long distances along the surface compared to the skin depth of the metal. Whereas light can usually penetrate on the order of tens of nanometers into a metal, these charge-density waves can often propagate over hundreds of microns.

These charge-density waves are known as *surface plasmons* (SPs), for reasons that will be explained as the chapter unfolds. Research into SPs exploded after 1998, when it was demonstrated that SPs can increase the amount of light transmitted through an array of subwavelength-sized holes in a silver plate [1]. Since then, the physics and application of SPs have become their own subfield of optical research, known as *plasmonics*.

Despite their name, which implies a quantum-mechanical nature (just as the 'photon' is the quantum of light), SPs appear naturally in Maxwell's equations. In this chapter, we consider the classical description of SPs and indicate their useful properties. We begin by considering the plasma model of a metal, which serves as a fair approximation for describing light–metal interactions.

11.1 Light propagation in a plasma

A plasma is an ionized gas possessing equal amounts of positive and negative charge, resulting in an overall net charge of zero. In calling it a gas, it is implied that at least one species of charge is unbound and is free to move in the presence of an external force. The surface of the Sun is a plasma, with the atoms ionized by the extreme temperature; the ionosphere of the Earth is a plasma, with the atoms ionized by solar radiation. The example most relevant to our interests, however, is a metallic

doi:10.1088/978-0-7503-6064-7ch11

conductor such as silver or gold, where electrons are very weakly bound and can be moved by an external electric field.

To characterize the basic optical properties of a plasma, we use a simplified version of the Lorentz model of the atom discussed in chapter 6; we now assume electrons experience a damping force $-g\dot{\mathbf{r}}$ but not a restoring force because they are unbound. The ions are tightly bound and form the structure of the metal; therefore, they are ignored in this model. Furthermore, interactions between electrons are ignored, which is a big assumption, but the results we will get agree well enough with experiment to justify it. This model is often referred to as the *Drude model*, after the work of German physicist Paul Drude, who used it to describe electron transport in metals [2].

Under our assumptions, the equation of motion for an individual electron is given by the expression

$$m\ddot{\mathbf{r}} = -g\dot{\mathbf{r}} - e\mathbf{E}_0 e^{-i\omega t}, \tag{11.1}$$

where again m is the mass of the electron, e is the magnitude of the electron charge, and $\mathbf{E}_0$ is the amplitude of an external monochromatic electric field. As in the Lorentz model, it is assumed that the motion of the electron is much smaller than the wavelength of the external field so that the field can be treated as uniform.

We again take

$$\mathbf{r}(t) = \mathbf{r}_0 e^{-i\omega t}, \tag{11.2}$$

and find that the time-harmonic position vector of the electron is of the form

$$\mathbf{r}_0 = \frac{e\mathbf{E}_0}{m\omega^2 - i\omega g}. \tag{11.3}$$

The displacement of an electron from equilibrium leaves a positive charge in its place, forming a dipole. If we assume that there are N electrons per unit volume, then the time-harmonic polarization density $\mathbf{P}_0$ can be written as

$$\mathbf{P}_0 = -Ne\mathbf{r}_0. \tag{11.4}$$

On substituting from equation (11.3), we find that the polarization induced in the plasma is

$$\mathbf{P}_0 = -\frac{eN\mathbf{E}_0}{m\omega^2 - i\omega g}. \tag{11.5}$$

We can convert this into a dielectric constant by recalling the relation

$$\mathbf{D} = \epsilon\mathbf{E} = \epsilon_0\mathbf{E} + \mathbf{P}. \tag{11.6}$$

If we substitute from equation (11.5) into this relation, we can eliminate $\mathbf{E}$ and find

$$\epsilon(\omega) = \epsilon_0\left[1 - \frac{Ne^2}{\epsilon_0(m\omega^2 - i\omega g)}\right]. \tag{11.7}$$

This expression can be written in a more elegant form as

$$\epsilon(\omega) = \epsilon_0 \left[1 - \frac{\omega_p^2}{\omega^2 - i\omega\omega_\tau} \right], \tag{11.8}$$

where ω_p is called the *plasma frequency*, defined as

$$\omega_p = \sqrt{\frac{Ne^2}{m\epsilon_0}}, \tag{11.9}$$

and $\omega_\tau = g/m$ is called the *relaxation frequency*. The relaxation frequency is a measure of the rate at which collision processes return electrons to their equilibrium states. Values of these frequencies for several metals, obtained from fitting to experimental data, are given in table 11.1. We note that the high end of the visible light spectrum has a frequency of 700 THz, which is equivalent to an angular frequency $\omega = 4.40 \times 10^{15}$ s^{-1}; therefore, the plasma frequency is higher than the highest frequency of the visible spectrum for all the metals listed.

Let us neglect the damping force in equation (11.8); we do this for simplicity, but it is also a good approximation as long as ω is sufficiently large. We may now rewrite our expression for ϵ as

$$\epsilon(\omega) = \epsilon_0 \left[1 - \frac{\omega_p^2}{\omega^2} \right]. \tag{11.10}$$

Looking this equation, we can see that ω_p represents a critical frequency at which the optical properties of the material change dramatically. When $\omega > \omega_p$, we have $\epsilon(\omega) > 0$; we therefore have a pure nonabsorbing dielectric with the wavenumber

$$k(\omega) = \omega\sqrt{\epsilon(\omega)\mu_0} = \frac{\omega}{c}\sqrt{1 - \frac{\omega_p^2}{\omega^2}}. \tag{11.11}$$

It is straightforward to calculate the phase velocity v_p and the group velocity v_g in this case; we have

$$v_p = \frac{c}{\sqrt{1 - \frac{\omega_p^2}{\omega^2}}} > c, \tag{11.12}$$

Table 11.1. Fitted values of the plasma frequency ω_p and the relaxation frequency ω_τ for several metals, after Ordal *et al* [3].

Metal	ω_p ($\times 10^{15}$ s^{-1})	ω_τ ($\times 10^{13}$ s^{-1})
Silver	13.7	2.73
Gold	13.7	4.07
Copper	12.0	5.24
Tungsten	9.12	8.16

$$v_g = c\sqrt{1 - \frac{\omega_p^2}{\omega^2}} < c. \tag{11.13}$$

When light with a frequency higher than the plasma frequency passes through a plasma, the phase velocity is always greater than the vacuum speed of light, and the group velocity is always less than the vacuum speed of light. This is a case where the group velocity provides an excellent measure of the speed of light in a medium.

For $\omega < \omega_p$, we now have $\epsilon(\omega) < 0$; this results in a purely imaginary wavenumber of the form

$$k = \pm\frac{i}{c}\sqrt{\omega_p^2 - \omega^2} \equiv \pm i\alpha, \tag{11.14}$$

where the $\pm$ is dictated by the direction of wave propagation. We expect this imaginary wavenumber to indicate the absorption of the electromagnetic wave; for a plane wave propagating in the $+z$-direction, we therefore choose the plus sign, and the electric field of the wave takes the form

$$\mathbf{E}(z) = \mathbf{E}_0 e^{ikz} = \mathbf{E}_0 e^{-\alpha z}, \tag{11.15}$$

where the wave decays exponentially as it propagates. The magnetic field can be found from Maxwell's equations:

$$\mathbf{H}(z) = -\frac{i\omega\mu_0}{\alpha}\hat{\mathbf{z}} \times \mathbf{E}_0 e^{-\alpha z}. \tag{11.16}$$

It should be noted that, due to the presence of the 'i' in the expression for the magnetic field, it is always ninety degrees out of phase with the electric field. This, in turn, means that there is no power flow in the system, as the Poynting vector $\mathbf{S}(z)$ demonstrates:

$$\mathbf{S}(z) = \frac{1}{2}\mathrm{Re}\{\mathbf{E}(z) \times \mathbf{H}^*(z)\} = -\frac{1}{2}\mathrm{Re}\left\{\frac{i\omega\mu_0}{\alpha}|\mathbf{E}_0|^2\right\} = 0. \tag{11.17}$$

What causes this dramatic change in the behavior of the plasma's optical properties at the plasma frequency? Because the electrons have mass, they do not move instantaneously in response to an applied electric field. If the frequency is sufficiently high, i.e. above the plasma frequency, the field switches direction too fast for the electrons to respond; the electrons effectively stand still, and no energy is transferred to them. If the frequency is below the plasma frequency, the field oscillates slowly enough that the electrons can respond to the field, and part of the energy of the field goes into setting the electrons into motion. It should be noted that the plasma frequency is inversely related to the mass m of the electrons, indicating that heavier particles have a lower plasma frequency.

If we maintain the complex permittivity of equation (11.8), we will find that there is some energy propagation into the metal. The case $\omega_\tau = 0$ is effectively a perfect metal approximation, where the electrons are undamped; with nonzero damping, the electrons can only imperfectly obstruct the propagation of light.

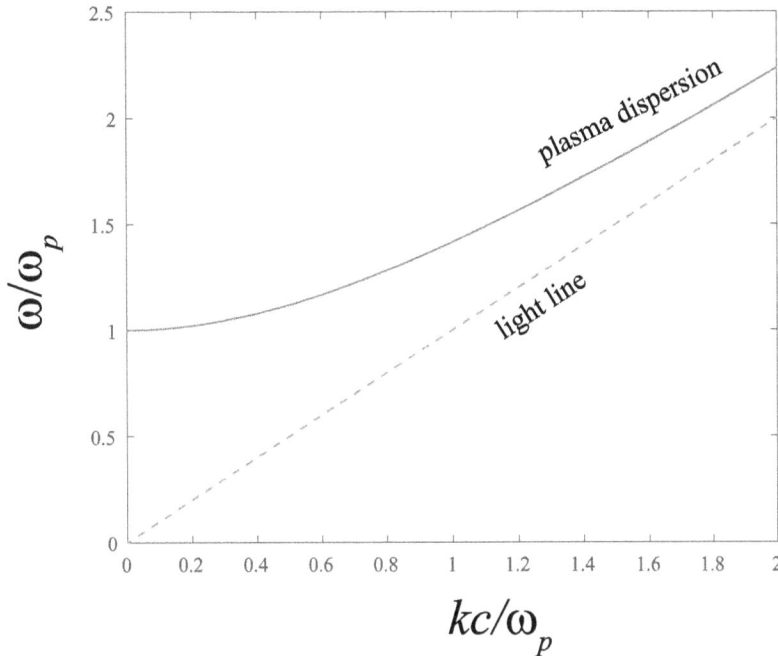

Figure 11.1. The dispersion curve for light in a plasma, with the light line $k = \omega/c$ included for comparison. We have taken $\omega_p = 13.7 \times 10^{15}$ s^{-1}.

It is of interest to introduce what is known as the *dispersion curve* of light in the plasma, or a plot of the dispersion relation showing how $\omega(k)$ is related to k. In quantum physics, the energy of a photon is interpreted as $\hbar\omega$, and the momentum of a photon is interpreted as $\hbar k$, so that the dispersion relation may also be viewed as characterizing the energy–momentum relationship of light in matter. (This interpretation must be used with caution, however, as the aforementioned Abraham–Minkowski controversy of section 7.4 has shown.)

The dispersion curve of light in the plasma is shown in figure 11.1. This plot clearly shows that there are no solutions for frequencies below the plasma frequency ω_p. Furthermore, it shows that as the frequency increases, the plasma curve approaches the light line; at high frequencies, the electrons are almost completely stationary, and the light propagates through the plasma as if nothing were there at all.

We also note that at any particular frequency, the wave in plasma has a lower k-value, or less momentum, than light in free space. This will become notable as we turn to SPs, for which the opposite is true and which plays a significant role in their usefulness.

11.2 Plasma oscillations

From the preceding discussion, it appears that the plasma frequency is the natural resonance frequency of the electron plasma. We can show this explicitly through Maxwell's equations, demonstrating that the plasma itself can undergo oscillations of the electron density even in the absence of an applied electromagnetic wave. These oscillations are typically referred to as *plasma oscillations*. They were first proposed by Tonks and Langmuir in 1929 [4] to explain unusually high electron speeds observed by Langmuir in low-pressure mercury arcs; they are therefore sometimes also called Langmuir waves. We will take a few moments to discuss the properties of plasma oscillations, which provide an excellent stepping stone to SPs.

We begin by using a simple model that, like the Drude model, ignores direct interactions between electrons; our discussion follows that of Bernstein and Trehan [5]. We assume that the background electron density is N and that this background is perturbed by a density $n(\mathbf{r}, t)$. The motion of the total electron density is described by a velocity field $\mathbf{v}(\mathbf{r}, t)$. We may therefore write the charge density as $\rho(\mathbf{r}, t) = -eN - en(\mathbf{r}, t)$ and the current density as $\mathbf{J}(\mathbf{r}, t) = -eN\mathbf{v}(\mathbf{r}, t)$. If we use these in the continuity equation, i.e. equation (2.57), we obtain

$$\frac{\partial n(\mathbf{r}, t)}{\partial t} + N\nabla \cdot \mathbf{v}(\mathbf{r}, t) = 0. \tag{11.18}$$

At a particular point $\mathbf{r}$, the motion of an electron is dictated by the force equation,

$$m\frac{\partial \mathbf{v}(\mathbf{r}, t)}{\partial t} = -e\mathbf{E}(\mathbf{r}, t), \tag{11.19}$$

where $\mathbf{E}(\mathbf{r}, t)$ is not an external field but is instead the internal electric field that arises from spatial variations of the charge density. This field can be determined from Gauss's law,

$$\nabla \cdot \mathbf{E}(\mathbf{r}, t) = -en(\mathbf{r}, t)/\epsilon_0. \tag{11.20}$$

Let us take the divergence of equation (11.19) and substitute from equations (11.18) and (11.20). We arrive at the equation

$$-\frac{m}{N}\frac{\partial^2 n(\mathbf{r}, t)}{\partial t^2} = \frac{e^2 n(\mathbf{r}, t)}{\epsilon_0}, \tag{11.21}$$

which may be rewritten as

$$\frac{\partial^2 n(\mathbf{r}, t)}{\partial t^2} + \frac{Ne^2}{m\epsilon_0}n(\mathbf{r}, t) = 0. \tag{11.22}$$

This equation represents simple harmonic motion of the electron density, and the oscillation frequency is equal to the plasma frequency ω_p, given by equation (11.9). This is the basic model of plasma oscillation, in which the electron density can oscillate even in the absence of an external field.

We have derived this result classically, but just as the electromagnetic field can be quantized to describe quanta of light called photons, plasma oscillations can be quantized—the fundamental quantum is called a *plasmon*. It turns out that 'plasmon' is a much less unwieldy term than 'plasma oscillation,' so plasmon has become the common name for both the classical and quantum descriptions of electron density oscillations. The volumetric oscillations of the electron density are often called *bulk plasmons* to distinguish them from the SPs we will describe momentarily.

It should be noted from equation (11.22) that our derived plasma oscillation has no spatial dependence whatsoever; it is a bulk oscillation of the entire electron plasma, with an effectively infinite wavelength. This suggests immediately that such a plasma oscillation cannot be excited by a light wave, as such a coupling would require that the light wave match both the frequency (energy) and wavelength (momentum) of the plasma oscillation.

The infinite wavelength is the result of our oversimplified model of the electron plasma; the model may be refined to include the pressure induced by the random thermal motion of the electrons. This model can also be derived in the context of Maxwell's equations, but this requires the use of some results from fluid mechanics and therefore falls somewhat outside the scope of this text. We refer the reader to the derivation of Bernstein and Trehan [5]; the resulting dispersion relation is of the form

$$\omega^2 = \omega_p^2 + \frac{3k_B T}{m}k^2, \tag{11.23}$$

where k_B is the Boltzmann constant and T is the temperature of the electron plasma. This result was first derived by Landau [6] in 1946, but due to Cold War isolation, it was not known in the West for many years. Bohm and Gross independently derived the formula in 1949 [7]. Because they have a finite group velocity, the plasma oscillations as derived by Landau and later by Bohm and Gross are genuine traveling electron density waves.

11.3 What is a surface plasmon?

In conventional electronics, energy is transported by means of electrical currents in conductors, where an electric potential is applied to create a flow of electrons in wires. In direct current systems, the current always flows in one direction, and this is typically used for small electronic devices. In alternating current systems, the current has a sinusoidal modulation, causing it to switch direction periodically. Alternating currents can transport energy over large distances without significant loss, so it is the method by which power is delivered to homes and businesses. The most common standard of frequency for AC operation is 50 or 60 Hz. In both of these cases, however, there is transport of electrons in the process, as roughly illustrated in figure 11.2(a).

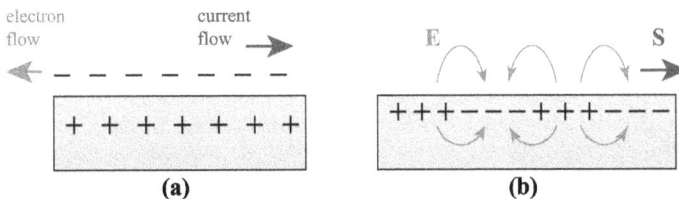

Figure 11.2. Rough illustration of (a) a direct current in an electrical conductor, and (b) a surface plasmon wave on the surface of a conductor.

DC current, for example, could be considered analogous to water flowing downstream in a river.

For water, of course, there is another way to transport energy, in the form of waves propagating along the surface of the river. Water waves do not involve a net transport of water itself but rather involve oscillations of water molecules that carry energy. The term 'wave' itself can be broadly defined as 'oscillations of a medium that transport energy without a net transport of the medium itself.'

An analogous phenomenon can occur in electrical conductors; waves can be excited in the electron density at the surface of the metal, and these waves can carry energy over surprisingly long distances along the surface. These waves are SPs, and a rough illustration of them is shown in figure 11.2(b). Because the waves are oscillations in electron density, they have local regions of net positive and net negative charge, which also means they have an electromagnetic wave associated with them. This electromagnetic wave, as we will see, is doubly evanescent, meaning that SPs are truly confined to the surface and are nonradiative; as long as they are propagating on a smooth, flat surface, they do not couple back into electromagnetic plane waves. This nonradiative feature also implies that the effective wavelength of SPs is smaller than that of light at the same frequency. Because their energy is localized to a two-dimensional surface, SPs tend to have a large field intensity near the surface, as we will see.

When surface plasmons are traveling over a long distance on a planar surface (here, 'long' is relative and means millimeters at most), then they are typically called surface plasmon polaritons (SPPs). A polariton is a hybrid particle involving a photon strongly coupled to an electric dipole, and a plasmon fits this description. SPs can also be excited on localized surfaces, such as nanoscale metal particles, and in such cases they tend to form standing waves and are called localized surface plasmons (LSPs) or surface plasmon resonances (SPRs). We will stick to the simple term 'surface plasmon,' or even 'plasmon,' for brevity. We must keep in mind, however, that bulk plasmons exist, as described in the previous section; we will only use the term 'plasmon' to describe SPs when there is no risk of ambiguity.

Before we delve into the mathematics of SPs, it is worth looking at the history of their discovery and appearance in a number of useful phenomena; these phenomena eventually propelled SPs into becoming an intense field of study. The concept of a surface plasmon was first discussed by Ritchie in 1957 [8] to explain experimental observations of anomalous electron energy loss through thin metal films. In short, Ritchie's analysis indicated that electrons lose more energy than expected because energy is coupled into SPs and that the characteristic frequency of surface plasma oscillations for a thin film is $\omega_p/\sqrt{2}$, lower than the bulk plasma frequency. In 1960, Stern and Ferrell further analyzed the problem, confirmed Ritchie's conclusions, and coined the term 'surface plasmon' [9]. From that point, SPs became a regular topic of study, though a relatively obscure one for quite some time.

SPs would eventually be found to play a significant role in several other optics mysteries. In the 1920s, Indian physicist Raman discovered the inelastic scattering of photons by matter, resulting in a change of direction and energy of the photon; this effect, for which Raman would win the 1930 Nobel Prize in Physics, is now called the Raman effect [10]. Raman scattering is the result of the inelastic scattering of light by vibrational modes of molecules, in which the molecule loses or gains energy depending on whether it emits or absorbs part of the energy from the photon, respectively. The vibrational modes of different species of molecules are highly unique, serving as a 'fingerprint' of the molecule, making Raman scattering a possible way to optically identify chemical samples. However, Raman scattering is typically very weak, having a cross section per molecule ranging from 10^{-30} cm^2 to 10^{-25} cm^2, orders of magnitude weaker than fluorescence, which has cross sections between 10^{-17} cm^2 to 10^{-16} cm^2 [11]. The weakness of the Raman signal greatly limited its usefulness as a laboratory tool.

In 1974, however, Fleischmann, Hendra, and McQuillan started applying Raman scattering to study molecules adsorbed at a silver electrode [12], finding unexpectedly strong Raman signals that they attributed to surface area effects. In attempting to replicate these results several years later, Jeanmaire and Van Duyne discovered that the strength of the Raman signal was orders of magnitude greater than expected for molecules on the metal surface [13] and well beyond what could be explained by geometry alone. Enhancements of up to 10^{14} would eventually be measured, largely mitigating the disadvantage of the weak signal produced by Raman scattering. The phenomenon became known as surface-enhanced Raman scattering (SERS). Naturally, the mechanism of this incredible advancement was immediately sought, and several candidates for enhancement were proposed. In 1979, Tsang, Kirtley, and Bradley showed that this enhancement appeared to be connected with the presence of surface plasmon modes in silver [14]. It was proposed that the strong enhancement of electric fields due to SPs, especially on rough surfaces, in turn enhances Raman scattering. Today, however, several mechanisms are thought to play a role in SERS, and it is still debated how much each mechanism, including SPs, contributes.

In spite of these discoveries, SPs remained a specialized topic for two decades. This changed in 1998 when Ebbesen *et al* announced the experimental observation of 'extraordinary optical transmission through subwavelength hole arrays' [1]. It had long been known that light transmission through holes smaller than the wavelength of light is very inefficient, and less light is transmitted than is even geometrically incident upon the hole. In a pioneering work published by Bethe in 1944 [15], he showed theoretically that the amount of light transmitted by a single subwavelength-sized hole depends on the hole diameter a as a^6, and inversely with wavelength λ as λ^{-4}. Thus, it was known that very little light passes through such an aperture, imposing a strong practical limit on applications such as near-field optics, where superresolution is achieved by using a subwavelength aperture to transmit or collect light near a surface [16].

Ebbesen and collaborators studied the transmission of light through a periodic array of subwavelength holes in a thin silver plate; remarkably, they found that the amount of light transmitted was about ten times greater than expected. They provided evidence that this enhancement, as in SERS, was due to the presence of SPs, and the phenomenon has become known as extraordinary optical transmission (EOT). However, as in the case of SERS, a lively debate over the physical mechanism of enhancement ensued, with various researchers arguing that the effect could be explained by diffracted evanescent waves [17] or waveguide mode resonances [18]; the latter paper even showed experimental evidence that SPs *suppress* the transmission of light.

The controversy was clarified somewhat by theoretical and experimental work by Schouten *et al* [19]; using a Young's double-slit configuration, they showed that both enhancement and suppression are possible with SPs and that these effects can be interpreted as plasmon interference between the slits. Though the other aforementioned effects can also play a role, it is now appreciated that SPs can cause dramatic changes to the transmission spectrum of subwavelength hole arrays. The work of Ebbesen *et al* truly launched the study of plasmonics as a major field of research in optics.

11.4 Surface plasmons in Maxwell's equations

We now turn to deriving the simplest solution for a surface plasmon, namely the solution at an interface between a dielectric with permittivity ϵ_1 and a metal with permittivity ϵ_2. The geometry is illustrated in figure 11.3, where again we use a curved line to indicate an evanescent wave. We will look for solutions to Maxwell's equations that are evanescent at both $z > 0$ and $z < 0$. Recall that, in solving the problem of refraction at an interface, we had three fields—incident, reflected, and refracted—and that this produced a unique solution. We expect that, with only two fields present in our plasmon derivation, there will be additional restrictions on the solution, in particular on the material properties of the media, and our goal is to determine what those restrictions are.

We begin, for once, with the magnetic field and define the magnetic field in each region as

$$\mathbf{H}_j(\mathbf{r}) = \hat{\mathbf{y}} H_j e^{i(k_{xj}x + k_{zj}z)}, \tag{11.24}$$

where $j = 1, 2$ and k_{xj} and k_{zj} are the x-component and the z-component, respectively, of the wave vector in each region. We therefore have a p-polarized, or TM, mode.

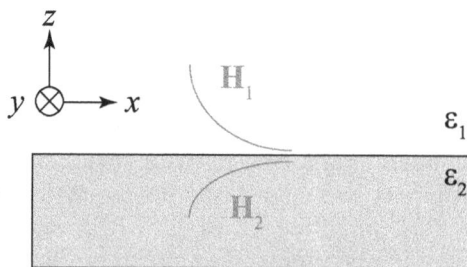

Figure 11.3. Rough illustration of the geometry and notation used to derive SPs at a single interface.

We can readily determine the electric field from the Maxwell–Ampère law in the form

$$\mathbf{E}_j(\mathbf{r}) = \frac{H_j}{\omega \epsilon_j}[\hat{\mathbf{x}}k_{zj} - \hat{\mathbf{z}}k_{xj}]e^{i(k_{xj}x + k_{zj}z)}. \tag{11.25}$$

You might wonder where the free current $\mathbf{J}_f(\mathbf{r})$ has gone in our application of the Maxwell–Ampère law, considering that a surface plasmon involves the motion of electric charges, i.e. a localized current. We recall from section 5.4 that for monochromatic fields, we can use Ohm's law to absorb the induced free current into a complex permittivity, which is the approach that we take here.

Following a procedure similar to the one we used for reflection and refraction, we now apply the boundary conditions

$$E_{x1} = E_{x2}, \quad H_1 = H_2, \tag{11.26}$$

which are the tangential boundary conditions. We first note that these conditions imply that $k_{x1} = k_{x2} \equiv k_x$. We then use equation (11.25) to write the electric field condition in terms of the magnetic field, i.e.

$$\frac{k_{z1}}{\epsilon_1}H_1 = \frac{k_{z2}}{\epsilon_2}H_2. \tag{11.27}$$

We therefore have two equations for the magnetic field components H_1 and H_2,

$$H_1 - H_2 = 0, \tag{11.28}$$

$$\frac{k_{z1}}{\epsilon_1}H_1 - \frac{k_{z2}}{\epsilon_2}H_2 = 0. \tag{11.29}$$

These can be written in matrix form as

$$\begin{bmatrix} \dfrac{k_{z1}}{\epsilon_1} & -\dfrac{k_{z2}}{\epsilon_2} \\ 1 & -1 \end{bmatrix} \begin{bmatrix} H_1 \\ H_2 \end{bmatrix} = \begin{bmatrix} 0 \\ 0 \end{bmatrix}. \tag{11.30}$$

This homogeneous matrix equation can only be satisfied if the determinant of the matrix is zero, or

$$\frac{k_{z1}}{\epsilon_1} - \frac{k_{z2}}{\epsilon_2} = 0. \tag{11.31}$$

Let us now take advantage of the fact that we may write

$$k_{zj}^2 = \epsilon_j k^2/\epsilon_0 - k_x^2, \tag{11.32}$$

where $k = \omega/c$ is again the free-space wavenumber and $j = 1, 2$. Let us move one component of equation (11.31) to the other side of the equals sign and then square the expression; on substituting from equation (11.32) twice into the resulting formula, we obtain

$$\frac{\epsilon_1 k^2}{\epsilon_1^2} - \frac{k_x^2}{\epsilon_1^2} = \frac{\epsilon_2 k^2}{\epsilon_0 \epsilon_2^2} - \frac{k_x^2}{\epsilon_0 \epsilon_2^2}. \tag{11.33}$$

We solve this expression for k_x, and after some manipulation arrive at the important result

$$k_x = k \sqrt{\frac{\epsilon_1 \epsilon_2}{\epsilon_0 (\epsilon_1 + \epsilon_2)}}. \tag{11.34}$$

If we introduce the relative permittivity $\tilde{\epsilon}_j = \epsilon_j / \epsilon_0$, we may write this in its standard form,

$$k_x = k \sqrt{\frac{\tilde{\epsilon}_1 \tilde{\epsilon}_2}{\tilde{\epsilon}_1 + \tilde{\epsilon}_2}}. \tag{11.35}$$

This expression, which quantifies the wave vector of our hypothetical plasmon wave along the surface, is the formula for the surface plasmon wavenumber. Our first task is to determine under what conditions this wavenumber is real-valued, which implies a free-propagating wave. This is easy to determine if we first assume that ϵ_1 and ϵ_2 are real-valued and that medium 1 is a dielectric with $\epsilon_1 > 0$. We readily see that the quantity under the square root will be positive, and the wavenumber real, if

$$\epsilon_2 < -\epsilon_1. \tag{11.36}$$

Neglecting the imaginary part of the permittivity for the moment, this condition is achieved for metals over much of the visible light range. Table 11.2 lists the relative permittivities of metals at several wavelengths. All of these examples have a negative real part, easily satisfying the plasmon condition if medium 1 is air or vacuum. It should be noted that silver and gold have the smallest imaginary parts, implying the lowest absorption; because of this, these are the usual metals used in plasmonics.

Table 11.2. Complex relative permittivities of several metals.

Material	Wavelength (nm)	Relative permittivity	[Ref]
Copper	617	$-10.1820 + 1.9230i$	[20]
Copper	1240	$-71.3841 + 7.3264i$	[21]
Silver	617	$-17.2355 + 0.4982i$	[20]
Silver	1240	$-71.9719 + 5.5864i$	[21]
Gold	617	$-10.6619 + 1.3742i$	[20]
Gold	1240	$-76.7745 + 6.5249i$	[21]
Aluminum	619.9	$-54.2604 + 19.4480i$	[21]
Aluminum	1240	$-153.56 + 29.880i$	[21]

One important consequence of condition (11.36), found by looking back at equation (11.10), is that SPs can only exist at frequencies below the plasma frequency, where light waves cannot freely propagate.

We now look at the consequences of condition (11.35) for the rest of the wave behavior. If we substitute from this condition back into equation (11.32), we find after a little manipulation that

$$k_{zj}^2 = k^2 \frac{\epsilon_j^2}{\epsilon_0(\epsilon_1 + \epsilon_2)}. \tag{11.37}$$

This is negative for both $j = 1, 2$ and demonstrates that our surface plasmon wave is doubly evanescent. By physical reasoning, we argue that the sign of k_{z1} must be positive to produce exponential decay in the $+z$-direction; then, equation (11.31), with the recognition that $\epsilon_2 < 0$, indicates that the sign of k_{z2} must be negative, producing exponential decay in the $-z$-direction.

Our formula (11.35) did not, in fact, assume that $\tilde{\epsilon}_2$ is real-valued, and so it can also be used to determine the complex plasmon wavenumber, which takes into account material absorption. Let us write

$$\tilde{\epsilon}_2 = \tilde{\epsilon}_2' + i\tilde{\epsilon}_2'', \tag{11.38}$$

where $\tilde{\epsilon}_2'$ and $\tilde{\epsilon}_2''$ are real numbers. The real and imaginary parts of k_x can, in fact, be determined exactly, but the exact formulas are very complicated and difficult to interpret. However, if we assume that $|\tilde{\epsilon}_2'| \gg \tilde{\epsilon}_2''$, which we can see applies to the values for metals given above, we may approximate the solutions as

$$k_x' = k \left(\frac{\tilde{\epsilon}_1 \tilde{\epsilon}_2'}{\tilde{\epsilon}_1 + \tilde{\epsilon}_2'} \right)^{1/2}, \tag{11.39}$$

$$k_x'' = k \left(\frac{\tilde{\epsilon}_1 \tilde{\epsilon}_2'}{\tilde{\epsilon}_1 + \tilde{\epsilon}_2'} \right)^{3/2} \frac{\tilde{\epsilon}_2''}{2(\tilde{\epsilon}_2')^2}. \tag{11.40}$$

The real part of the surface plasmon wavenumber, and the corresponding surface plasmon wavelength, can therefore be determined from the real part of $\tilde{\epsilon}_2$.

We are now in a good position to calculate the most important lengths associated with SPs; we use silver at a wavelength of 617 nm, with permittivity $\tilde{\epsilon}_2 = -17.2355 + 0.4982i$, adjacent to air, as a typical example. From equation (11.39), we can find the plasmon wavelength along the surface to be

$$\lambda_{SP} = \frac{2\pi}{k_x'} = 599 \text{ nm}, \tag{11.41}$$

which, we note, is smaller than the free-space wavelength. The propagation distance along the surface, i.e. the distance over which the plasmon intensity decays by $1/e$, is given by

$$l_{SP} = \frac{1}{2k_x''} = 54 \ \mu m, \tag{11.42}$$

which appears to be a very short propagation distance indeed. We may compare this, however, to the penetration depth to which a light wave propagates in the metal at normal incidence; in particular, the distance z at which a light wave's intensity decays by $1/e$,

$$z = \frac{1}{2k\,\mathrm{Im}\{\sqrt{\tilde{\epsilon}_2}\}} = 12 \ nm. \tag{11.43}$$

Relative to penetration depth, the plasmon propagation distance is 'long range.' However, we will soon see that plasmons can propagate over much longer distances in thin films.

Finally, we look at the extension d of the evanescent wave into both air and the metal; in other words, the $1/e$ distance of intensity for both:

$$d_j = \frac{1}{2\,\mathrm{Im}\{k_{zj}\}} = \frac{1}{2k}\mathrm{Im}\left\{\sqrt{\frac{\tilde{\epsilon}_1 + \tilde{\epsilon}_2}{\tilde{\epsilon}_j^2}}\right\}. \tag{11.44}$$

This results in $d_1 = 0.2 \ \mu m$ for air and $d_2 = 11$ nm for the metal.

We therefore find that a surface plasmon is a wave that has a wavelength shorter than light of the same frequency, which propagates over distances significantly longer than the penetration depth of the metal along the surface and is tightly localized to that surface.

We may determine the dispersion relation for plasmons by substituting from equation (11.10) into equation (11.39); if we consider the case in which the first medium is air, we readily find

$$k_x^2 = \frac{\omega^2}{c^2}\left[1 - \frac{\omega^2}{2\omega^2 - \omega_p^2}\right]. \tag{11.45}$$

The resulting dispersion curve for a surface plasmon is shown in figure 11.4 along with the plasma dispersion curve and the light line.

There are several things to note in this figure. First, SPs exist at low frequencies where light cannot propagate in a plasma. Second, we note that at high wavenumbers, the surface plasmon frequency asymptotically approaches $\omega_p/\sqrt{2}$. Finally, we note that the surface plasmon curve never crosses the light line. In order for light to couple into a surface plasmon, it has to match both the frequency and wavenumber of the surface plasmon, i.e. there must be an intersection of the dispersion curves. The lack of intersection indicates that light illuminating a smooth metal surface cannot excite a surface plasmon (electrons, with a very different dispersion curve, are a different story). We will introduce methods of exciting SPs with light in the next section.

We may also reverse this observation and note that a surface plasmon cannot spontaneously radiate into light while propagating on a smooth surface. This is one

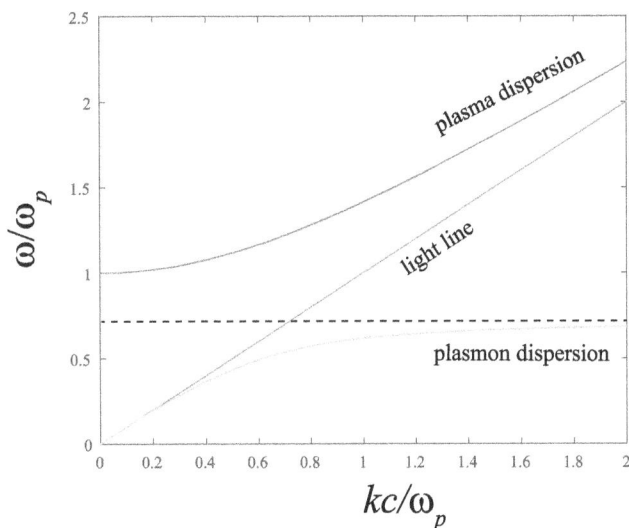

Figure 11.4. The dispersion curve for SPs, with both the light line $k = \omega/c$ and the light in plasma line included for comparison. The dashed line indicates $\omega_p/\sqrt{2}$.

of the properties that makes SPs appealing as an approach for making extremely small-scale optical elements.

It should be noted that there is another way to derive the surface plasmon dispersion relation, or, in fact, any surface wave dispersion relation. If we consider the Fresnel formulas for transmission and reflection at an interface, surface waves are found as singularities of the reflection amplitude as a function of k_x. We will leave this demonstration as an exercise but will use this approach to study other surface effects later in the text.

11.5 Optical excitation of surface plasmons

Because SPs do not couple to freely propagating light at a smooth surface, some sort of intermediate coupling mechanism must be used to generate plasmons from light. We briefly outline the most important of these.

In section 9.4.4, we noted that an evanescent wave produced in total internal reflection has a transverse wavenumber greater than the total wavenumber of light in the rarer medium. If the denser medium has a refractive index of n_1 and is incident upon the interface at an incident angle of θ_i, then the transverse wavenumber is

$$k_x = k_0 n_1 \sin \theta_i, \tag{11.46}$$

which will be greater than $k_0 n_2$ if $\theta_i > \theta_c$. If this wavenumber is matched to the surface plasmon wavenumber, the evanescent wave can couple directly to a surface plasmon on a smooth surface.

The reasonableness of this approach can be seen by looking at the case of silver at a wavelength $\lambda = 617$ nm and a surface plasmon wavelength $\lambda_{\text{SP}} = 599$ nm. If total internal reflection occurs in a glass prism of index $n_1 = 1.55$ at an interface with air of $n_2 = 1.0$, we find that our evanescent wave will match the surface plasmon wavelength if

$$\sin \theta_i = \frac{\lambda}{n_1 \lambda_{SP}}, \tag{11.47}$$

or if $\theta_i = 41.65°$. This angle is just above the critical angle of $\theta_c = 40.18°$.

There are two standard configurations for exciting SPs, named after their creators and introduced in the same year, which are illustrated in figure 11.5. In the Otto configuration, a prism is brought close to the surface of a plasmon-supporting metal, with an air gap in between, and the plasmons are created on the surface facing the prism [22]. In the Kretschmann–Raether configuration, a thin film of metal is deposited directly on the prism, and plasmons are excited on the metal–dielectric interface opposite the prism [23]. It is important to note that the incident field must be TM polarized in order to excite a surface plasmon, as an electric field along the direction of propagation is required.

A second method for exciting SPs is to use a grating on the surface of the metal to couple light into a surface plasmon. Crudely analogous to a race runner using a starting block, the surface plasmon uses the grating to give itself leverage to launch itself. If the grating has a spatial period a, then the incident wave can couple to wavenumbers

$$k_x = k \sin \theta_i + 2\pi m / a, \tag{11.48}$$

where m is an integer.

We can estimate the needed grating period for silver, again at $\lambda = 617$ nm, and take $\theta_i = 45°$ and $m = 1$. Solving equation (11.48) for a, we find that $a = 1.91 \mu$m, which is comparable to the periods used in ordinary diffraction gratings at optical wavelengths.

A third method for exciting SPs is to use a roughened surface. The roughened surface can be treated, at least approximately, as a superposition of diffraction gratings with a spatial power spectrum $S(\mathbf{k})$, where $\mathbf{k}$ represents the wavenumber of the gratings. An incident plane wave can therefore excite SPs provided the power spectrum possesses a nonzero $\mathbf{k}$ that satisfies equation (11.48). This process can also work in reverse; SPs excited on a rough surface can couple back into light;

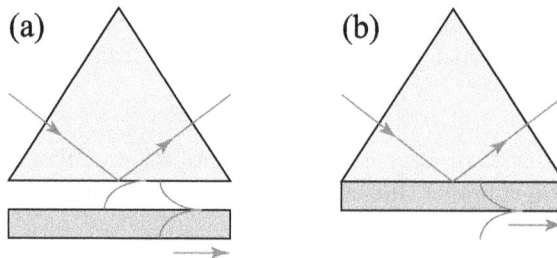

Figure 11.5. Illustration of the methods of evanescent wave excitation of SPs. (a) Otto configuration, (b) Kretschmann–Raether configuration.

Kretschmann provides an early analysis of this process [24]. In fact, all of the methods presented here for coupling light into SPs can be reversed. This straightforward ability to convert from SPs into light and vice versa, as well as the subwavelength nature of SPs, is part of what makes plasmons appealing in the design of optical interconnects.

11.6 Field enhancement of surface plasmons

We have noted that the electric field of an SP is enhanced with respect to the free-propagating field that excited it; this is one part of the plasmonic contribution to SERS. Ideally, we could calculate this enhancement analytically by solving Maxwell's equations for one of the coupling mechanisms of the previous section; this approach, however, does not match the experimental approach typically used for SERS. As an alternative, we can estimate the maximum field enhancement of an incident plane wave by use of a clever method introduced by Weber and Ford in 1981 [25], which is a simple application of Maxwell's equations. We present this method in this section.

The method is based on looking for the equilibrium condition between the power a plane wave deposits into a smooth metal surface and the power dissipated in the metal by a surface plasmon. We do not, in general, expect 100% efficiency in coupling, so this method imposes an upper limit on the amount by which the electric field is enhanced.

From section 9.4.3, we can write the component of the power per unit area normally incident upon the dielectric–metal interface as

$$S_{PW} = \frac{|\mathbf{E}_0|^2}{2} \sqrt{\frac{\epsilon_1}{\mu_0}} \cos \theta_i [1 - |r|^2], \tag{11.49}$$

where PW denotes 'plane wave,' $\mathbf{E}_0$ is the vector amplitude of the incident wave, θ_i is the angle of incidence, and r is the reflected amplitude. This equation, in essence, says: 'incident power minus reflected power equals power transmitted into the metal.'

Turning to SPs, we can use equations (11.24) and (11.25), together with our definition of the Poynting vector, equation (7.26a), to write the complex component of the Poynting vector along the surface as

$$S_{\text{SP}}(z) = |\mathbf{H}_{\text{SP}}|^2 \frac{k_x}{2\omega\epsilon_j} e^{-2\alpha_j z} e^{-2k_x'' x}, \tag{11.50}$$

where $j = 1, 2$ and depends on the medium we are in, $\mathbf{H}_{\text{SP}}$ is the amplitude of the magnetic field of the surface plasmon, and α_j is the imaginary part of k_{zj}, given by equation (11.37). We recall that the magnetic field amplitude is continuous across the interface, so the same amplitude $\mathbf{H}_{\text{SP}}$ applies on either side of the boundary.

We now integrate over z to give the total power flow per unit length along y, i.e.

$$P_{\text{SP}} = \int_{-\infty}^{\infty} \text{Re}\{S_{\text{SP}}\} dz, \tag{11.51}$$

which gives us

$$P_{\text{SP}} = \frac{|\mathbf{H}_{\text{SP}}|^2}{2\omega}\text{Re}\left\{\frac{k_x}{2\tilde{\alpha}_2\epsilon_2} + \frac{k_x}{2\tilde{\alpha}_1\epsilon_1}\right\}e^{-2k_x''x}, \tag{11.52}$$

where we have assumed $\alpha_1 > 0$, $\alpha_2 < 0$ and let $\tilde{\alpha}_j = |\alpha_j|$.

We may use the Ampère–Maxwell law to write the magnetic field amplitude $|\mathbf{H}_{\text{SP}}|^2$ in terms of the electric field amplitude $|\mathbf{E}_{\text{SP}}|^2$. We note that because the $\mathbf{H}$-field is continuous across the boundary, we may use either the value just within medium 1 or just within medium 2. We use medium 1, so we may write

$$|\mathbf{E}_{\text{SP}}|^2 = \frac{|k_1|^2}{\omega^2\epsilon_1^2}|\mathbf{H}_{\text{SP}}|^2, \tag{11.53}$$

which gives us the Poynting vector

$$S_{\text{SP}}(z) = |\mathbf{E}_{\text{SP}}|^2\frac{\omega\epsilon_1}{4k_1^2}\text{Re}\left\{k_x\frac{\tilde{\alpha}_2\epsilon_2 + \tilde{\alpha}_1\epsilon_1}{\tilde{\alpha}_1\tilde{\alpha}_2\epsilon_2}\right\}e^{-2k_x''x}. \tag{11.54}$$

For simplicity at this point, we are assuming that ϵ_1 is real-valued.

We finally note that the power dissipated in the metal, which is power per unit area, is of the form

$$\Delta P = -\frac{dS_{\text{SP}}}{dx} = 2k_x''S_{\text{SP}}, \tag{11.55}$$

which finally leads us to an expression for the dissipated power,

$$\Delta P = |\mathbf{E}_{\text{SP}}|^2\frac{\omega\epsilon_1}{4k_1^2}k_x''\text{Re}\left\{k_x\frac{\tilde{\alpha}_2\epsilon_2 + \tilde{\alpha}_1\epsilon_1}{\tilde{\alpha}_1\tilde{\alpha}_2\epsilon_2}\right\}. \tag{11.56}$$

We now set $x = 0$, as we are interested in the maximum power dissipated by the metal.

From this point, we assume that the plane wave power incident upon the metal is equal to the power dissipated in the surface, or $\Delta P = S_{PW}$, which allows us to solve for the ratio of electric fields $|\mathbf{E}_{\text{SP}}|^2/|\mathbf{E}_0|^2$. The algebra is surprisingly difficult but eventually leads to an expression of the form

$$\frac{|\mathbf{E}_{\text{SP}}|^2}{|\mathbf{E}_0|^2} = 2\cos\theta_i\frac{\epsilon_2'^2}{\epsilon_1^{1/2}}\frac{\sqrt{-(\epsilon_1 + \epsilon_2')}}{(\epsilon_1 - \epsilon_2')}\frac{1}{\epsilon_2''}[1 - |r|^2], \tag{11.57}$$

where we have taken advantage of the fact that $|\epsilon_2''| \ll |\epsilon_2'|$ for a plasmonic metal and applied equations (11.39) and (11.40). It should be noted that $|\mathbf{E}_{\text{SP}}|^2/|\mathbf{E}_0|^2 > 0$ as long as the usual plasmon conditions are satisfied.

The plasmon enhancement factor is plotted in figure 11.6 for several characteristic metals using experimental values of the permittivity. We follow the approach of Weber and Ford [25] and assume perfect coupling, $|r| = 0$, as well as first-order

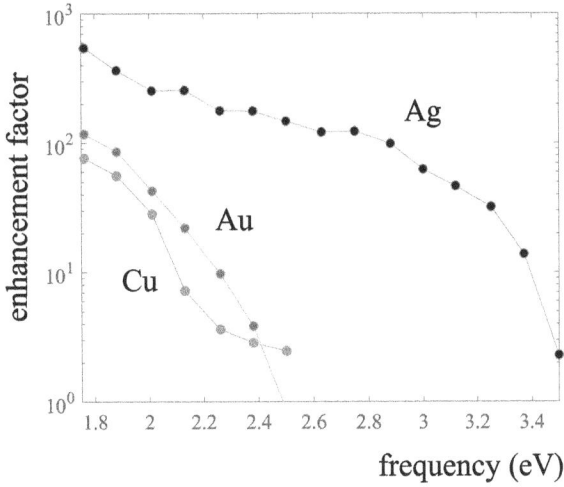

Figure 11.6. Calculated plasmon enhancement factor $|E_{SP}|^2/|E_0|^2$ for silver, copper, and gold. Optical data taken from [20].

grating coupling with a grating of period 800 nm, which implies that θ_i must satisfy the equation

$$\sin \theta_i = \left[\frac{\epsilon_2'}{1 + \epsilon_2'} \right]^{1/2} - \lambda/d. \tag{11.58}$$

We also follow the Weber and Ford approach of plotting the frequency of light in electron volts (eV). Optical constants for the metals are taken from the experimental data of Johnson and Christy [20], and results are only plotted for those frequencies at which plasmons are supported.

It can be seen that the field enhancement can be three orders of magnitude for silver and up to two orders of magnitude for copper and gold. These factors are not enough to account for the strong enhancement of SERS, and in fact, Weber and Ford argue that rough metal surfaces provide significantly less enhancement. These results demonstrate, however, that SPs can produce strong field enhancement under certain conditions.

The physical origin of this enhancement can be understood dimensionally. A plane wave is a three-dimensional volume wave, and the energy in a given volume is converted into what amounts effectively to a two-dimensional surface wave. The energy density is increased because of compression of the energy along one dimension.

11.7 Surface plasmons in thin films

SPs can, in principle, be excited at both interfaces of a metal film of finite thickness. For a sufficiently thick film, these plasmonic modes are effectively independent. When the thickness of the film becomes thin enough for the plasmon fields to reach from one interface to the other, the modes become coupled and split into a symmetric and an antisymmetric mode. The symmetric mode is particularly useful, as its propagation range increases as the thickness of the film decreases. In this section, we derive the properties of the thin-film modes as best we can. As we will see, analytic solutions can only be found with the use of significant approximations.

The geometry of our system is illustrated in figure 11.7; there are three materials, with the thin film possessing permittivity ϵ_2 and thickness d. We consider monochromatic excitations, and now have three magnetic fields to concern ourselves with, which we write as

$$\mathbf{H}_1(\mathbf{r}) = \hat{\mathbf{y}} H_1 e^{ik_x x} e^{\alpha_1 z}, \tag{11.59}$$

$$\mathbf{H}_2(\mathbf{r}) = \hat{\mathbf{y}} H_2^- e^{ik_x x} e^{-\alpha_2 z} + \hat{\mathbf{y}} H_2^+ e^{ik_x x} e^{\alpha_2(z-d)}, \tag{11.60}$$

$$\mathbf{H}_3(\mathbf{r}) = \hat{\mathbf{y}} H_3 e^{ik_x x} e^{-\alpha_3(z-d)}. \tag{11.61}$$

We already assume that the z-dependence will largely involve exponential decay and have chosen the signs of the exponentials appropriately. Furthermore, we have shifted the origins of the latter two exponentials so that H_2^+ and H_3 represent the field amplitudes at the surface $z = d$; this will make the analysis easier going forward. We know from our boundary conditions that all three fields have the same x-dependence. From the Ampère–Maxwell law, we may determine that the tangential components of the electric fields are

$$E_{x1} = \frac{H_1 \alpha_1}{i\omega\epsilon_1} e^{ik_x x} e^{\alpha_1 z}, \tag{11.62}$$

$$E_{x2} = \frac{1}{i\omega\epsilon_2} e^{ik_x x} \left[-\alpha_2 H_2^- e^{-\alpha_2 z} + \alpha_2 H_2^+ e^{\alpha_2(z-d)} \right], \tag{11.63}$$

$$E_{x3} = -\frac{1}{i\omega\epsilon_3} H_3 e^{ik_x x} \alpha_3 e^{-\alpha_3(z-d)}. \tag{11.64}$$

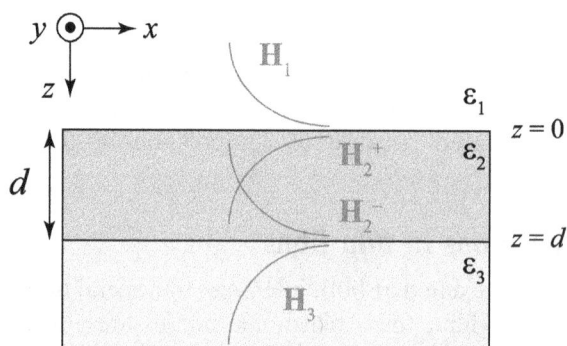

Figure 11.7. Illustration of the geometry and notation used to derive SPs on a thin metal film.

The magnetic field and the tangential component of the electric field must be continuous at each boundary, which leads us to a set of four boundary conditions involving H_1, H_2^+, H_2^-, and H_3. These conditions may be written in matrix form as

$$
\begin{bmatrix}
1 & -1 & -e^{-\alpha_2 d} & 0 \\
\dfrac{\alpha_1}{\epsilon_1} & \dfrac{\alpha_2}{\epsilon_2} & -\dfrac{\alpha_2}{\epsilon_2}e^{-\alpha_2 d} & 0 \\
0 & e^{-\alpha_2 d} & 1 & -1 \\
0 & -\dfrac{\alpha_2}{\epsilon_2}e^{-\alpha_2 d} & \dfrac{\alpha_2}{\epsilon_2} & \dfrac{\alpha_3}{\epsilon_3}
\end{bmatrix}
\begin{bmatrix}
H_1 \\ H_2^- \\ H_2^+ \\ H_3
\end{bmatrix}
=
\begin{bmatrix}
0 \\ 0 \\ 0 \\ 0
\end{bmatrix}.
\tag{11.65}
$$

This matrix equation is satisfied if and only if the determinant of the matrix vanishes. The determinant takes some effort to calculate; with some significant but straightforward manipulations (i.e. no tricks involved), we can write the result in the intermediate form

$$
\left[\frac{\alpha_2 \alpha_3}{\epsilon_2 \epsilon_3} + \frac{\alpha_2^2}{\epsilon_2^2} + \frac{\alpha_1 \alpha_3}{\epsilon_1 \epsilon_3} + \frac{\alpha_1 \alpha_2}{\epsilon_1 \epsilon_2}\right]e^{\alpha_2 d}
$$

$$
+ \left[\frac{\alpha_2 \alpha_3}{\epsilon_2 \epsilon_3} - \frac{\alpha_2^2}{\epsilon_2^2} - \frac{\alpha_1 \alpha_3}{\epsilon_1 \epsilon_3} + \frac{\alpha_1 \alpha_2}{\epsilon_1 \epsilon_2}\right]e^{-\alpha_2 d} = 0.
\tag{11.66}
$$

If we write the exponentials in terms of hyperbolic sine and cosine functions, we can express this in the significantly more compact form

$$
\tanh(\alpha_2 d) = -\frac{\left(\dfrac{\alpha_2 \alpha_3}{\epsilon_2 \epsilon_3} + \dfrac{\alpha_1 \alpha_2}{\epsilon_1 \epsilon_2}\right)}{\left(\dfrac{\alpha_2^2}{\epsilon_2^2} + \dfrac{\alpha_1 \alpha_3}{\epsilon_1 \epsilon_3}\right)}.
\tag{11.67}
$$

This is our surface plasmon dispersion relation for a thin metal film, as k_x and $k_0 = \omega/c$ are buried within the constants α_j. As one can see, we will not be able to get a simple analytic result for k_x due to the presence of the hyperbolic tangent function.

We can use this compact form to perform a consistency check on this formula. In the limit that the film becomes infinitely thick, we expect the SPs on the two surfaces to uncouple and take on the form of the single interface plasmons. When $d \to \infty$, we have $\tanh \to 1$, and we may rewrite our dispersion relation in the simplified form

$$
\left(\frac{\alpha_1}{\epsilon_1} + \frac{\alpha_2}{\epsilon_2}\right)\left(\frac{\alpha_2}{\epsilon_2} + \frac{\alpha_3}{\epsilon_3}\right) = 0.
\tag{11.68}
$$

If we compare this to equation (11.31), taking into account the different sign conventions we have used in this section, we see that we have two solutions to this equation, one that represents the surface plasmon at the 1–2 interface and the other that represents the surface plasmon at the 2–3 interface. Our new dispersion relation is at least consistent with what we have done previously. Furthermore, this result should convince us that there are two distinct solutions to equation (11.67).

It is hard to progress further analytically without making some simplifications. Let us consider the case where $\epsilon_1 = \epsilon_3$, e.g. the thin film is suspended in air. We may then simplify equation (11.67) to the form

$$\left(\frac{\alpha_1}{\epsilon_1} + \frac{\alpha_2}{\epsilon_2}\right)^2 - \left(\frac{\alpha_1}{\epsilon_1} - \frac{\alpha_2}{\epsilon_2}\right)^2 e^{-2\alpha_2 d} = 0, \tag{11.69}$$

which is now in a form that can be factorized,

$$\left[\left(\frac{\alpha_1}{\epsilon_1} + \frac{\alpha_2}{\epsilon_2}\right) + \left(\frac{\alpha_1}{\epsilon_1} - \frac{\alpha_2}{\epsilon_2}\right)e^{-\alpha_2 d}\right]$$
$$\times \left[\left(\frac{\alpha_1}{\epsilon_1} + \frac{\alpha_2}{\epsilon_2}\right) - \left(\frac{\alpha_1}{\epsilon_1} - \frac{\alpha_2}{\epsilon_2}\right)e^{-\alpha_2 d}\right] = 0. \tag{11.70}$$

Each of the terms in square brackets now represents a solution of the dispersion relation, i.e.

$$\left(\frac{\alpha_1}{\epsilon_1} + \frac{\alpha_2}{\epsilon_2}\right)e^{\alpha_2 d/2} + \left(\frac{\alpha_1}{\epsilon_1} - \frac{\alpha_2}{\epsilon_2}\right)e^{-\alpha_2 d/2} = 0, \tag{11.71}$$

$$\left(\frac{\alpha_1}{\epsilon_1} + \frac{\alpha_2}{\epsilon_2}\right)e^{\alpha_2 d/2} - \left(\frac{\alpha_1}{\epsilon_1} - \frac{\alpha_2}{\epsilon_2}\right)e^{-\alpha_2 d/2} = 0. \tag{11.72}$$

With hyperbolic trigonometric identities, these can be simplified to

$$\frac{\alpha_1}{\epsilon_1} + \frac{\alpha_2}{\epsilon_2}\tanh(\alpha_2 d/2) = 0, \tag{11.73}$$

$$\frac{\alpha_1}{\epsilon_1} + \frac{\alpha_2}{\epsilon_2}\coth(\alpha_2 d/2) = 0. \tag{11.74}$$

Our system is symmetric along the z-axis through the center of the film; this suggests that the two solutions must be of a symmetric and antisymmetric form. To determine which is which, let us look for a symmetric solution that satisfies equation (11.65). We take $H_1 = H_3 = H_o$ (for 'H outer') and $H_2^+ = H_2^- = H_i$ (for 'H inner'). On substitution into our matrix equation, we find that we obtain only two distinct equations of the form

$$H_o - H_i = H_i e^{-\alpha_2 d}, \tag{11.75}$$

$$\frac{\alpha_1}{\epsilon_1}H_o + \frac{\alpha_2}{\epsilon_2}H_i = \frac{\alpha_2}{\epsilon_2}H_i e^{-\alpha_2 d}. \tag{11.76}$$

If we eliminate H_i from these equations, we find that

$$\frac{\alpha_1}{\epsilon_1} + \frac{\alpha_2}{\epsilon_2}\tanh(\alpha_2 d/2) = 0, \tag{11.77}$$

which is equation (11.73), which we have now demonstrated represents the dispersion relation of the symmetric mode. By taking $H_1 = H_o$, $H_3 = -H_o$, $H_2^- = -H_i$, $H_2^+ = H_i$, we can show that the antisymmetric mode satisfies equation (11.74).

The only remaining problem is to use the dispersion relations to determine the propagation behaviors of the symmetric and antisymmetric modes. Here, we are significantly limited by the transcendental nature of equations (11.73) and (11.74). Let us consider the case of an extremely thin metal film, significantly smaller than the wavelength of light. Then it is reasonable to assume that the argument of the hyperbolic tangent function is small, and we can use the linear approximation $\tanh(x) \approx x$. Our dispersion relations are then of the form

$$\frac{\alpha_1}{\epsilon_1} = -\frac{\alpha_2}{\epsilon_2}\frac{\alpha_2 d}{2}, \quad \text{(symmetric)}, \tag{11.78}$$

$$\frac{\alpha_2 d}{2}\frac{\alpha_1}{\epsilon_1} = -\frac{\alpha_2}{\epsilon_2}, \quad \text{(antisymmetric)}. \tag{11.79}$$

Let us square each of these equations and solve for k_x, the complex surface plasmon wavenumber. Beginning with the antisymmetric case, we find that we can eliminate the factor of α_2^2 on each side; then, using the definition of α_1 adapted from equation (11.32),

$$\alpha_j^2 = k_x^2 - \epsilon_j k^2/\epsilon_0, \tag{11.80}$$

we may readily write

$$k_x^{AS} = k_0\sqrt{\frac{\epsilon_1}{\epsilon_0} + \left(\frac{2\epsilon_1}{dk_0\epsilon_2}\right)^2}, \tag{11.81}$$

where we write k_x^{AS} for the 'antisymmetric' plasmon wavenumber. In the limit $d \to 0$, the second term dominates the expression, and we find that

$$k_x^{AS} \approx \frac{2\epsilon_1}{\epsilon_2 d}. \tag{11.82}$$

Because of the permittivity of the metal and the thickness appearing in the denominator, the imaginary part of this expression ends up very large, resulting in high absorption. This antisymmetric mode therefore becomes strongly damped for thin films.

We may do a similar analysis for the symmetric relation, equation (11.78), though some additional steps are needed. We first note that we may write

$$\alpha_1^2 = \alpha_2^2 - \left(\frac{\epsilon_1}{\epsilon_0} - \frac{\epsilon_2}{\epsilon_0}\right)k_0^2, \tag{11.83}$$

which follows from the definition of α_j. Upon substitution into equation (11.78), we get a quadratic equation for α_2^2,

$$\alpha_2^4 - \frac{4\epsilon_2^2}{d^2\epsilon_1^2}\alpha_2^2 + \frac{4\epsilon_2^2}{d^2\epsilon_1^2}\left(\frac{\epsilon_1}{\epsilon_0} - \frac{\epsilon_2}{\epsilon_0}\right)k_0^2 = 0. \tag{11.84}$$

This equation can be solved using the quadratic formula of the form

$$\alpha_2^2 = \frac{2\epsilon_2^2}{d^2\epsilon_1^2}\left[1 \pm \sqrt{1 - \frac{d^2\epsilon_1^2}{\epsilon_2^2}\left(\frac{\epsilon_1}{\epsilon_0} - \frac{\epsilon_2}{\epsilon_0}\right)k_0^2}\right]. \tag{11.85}$$

We now again assume that d is extremely small, which means we can approximate the square root using the first three terms of its binomial expansion,

$$\sqrt{1 - x} \approx 1 - \frac{1}{2}x - \frac{1}{8}x^2. \tag{11.86}$$

If we take the negative root of our quadratic formula, which allows us to satisfy the unsquared dispersion relation, equation (11.78), and use the definition of α_j^2, we finally arrive at the result

$$k_x^S = k_0\sqrt{\frac{\epsilon_1}{\epsilon_0} + \frac{k_0^2 d^2 \epsilon_1^2}{\epsilon_0^2}\left(1 - \frac{\epsilon_1}{\epsilon_2}\right)^2}. \tag{11.87}$$

For this symmetric mode, we can see that in the limit $d \to 0$, the permittivity of the metal becomes increasingly less significant due to the factor of d^2 in the numerator, leaving us with a purely real-valued k_x^S (assuming that ϵ_1 is real-valued itself). Therefore, the propagation distance of the symmetric mode becomes increasingly large as the metal film becomes thinner.

What is the origin of this long-range behavior? Looking back at equation (11.63), we can see that a mode that is symmetric in H_y is antisymmetric in E_x. Thus, within the film itself, the component of the electric field that drives electrons along the film and consequently causes a loss of energy into the metal becomes smaller as the thickness of the film decreases. In short: for the symmetric mode, the longitudinal electric field is largely pushed out of the lossy metal, whereas the opposite is true for the antisymmetric mode. This disappearance of damping may seem too good to be true, and in a sense, it is: as the film becomes thinner, it can be shown that the plasmon field extends over a larger range outside the metal, in regions 1 and 3.

It should be noted that we have described the modes as 'symmetric' and 'antisymmetric' based on the behavior of the **H**-field. Other researchers may label these modes in the opposite sense by basing their description on E_x; it is therefore important to specify what is being referred to when discussing the symmetry of a mode.

The first work to recognize the long-range thin-film plasmon mode appears to have been a study by Sarid [26], who used an analysis very different from ours. Other worthwhile references about thin-film plasmonics are the reviews by Berini [27]; Han and Bozhevolnyi [28]; and the monograph by Li [29].

11.8 Extraordinary optical transmission

We have already described how the observation in 1998 [1] of extraordinary optical transmission through an array of subwavelength holes in a metal plate really launched the field of plasmonics; it is worthwhile to spend a few minutes elucidating the physics of this extraordinary optical transmission.

As we have seen, a surface plasmon is a wave phenomenon, and it therefore has interference, diffraction, and resonance effects that are strongly dependent on wavelength. If we consider a periodic array of subwavelength holes, we can already imagine that the transmission can depend on the size of the wavelength relative to the hole diameter, the period of the array, and even the thickness of the metal film. It is not possible to construct an exact analytic model for extraordinary optical transmission, so here we consider a toy model introduced by Schouten *et al* [19] to demonstrate how SPs can affect light transmission in Young's double-slit experiment.

The basic principle is illustrated in figure 11.8. We consider two long slits, separated by a distance d, in a metal plate that can support plasmons. We imagine a plane wave of amplitude E_0 normally incident upon the slits with polarization in the plane of the figure (the only polarization that can excite SPs).

We consider the transmission of light through the right slit; because of the symmetry of this problem, the transmission through the left slit is the same. There are two possible ways in which light can be transmitted through the right slit. First, light can be directly transmitted through the slit, with a real-valued transmission amplitude α. Second, light can be incident upon the left slit, be coupled into a surface plasmon, propagate to the right slit, and there be coupled back into light; we consider the overall amplitude and phase of this transmission process to be given by $\alpha\beta \exp[i\Phi]$, with β and Φ real-valued.

The total electric field amplitude E_r emitted by the right slit can then be written as

$$E_r(d) = \alpha E_0[1 + \beta e^{i(k_{\mathrm{SP}}d+\Phi)}], \tag{11.88}$$

where k_{SP} is the plasmon wavenumber. The total transmitted intensity, including the contributions from both slits, is therefore

$$I(d) = 2|E_r(d)|^2 = \alpha^2[1 + \beta^2 + 2\beta \cos(k_{\mathrm{SP}}d + \Phi)]. \tag{11.89}$$

We predict that the transmission through the slits should vary sinusoidally with the slit separation d, or equivalently (neglecting dispersion) with the surface plasmon wavenumber k_{SP}. Figure 11.9(a) shows the experimentally measured transmission

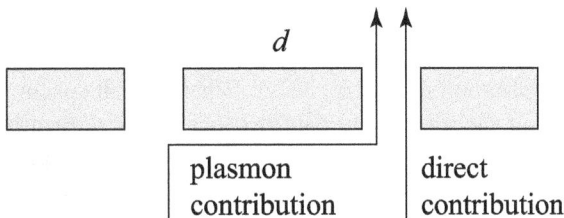

Figure 11.8. Basic principle of the toy model of Schouten *et al* [19].

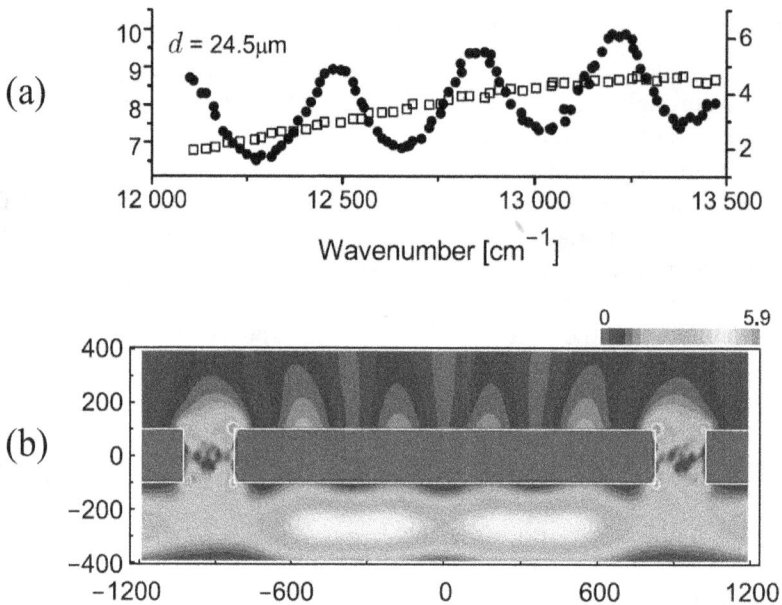

Figure 11.9. (a) Experimentally measured transmission in the plasmonic version of Young's double-slit experiment, with $d = 24.5\mu$m. The black circles indicate transverse magnetic (TM) (plasmon-exciting) polarization, while the white squares indicate transverse electric (TE) (no plasmon) polarization. (b) Numerical simulation of the electric field at a maximum of transmission, where the slit separation is $5\lambda_{SP}/2$. Parts (a) and (b) reprinted with permission from [19], Copyright (2005) by the American Physical Society.

for a slit separation $d = 24.5\mu$m. One can see that there are strong oscillations of the transmission when plasmons are present (black circles); for the orthogonal polarization where plasmons are not present, the oscillations are absent (white circles). The slit separation is many multiples of the free-space wavelength and much larger than the skin depth of the metal; this effectively rules out the possibility that the enhancement is due to short-range evanescent wave coupling between the slits.

Figure 11.9(b) shows an exact solution of Maxwell's equations using the domain integral equation method (to be discussed in chapter 16). The slit separation is taken to be $5\lambda_{SP}/2$, a case where it was found that the transmission is at a maximum. One can see from the figure that the antinodes of the plasmon standing wave on both the dark and light sides of the metal plate coincide with the slit positions; when the transmission is at a minimum, it is found that the nodes of the standing wave coincide with the slit positions.

From these results, we may conclude that extraordinary optical transmission is, in large part, a surface plasmon resonance effect. When the slit spacing is such that there is field enhancement at the slits, the transmitted field is consequently enhanced. These results also illustrate that, for certain slit separations, SPs can suppress the transmission of light, explaining some early results that suggested a suppression effect [18].

It should be noted that other effects can certainly contribute to extraordinary optical transmission, including direct evanescent coupling of apertures when they are close together, as well as waveguide resonances existing in individual holes. The double-slit results suggest, however, that the primary cause of EOT is light–plasmon wave interference.

Though EOT appears to be strongest for an array of subwavelength-sized holes, researchers have also demonstrated that it is possible to use plasmons to enhance the transmission of light through a single aperture by applying a pattern of grooves around the aperture [30]. Such results open the possibility of using SPs for applications such as near-field optical imaging.

11.9 Zenneck waves

Though we have focused on SPs in this chapter, it should be noted that there are a number of other surface wave and quasi-surface wave phenomena that can arise at interfaces. In this section, we discuss the surface plasmon's evil twin, known as a Sommerfeld–Zenneck wave or simply a Zenneck wave, after the two researchers who first discussed them in the early 1900s [31, 32]. These discoveries were made in the early days of radio communication, and the researchers were interested in the ways in which a dipole antenna close to the surface of the Earth could transmit radiation over long distances, even beyond the visible horizon. Zenneck and Sommerfeld's solutions suggested that radio waves could propagate along the surface in a form of what we loosely refer to as a surface wave.

These results were almost immediately mired in controversy, including a claim that Sommerfeld had made a sign error in his calculation that made spurious surface effects appear. (These claims were later refuted.) Since then, however, there has been an ongoing struggle to demonstrate experimentally that Zenneck waves exist at all. As we will see, this effort is complicated by the observation that Zenneck waves satisfy the same dispersion formula as SPs. Early work on SPs sometimes conflated the two phenomena, though today most researchers appear to agree that they are distinct.

We do not attempt to give a definitive explanation of Zenneck waves but only present an introduction to the concepts and provide references for readers to investigate further.

To begin, let us note that, as our short historical review indicates, we are not considering waves at a dielectric/metal interface. In this case, we are interested in a dielectric/lossy dielectric interface and want to see whether some sort of surface wave is possible in this case.

Let us consider again the p-polarization reflection amplitude in its most general (nonmagnetic) form, from equation (9.144),

$$r_\parallel = \frac{\dfrac{n_2}{n_1}k_{iz} - \dfrac{n_1}{n_2}k_{tz}}{\dfrac{n_2}{n_1}k_{iz} + \dfrac{n_1}{n_2}k_{tz}}, \tag{11.90}$$

where $k_{iz} = \sqrt{n_1^2 k_0^2 - k_x^2}$ and $k_{tz} = \sqrt{n_2^2 k_0^2 - k_x^2}$. We noted at the end of section 11.4 that SPs may be viewed as poles of the reflection amplitude, i.e. those values of

k_x such that the denominator vanishes. If we substitute the values of k_{iz} and k_{tz}, this results in the surface plasmon wavenumber, equation (11.35), i.e.

$$k_x = k\sqrt{\frac{\tilde{\epsilon}_1\tilde{\epsilon}_2}{\tilde{\epsilon}_1 + \tilde{\epsilon}_2}}, \qquad (11.91)$$

where again the tilde indicates that these are the relative permittivities.

A surface plasmon is a wave with only two components, one in each medium. We may crudely interpret the singularity of the reflection coefficient as representing a case where there is only a reflected and transmitted wave: the zero amplitude of the incident wave is countered by the infinite magnitude of the reflection coefficient.

There is, however, another situation in which the field at an interface only has two components: when the reflection coefficient is equal to zero, i.e. a wave is incident at the Brewster angle. In this case, we formally have only an incident wave and a transmitted wave, with no reflected wave. This is the situation that is associated with the Zenneck wave. Because the numerator and denominator of the reflection amplitude only differ by a sign, and we must square our expressions to derive the wavenumber, we can readily see that our Zenneck wave satisfies equation (11.91) as well.

Let us consider for a moment what a hypothetical Zenneck wave would look like with a dielectric/lossy dielectric interface using equation (11.91). We take $\tilde{\epsilon}_1$ to be real and write $\tilde{\epsilon}_2 = \tilde{\epsilon}_2' + i\tilde{\epsilon}_2''$, where $\tilde{\epsilon}_2'$, $\tilde{\epsilon}_2''$ are real-valued. If we assume that the imaginary part is smaller than the real part, we may reuse equation (11.39) to write the real part of k_x as

$$k_x' = k\left(\frac{\tilde{\epsilon}_1\tilde{\epsilon}_2'}{\tilde{\epsilon}_1 + \tilde{\epsilon}_2'}\right)^{1/2}. \qquad (11.92)$$

In this case, with both permittivities positive, if $\tilde{\epsilon}_1 \approx 1$, we find that $k_x < k$, which implies that the effective refractive index of this wave would be less than unity. This is in contrast to a surface plasmon, where the effective index is greater than one. A Zenneck wave is therefore expected to be a 'fast' surface wave, while a surface plasmon is a 'slow' surface wave.

The description of a Zenneck wave as a zero of the reflection coefficient is somewhat unsatisfying because we have stressed that surface waves are typically found as *singularities* of the reflection coefficient. Here, we note a curious mathematical wrinkle: in terms of a complex-valued k_x, the function $r_\parallel$ is a Riemann surface with four distinct sheets. The existence or nonexistence of surface effects associated with this reflection amplitude depends crucially on which sheet we have 'excited.' We may summarize the possibilities as

$$r_\parallel = \pm_a \frac{\dfrac{n_2}{n_1}k_{iz} \mp_b \dfrac{n_1}{n_2}k_{tz}}{\dfrac{n_2}{n_1}k_{iz} \pm_b \dfrac{n_1}{n_2}k_{tz}}, \qquad (11.93)$$

where $\pm_a$ and $\pm_b$ represent independent sign possibilities. On two of the sheets, the zero of the denominator represents a surface plasmon, and on the other two sheets, the zero represents a Zenneck wave. The trick, then, is to determine a method of exciting the Zenneck wave on a different sheet of the Riemann surface.

If the wave is excited, what do we expect its properties to be? In this case, we fully expect all of k_x, k_{iz}, and k_{tz} to be complex, even in air. We expect the wave to be exponentially damped in the lossy dielectric. For a surface plasmon, k_{iz} and k_{tz} are defined such that the plasmon exponentially decays away from the interface. However, based on the relationship between the surface plasmon condition and the Zenneck condition, we naively expect that the wave in air exponentially *increases* away from the interface.

This goes against our physical intuition for waves—how can they increase without limit in unbounded space? This is, however, the manifestation of what is known as an improper *leaky wave* [33]. A leaky wave is a wave that is imperfectly guided and thus gradually loses energy through 'leaks' into propagating waves. The classic example is a rectangular metal waveguide that has an open slit cut along the length of one of its boundaries. In a closed waveguide, a wave is perfectly guided; the slit, however, allows energy to gradually escape. The seeming exponential growth is a mathematical approximation of the escaping wave and turns out to be valid only in a finite region near the waveguide itself.

This interpretation of a Zenneck wave as an improper leaky surface wave seems to be the currently accepted view [34]. There has been a lot of discussion about the relationship between SPs and Zenneck waves and how to define them; see, for example, Sarkar *et al* [35]. Zenneck waves have apparently remained elusive due to the difficulty in exciting them, as simple dipole radiation sources are ineffective [36]. There have been claims of experimental detection, the most recent appearing in 2020 [37]. For those interested in learning more about leaky waves, the earliest study appears to be a 1959 investigation by Marcuvitz [38].

11.10 Dyakonov waves

Let us conclude this chapter by looking at a surface wave of a very different nature, one that can appear under certain circumstances at the interface between an isotropic medium and an anisotropic medium, or between two anisotropic media. These waves were theoretically predicted by M.I. Dyakonov in 1987 and are therefore referred to as *Dyakonov waves* [39]; the quantized versions have been given the name Dyakonons. These waves are a good opportunity to test our experience and intuition about anisotropic media, in addition to being an interesting physical phenomenon.

Let us consider the interface between a dielectric with normalized permittivity $\tilde{\varepsilon}_1$ (region 1) and a uniaxial crystal (region 2), the latter of which has normalized permittivities $\tilde{\varepsilon}_o$, $\tilde{\varepsilon}_e$. The optic axis—the axis of the extraordinary permittivity—is taken to lie in the xy-plane at an angle ϕ from the x-axis, as illustrated in figure 11.10(a). Our approach will be similar to that used to derive the surface plasmon wavenumber— we will introduce waves on both sides of the interface and match boundary conditions

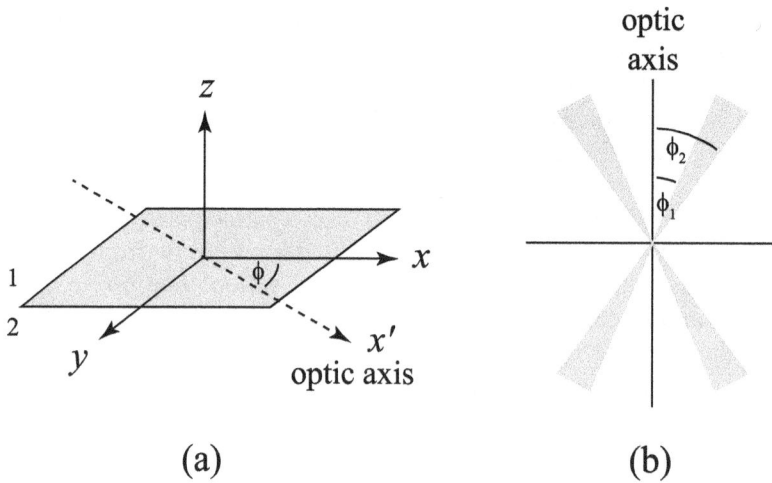

Figure 11.10. (a) Illustration of the geometry for Dyakonov waves. (b) Illustration of the allowed propagation directions for Dyakonov waves.

to find the wavenumber. A gentle warning: the calculation is quite involved, so we give the major steps only and leave out the simpler algebraic manipulations.

The surface wave is taken to propagate in the xz-plane and be evanescent on both sides of the interface. On the air side, we have the **k**-vector

$$\mathbf{k}_1 = k_x \hat{\mathbf{x}} + iq\hat{\mathbf{z}}, \tag{11.94}$$

while in the crystal, we have two components,

$$\mathbf{k}_2 = k_x \hat{\mathbf{x}} - iq_o\hat{\mathbf{z}}, \tag{11.95}$$

$$\mathbf{k}_3 = k_x \hat{\mathbf{x}} - iq_e\hat{\mathbf{z}}, \tag{11.96}$$

where q_o and q_e refer to the ordinary and extraordinary waves, respectively. In the dielectric, we may readily write

$$k_x^2 - q^2 = k_0^2 \tilde{\epsilon}_1, \tag{11.97}$$

which provides a relationship between k_x and q. To determine relationships for q_o and q_e, we turn to the Fresnel formula for wave normals, equation (8.48), which can be written as

$$\left(\frac{1}{n^2} - \frac{1}{\tilde{\epsilon}_o} \right) \left[s_x'^2 \left(\frac{1}{n^2} - \frac{1}{\tilde{\epsilon}_o} \right) + s_y'^2 \left(\frac{1}{n^2} - \frac{1}{\tilde{\epsilon}_e} \right) + s_z'^2 \left(\frac{1}{n^2} - \frac{1}{\tilde{\epsilon}_e} \right) \right] = 0, \tag{11.98}$$

where we have used $v_p = c/n$ and $v_o^2 = c^2/\tilde{\epsilon}_o$, while s_x', s_y', and s_z' represent the components of the unit vector of the wave in the crystal, measured with respect to the optic axis. We factor out the first parenthetical term, which has a solution

$$n^2 = \tilde{\epsilon}_o \tag{11.99}$$

that represents the index of the ordinary wave. If we multiply both sides by k_0^2 and note that we must have $k_0 n^2 = k_x^2 - q_o^2$, we may write

$$k_x^2 - q_o^2 = k_0^2 \tilde{\epsilon}_o, \tag{11.100}$$

which becomes our second relationship.

For the remaining part of the Fresnel formula, we can rearrange and use $s_x'^2 + s_y'^2 + s_z'^2 = 1$ to get

$$\frac{1}{n^2} = \frac{s_x'^2}{\tilde{\epsilon}_o} + \frac{s_y'^2 + s_z'^2}{\tilde{\epsilon}_e}. \tag{11.101}$$

We now multiply both sides of this expression by $k_0^2 n^2$ and note that $k_0^2 n^2 s_x'^2 = k_x'^2$, $k_0^2 n^2 s_y'^2 = k_y'^2$, and $k_0^2 n^2 s_z'^2 = -q_e^2$. In this system, we have $k_x' = k_x \cos\phi$ and $k_y' = k_x \sin\phi$. Finally, we have

$$\frac{k_x^2 \cos^2\phi}{\tilde{\epsilon}_o} + \frac{k_x^2 \sin^2\phi - q_e^2}{\tilde{\epsilon}_e} = k_0^2. \tag{11.102}$$

This becomes a third equation relating k_x, q, q_o, and q_e. There are, however, four unknowns, which means we need an additional constraint to solve for the surface wave parameters. We get this constraint by matching the transverse boundary conditions of the electric and magnetic fields at the interface. We thus combine *four* waves: the s- and p-polarizations in air, and the ordinary and extraordinary waves in the medium.

In air, let us take

$$\mathbf{E}_s(\mathbf{r}) = E_s \hat{\mathbf{y}} e^{i\mathbf{k}_1 \cdot \mathbf{r}}, \tag{11.103}$$

$$\mathbf{H}_s(\mathbf{r}) = \frac{E_s}{i\omega\mu_0} [q\hat{\mathbf{x}} + ik_x\hat{\mathbf{z}}] e^{i\mathbf{k}_1 \cdot \mathbf{r}} \tag{11.104}$$

for the s-polarization state, and

$$\mathbf{E}_p(\mathbf{r}) = E_p \left[\hat{\mathbf{x}} + \frac{ik_x}{q}\hat{\mathbf{z}} \right] e^{i\mathbf{k}_1 \cdot \mathbf{r}}, \tag{11.105}$$

$$\mathbf{H}_p(\mathbf{r}) = \frac{E_p}{i\omega\mu_0} \frac{\tilde{\epsilon}_1 k_0^2}{q} e^{i\mathbf{k}_1 \cdot \mathbf{r}} \tag{11.106}$$

for the p-polarization state. The **H**-fields are, of course, found through Faraday's law.

Deriving the fields in the crystal is more involved. First, let us note that we must work with the **D**-field in the crystal and that the ordinary wave must be entirely perpendicular to both $\mathbf{k}_2$ and the optic axis, $\mathbf{o} \equiv \hat{\mathbf{x}} \cos\phi + \hat{\mathbf{y}} \sin\phi$. However, because the **D**-vector lies entirely in an ordinary permittivity direction, it is parallel to **E** and we can work directly with **E**. Setting up a system of equations, we find

$$\mathbf{E}_o = E_o \left[\cos\phi\hat{\mathbf{x}} - \cos\phi\hat{\mathbf{y}} - \frac{ik_x}{q_o}\sin\phi\hat{\mathbf{z}} \right] e^{i\mathbf{k}_2 \cdot \mathbf{r}}, \qquad (11.107)$$

$$\mathbf{H}_o = \frac{E_o}{i\omega\mu_0} \left[q_o \sin\phi\hat{\mathbf{x}} - \frac{k_0^2 \tilde{\epsilon}_o}{q_o}\sin\phi\hat{\mathbf{y}} - ik_x \cos\phi\hat{\mathbf{z}} \right] e^{i\mathbf{k}_2 \cdot \mathbf{r}}. \qquad (11.108)$$

It should be noted that we have used our relations between q, q_o, and k_x to simplify the preceding formulas.

The extraordinary wave is the trickiest to calculate. We must start with the **D**-field because it is the only wave that is perpendicular to $\mathbf{k}_3$. To start, we note that for any direction of propagation, there is an ordinary wave and an extraordinary wave. For the direction $\mathbf{k}_3$, we can first define an ordinary wave $\mathbf{D}_2(\mathbf{r})$ using the same form as equation (11.107):

$$\mathbf{D}_2(\mathbf{r}) = D_2 \left[\sin\phi\hat{\mathbf{x}} - \cos\phi\hat{\mathbf{y}} - \frac{ik_x}{q_e}\sin\phi\hat{\mathbf{z}} \right] e^{i\mathbf{k}_3 \cdot \mathbf{r}}. \qquad (11.109)$$

We then introduce our extraordinary $\mathbf{D}_e(\mathbf{r})$ as being orthogonal to this wave and the direction $\mathbf{k}_3$, which results in

$$\mathbf{D}_e(\mathbf{r}) = \left[\cos\phi\hat{\mathbf{x}} + \sin\phi\left(1 - \frac{k_x^2}{q_e^2}\right)\hat{\mathbf{y}} - \frac{ik_x}{q_e}\cos\phi\hat{\mathbf{z}} \right] e^{i\mathbf{k}_3 \cdot \mathbf{r}}. \qquad (11.110)$$

This expression must now be converted into an electric field. We know that $\mathbf{E} = \epsilon^{-1}\mathbf{D}$; in the principal axis coordinate system, we have

$$\epsilon^{-1} = \begin{bmatrix} 1/\tilde{\epsilon}_e & 0 & 0 \\ 0 & 1/\tilde{\epsilon}_o & 0 \\ 0 & 0 & 1/\tilde{\epsilon}_o \end{bmatrix}. \qquad (11.111)$$

It is a straightforward calculation to show that in our problem's xyz-system, we have

$$\epsilon^{-1} = \begin{bmatrix} \cos^2\phi/\tilde{\epsilon}_e + \sin^2\phi/\tilde{\epsilon}_o & \left(\dfrac{1}{\tilde{\epsilon}_e} - \dfrac{1}{\tilde{\epsilon}_o}\right)\sin\phi\cos\phi & 0 \\ \left(\dfrac{1}{\tilde{\epsilon}_e} - \dfrac{1}{\tilde{\epsilon}_o}\right)\sin\phi\cos\phi & \sin^2\phi/\tilde{\epsilon}_e + \cos^2\phi/\tilde{\epsilon}_o & 0 \\ 0 & 0 & 1 \end{bmatrix}. \qquad (11.112)$$

This results in an electric field and a corresponding magnetic field of the form

$$\mathbf{E}_e(\mathbf{r}) = E_e \left[\frac{\cos\phi}{q_e^2}\left(k_x^2/\tilde{\epsilon}_o - k_0^2\right)\hat{\mathbf{x}} - \frac{k_0^2 \sin\phi}{q_e^2}\hat{\mathbf{y}} - \frac{ik_x}{q_e \tilde{\epsilon}_o}\cos\phi\hat{\mathbf{z}} \right] e^{i\mathbf{k}_3 \cdot \mathbf{r}}, \qquad (11.113)$$

$$\mathbf{H}_e(\mathbf{r}) = \frac{E_e}{i\omega\mu_0}\left[\frac{k_0^2\sin\phi}{q_e}\hat{\mathbf{x}} - \frac{k_0^2\cos\phi}{q_e}\hat{\mathbf{y}} - \frac{ik_xk_0^2\sin\phi}{q_e^2}\hat{\mathbf{z}}\right]e^{ik_3\cdot\mathbf{r}}. \quad (11.114)$$

We now require the tangential components of the electric and magnetic fields to be continuous across the boundary. This amounts to the sum of the tangential s- and p-polarized fields being continuous with the sum of the o- and e-polarized fields. This leads to a fourth-order matrix equation,

$$\begin{bmatrix} 0 & 1 & -\sin\phi & -\dfrac{q_o^2}{\tilde{\epsilon}_o q_e^2}\cos\phi \\ 1 & 0 & -q_o\cos\phi & -\dfrac{k_0^2}{q_e}\sin\phi \\ q & 0 & -q_o\cos\phi & -\dfrac{k_0^2}{q_e}\sin\phi \\ 0 & \dfrac{\tilde{\epsilon}_1 k_0^2}{q} & \dfrac{k_0^2\tilde{\epsilon}_o}{q_o}\sin\phi & \dfrac{k_0^2}{q_e}\cos\phi \end{bmatrix}\begin{bmatrix} E_s \\ E_p \\ E_o \\ E_e \end{bmatrix} = 0. \quad (11.115)$$

This system of equations can only be solved if the determinant of the matrix vanishes (as was the case in our surface plasmon derivation). Here, we have a fourth-order determinant, but the presence of zeros in the matrix makes the calculation easier. With some algebraic effort, we can reduce this determinant to

$$q_o(q_o + q)(\tilde{\epsilon}_o q q_e + \tilde{\epsilon}_1 q_o^2)\cos^2\phi - k_0^2\tilde{\epsilon}_o(q + q_e)(\tilde{\epsilon}_o q + \tilde{\epsilon}_1 q_o)\sin^2\phi = 0. \quad (11.116)$$

Finally, this equation can be manipulated, with the help of equations (11.97), (11.100), and (11.102) and some 'fairly tedious algebraic transformations' [39], into an angle-independent form,

$$(q + q_e)(q + q_o)(\tilde{\epsilon}_1 q_o + \tilde{\epsilon}_o q_e) = (\tilde{\epsilon}_e - \tilde{\epsilon}_1)(\tilde{\epsilon}_1 - \tilde{\epsilon}_o)q_o k_0^2. \quad (11.117)$$

From our definitions of q, q_o, and q_e, we only have a surface wave if all three are positive; equation (11.117) implies that this is the case if $\tilde{\epsilon}_e > \tilde{\epsilon}_1 > \tilde{\epsilon}_o$. This implies, incidentally, that Dyakonov waves cannot arise at an air/crystal interface, for which we expect $\tilde{\epsilon}_1$ to be the smallest permittivity in the system.

Further analysis of the four equations that define k_x, q, q_o, and q_e indicates that Dyakonov waves will exist for limited angular ranges with respect to the optic axis; these angles can be found to be of the form

$$\sin^2\phi_1 = \frac{\xi}{2}\{1 - \eta\xi + [(1 - \eta\xi)^2 + 4\eta]^{1/2}\}, \quad (11.118)$$

$$\sin^2\phi_2 = \frac{(1 + \eta)^3\xi}{(1 + \eta)^2(1 + \eta\xi) - \eta^2(1 - \xi)^2}, \quad (11.119)$$

with

$$\eta = \frac{\tilde{\epsilon}_e}{\tilde{\epsilon}_o} - 1, \quad \xi = \frac{\tilde{\epsilon}_1 - \tilde{\epsilon}_o}{\tilde{\epsilon}_e - \tilde{\epsilon}_o}. \tag{11.120}$$

For natural anisotropic materials, we expect this angular range to be quite small. In fact, it took decades for Dyakonov waves to be observed experimentally, which was finally accomplished in 2009 [40]. The experiment used a potassium titanyl phosphate (KTP) biaxial crystal, with $n_x = 1.7619$, $n_y = 1.7712$, and $n_z = 1.8648$ at $\lambda = 632.8$ nm. An index-matching liquid was used as the dielectric medium with $n = 1.7868$. The Dyakonov waves were predicted to occur between angles of $30.5°$ and $30.6°$ and experimentally observed in this range.

To date, there has been relatively little work on Dyakonov waves, at least compared to SPs. However, the introduction of metamaterials, to be discussed in the next chapter, has increased the variety of optical materials that can be used to support them. In particular, it has been shown that materials designed with form birefringence (see section 12.7) can support not only Dyakonov waves but also hybrid plasmon–Dyakonov waves [41], thus unifying many of the topics we have discussed in this chapter.

11.11 Exercises

1. The refractive index of silver at 500 nm is given by $n = 0.05 + i2.87$. Using only the real part of ϵ that arises from this, calculate the surface plasmon wavelength at an air–silver interface. Also, determine how far the electric field of the plasmon penetrates into the silver, i.e. what is the depth at which the electric field amplitude reduces to $1/e$ of its surface value?

2. The refractive index of gold at 600 nm is given by $n = 0.21 + i3.27$. Using only the real part of ϵ that arises from this, calculate the surface plasmon wavelength at an air–gold interface. Also, determine how far the electric field of the plasmon penetrates into the gold, i.e. what is the depth at which the electric field amplitude reduces to $1/e$ of its surface value?

3. At an air–gold interface at a free-space wavelength of 775 nm, a surface plasmon wavelength of 758 nm is measured. Assuming that it is real-valued, determine the permittivity of gold at that wavelength.

4. The surface plasmon dispersion relation can be alternatively derived by viewing a surface plasmon as a singularity of the reflection amplitude. We have seen that for light incident from region 1 to region 2, with complex refractive indices n_1 and n_2, the reflection amplitudes are

$$r_\parallel = \frac{\frac{n_2}{n_1}k_{iy} - \frac{n_1}{n_2}k_{ty}}{\frac{n_2}{n_1}k_{iy} + \frac{n_1}{n_2}k_{ty}},$$

$$r_\perp = \frac{k_{iy} - k_{ty}}{k_{iy} + k_{ty}},$$

where y is the normal to the interface. Show that the SP dispersion relation can be derived from $r_{\parallel}$ and that no such solution exists for $r_{\perp}$.

5. The thin-film surface plasmon dispersion relation can be alternatively derived by viewing a surface plasmon as a singularity of the reflection amplitude. For a 1-2-3 layer system normal to the y-axis, the reflection amplitude can be written as

$$r = \frac{r_{12} + r_{23}e^{2i\beta}}{1 + r_{12}r_{23}e^{2i\beta}},$$

where r_{ij} represents the reflection amplitude going from medium 1 to medium 2 and $\beta = n_2 k_{2y} d_2$. Furthermore, we know that the reflection amplitude for p-polarization may be written as

$$r_{12}^{\parallel} = \frac{\dfrac{n_2}{n_1}k_{1y} - \dfrac{n_1}{n_2}k_{2y}}{\dfrac{n_2}{n_1}k_{1y} + \dfrac{n_1}{n_2}k_{2y}},$$

with a similar expression for $r_{23}^{\parallel}$. Show that we can derive the thin-film dispersion relation,

$$\tanh(\alpha_2 d) = -\frac{\left(\dfrac{\alpha_2\alpha_3}{\epsilon_2\epsilon_3} + \dfrac{\alpha_1\alpha_2}{\epsilon_1\epsilon_2}\right)}{\left(\dfrac{\alpha_2^2}{\epsilon_2^2} + \dfrac{\alpha_1\alpha_3}{\epsilon_1\epsilon_3}\right)},$$

where $\alpha_j = i|k_{jy}|$, by locating the singularity of r.

6. Let us consider SPs on free-standing thin silver films at a wavelength of 617 nm, for which the permittivity is $\epsilon_2 = -17.2355 + 0.4982i$. Calculate the surface plasmon wavelength and the propagation distance over which the electric field amplitude decays by $1/e$ for the symmetric and antisymmetric modes of the film for thicknesses of 100 nm, 50 nm, 25 nm, and 10 nm.

7. Use matrix methods to study the Otto configuration used to excite SPs, treating it as a four-material system. Light enters through glass ($n_1 = 1.52$), passes through an air gap ($n_2 = 1$, $d_2 = 200$ nm), enters a copper metal film ($n_3 = -10.1820 + 1.9230i$, $d_3 = 1000$ nm), and can end in air ($n_4 = 1$). Plot the reflected power R as a function of the incident angle for both s-polarization and p-polarization, and locate the plasmon resonance. Does the angle where the resonance appears agree with the simple theoretical prediction? If not, can you think of possible reasons why?

8. Use matrix methods to study the Kretschmann configuration used to excite SPs, treating it as a three-material system. Light enters through glass ($n_1 = 1.52$), enters a silver metal film ($n_3 = -17.2355 + 0.4982i$, $d_3 = 50$ nm), and ends in air ($n_3 = 1$). Plot the reflected power R as a function of the incident angle for both s-polarization and p-polarization, and locate the plasmon resonance. Does the angle where the resonance appears

Table 11.3. Optical properties of aluminum for selected energies.

Energy	n_R	n_I
1.8	1.830	8.060
1.9	1.572	7.735
2.0	1.367	7.405
2.20	1.073	6.784
2.40	0.873	6.242
2.60	0.728	5.778
2.80	0.608	5.368
3.00	0.521	5.000
3.40	0.399	4.396

agree with the simple theoretical prediction? If not, can you think of possible reasons why?

9. Depending on the desired wavelength of operation, different materials may be used for an ideal surface plasmon resonance. The refractive index of aluminum as a function of energy (electron volts, eV) is given in table 11.3.

   Calculate and plot the plasmonic field enhancement as a function of eV using the method of Weber and Ford, assuming perfect coupling ($r = 0$) and a coupling grating with a period of 800 nm. What would be the optimal energy for field enhancement, and how does the field enhancement compare to silver, gold, and copper?

References

[1] Ebbesen T W, Lezec H J, Ghaemi H F, Thio T and Wolff P A 1998 Extraordinary optical transmission through sub-wavelength hole arrays *Nature* **391** 667–9
[2] Drude P 1900 Zur Elektronentheorie der Metalle *Ann. Phys., Lpz.* **306** 566–613
[3] Ordal M A *et al* 1983 Optical properties of the metals Al, Co, Cu, Au, Fe, Pb, Ni, Pd, Pt, Ag, Ti, and W in the infrared and far infrared *Appl. Opt.* **22** 1099–119
[4] Tonks L and Langmuir I 1929 Oscillations in ionized gases *Phys. Rev.* **33** 195–210
[5] Bernstein I B and Trehan S K 1960 Plasma oscillations (I) *Nucl. Fusion* **1** 3
[6] Landau L D 1965 61 - On the vibrations of the electronic plasma *Collected Papers of L.D. Landau* ed D T Haar (Oxford: Pergamon) 445–60
[7] Bohm D and Gross E P 1949 Theory of plasma oscillations. A. Origin of medium-like behavior *Phys. Rev.* **75** 1851–64
[8] Ritchie R H 1957 Plasma losses by fast electrons in thin films *Phys. Rev.* **106** 874–81
[9] Stern E A and Ferrell R A 1960 Surface plasma oscillations of a degenerate electron gas *Phys. Rev.* **120** 130–6
[10] Raman C V 1928 A new radiation *Indian J. Phys.* **2** 387–98
[11] Kneipp K 2007 Surface-enhanced Raman scattering *Phys. Today* **60** 40–6
[12] Fleischmann M, Hendra P J and McQuillan A J 1974 Raman spectra of pyridine adsorbed at a silver electrode *Chem. Phys. Lett.* **26** 163–6

[13] Jeanmaire D L and Van Duyne R P 1977 Surface Raman spectroelectrochemistry: Part I. Heterocyclic, aromatic, and aliphatic amines adsorbed on the anodized silver electrode *J. Electroanal. Chem. Interfacial Electrochem* **84** 1–20

[14] Tsang J C, Kirtley J R and Bradley J A 1979 Surface-enhanced Raman spectroscopy and surface plasmons *Phys. Rev. Lett.* **43** 772–5

[15] Bethe H A 1944 Theory of diffraction by small holes *Phys. Rev.* **66** 163–82

[16] Novotny L and Hecht B 2012 *Principles of Nano-Optics* 2nd edn (Cambridge: Cambridge University Press)

[17] Lezec H J and Thio T 2004 Diffracted evanescent wave model for enhanced and suppressed optical transmission through subwavelength hole arrays *Opt. Exp.* **12** 3629–51

[18] Cao Q and Lalanne P 2002 Negative role of surface plasmons in the transmission of metallic gratings with very narrow slits *Phys. Rev. Lett.* **88** 057403

[19] Schouten H F *et al* 2005 Plasmon-assisted two-slit transmission: Young's experiment revisited *Phys. Rev. Lett.* **94** 053901

[20] Johnson P B and Christy R W 1972 Optical constants of the noble metals *Phys. Rev.* **6** 4370–9

[21] 1985 *Handbook of the Optical Constants of Solids* **vol 1** ed E D Palik (San Diego, CA: Academic)

[22] Otto A 1968 Excitation of nonradiative surface plasma waves in silver by the method of frustrated total reflection *Z. Phys.* **216** 398–410

[23] Kretschmann E and Raether H 1968 Radiative decay of non radiative surface plasmons excited by light *Z. Nat.forsch.* A **23** 2135–6

[24] Kretschmann E 1972 The angular dependence and the polarisation of light emitted by surface plasmons on metals due to roughness *Opt. Commun.* **5** 331–6

[25] Weber W H and Ford G W 1981 Optical electric-field enhancement at a metal surface arising from surface-plasmon excitation *Opt. Lett.* **6** 122–4

[26] Sarid D 1981 Long-range surface-plasma waves on very thin metal films *Phys. Rev. Lett.* **47** 1927–30

[27] Berini P 2009 Long-range surface plasmon polaritons *Adv. Opt. Photon.* **1** 484–588

[28] Han Z and Bozhevolnyi S I 2012 Radiation guiding with surface plasmon polaritons *Rep. Prog. Phys.* **76** 016402

[29] Li Y 2017 *Plasmonic Optics: Theory and Applications* (Bellingham, WA: SPIE Press)

[30] Caglayan H, Bulu I and Ozbay E 2005 Extraordinary grating-coupled microwave transmission through a subwavelength annular aperture *Opt. Express* **13** 1666–71

[31] Zenneck J 1907 Über die Fortpflanzung ebener elektromagnetischer Wellen längs einer ebenen Leiterfläche und ihre Beziehung zur drahtlosen Telegraphie *Ann. Phys., Lpz.* **328** 846–66

[32] Sommerfeld A 1909 Über die Ausbreitung der Wellen in der drahtlosen Telegraphie *Ann. Phys., Lpz.* **333** 665–736

[33] Monticone F and Alù A 2015 Leaky-wave theory, techniques, and applications: from microwaves to visible frequencies *Proc. IEEE* **103** 793–821

[34] Jackson D R, Mesa F, Michalski K A and Mosig J R 2022 Reflections on the Zenneck wave *2022 IEEE Int. Symp. on Antennas and Propagation and USNC-URSI Radio Science Meeting (AP-S/URSI)* 859–60

[35] Sarkar T K, Abdallah M N, Salazar-Palma M and Dyab W M 2017 Surface plasmons-polaritons, surface waves, and Zenneck waves: clarification of the terms and a description of the concepts and their evolution *IEEE Antennas Propag. Mag.* **59** 77–93

[36] Mesa F and Jackson D R 2020 Excitation of the Zenneck wave by a tapered line source above the earth or ocean *IEEE Trans. Antennas Propag.* **68** 4848–59

[37] Oruganti S K *et al* 2020 Experimental realization of Zenneck type wave-based non-radiative, non-coupled wireless power transmission *Sci. Rep.* **10** 925

[38] Marcuvitz N 1956 On field representations in terms of leaky modes or eigenmodes *IRE Trans. Antennas Propag.* **4** 192–4

[39] Dyakonov M I 1988 New type of electromagnetic wave propagating at an interface *Sov. Phys. JETP* **67** 714–6

[40] Takayama O, Crasovan L, Artigas D and Torner L 2009 Observation of dyakonov surface waves *Phys. Rev. Lett.* **102** 043903

[41] Takayama O, Artigas D and Torner L 2012 Practical dyakonons *Opt. Lett.* **37** 4311–3

Chapter 12

Metamaterials

In recent decades, the study of optics has been transformed by the recognition that it is possible to construct materials with optical properties not found in nature. These materials, now known as *metamaterials*, are fabricated out of 'meta-atoms': structures made of many atoms but still subwavelength in size. These subwavelength structures can produce nontrivial and even counterintuitive optical effects in bulk.

In this chapter, we introduce the basic concepts of metamaterials. The subject is already too broad to cover comprehensively here, so we focus on some of its most surprising predictions, such as negative refraction and perfect lenses.

12.1 Background

In section 9.8, we showed that it is possible to create materials with magnetic properties that suppress reflection at an interface for a particular direction of illumination. For normal incidence, this can be achieved if the electromagnetic impedance, $\eta = \sqrt{\mu/\epsilon}$, is the same for the materials on either side of the interface. Materials with a nontrivial magnetic response, i.e. $\mu \neq \mu_0$, are rare at visible wavelengths, and so optics researchers historically did not focus much attention on them.

The significance of impedance matching was not lost on early electromagnetics researchers, however. In the 1964 English translation of his book *Optics*, optical physicist Arnold Sommerfeld reflected upon his work during World War II,

> During the war the problem arose to find, as a counter-measure against allied radar, a largely non-reflecting ('black') surface layer of small thickness. This layer was to be particularly non-reflecting for perpendicular or almost perpendicular incidence of the radar wave.
>
> ...
>
> The criterion is, thus, not the index of refraction n but the ratio of wave resistances m. In order to 'camouflage' an object against radar

waves, one must cover it with a layer for which this ratio of wave resistances has the value 1 in the region of centimeter waves.

Sommerfeld and his colleagues apparently did not make much progress in creating their radar-proof camouflage, but their efforts might be considered an example of scientific foreshadowing. In the 1990s, a British company called Marconi Materials Technology succeeded in developing a radar-absorbing carbon material that could be painted onto ships; the only problem was that the company did not know why it worked! They enlisted physicist John Pendry to explain the properties of the material, and Pendry made a striking discovery: the radar-absorbing properties of the material came not from the chemical properties of carbon but from the subwavelength structure of the carbon, which was drawn into thin fibers.

This led Marconi and Pendry to wonder what other optical effects could be created by manipulating the structure of matter on a subwavelength scale. Their next challenge was to create a magnetic response in a nonmagnetic material [1]. To achieve this, they opted to fashion a bulk material out of subwavelength metallic structures known as *split ring resonators*, which simultaneously have capacitance and inductance because of their structure; these are illustrated in figure 12.1. These structures would be smaller than a wavelength but constructed of many physical atoms and would later be referred to as meta-atoms. Because of their subwavelength size, a collection of meta-atoms would act as a bulk material with optical properties dictated by the meta-atom structure.

Materials made of such meta-atoms, with optical properties potentially not found in natural materials, were later referred to as *metamaterials* ('meta' = 'beyond'). In principle, metamaterials can be designed to have any value of ϵ and μ at a given frequency. The idea of metamaterials did not get much attention until 2000, when Pendry published a paper arguing that it is possible to make a material with a *negative* index of refraction, and furthermore, that such a negative-refractive-index material can be used to make a lens with no resolution limit—a perfect lens [2]. Pendry was building upon the idea of a negative refractive index first introduced by Victor Vesalago in the 1960s [3]. The introduction of metamaterials has

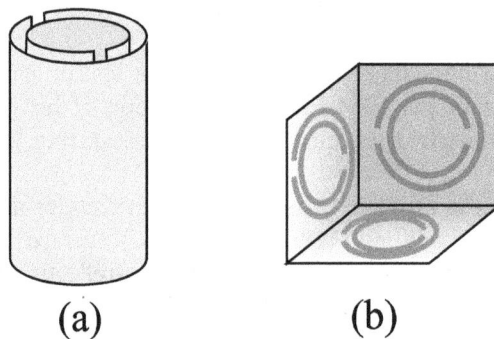

(a)   (b)

Figure 12.1. Illustration of (a) a cylindrical split ring resonator, producing anisotropic material properties in bulk, and (b) thin split ring resonators, designed to produce isotropic properties in bulk.

fundamentally changed optics: we no longer only ask what light can do but also ask how we can make light do whatever we want.

There is one significant problem with metamaterials at optical frequencies, however: we currently do not have efficient and reliable ways to make three-dimensional metamaterials at those frequencies. If we consider light at a wavelength of around 500 nm, the building blocks used to construct a metamaterial must be of a size of 50 nm or smaller. Imagine constructing a metamaterial out of building blocks 50 nm on a side and putting together enough of them to make a macroscopic object that is 1 mm on a side! Nobody is quite sure how to do this efficiently in a macroscopic volume, so much of the effort in optics in recent years has focused on fabricating two-dimensional metasurfaces as thin optical elements [4]. It is probably only a matter of time, though, before new techniques allow the construction of a variety of three-dimensional optical metamaterials.

12.2 Negative refraction

The possibility of negative refraction results in many counterintuitive properties of light waves. Here, we look at how Maxwell's equations are consistent with the possibility of negative refraction.

Let us begin by considering how Snell's law changes when light passes into a medium with a negative refractive index:

$$n_1 \sin \theta_i = n_2 \sin \theta_t. \tag{12.1}$$

If $n_2 < 0$, then, in principle, we expect that $\theta_t < 0$ as well, and we expect the transmitted wave to be refracted on the opposite side of the surface normal, as shown in figure 12.2.

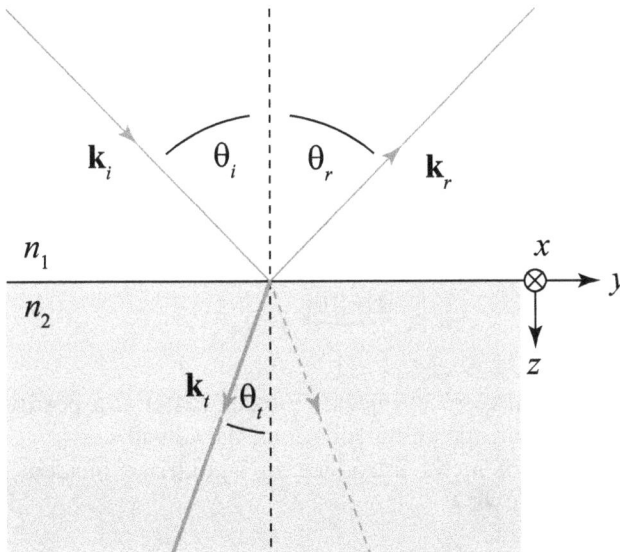

Figure 12.2. Illustration of negative refraction.

But it is not obvious that we are allowed to interpret Snell's law in that way, so we must look more carefully at whether a negative refractive index does, in fact, produce negative refraction and whether it is consistent with Maxwell's equations. Our first step is our definition of the refractive index,

$$n = \sqrt{\frac{\epsilon\mu}{\epsilon_0\mu_0}}.$$ (12.2)

Let us view ϵ and μ as complex numbers,

$$\epsilon = r_\epsilon e^{i\phi_\epsilon}, \quad \mu = r_\mu e^{i\phi_\mu}.$$ (12.3)

Then the refractive index is given by

$$n = \sqrt{r_\epsilon r_\mu}\, e^{i(\phi_\epsilon + \phi_\mu)/2}.$$ (12.4)

If we let ϵ and μ be real-valued and negative, then $\phi_\epsilon = \phi_\mu = \pi$. This results in

$$n = \sqrt{r_\epsilon r_\mu}\, e^{i(2\pi)/2} = -\sqrt{r_\epsilon r_\mu},$$ (12.5)

and the refractive index is apparently negative if the permittivity and permeability are simultaneously negative.

This argument is still not a proof that Maxwell's equations are consistent with the idea of a negative refractive index. We will need to derive a new version of the Fresnel formulas *again* to explore this case. In principle, equations (9.175)–(9.178) should still apply, but the introduction of a negative index introduces some ambiguity into our definitions of a plane wave. For the expression $\exp[ik_0 n \mathbf{s} \cdot \mathbf{r}]$, we have previously had a reasonable definition of $\mathbf{s}$, based on the fact that n was positive. If we consider $n < 0$, it is unclear what happens to $\mathbf{s}$, and this is what we need to determine. We will find that we need to go a bit beyond Maxwell's equations to answer this question.

We recall that a monochromatic field in any uniform medium satisfies the Helmholtz equation,

$$[\nabla^2 + n^2 k_0^2]\mathbf{E} = 0.$$ (12.6)

For convenience, we will *by definition* consider the refractive index that appears in this equation to be positive, i.e. we have

$$n \equiv \sqrt{\frac{|\epsilon||\mu|}{\epsilon_0\mu_0}} > 0.$$ (12.7)

It may seem paradoxical to give a negative index material a positive index, but it removes ambiguity in our upcoming equations, as we will see.

With this definition of n, we introduce an s-polarized incident, reflected, and transmitted wave as follows:

$$\mathbf{E}_i(\mathbf{r}) = \hat{\mathbf{x}} E_{0i} e^{ik_0 n_1 \mathbf{s}_i \cdot \mathbf{r}},$$ (12.8)

$$\mathbf{E}_r(\mathbf{r}) = \hat{\mathbf{x}} E_{0r} e^{ik_0 n_1 \mathbf{s}_r \cdot \mathbf{r}}, \tag{12.9}$$

$$\mathbf{E}_t(\mathbf{r}) = \hat{\mathbf{x}} E_{0t} e^{ik_0 n_2 \mathbf{s}_t \cdot \mathbf{r}}, \tag{12.10}$$

where $\mathbf{s}_i$, $\mathbf{s}_r$, and $\mathbf{s}_t$ are the unit vectors of wave propagation for the incident, reflected, and transmitted waves, respectively. From our boundary conditions, we may introduce the law of reflection and Snell's law as

$$n_1 s_{iy} = n_1 s_{ry} = n_2 s_{ty}. \tag{12.11}$$

We use these relations to write the expressions for the unit vectors explicitly as

$$\mathbf{s}_i = \hat{\mathbf{y}} s_{iy} + \hat{\mathbf{z}} s_{iz}, \tag{12.12}$$

$$\mathbf{s}_r = \hat{\mathbf{y}} s_{iy} - \hat{\mathbf{z}} s_{iz}, \tag{12.13}$$

$$\mathbf{s}_t = \hat{\mathbf{y}} s_{ty} + \hat{\mathbf{z}} s_{tz}. \tag{12.14}$$

It is important to note that s_{ty} points in the same direction as s_{iy}, as we have defined $n_2 > 0$. Furthermore, we have already used the usual sign for the z-component of the reflected component, but for the moment, we leave the sign of s_{tz} unspecified; this will be key in demonstrating negative refraction.

We now incorporate our boundary conditions. First, we have

$$\mathbf{n} \times (\mathbf{E}_1 - \mathbf{E}_2) = 0, \tag{12.15}$$

which leads us to our usual expression in terms of normalized fields,

$$1 + r = t. \tag{12.16}$$

We also have

$$\mathbf{n} \times (\mathbf{H}_1 - \mathbf{H}_2) = 0, \tag{12.17}$$

with

$$\mathbf{H} = \frac{1}{i\omega\mu} \nabla \times \mathbf{E}. \tag{12.18}$$

Our magnetic field boundary condition ends up in the form

$$\frac{n_1 s_{iz}}{\mu_1}(1 - r) = \frac{n_2 s_{tz}}{\mu_2} t. \tag{12.19}$$

We solve our two boundary condition equations to get the reflection and transmission in the following forms:

$$r_\perp = \frac{\dfrac{n_1 s_{iz}}{\mu_1} - \dfrac{n_2 s_{tz}}{\mu_2}}{\dfrac{n_1 s_{iz}}{\mu_1} + \dfrac{n_2 s_{tz}}{\mu_2}}, \tag{12.20}$$

$$t_\perp = \frac{2\dfrac{n_1 s_{iz}}{\mu_1}}{\dfrac{n_1 s_{iz}}{\mu_1} + \dfrac{n_2 s_{tz}}{\mu_2}}. \tag{12.21}$$

We finally see a hint of a difference for an $\epsilon < 0$, $\mu < 0$ case. If $\mu_2 < 0$, then there is the possibility of the denominator becoming zero, and the reflection and transmission becoming *infinite*, unless we also take $s_{tz} < 0$. But this, in itself, is not justification enough to make that sign choice: there is always the possibility that such an infinite case does not arise for some reason that is not obvious in the equations so far.

However, we may use an additional tool to answer the question: energy conservation. We again use the definitions

$$R \equiv \frac{|\mathbf{S}_r \cdot \hat{\mathbf{z}}|}{|\mathbf{S}_i \cdot \hat{\mathbf{z}}|}, \tag{12.22}$$

$$T \equiv \frac{|\mathbf{S}_t \cdot \hat{\mathbf{z}}|}{|\mathbf{S}_i \cdot \hat{\mathbf{z}}|} \tag{12.23}$$

and note that the definition of the Poynting vector is apparently unchanged for negative-refractive-index materials:

$$\mathbf{S} = \frac{1}{2}\mathrm{Re}\{\mathbf{E} \times \mathbf{H}^*\}. \tag{12.24}$$

With some effort, we find that

$$R = |r_\perp|^2, \tag{12.25}$$

$$T = \frac{n_2}{n_1}\left|\frac{\mu_1}{\mu_2}\right|\left|\frac{s_{tz}}{s_{iz}}\right||t_\perp|^2. \tag{12.26}$$

Although we have negative ϵ and μ in the second region, we have a real-valued refractive index. We do not expect absorption, and thus expect $R + T = 1$. In fact, if we add together these quantities, we may rearrange the results into the form

$$R + T = 1 + 4n_1 n_2 \frac{\dfrac{|s_{iz}s_{tz}|}{|\mu_1\mu_2|} - \dfrac{s_{iz}s_{tz}}{\mu_1\mu_2}}{\left|\dfrac{n_1 s_{iz}}{\mu_1} + \dfrac{n_2 s_{tz}}{\mu_2}\right|^2}. \tag{12.27}$$

With $\mu_1 > 0$, $\mu_2 < 0$, and $s_{iz} > 0$, the only way we get $R + T = 1$ is if $s_{tz} < 0$. This remarkably means that the wave is propagating *toward* the interface in the transmitted region, as shown in figure 12.3.

We may connect this back to the angle of refraction by noting that $s_{ty}/s_{tz} < 0$, which means that $\tan\theta_t < 0$, which means that $\theta_t < 0$.

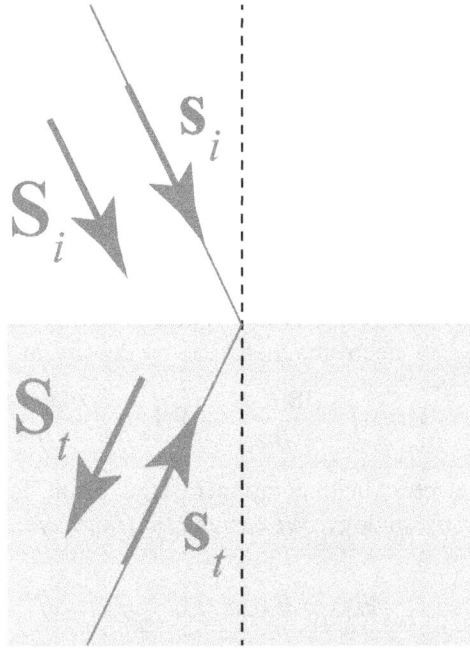

Figure 12.3. Directions of wave propagation (**s**) and the Poynting vector (**S**) in negative refraction.

The idea that the wave is oscillating towards the interface in the transmitted region is troubling, but we may also calculate the Poynting vector explicitly:

$$\mathbf{S}_t = \frac{1}{2}|E_{0t}|^2 \frac{k_0 n_2}{\omega \mu_2}[-\hat{\mathbf{z}}|s_{tz}| + \hat{\mathbf{y}}s_y]. \tag{12.28}$$

Since $\mu_2 < 0$, the energy flow is in the positive z-direction, i.e. away from the interface.

Our proof, which goes beyond Maxwell's equations and further adds the assumption of energy conservation, can still be criticized. In fact, the existence of negative refraction was fiercely debated in the years immediately following the introduction of Pendry's perfect lens. However, there is now a broad collection of experimental work that appears to show negative refraction. The first results were demonstrated at microwave frequencies [5–7], followed by results in the near-infrared [8] and optical frequencies [9]. Using plasmonic materials, negative-index materials have even been tested in the ultraviolet regime [10]. Overall, the scientific community appears to have come to the consensus that negative refraction exists.

We have noted that the Poynting vector provides intuitively reasonable results in a negative-index material, but the same cannot be said for the energy density at a single frequency, i.e.

$$U(\mathbf{r}, \omega) = \frac{1}{4}\mathbf{D}^*(\mathbf{r}, \omega) \cdot \mathbf{E}(\mathbf{r}, \omega) + \frac{1}{4}\mathbf{H}^*(\mathbf{r}, \omega) \cdot \mathbf{B}(\mathbf{r}, \omega). \tag{12.29}$$

With $\mathbf{D} = \epsilon\mathbf{E}$ and $\mathbf{B} = \mu\mathbf{H}$, a negative-index material would appear to be associated with a negative energy density, truly a peculiar conclusion!

The most common explanation for this result is analogous to our distinction between phase velocity and group velocity in section 6.4. Though the phase velocity is often a reasonable measure in regions of low dispersion, in regions of significant dispersion we must take into account the finite bandwidth of the signal. Analogously, negative-index materials typically rely on resonance effects in order to produce a negative index, and in such a resonance region, we must account for the finite bandwidth in order to properly analyze the energy density.

To see this, let us return to equation (7.9) and note that in the time domain, we can describe the change in electromagnetic energy density in a volume as

$$\frac{\partial U(\mathbf{r}, t)}{\partial t} = \mathbf{H}(\mathbf{r}, t) \cdot \frac{\partial \mathbf{B}(\mathbf{r}, t)}{\partial t} + \mathbf{E}(\mathbf{r}, t) \cdot \frac{\partial \mathbf{D}(\mathbf{r}, t)}{\partial t}. \tag{12.30}$$

We focus on the electric field component; the result for the magnetic field component will follow by analogy. We assume that we have a narrowband electric field of the form

$$\mathbf{E}(t) = \mathbf{E}_s(t)e^{-i\omega_s t}, \tag{12.31}$$

where $\mathbf{E}_s(t)$ is a very slowly varying function of t and ω_s is the central frequency of our finite-bandwidth signal. For brevity, we leave out the $\mathbf{r}$-dependence in our formulas, but they do not affect the outcome. We are explicitly using a complex electric field in this formula; the real-valued physical field can be found by taking the real part.

By Fourier analysis, we may readily find that

$$\tilde{\mathbf{E}}(\omega) = \frac{1}{2\pi} \int_{-\infty}^{\infty} \mathbf{E}_s(t)e^{i(\omega - \omega_s)t}dt = \tilde{\mathbf{E}}_s(\omega - \omega_s), \tag{12.32}$$

with $\tilde{\mathbf{E}}_s(\omega)$ the Fourier transform of $\mathbf{E}_s(t)$. We may now write the displacement field $\mathbf{D}(\omega)$ as

$$\hat{\mathbf{D}}(\omega) = \epsilon(\omega)\tilde{\mathbf{E}}_s(\omega - \omega_s), \tag{12.33}$$

or, in the time domain, we have

$$\mathbf{D}(t) = \int_{-\infty}^{\infty} \epsilon(\omega)\tilde{\mathbf{E}}_s(\omega - \omega_s)e^{-i\omega t}d\omega. \tag{12.34}$$

We now assume that our signal is narrowband, so that the only significant values of the integration range lie close to ω_s. We may then approximate $\epsilon(\omega)$ by the first two terms of its Taylor expansion,

$$\epsilon(\omega) \approx \epsilon(\omega_s) + \frac{d\epsilon}{d\omega}\bigg|_{\omega_s}(\omega - \omega_s). \tag{12.35}$$

With this, we have

$$\mathbf{D}(t) = \epsilon(\omega_s)\mathbf{E}_s(t)e^{-i\omega_s t} + \frac{d\epsilon}{d\omega}\Big|_{\omega_s} \int_{-\infty}^{\infty} (\omega - \omega_s)\tilde{\mathbf{E}}_s(\omega - \omega_s)e^{-i\omega t}d\omega. \quad (12.36)$$

The $(\omega - \omega_s)$ term may be written as a time derivative of the exponential, so that we have

$$\mathbf{D}(t) = \epsilon(\omega_s)\mathbf{E}_s(t)e^{-i\omega_s t} + \frac{d\epsilon}{d\omega}\Big|_{\omega_s} e^{-i\omega_s t} \int_{-\infty}^{\infty} \tilde{\mathbf{E}}_s(\omega - \omega_s)i\frac{d}{dt}e^{-i(\omega-\omega_s)t}d\omega. \quad (12.37)$$

Pulling out the time derivative, we may then write

$$\mathbf{D}(t) = \epsilon(\omega_s)\mathbf{E}_s(t)e^{-i\omega_s t} + i\frac{d\epsilon}{d\omega}\Big|_{\omega_s} e^{-i\omega_s t}\frac{d\mathbf{E}_s(t)}{dt}. \quad (12.38)$$

We now take the time derivative of this expression; we find with some simple manipulations that

$$\frac{d\mathbf{D}(t)}{dt} = \left\{\frac{d[\epsilon(\omega)\omega]}{d\omega}\Big|_{\omega_s} \frac{d\mathbf{E}_s(t)}{dt} - i\omega_s \epsilon(\omega_s)\mathbf{E}_s(t)\right\}e^{-i\omega_s t}, \quad (12.39)$$

where we have dropped a second derivative of $\mathbf{E}_s(t)$ with respect to time. Finally, we substitute from this expression into equation (12.30) using the real fields, e.g. $(\mathbf{E} + \mathbf{E}^*)/2$ and so forth. We have

$$\frac{dU_e(t)}{dt} = \frac{1}{4}[\mathbf{E}(t) + \mathbf{E}^*(t)] \cdot \frac{d}{dt}[\mathbf{D}(t) + \mathbf{D}^*(t)]. \quad (12.40)$$

If we take a time average over a significant number of cycles, we are left with

$$\left\langle \frac{dU_e(t)}{dt} \right\rangle = \frac{1}{4}\left[\mathbf{E}(t) \cdot \frac{d\mathbf{D}^*(t)}{dt} + \mathbf{E}^*(t) \cdot \frac{d\mathbf{D}(t)}{dt}\right]. \quad (12.41)$$

Substituting from equation (12.39) into this expression, we obtain

$$\left\langle \frac{dU_e(t)}{dt} \right\rangle = \frac{1}{4}\left\{\frac{d[\epsilon^*(\omega)\omega]}{d\omega}\Big|_{\omega_s} \mathbf{E}_s(t) \cdot \frac{d\mathbf{E}_s^*(t)}{dt} + \frac{d[\epsilon(\omega)\omega]}{d\omega}\Big|_{\omega_s} \mathbf{E}_s^*(t) \cdot \frac{d\mathbf{E}_s(t)}{dt}\right\}. \quad (12.42)$$

There is one final simplification. If we assume that the material is nonabsorbing (and we are particularly interested in ideal negative-index materials, which possess no absorption), we have $\epsilon^* = \epsilon$, and the final result, including the magnetic part as well, is

$$\langle U(t) \rangle = \frac{1}{4}\left\{\frac{d[\epsilon(\omega)\omega]}{d\omega}\Big|_{\omega_s} |\mathbf{E}_s(t)|^2 + \frac{d[\mu(\omega)\omega]}{d\omega}\Big|_{\omega_s} |\mathbf{H}_s(t)|^2\right\}. \quad (12.43)$$

This expression for the energy density can be compared to equation (12.29). For light of finite bandwidth, the energy density depends on derivatives of ϵ and μ. It can be shown that these quantities are always positive due to causality, though we refer the reader to Milonni (see section 7.3 of [11]) for the details.

Because the energy of negative-index materials can only be derived by taking into account the finite bandwidth of light, such materials are often referred to as *inherently dispersive*: the dispersion properties must always be considered, much as dispersion properties must be taken into account when defining the velocity of light. It should be noted that at least some authors have also argued that a negative electromagnetic energy density can be interpreted in terms of energy temporarily 'borrowed' from the material [12].

In a negative-index material, the vector triplet (**E**, **H**, **s**) satisfies a left-hand rule of multiplication, in that one can determine the cross-product direction of any two of the vectors by using one's left hand, instead of one's right. For this reason, such materials are often referred to as *left-handed materials*.

It is of interest to note that effective negative refraction can also be achieved with metallic photonic crystals, as demonstrated in 2004 [13]. In this case, the mechanism of negative refraction is distinct from the metamaterial effect, as the periodic structure of a photonic crystal is comparable to a wavelength rather than the subwavelength structure of a metamaterial.

12.3 The perfect lens

Negative refraction introduces one other remarkable feature: let us suppose that we impedance match the negative index medium to the external medium, i.e. we take $\mu_2 = -\mu_1$ and $\epsilon_2 = -\epsilon_1$, which gives us $\eta_1 = \eta_2$. We then have $\theta_t = -\theta_i$, but $\cos \theta_t = \cos \theta_i$. If we return to our Fresnel formulas for magnetic materials (we use equation (9.175) for convenience), we readily find that $r_\perp = 0$, and we have perfect transmission for all angles θ_i! We therefore have a reflectionless material which nevertheless refracts light, because the refractive index $n_2 = -n_1$.

This is the mechanism by which Pendry's aforementioned 'perfect' lens functions. In air, we imagine a slab of material of thickness $2d$ with $\epsilon = -\epsilon_0$ and $\mu = -\mu_0$. All the light incident on the negative-index slab is transmitted, and furthermore, we can demonstrate geometrically that any light rays emanating from a point at a distance d from the slab are refocused at a distance d after the slab, as illustrated in figure 12.4.

Pendry went further and argued that this imaging effect also applies to evanescent waves, thus justifying the term 'perfect.' We may explicitly demonstrate this ourselves using the matrix method of chapter 10. Let us consider a three-layer system: the first layer is an air layer of thickness d; the second layer is a negative-index layer with $\epsilon = -\epsilon_0$, $\mu = -\mu_0$, and thickness $2d$; and the third layer is another air layer of thickness d.

To demonstrate perfect imaging, we simply need to multiply the three characteristic matrices together. The two outer matrices, $\mathbf{M_1}$ and $\mathbf{M_3}$, are identical and represent propagation in air:

$$\mathbf{M_1} = \mathbf{M_3} = \begin{bmatrix} \cos(k_z d) & -(i/p)\sin(k_z d) \\ -ip\sin(k_z d) & \cos(k_z d) \end{bmatrix}.$$

The middle layer requires a little more thought. Its thickness is $2d$; the magnitude of k_z is the same as in air because it has the same magnitude of refractive index, but the

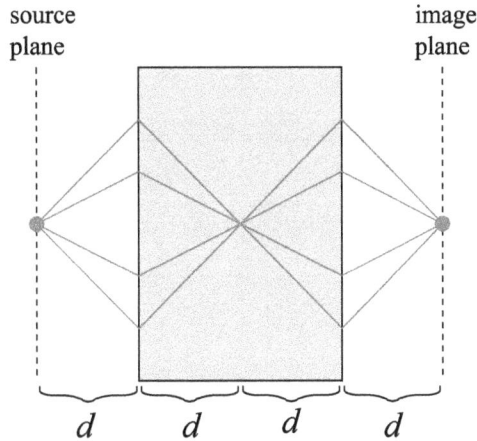

Figure 12.4. Illustration of the ray picture of the perfect lens.

sign is reversed: recall that waves flow *toward* the interface in negative-index materials. Referring to equation (10.21), the quantity p remains positive. The form of the second matrix is therefore

$$\mathbf{M}_2 = \begin{bmatrix} \cos(2k_z'd) & -(i/p)\sin(2k_z'd) \\ -ip\sin(2k_z'd) & \cos(2k_z'd) \end{bmatrix},$$

where k_z' is the component of the wave vector in the second medium. Let us, for brevity, define $\beta = k_z d$ and $\beta' = 2k_z'd$. We then multiply all three matrices together, and use familiar trigonometric identities such as

$$\cos(A \pm B) = \cos A \cos B \mp \sin A \sin B.$$

We readily find that

$$\mathbf{M}_1\mathbf{M}_2\mathbf{M}_3 = \begin{bmatrix} \cos(2\beta + \beta') & -(i/p)\sin(2\beta + \beta') \\ -ip\sin(2\beta + \beta') & \cos(2\beta + \beta') \end{bmatrix}.$$

For both evanescent waves and plane waves, $k_z' = -k_z$; this means that for any incident wave, the characteristic matrix reduces to

$$\mathbf{M}_1\mathbf{M}_2\mathbf{M}_3 = \begin{bmatrix} 1 & 0 \\ 0 & 1 \end{bmatrix}, \tag{12.44}$$

which is the identity matrix! Therefore, any plane wave at the entrance plane of the first air layer exits the second air layer in exactly the same state. An object field at the entrance plane is therefore perfectly reconstructed at the exit plane: the lens produces a perfect image of the object field without any curvature, unlike a traditional lens.

It is natural to wonder how the lens can reconstitute the evanescent waves of the object field, which are, of course, exponentially decaying. In the negative-index medium, however, k_z has a negative sign, which indicates exponential growth in the medium. Figure 12.5 shows how this works. An exponentially decaying wave

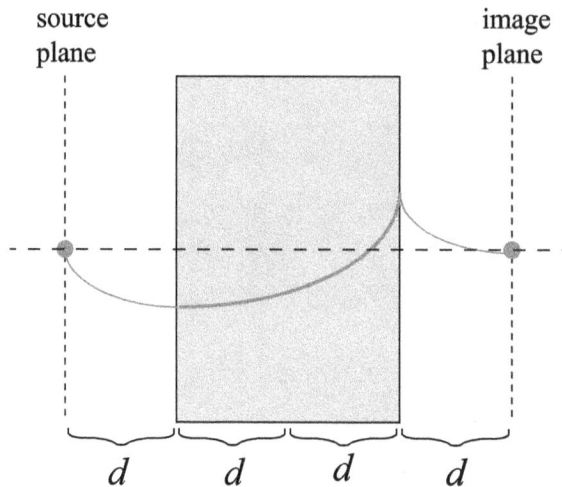

Figure 12.5. Illustration of how evanescent waves are enhanced for perfect imaging.

experiences exponential growth in the medium, sufficient to compensate for both the original decay and the additional exponential decay it acquires after leaving the medium again.

This seeming amplification of evanescent waves was one reason that some researchers suggested early in the metamaterials era that the perfect lens could not function in practice. It was argued that the high electric field intensities could cause a breakdown in the metal that would ruin the imaging effect. In 2005, however, a crude experimental test of a 'superlens' was performed that demonstrated super-resolved imaging [14]. This lens was not a metamaterial but a silver slab with a negative permittivity and a subwavelength thickness; Pendry had already predicted that in this quasi-static limit, TM waves would be affected purely by the negative permittivity [2], alleviating the need for negative permeability. The amplification can, in fact, be explained as the excitation of a surface plasmon (SP) on the shadow side of the perfect lens; in a steady state, this SP resonance provides the high electric field needed to enhance resolution [15].

The construction of an actual negative-index perfect lens has remained elusive, though a microwave version of such a lens was reported by Aydin *et al* in 2007 [16]. There remain significant practical and conceptual difficulties in constructing perfect lenses, and debate about their existence and effectiveness will probably linger until someone is able to experimentally construct one that satisfies all the conditions to be considered an ideal negative-index material—and sees whether it works.

It should be noted that even an ideal metamaterial lens would still fall short of perfection; absorption in the metamaterial would reduce or even destroy the resolution, and high-spatial-frequency evanescent fields would become sensitive to the arrangement of the meta-atoms themselves and would no longer 'see' the effective negative index; see the discussion at the end of section 9.4.4. Furthermore, such a lens would not provide any magnification, making it of limited use in applications such as microscopy. In 2002, Pendry and Ramakrishna introduced a modified lens that trades perfection for magnification [17].

The perfect lens is a good demonstration that some long-established fundamental limits in optical system performance are more guidelines than rules. The Rayleigh resolution criterion, which applies to the resolution of imaging systems with conventional lenses, can be circumvented by a metamaterial lens; however, new resolution limits are encountered beyond the Rayleigh criterion. We already encountered a similar situation when we discussed the use of evanescent waves to image subwavelength features in near-field optics in section 9.4.4.

12.4 Epsilon-near-zero materials

The introduction of metamaterials naturally led to the question: what other unusual optical properties can be produced using them? Researchers very quickly realized that metamaterials had the potential to allow permittivities very close to zero; such materials are known as *epsilon-near-zero* (ENZ) materials. These materials can come in several varieties and can possess a number of interesting properties, which we summarize here.

The earliest discussion of ENZ materials was presented in 2002 by Garcia, Ponizovskaya, and Xiao [18]. They considered the effects of a nonmagnetic medium with $n_2 = 0$. It can readily be seen from equations (9.68) and (9.78) that $R = 1$ for both transverse electric (TE) and transverse magnetic (TM) polarizations; the ENZ medium acts as a perfectly reflective medium. Furthermore, provided there is no significant dispersion, the medium reflects perfectly over a range of wavelengths; it acts as a photonic bandgap medium similar to the dielectric mirrors of section 10.6.

It should be noted that a zero refractive index does not necessarily require metamaterials; we saw in section 11.1 that the permittivity of an ideal plasma equals zero when the frequency of light equals the plasma frequency, as shown in equation (11.10). However, the noble metals all have plasma frequencies in the UV range. For example, for silver, $\omega_p = 1.36 \times 10^{16}$ s^{-1}, and for copper, we have $\omega_p = 2.24 \times 10^{16}$ s^{-1} [19]; these lie firmly in the ultraviolet regime. We would like an ENZ material with zero permittivity for optical frequencies.

The approach of Garcia *et al* is to introduce a metamaterial consisting of metal rods with a permittivity of ϵ_m in a dielectric background with a permittivity of ϵ_d, as illustrated in figure 12.6. Provided the relevant size scales (size of rods, rod spacing) are significantly smaller than the wavelength, the effective permittivity $\bar{\epsilon}$ can be written approximately as

$$\bar{\epsilon} = f\epsilon_m + (1 - f)\epsilon_d, \tag{12.45}$$

where f is the volume filling factor of the metal (the fraction of the total volume occupied by the metal). Using an appropriate choice of f, ϵ_m, and ϵ_d, the effective permittivity can be tuned to reach zero at a particular frequency or, if dispersion effects are small, over a range of frequencies.

Even if dispersion is neglected, the material only responds as a medium with effective permittivity $\bar{\epsilon}$ if the wavelength is much larger than the relevant size scales.

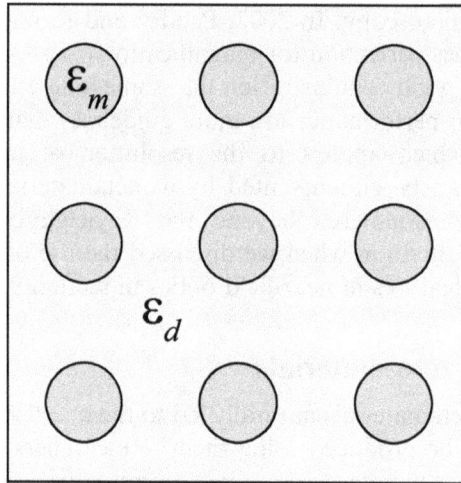

Figure 12.6. Top-down illustration of the structure of Garcia *et al*'s zero-index metamaterial.

This means that any bandgap created is a low-frequency bandgap, and once the frequency increases such that the wavelength is comparable to the size and spacing of the metal rods, the effect breaks down.

A nonmagnetic zero-index material acts as a perfect reflector and has no transmission; if we consider a system where ϵ is close to but not quite zero, other interesting refractive properties manifest. In 2002, Enoch *et al* [20] proposed that a low-index metamaterial could produce highly directional light emission. The premise used here can be understood directly from Snell's law; if a plane wave passes from a medium with low index n_{LI} into a medium n_2, Snell's law indicates that

$$\sin(\theta_t) = \frac{n_{LI}}{n_2} \sin(\theta_i). \tag{12.46}$$

For $n_{LI} \approx 0$, we have $\theta_t \approx 0$, nearly independent of the angle of incidence θ_i.

The system envisioned by Enoch *et al* is illustrated in figure 12.7. An antenna is situated within an ENZ material, and due to Snell's law, even plane waves at grazing incidence are refracted at an angle very close to the normal. The researchers experimentally realized their metamaterial using layers of copper mesh sandwiched between slices of foam with a permittivity very close to unity at microwave frequencies. The resulting structure had a plasma frequency of 14.5 GHz, and experiments showed high directionality of the emitted radiation near this frequency.

The refractive index does not even need to be very close to zero to exhibit unusual effects. As studied by Schwartz and Piestun [21], any material with a refractive index below unity exhibits total *external* reflection in air, governed again by Snell's law. Again, such a medium was realized by arrays of metal cylinders in a dielectric background medium.

The perfect reflectivity of a dielectric ENZ material can be a benefit or a limitation, depending on the application. It can be eliminated, in principle, by the

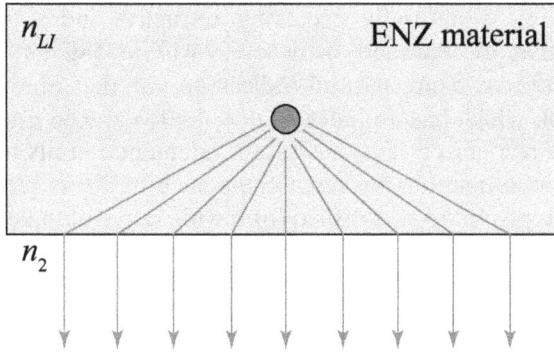

Figure 12.7. Illustration of Enoch *et al*'s directional ENZ emitter.

use of a metamaterial with zero permeability, $\mu = 0$, as well as zero permittivity. The dispersion characteristics of the metamaterial must be tailored such that the impedance of the medium $\eta = \sqrt{\mu/\epsilon}$ matches the impedance of free space, $\eta_0 = \sqrt{\mu_0/\epsilon_0}$. The transmission is not perfect except at normal incidence, as discussed in section 9.8, but it is higher than zero. This possibility was first discussed by Ziolkowski in 2004 [22], who demonstrated that a source in this impedance-matched medium produces a highly directional beam, as in the purely dielectric case discussed by Enoch *et al* previously.

One other property of note of ENZ materials was introduced by Alù *et al* in 2007 [23]. Because the wavelength of light becomes λ/n within a medium, in an ENZ material, the wavelength becomes extremely large, in principle larger than any size of the ENZ object itself. The phase within the object is then constant; if the output face of the object is curved, the phase front coming out of the object matches the curvature. Such a device, in principle, produces perfect wave front shaping; a wave with arbitrary wave front curvature entering the device exits the device with a wave front determined by the exit face curvature. The original analysis of Alù *et al* considered only purely dielectric devices, which, as we have seen, can have small throughput, but the same principle should hold for an impedance-matched material as well.

Though three-dimensional metamaterials that work at visible light frequencies are generally difficult to fabricate, Mass *et al* demonstrated such a material in 2013 [24]. It was fabricated from alternating layers of Ag and SiN of subwavelength thickness, which resulted in an ENZ condition being satisfied at wavelengths $\lambda = 662, 545,$ and 428 nm, with good impedance matching. The desired permittivity was determined by use of equation (12.45), which relates the filling factor of the metal to the effective permittivity of the metamaterial.

12.5 High-index metamaterials

Having spent a significant amount of time on epsilon-near-zero metamaterials, it is worthwhile to at least briefly consider the other extreme: metamaterials designed to have refractive indices significantly higher than any natural material. At optical frequencies, the highest naturally occurring refractive index for a transparent material is near that of diamond, with $n = 2.417$ at 589 nm. For transparent materials in the infrared band, natural indices appear that are somewhat higher, such as germanium, which has an index of $n = 4.0043$ at 8.66 μm [25].

It has long been recognized that there are fundamental limits to refractive index values, and early investigations by researchers such as Moss [26] found heuristic relationships that appeared to be satisfied for a wide range of materials. Much more recently, Shim *et al* [27] argued that the refractive index is limited by electron density, operating frequency, and the dispersion characteristics of materials. But dispersion, in particular, is one aspect that can be greatly changed through the use of metamaterial structures, and so researchers have looked at a variety of ways to boost refractive index values.

One of the first approaches was suggested by Shen *et al* [28] for infrared wavelengths. By introducing a periodic array of subwavelength slits of width a and period d in an approximately perfectly conducting metal, they found that TM incident light responded to the metal as if it were a dielectric of refractive index $n = d/a$. They simulated materials with an index as high as $n = 16$. Because of the slit structure, the material had anisotropic optical properties.

We have noted that natural materials typically have a permeability nearly equal to μ_0 at optical frequencies. A natural strategy for increasing the refractive index is therefore to construct a material out of meta-atoms that simultaneously yield high permittivity and permeability; however, it has been found that simple meta-atoms with large permittivities typically also have a strong diamagnetic response that lowers the permeability below μ_0, leaving only moderate gains in refractive index comparable to natural values. In 2009, Shin *et al* introduced [29] an antenna-like meta-atom structure that restricted the induced currents that cause diamagnetism; their simulations in the 3–6 μm range showed an effective index between 5.5 and 7. This material, in principle, had the advantage of also being broadband, as its electromagnetic response was determined primarily by the shape of the meta-atoms and not their material properties.

An experimental demonstration of a high-index material was performed by Choi *et al* in 2009 at terahertz frequencies [30]. Using a thin meta-atom in the shape of an 'H,' they suppressed the diamagnetic response and were able to produce a quasi-three-dimensional metamaterial with an index of 33.22 at a frequency of 0.851 THz. Here, 'quasi-three-dimensional' means that five layers of meta-atoms were stacked to produce a very thin but bulk metamaterial structure.

Dielectric materials may also be used. In 2019, Nguyen *et al* [31] used a pair of strontium titanate cubes of slightly different sizes as a meta-atom. Mie-like resonances (to be discussed in section 15.7) within the cubes provided the strong electromagnetic response needed, and the cube sizes were tuned to provide simultaneously high permittivity and permeability in the X-band (7.0–11.2 GHz) of the electromagnetic spectrum. The permittivity and permeability were also designed to provide impedance matching to minimize reflections.

A high refractive index was most recently demonstrated by Gao *et al* [32], who etched both sides of a dielectric slab with metasurfaces to produce a refractive index as high as 27 in the frequency range 0.39–0.65 THz. This material was argued to be broadband and to have low dispersion characteristics.

As is typical for metamaterials, all of these demonstrations so far are in the longer-wavelength regime and involve, at best, relatively thin slabs of material. It remains to be seen if, how, and when these properties can be scaled up for use in bulk optics.

12.6 Spoof surface plasmons

Metamaterial concepts may also be applied to manipulate other types of electromagnetic waves, most notably the SPs discussed in chapter 11. In that chapter, we noted that, in theory, SPs can be generated at an air–metal interface at any frequency below $\omega_p/\sqrt{2}$, where ω_p is the plasma frequency.

The value $\omega_p/\sqrt{2}$ is an upper limit on the frequency of SPs; however, there is also, in practice, a lower limit. We can see this by looking back at the permittivity for a plasma in the Drude model,

$$\epsilon(\omega) = \epsilon_0 \left[1 - \frac{\omega_p^2}{\omega^2} \right]. \tag{12.47}$$

As $\omega \to 0$, we find that $\epsilon \to -\infty$. A large negative permittivity is the mathematical form of a perfect conductor, and we may present several arguments that SPs are not sustained in this limit. First, using equation (11.35), the SP dispersion relation, we see that $k_{\mathrm{SP}} \to k_0$ in the perfect conductor limit. The plasmon wavenumber becomes so close to the free-space wavenumber that the plasmon is only weakly bound at best and easily coupled out of the metal. Second, if we look at equation (11.44), which describes the penetration depth of the plasmon in air and the metal, we find that the plasmon becomes unbounded in air in the perfect conductor limit. Finally, from this same equation, we find that the field does not penetrate the metal at all in this limit and thus cannot support SPs. The resulting surface wave is better described as a surface current and not an SP.

In practice, this means that SPs cannot be excited by ordinary materials at microwave frequencies or below. However, by manipulating the structure of the metal, we can create a medium whose effective permittivity and permeability allow these waves. The approach is roughly analogous to the effective medium approach of section 12.4 but accounts for the wave properties of light more explicitly. The surface waves created on our structured surface exhibit mathematical properties similar to those of ordinary SPs and are known as 'spoof' SPs.

Spoof SPs were first proposed by Pendry et al in 2004 [33]; their initial rough argument was soon followed up with more rigorous calculations [34]. Curiously, the original paper did not use the term 'spoof plasmon,' which was coined by the authors but removed at the request of the journal editor; it is nevertheless now the common term for the phenomenon. We will follow the derivation of the latter paper, which is also an excellent demonstration of solving problems with a mixture of exact formulas and careful approximations. Our approach will be to determine a formula

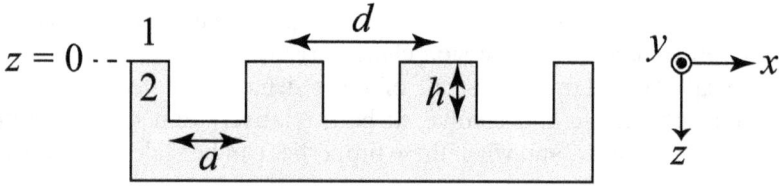

Figure 12.8. Illustration of the geometry of grooves in a metal plate used to create 'spoof' SPs.

for the reflected field; the surface resonance will then be a singularity of this reflected field.

We consider a one-dimensional periodic array of subwavelength-sized grooves patterned in a metal surface, as illustrated in figure 12.8. The holes are taken to have a width a, the period of the array is d, and the holes have a depth h. The quantities a, d, and h are assumed to be significantly smaller than the vacuum wavelength.

The metal is taken to be a perfect conductor; however, the holes allow electromagnetic waves to penetrate into the surface, much like an SP does. Because the holes are subwavelength in size, the electromagnetic field experiences the structure as a continuous medium of finite conductivity.

Let us assume that we have an incident TM plane wave, with electric and magnetic fields of the form

$$\mathbf{E}_i(\mathbf{r}) = \frac{E_0}{k_0}(k_z\hat{\mathbf{x}} - k_x\hat{\mathbf{z}})e^{ik_x x}e^{ik_z z}, \qquad (12.48)$$

$$\mathbf{H}_i(\mathbf{r}) = \frac{E_0}{\mu_0 c}\hat{\mathbf{y}}e^{ik_x x}e^{ik_z z}, \qquad (12.49)$$

where k_x and k_z are the components of the plane wave incident from air in region 1. Because our array of grooves is effectively a diffraction grating, we may write the reflected fields as

$$\mathbf{E}_r(\mathbf{r}) = \frac{E_0}{k_0}\sum_{n=-\infty}^{\infty}(-k_{zn}\hat{\mathbf{x}} - k_{xn}\hat{\mathbf{z}})e^{ik_{xn}x}e^{-ik_{zn}z}, \qquad (12.50)$$

$$\mathbf{H}_r(\mathbf{r}) = \frac{E_0}{\mu_0 c}\hat{\mathbf{y}}\sum_{n=-\infty}^{\infty}r_n e^{ik_{xn}x}e^{-ik_{zn}z}, \qquad (12.51)$$

where we have defined $k_{xn} = k_x + 2\pi n/d$, n is an integer, $k_{zn} = \sqrt{k_0^2 - k_{xn}^2}$, and r_n represents the reflection amplitude of the nth diffraction order. (We have not discussed diffraction gratings explicitly in this book, though we alluded to their properties in section 11.5.)

We now consider the fields within the grooves. A groove acts as a waveguide, and because we are assuming that the groove is subwavelength in size, only the lowest-order TEM mode of the waveguide is significant; this mode may be written as

$$\mathbf{E}_\pm(\mathbf{r}) = E_0 e^{\pm ik_0 z}\hat{\mathbf{x}}, \qquad (12.52)$$

$$\mathbf{H}_{\pm}(\mathbf{r}) = \pm \frac{E_0}{\mu_0 c} e^{\pm i k_0 z} \hat{\mathbf{y}}. \tag{12.53}$$

This result will be derived explicitly in section 13.2.1; for now, we take it as a given. The $\pm$ represents waves propagating into or out of the groove, respectively.

We may write the total electric fields in region 1 and region 2 as

$$\mathbf{E}_1(\mathbf{r}) = \mathbf{E}_i(\mathbf{r}) + \mathbf{E}_r(\mathbf{r}), \tag{12.54}$$

$$\mathbf{E}_2(\mathbf{r}) = C_+ \mathbf{E}_+(\mathbf{r}) + C_- \mathbf{E}_-(\mathbf{r}). \tag{12.55}$$

Our approach is to treat this very much like the Fresnel formulas: we want to match the boundary conditions for the electric and magnetic fields at the interface $z = 0$ and at the bottom of the groove, $z = h$, in order to find the coefficients r_n and $C_{\pm}$. We begin with the electric field condition at $z = h$; because $E_x = 0$ at the surface of a perfect conductor, we find that

$$\mathbf{E}_2(h) = C_+ E_0 e^{i k_0 h} + C_- E_0 e^{-i k_0 h} = 0, \tag{12.56}$$

or $C_- = -C_+ \exp[2i k_0 h]$. We also take $C_+ \equiv C$, so that we may write

$$\mathbf{E}_2(\mathbf{r}) = C E_0 [e^{i k_0 z} - e^{-i k_0 z} e^{2i k_0 h}] \hat{\mathbf{x}}, \tag{12.57}$$

$$\mathbf{H}_2(\mathbf{r}) = \frac{C E_0}{\mu_0 c} [e^{i k_0 z} + e^{-i k_0 z} e^{2i k_0 h}] \hat{\mathbf{y}}. \tag{12.58}$$

We now attempt to match E_x at the boundary $z = 0$, requiring it to be continuous over a single period of the grating. The fields in each region are thus

$$E_{1x}(x) = E_0 \frac{k_z}{k_0} e^{i k_x x} - E_0 \sum_{n=-\infty}^{\infty} \frac{k_{zn}}{k_0} r_n e^{i k_{xn} x}, \tag{12.59}$$

$$E_{2x}(x) = \begin{cases} C E_0 [1 - e^{2i k_0 h}], & |x| \leqslant a/2, \\ 0, & |x| > a/2. \end{cases} \tag{12.60}$$

To match this condition, we use a Fourier series expansion of each field over a period d; that is, we write

$$E_{1x}(x) = \sum_{n=-\infty}^{\infty} e_{1n} e^{i k_{xn} x}, \tag{12.61}$$

$$E_{2x}(x) = \sum_{n=-\infty}^{\infty} e_{2n} e^{i k_{xn} x}, \tag{12.62}$$

with

$$e_{1n} = \frac{1}{d} \int_0^d E_{1x}(x') e^{-i k_{xn} x'} dx', \tag{12.63}$$

$$e_{2n} = \frac{1}{d} \int_0^d E_{2x}(x')e^{-ik_{xn}x'}dx'. \tag{12.64}$$

It is straightforward to show that we get

$$e_{1n} = E_0 \left[\frac{k_z}{k_0} \delta_{n0} - \frac{k_{zn}}{k_0} r_n \right], \tag{12.65}$$

$$e_{2n} = -\frac{2iaCE_0}{d} e^{ik_0h} \sin(k_0h) \mathrm{sinc}[k_{xn}a/2], \tag{12.66}$$

where $\mathrm{sinc}(x) = \sin(x)/x$ is the 'sinc' function. If we require the field to be continuous for each Fourier component, we get the expression

$$\frac{k_{mz}}{k_0} r_m = \frac{k_z}{k_0} \delta_{m0} + \frac{2iaC}{d} e^{ik_0h} \sin(k_0h) \mathrm{sinc}[k_{xm}a/2]. \tag{12.67}$$

Now, ideally, we would like to do a similar matching of the boundary condition for the magnetic field. Here, however, we have a problem: we have already assumed that the field in the grooves is produced by a single TEM mode. While this can be approximately true, it is not exactly true, because even for grooves of subwavelength width, higher-order evanescent modes are present. We therefore cannot perfectly match the magnetic field boundary condition with our approach.

Our solution is to require the average TM field H_y, integrated over the groove width, to be continuous. For a subwavelength groove, this roughly equates the intensities of the fields in region 1 and region 2; it should be noted, however, that this approach fails for a very large groove because then the average over the oscillating field approaches zero in region 1. Our fields in the two regions are

$$H_{1y}(x) = \frac{E_0}{\mu_0 c} \left[e^{ik_x x} + \sum_{n=-\infty}^{\infty} r_n e^{ik_{xn}x} \right], \tag{12.68}$$

$$H_{2y}(x) = \frac{CE_0}{\mu_0 c} [1 + e^{2ik_0h}]. \tag{12.69}$$

Let us define the magnetic field average in each region as

$$\frac{1}{a} \int_{-a/2}^{a/2} H_y(x')dx', \tag{12.70}$$

which readily leads us to the boundary condition

$$\mathrm{sinc}(k_x a/2) + \sum_{m=-\infty}^{\infty} r_m \mathrm{sinc}(k_{xm}a/2) = 2Ce^{ik_0h} \cos(k_0h). \tag{12.71}$$

If we substitute for r_m from equation (12.67) into this expression, we can find an expression for C. Then, substituting back into equation (12.67), we finally get an expression for the reflection coefficients,

$$r_m = \delta_{m0} + \frac{2i \tan(k_0 h) S_0 S_m k_0 / k_{zm}}{1 - i \tan(k_0 h) \sum_{n=-\infty}^{\infty} S_n^2 k_0 / k_{zn}}, \tag{12.72}$$

where we have introduced

$$S_n \equiv \sqrt{\frac{a}{d}} \operatorname{sinc}(k_{xn} a / 2) \tag{12.73}$$

for brevity.

As noted at the end of section 11.4, surface waves—in this case, spoof SPs—are defined by the singularities of r_m, or the zeros of the denominator. This is still a bit too difficult to evaluate; in the limit that the grooves are subwavelength in size, however, we expect only the zeroth-order reflected field to be significant. We therefore only keep the zeroth-order term of the sum in the denominator and obtain

$$r_0 \approx \frac{1 + i \tan(k_0 h) S_0^2 k_0 / k_z}{1 - i \tan(k_0 h) S_0^2 k_0 / k_z}. \tag{12.74}$$

This denominator vanishes if

$$\frac{\sqrt{k_0^2 - k_x^2}}{k_0} = i S_0^2 \tan(k_0 h). \tag{12.75}$$

We know that surface waves exist when $k_x > k_0$; we therefore assume this to be the case and finally arrive at the dispersion relation

$$k_x = k_0 \sqrt{1 + S_0^2 \tan^2(k_0 h)}. \tag{12.76}$$

For subwavelength structures, we can further take $S_0 \approx \sqrt{a/d}$, which gives us

$$k_x = k_0 \sqrt{1 + \frac{a}{d} \tan^2(k_0 h)}. \tag{12.77}$$

This is the standard form of the dispersion relation for spoof plasmons created by a one-dimensional array of grooves. From the formula, we can see that the dispersion curve, like that for ordinary SPs, always lies to the right of the light line. Furthermore, as $k_x \to \infty$, we can see that the angular frequency approaches $\omega = \pi c / 2h$, which effectively serves the role of the plasma frequency $\omega_p / \sqrt{2}$. If we view a groove as an open cavity of length h, we can see that this frequency corresponds to the smallest resonant frequency of the cavity, with $h = \lambda/4$.

The dispersion curve is plotted in figure 12.9 for several choices of the groove parameters. It can be seen that both the slope of the curve and the limiting frequency can be tailored with an appropriate selection of parameters, and in this case, the structured surface is designed for microwave frequencies. This figure should be compared to figure 11.4, which shows the dispersion of natural SPs.

The groove system described here only works for a TM incident field, making the effective medium anisotropic. A similar calculation can be done in which a surface is

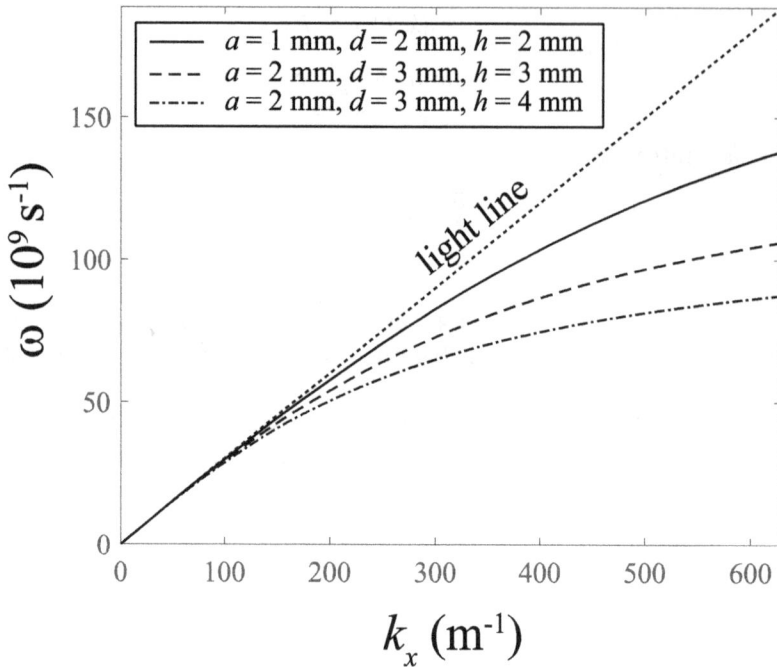

Figure 12.9. The dispersion curve of spoof SPs for several illustrative choices of groove parameters. The minimum wavelength shown in the figure is 1cm, which lies in the microwave band.

perforated by an array of square holes or dimples, making the effective medium isotropic with respect to the in-plane coordinates [34]. In the case of a hole array, it can be shown that the in-plane permittivity takes the form of the Drude model, equation (11.10), with the plasma frequency dictated by the geometry of the hole array.

The first experimental demonstration of spoof SPs was reported in 2005 [35]. The sample was an array of square brass tubes with a side length of $d = 9.525$ mm, a hole width of $a = 6.960$ mm, and a length of $h = 45$ mm. In principle, spoof SPs should only be observable via an evanescent incident wave; the researchers added an array of brass cylindrical rods to their surface to provide diffractive coupling to the surface mode (the grating approach for exciting SPs discussed in section 11.5). The signature of a resonant surface mode at 12.3GHz was clearly seen through reflection measurements.

There is much more that can be said about the physics and applications of spoof SPs; we refer the reader to a 2022 review article [36].

12.7 Form birefringence

It is often the case that significant scientific discoveries have a number of forerunners that anticipated their broad recognition. We have already noted this, for example, in the work of Pancharatnam in the 1950s, where his discovery of an anomalous phase in polarization transformations preceded the later work on geometric phase.

Metamaterials had their own precursor discoveries, most notably the phenomenon of *form birefringence*, in which isotropic materials were made to exhibit anisotropic behavior through subwavelength structuring. Such materials achieve their birefringence through their form, rather than the properties of the atoms and molecules they are constructed from.

Perhaps the earliest discussion of such an effect is due to Lord Rayleigh in 1892, who estimated the optical properties of a rectangular array of cylinders and found they exhibited birefringence [37]. Rayleigh was not interested in developing new materials as much as generalizing the Lorentz–Lorenz formula that we discussed in chapter 6. In 1912, Wiener wrote an extensive treatment of the optical properties of systems of particles and discussed the appearance of anisotropy [38].

In these cases, the bulk anisotropy arises from an array of nonspherical particles with similar orientation that are smaller than a wavelength in size but significantly larger than atoms or molecules; the anisotropy comes from the anisotropic shape of the particles. The similarity to discussions of metamaterials and meta-atoms should be obvious.

There is, however, a much easier way to create form birefringence: through a periodic multilayered thin film. In chapter 9, we saw that s- and p-polarized light incident upon an interface generally has different reflected and transmitted amplitudes; on passing through a multilayer thin film, these two polarizations consequently accrue different phase delays. If the period of the structure is taken to be significantly smaller than the wavelength of light, the two polarizations then effectively experience different phase velocities: the material exhibits birefringence. Using the matrix approach for stratified media from chapter 10, we can demonstrate this rigorously.

We consider, as in section 10.6, a periodic multilayer film, where each period consists of two layers of thickness a and b and indices n_1 and n_2, as illustrated in figure 12.10. The total thickness of one period is taken to be d. For mathematical simplicity, we take the number of total periods to be infinite, and we look for the natural eigenmodes that propagate in this periodic structure.

Let us write the field vector in the mth period of the medium, in the ith sublayer, as $|\mathbf{Q}\rangle_m^i$, where $i = 1, 2$ and the mth layer ranges from $md < z < (m + 1)d$. This field vector was introduced in section 10.2, and we take it to represent the value of the fields at the left side of the layer in question, e.g. $|\mathbf{Q}\rangle_m^1$ is the value of the fields at $z = md$. Using the characteristic matrices, we may readily write

$$|\mathbf{Q}\rangle_m^1 = \mathbf{M}_1\mathbf{M}_2|\mathbf{Q}\rangle_{m+1}^1, \tag{12.78}$$

where $\mathbf{M}_1$ and $\mathbf{M}_2$ are the characteristic matrices of the two sublayers.

We now turn to the Bloch–Floquet theorem, which indicates that the eigenfunctions for a given value of k_y for a component of the electric field must have the form

$$E_{k_y}(y, z) = E_{k_y}(z)e^{ik_y y}e^{ik_z z}, \tag{12.79}$$

where $E_{k_y}(z)$ is a periodic function in z. This theorem was originally developed by Floquet for one-dimensional periodic structures in 1883 [39] and was generalized by Bloch to three-dimensional periodic structures in 1929 [40].

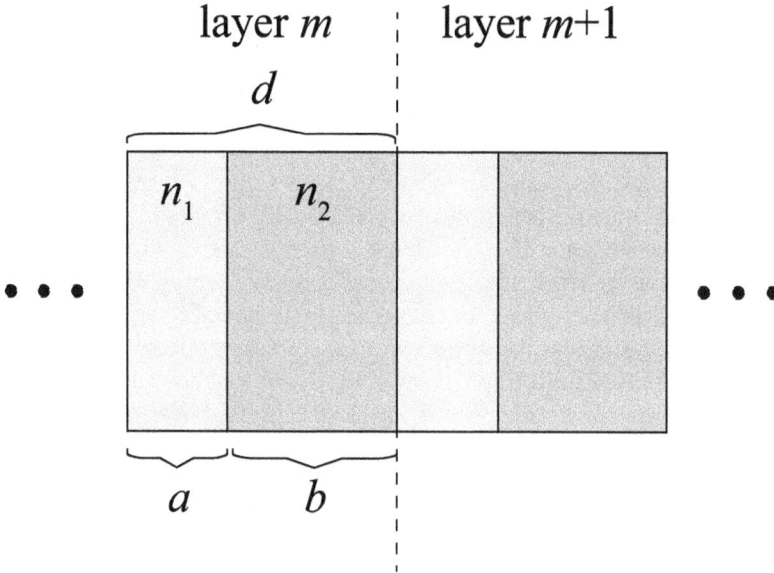

Figure 12.10. Illustration of the relevant geometry for form birefringence.

Our goal now is to determine the values of $E_{k_y}(z)$ and k_z for a given medium and value of k_y; it should be noted that k_z no longer has a simple relationship to k_y due to the periodic structure of the medium. However, determining the value is rather straightforward from the Bloch–Floquet theorem because it suggests that we must have

$$|\mathbf{Q}\rangle_{m+1}^1 = e^{ik_z d}|\mathbf{Q}\rangle_m^1, \tag{12.80}$$

or, comparing with equation (12.78), we may write an eigenequation of the form

$$\mathbf{M}|\mathbf{Q}\rangle_m^1 \equiv \mathbf{M}_1\mathbf{M}_2|\mathbf{Q}\rangle_m^1 = e^{-ik_z d}|\mathbf{Q}\rangle_m^1, \tag{12.81}$$

where we have written the matrix of a single period of the structure as $\mathbf{M}$. For simplicity, we temporarily write this matrix as

$$\mathbf{M} = \begin{bmatrix} A & B \\ C & D \end{bmatrix}, \tag{12.82}$$

and it is readily found that the eigenvalues Λ of this matrix are given by

$$\Lambda_\pm = \frac{(A + D) \pm \sqrt{A^2 + D^2 - 2AD + 4BC}}{2}. \tag{12.83}$$

This may be simplified further if we are working with a nonabsorbing medium, because then each matrix $\mathbf{M}_i$ and the combined matrix $\mathbf{M}$ must have a unit determinant, so $AD - BC = 1$. We then may write

$$\Lambda_\pm = \frac{A + D}{2} \pm \sqrt{\left(\frac{A + D}{2}\right)^2 - 1}. \tag{12.84}$$

This should equal $\exp[\pm ik_z d]$, representing left- and right-propagating waves. We may simplify even further by taking the sum of the two eigenfunctions, which readily leads to the result

$$\cos(k_z d) = \frac{A + D}{2}. \tag{12.85}$$

This expression is our dispersion relation that relates k_y, k_z, and ω implicitly through the values of A and D, with these values also depending on the polarization state of the wave.

Now we may take advantage of the symmetry of s- and p-polarization in our matrix system: the equations for the characteristic matrices $\mathbf{M}_i$ are structurally the same for the two polarizations, except we use p_i in the s-polarization case and q_i in the p-polarization case. Let us label the general case v_i, where v_i can be p_i or q_i. Then, with a little bit of effort, we can find that

$$\cos(k_z d) = \cos(k_{1z}a)\cos(k_{2z}b) - \frac{1}{2}\left[\frac{v_1}{v_2} + \frac{v_2}{v_1}\right]\sin(k_{1z}a)\sin(k_{2z}b), \tag{12.86}$$

where $k_{iz} = \sqrt{n_i^2 k_0^2 - k_y^2}$.

Equation (12.86) can be used to find the bandgaps of our infinite periodic multilayer film, which arise when the right-hand side of the equation is greater than unity and therefore no real-valued k_z can satisfy it. Bandgaps arise when the period of the structure is comparable to the wavelength; however, we are currently interested in subwavelength periodic structures.

We know we may write $p_i = k_{iz}/k_0\mu_i$ and $q_i = k_{iz}/k_0\epsilon_i$ to simplify our expression. Let us define $v_i = k_{iz}w_i/k_0$, where now $w_i = 1/\mu_i$ for s-polarization and $w_i = 1/\epsilon_i$ for p-polarization. We assume that a, b, and d are all significantly smaller than the wavelength, so that we may approximate each sine and cosine function by the lowest terms of their Taylor series, keeping only up to the quadratic terms. We arrive at the expression

$$k_0^2 = \frac{k_z^2 d^2 + k_y^2\left[a^2 + b^2 + \dfrac{w_1}{w_2}ab + \dfrac{w_2}{w_1}ab\right]}{n_1^2 a^2 + n_2^2 b^2 + n_1^2 ab\dfrac{w_1}{w_2} + n_2^2 ab\dfrac{w_2}{w_1}}. \tag{12.87}$$

This is very much in the form of a dispersion relation that relates k_0, k_z, and k_y. Let us make one final simplifying but reasonable assumption: that the materials are nonmagnetic and thus $\mu_i = \mu_0$. From this, we may use $w_i = 1$ for s-polarization and $w_i = 1/n_i^2$ for p-polarization. On substitution, and after performing some significant algebraic manipulations (including $a + b = d$), we find that we may write the dispersion relations for the two cases as

$$\frac{k_y^2}{n_o^2} + \frac{k_z^2}{n_o^2} = \frac{\omega^2}{c^2}, \; s-\text{polarization}, \tag{12.88}$$

$$\frac{k_y^2}{n_e^2} + \frac{k_z^2}{n_o^2} = \frac{\omega^2}{c^2}, \text{ p--polarization,} \tag{12.89}$$

where

$$n_o^2 = \frac{n_1^2 a}{d} + \frac{n_2^2 b}{d}, \tag{12.90}$$

$$\frac{1}{n_e^2} = \frac{a}{n_1^2 d} + \frac{b}{n_2^2 d}. \tag{12.91}$$

The system, therefore, behaves like a uniaxial crystal with ordinary axes pointing in the x and y directions. For normal incidence, of course, the two polarization states give the same effective refractive index, equivalent to propagating along the extraordinary axis. Our multilayer film, despite being made of isotropic materials, exhibits anisotropic behavior.

The analysis given above was performed by Yeh, Yariv, and Hong in a pair of back-to-back papers in 1977 [41, 42]. The Soviet physicist Rytov, however, had already demonstrated this phenomenon years earlier [43]. A noteworthy early experimental measurement of form birefringence was reported in 1976 by van der Ziel, Ilegems, and Mikulyak [44]. They used a structure consisting of fifteen alternating layers of 0.1235 μm thick AlAs and 0.1062 μm thick GaAs; at a wavelength of 1 μm, the refractive indices are $n_{AlAs} = 2.9474$ and $n_{GaAs} = 3.5039$. Using our expressions derived above, we estimate that $n_o = 3.2167$ and $n_e = 3.1694$, and the birefringence $\Delta n = n_e - n_o$ has the value $\Delta n = -0.0473$, which is in good agreement with their measured value of $\Delta n = -0.042$. This value is also significantly greater than those of many natural birefringent materials, such as quartz, which has a birefringence $\Delta n = 0.009$.

12.8 Exercises

1. One of the striking properties of the perfect lens is that it can be divided in half, a gap can be introduced between the two halves, and it can still retain its imaging properties. Assuming a perfect lens of thickness $2d$ is divided into two halves of thickness d, determine the object distance, image distance, and gap size that retains 'perfect' imaging. Provide calculations justifying your result.

2. Let us suppose light is going from a positive-index medium with ϵ_1, μ_1 into a negative-index medium with ϵ_2, μ_2. By looking for angles at which the reflected amplitude is zero, determine formulas for the Brewster angle in this case for both TE and TM waves. Write your answers in terms of the quantities $X = \epsilon_1 \mu_1 / \epsilon_2 \mu_2$ and $Y = \epsilon_2 \mu_1 / \epsilon_1 \mu_2$ (both of which are positive-valued).

3. Let us assume that we have magnetic materials on either side of an interface, i.e. we have permeabilities μ_1 and μ_2. Using the p-polarization reflection formula for such materials,

$$r_\parallel = \frac{k_{iz}\eta_1/n_1 - k_{tz}\eta_2/n_2}{k_{iz}\eta_1/n_1 + k_{tz}\eta_2/n_2},$$

determine the SP dispersion equation for k_x by looking for singularities in the reflection. Show that this formula reduces to the usual dispersion formula for nonmagnetic materials. Now imagine that we have a negative-index, impedance-matched system, with $\eta_1 = \eta_2$ and $n_2 = -n_1$; describe what happens to the allowed values of k_x.

4. Let us construct an ENZ material that operates at a wavelength of $\lambda_0 = 500$ nm. The substrate is fused silica with $\epsilon_d = 1.46$, and the metal is taken to be gold, with $\omega_p = 13.7 \times 10^{15}$ s^{-1}. (We neglect the absorption.) We assume that the material consists of metal rods of radius a that are periodically arranged $d = 100$ nm apart from each other. Find the radius a that results in the material having an index of $n = 0$ at λ_0.

5. Let us consider light propagating from air into a medium with $n = 0.5$. Determine the angle of total *external* reflection and write an expression for the reflected amplitude $r_\perp$ past this critical angle. Plot the reflected power $R_\perp$ as a function of angle, from $\theta_i = 0$ to $\theta_i = \pi/2$.

6. Total internal reflection can occur for light at the interface of a negative-index material. From Snell's law, determine the condition under which TIR occurs when light goes (a) from a positive-index material to a negative-index material, and (b) from a negative-index material to a positive-index material. Describe the spatial behavior of the electric field in the 'forbidden' region in both cases, and write the expression for the reflected amplitude for both cases.

7. Let us construct a form birefringent material using the materials that we used to construct a dielectric mirror in the chapter on stratified media. We use zinc sulfide ($n_1 = 2.3$) and cryolite ($n_2 = 1.35$), but each of these layers is now taken to have an optical path length of 1/20th of a wavelength for the wavelength $\lambda_0 = 546$ nm. Determine the effective ordinary and extraordinary refractive indices for this case.

References

[1] Pendry J B, Holden A J, Robbins D J and Stewart W J 1999 Magnetism from conductors and enhanced nonlinear phenomena *IEEE Trans. Microw. Theory Tech.* **47** 2075–84
[2] Pendry J B 2000 Negative refraction makes a perfect lens *Phys. Rev. Lett.* **85** 3966–9
[3] Vesalago V G 1968 The electrodynamics of substances with simultaneously negative values of ϵ and μ *Sov. Phys. Usp* **10** 509–14
[4] 2011 *Structured Surfaces as Optical Metamaterials* ed A I Maradudin (Cambridge: Cambridge University Press)
[5] Shelby R A, Smith D R and Schultz S 2001 Experimental verification of a negative index of refraction *Science* **292** 77–9
[6] Parazzoli C G, Greegor R B, Li K, Koltenbah B E C and Tanielian M 2003 Experimental verification and simulation of negative index of refraction using Snell's law *Phys. Rev. Lett.* **90** 107401

[7] Huangfu J *et al* 2004 Experimental confirmation of negative refractive index of a metamaterial composed of V-like metallic patterns *Appl. Phys. Lett.* **84** 1537–9

[8] Zhang S, Fan W, Panoiu N C, Malloy K J, Osgood R M and Brueck S R J 2005 Experimental demonstration of near-infrared negative-index metamaterials *Phys. Rev. Lett.* **95** 137404

[9] Valentine J *et al* 2008 Three-dimensional optical metamaterial with a negative refractive index *Nature* **455** 376–9

[10] Jin Q *et al* 2022 Negative index metamaterial at ultraviolet range for subwavelength photolithography *Nanophotonics* **11** 1643–51

[11] Milonni P W 2005 *Fast Light, Slow Light and Left-Handed Light* (Bristol: Institute of Physics Publishing)

[12] Shivanand W K J 2012 Electromagnetic field energy density in homogeneous negative index materials *Opt. Express* **20** 11370–81

[13] Parimi P V, Lu W T, Vodo P, Sokoloff J, Derov J S and Sridhar S 2004 Negative refraction and left-handed electromagnetism in microwave photonic crystals *Phys. Rev. Lett.* **92** 127401

[14] Fang N, Lee H, Sun C and Zhang X 2005 Sub-diffraction-limited optical imaging with a silver superlens *Science* **308** 534–7

[15] Pendry J B and Ramakrishna S A 2003 Refining the perfect lens *Physica* **338** 329–32 Proceedings of the Sixth International Conference on Electrical Transport and Optical Properties of Inhomogeneous Media

[16] Aydin K, Bulu I and Ozbay E 2007 Subwavelength resolution with a negative-index metamaterial superlens *Appl. Phys. Lett.* **90** 254102

[17] Pendry J B and Ramakrishna S A 2002 Near-field lenses in two dimensions *J. Phys.: Condens. Matter* **14** 8463

[18] Garcia N, Ponizovskaya E V and Xiao J Q 2002 Zero permittivity materials: band gaps at the visible *Appl. Phys. Lett.* **80** 1120–2

[19] Ordal M A *et al* 1983 Optical properties of the metals Al, Co, Cu, Au, Fe, Pb, Ni, Pd, Pt, Ag, Ti, and W in the infrared and far infrared *Appl. Opt.* **22** 1099–119

[20] Enoch S, Tayeb G, Sabouroux P, Guérin N and Vincent P 2002 A metamaterial for directive emission *Phys. Rev. Lett.* **89** 213902

[21] Schwartz B T and Piestun R 2003 Total external reflection from metamaterials with ultralow refractive index *J. Opt. Soc. Am.* B **20** 2448–53

[22] Ziolkowski R W 2004 Propagation in and scattering from a matched metamaterial having a zero index of refraction *Phys. Rev.* E **70** 046608

[23] Alù A, Silveirinha M G, Salandrino A and Engheta N 2007 Epsilon-near-zero metamaterials and electromagnetic sources: tailoring the radiation phase pattern *Phys. Rev.* **75** 155410

[24] Maas R, Parsons J, Engheta N and Polman A 2013 Experimental realization of an epsilon-near-zero metamaterial at visible wavelengths *Nat. Photon.* **7** 907–12

[25] Salzberg C D and Villa J J 1958 Index of refraction of germanium *J. Opt. Soc. Am.* **48** 579

[26] Moss T S 1950 A relationship between the refractive index and the infra-red threshold of sensitivity for photoconductors *Proc. Phys. Soc.* B **63** 167

[27] Shim H, Monticone F and Miller O D 2021 Fundamental limits to the refractive index of transparent optical materials *Adv. Mater.* **33** 2103946

[28] Shen J T, Catrysse P B and Fan S 2005 Mechanism for designing metallic metamaterials with a high index of refraction *Phys. Rev. Lett.* **94** 197401

[29] Shin J, Shen J T and Fan S 2009 Three-dimensional metamaterials with an ultrahigh effective refractive index over a broad bandwidth *Phys. Rev. Lett.* **102** 093903

[30] Choi M, Lee S H and Kim Y 2011 A terahertz metamaterial with unnaturally high refractive index *Nature* **470** 369–73

[31] Nguyen Q M, Anthony T K and Zaghloul A I 2019 Free-space-impedance-matched composite dielectric metamaterial with high refractive index *IEEE Antennas Wirel. Propag. Lett.* **18** 2751–5

[32] X G, Yu F L, Cai C L, Guan C Y, Shi J H and Hu F 2020 Terahertz metamaterial with broadband and low-dispersion high refractive index *Opt. Lett.* **45** 4754–7

[33] Pendry J B, Martín-Moreno L and Garcia-Vidal F J 2004 Mimicking surface plasmons with structured surfaces *Science* **305** 847–8

[34] Garcia-Vidal F J, Martín-Moreno L and Pendry J B 2005 Surfaces with holes in them: new plasmonic metamaterials *J. Opt. A: Pure Appl. Opt.* **7** S97

[35] Hibbins A P, Evans B R and Sambles J R 2005 Experimental verification of designer surface plasmons *Science* **308** 670–2

[36] Garcia-Vidal F J *et al* 2022 Spoof surface plasmon photonics *Rev. Mod. Phys.* **94** 025004

[37] Rayleigh L 1892 LVI. On the influence of obstacles arranged in rectangular order upon the properties of a medium London, Edinburgh Dublin *Phil. Mag. J. Sci.* **34** 481–502

[38] Wiener O 1912 Die theorie des Mischkörpers für das Feld der Stationären Strömung *Abh Sächs Ges Akad Wiss, Math-Phys Kl* **32** 509–604

[39] Floquet G 1883 Sur les équations différentielles linéaires àcoefficients périodiques *Annales scientifiques de l'École Normale Supérieure* **12** 47–88

[40] Bloch F 1929 Über die Quantenmechanik der Elektronen in Kristallgittern *Z. Phys.* **52** 555–600

[41] Yeh P, Yariv A and Hong C S 1977 Electromagnetic propagation in periodic stratified media. I. General theory *J. Opt. Soc. Am.* **67** 423–38

[42] Yariv A and Yeh P 1977 Electromagnetic propagation in periodic stratified media. II. Birefringence, phase matching, and x-ray lasers *J. Opt. Soc. Am.* **67** 438–47

[43] Rytov S M 1956 Electromagnetic properties of a finely stratified medium *Sov. Phys. JETP* **2** 466–75

[44] van der Ziel J P, Ilegems M and Mikulyak R M 1976 Optical birefringence of thin GaAs-AlAs multilayer films *Appl. Phys. Lett.* **28** 735–7

IOP Publishing

Electromagnetic Optics

Gregory J Gbur

Chapter 13

Guided waves

When light is confined in one or two dimensions and allowed to propagate freely along the other dimensions of the system, we refer to it as a *guided wave*, and the system that guides it is called a *waveguide*. We already briefly touched upon guided waves in section 9.4.4 on total internal reflection (TIR), where we noted that light can be guided by a stream of water or a glass fiber, provided it is propagating through the fiber in such a way that it is trapped by TIR.

We may broadly classify most waveguides by the method that they use to confine light. Dielectric waveguides, like glass fibers, confine light via TIR. Fiber optics has replaced electrical cables as the preferred method to transmit data over long distances; it offers higher communications bandwidths and lower loss. For communications, transmission wavelengths are chosen that coincide with low material absorption; the conventional band, for example, is from 1530 to 1565 nm, firmly in the infrared.

The other major class of waveguides is hollow metal waveguides, which confine light by strong reflection at the metal surfaces. Metal waveguides are used at radio and microwave frequencies to couple wireless signals to and from antennas and are also used in microwave ovens to couple microwaves from the magnetron source to the cooking area. Metal waveguides are most commonly used for long-wavelength applications and are inappropriate for optics; however, the physics of metal waveguides provides insight into light propagation through small holes in metal plates.

We should note a more recent class of waveguides known as photonic crystal fibers. These use a photonic crystal structure (discussed in section 10.6) to confine light to the fiber core. The advantage of such a photonic crystal fiber over a TIR fiber is that the core can be hollow, allowing light to travel mostly through air of low absorption [1]. An earlier fiber design similar in concept is the Bragg fiber, which uses a multilayered fiber to produce a photonic bandgap [2]. Unfortunately, we will not have time to discuss these classes in detail here.

doi:10.1088/978-0-7503-6064-7ch13

We have already encountered another class of waveguide in the form of surface plasmon polaritons in chapter 11. In this case, the wave is confined to the surface by the wave vector mismatch between the surface plasmon and light in free space.

We begin our discussion with some general observations relating to both dielectric and hollow metal waveguides and then turn to an analysis of hollow metal waveguides, which are mathematically simpler than the dielectric case. We then examine the properties of dielectric fiber optic waveguides.

13.1 General observations

We consider the propagation of light in a tube structure, with an arbitrary cross section and invariant in the longitudinal direction, as shown in figure 13.1. The waveguide is taken to be invariant in the z-direction, and we allow the possibility (for future consideration) that the waveguide consists of multiple transverse regions of different constant permittivity ϵ and permeability μ. It should be noted that this excludes gradient index (GRIN) fibers from our mathematical analysis, which have a continuous radial gradient of refractive index in their core.

Our goal is to search for traveling-wave solutions for the electric and magnetic fields of the form

$$\mathbf{E}(x, y, z, t) = \tilde{\mathbf{E}}_0(x, y)e^{i(k_l z - \omega t)}, \tag{13.1}$$

$$\mathbf{H}(x, y, z, t) = \tilde{\mathbf{H}}_0(x, y)e^{i(k_l z - \omega t)}, \tag{13.2}$$

where k_l represents the longitudinal component of the wavenumber in the region. The form of these equations represents traveling waves that do not change their spatial form upon propagation, aside from a possible change in overall amplitude due to absorption. It should be noted that k_l is not necessarily equal to the free-space

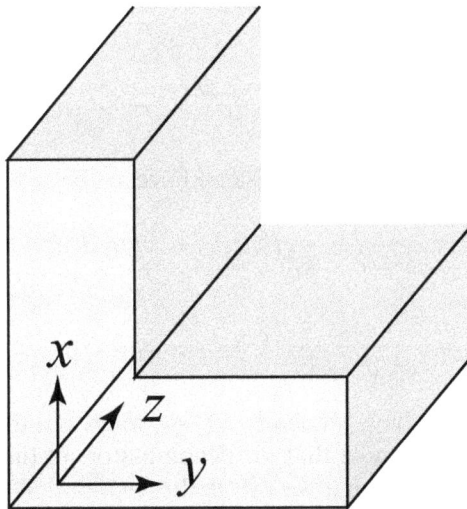

Figure 13.1. Illustration of the basic geometry for guided waves.

wavenumber ω/c, even in a hollow waveguide; part of our objective will be to determine the value of k_l. We also note that $\tilde{\mathbf{E}}_0$ and $\tilde{\mathbf{H}}_0$ are not necessarily transverse fields. In general, waves propagating in waveguides have a longitudinal component.

We assume that $\tilde{\mathbf{E}}_0$ and $\tilde{\mathbf{H}}_0$ have the forms

$$\tilde{\mathbf{E}}_0(x, y) = E_x\hat{\mathbf{x}} + E_y\hat{\mathbf{y}} + E_z\hat{\mathbf{z}}, \tag{13.3}$$

$$\tilde{\mathbf{H}}_0(x, y) = H_x\hat{\mathbf{x}} + H_y\hat{\mathbf{y}} + H_z\hat{\mathbf{z}}, \tag{13.4}$$

which leads us to three equations from Faraday's law,

$$\partial_y E_z - ik_l E_y = i\omega\mu H_x, \tag{13.5}$$

$$ik_l E_x - \partial_x E_z = i\omega\mu H_y, \tag{13.6}$$

$$\partial_x E_y - \partial_y E_x = i\omega\mu H_z, \tag{13.7}$$

and three equations from the Ampère–Maxwell law,

$$\partial_y H_z - ik_l H_y = -i\omega\epsilon E_x, \tag{13.8}$$

$$ik_l H_x - \partial_x H_z = -i\omega\epsilon E_y, \tag{13.9}$$

$$\partial_x H_y - \partial_y H_x = -i\omega\epsilon E_z, \tag{13.10}$$

where we have used ∂_x as a shorthand for the partial derivative with respect to x, for example. It should be noted that we have four equations, namely equations (13.5), (13.6), (13.8), and (13.9), that can be solved explicitly for E_x, E_y, H_x, and H_y. For the transverse magnetic (TM) fields, we have

$$H_x = \frac{i}{\omega^2/v^2 - k_l^2}[k_l\partial_x H_z - \omega\epsilon\partial_y E_z], \tag{13.11}$$

$$H_y = \frac{i}{\omega^2/v^2 - k_l^2}[k_l\partial_y H_z + \omega\epsilon\partial_x E_z], \tag{13.12}$$

and for the transverse electric (TE) fields, we have

$$E_x = \frac{i}{\omega^2/v^2 - k_l^2}[k_l\partial_x E_z + \omega\mu\partial_y H_z], \tag{13.13}$$

$$E_y = \frac{i}{\omega^2/v^2 - k_l^2}[k_l\partial_y E_z - \omega\mu\partial_x H_z]. \tag{13.14}$$

In these expressions, we have taken $v^2 = 1/\epsilon\mu$, where v is the speed of light in the medium at frequency ω. We note that the denominators of these equations indicate that, in general, $k_l \neq \omega/v$, though we will find exceptions! We also note that these equations all depend entirely on E_z and H_z. Apparently, if we calculate the longitudinal field components, we can determine all components of the field. How do we determine these longitudinal components? If we return to equations (13.7) and (13.10), we may substitute in from our solutions for the transverse field components. We find that some terms cancel, and we are left with

$$\left[\partial_x^2 + \partial_y^2 + (\omega/v)^2 - k_l^2 \right] E_z(x, y) = 0, \tag{13.15}$$

$$\left[\partial_x^2 + \partial_y^2 + (\omega/v)^2 - k_l^2 \right] H_z(x, y) = 0, \tag{13.16}$$

which represent uncoupled two-dimensional Helmholtz wave equations, which have established methods of solution.

The fact that these two equations are uncoupled does not necessarily imply that the solutions are independent. In order to properly assess the nature of the solutions, we must take into account any boundary conditions associated with the system in question. The different types of possible modes include TE modes, for which $E_z = 0$, TM modes, for which $H_z = 0$, mixed modes that combine longitudinal electric and magnetic fields, labeled EH and HE modes, as well as transverse electromagnetic (TEM) modes, for which the field is entirely longitudinal. Hollow rectangular metal waveguides of perfect conductivity only support TE and TM modes, and we can solve separately for the TE and TM modes; hollow slab waveguides support a TEM mode as well. We begin with these cases and work our way up to more complicated situations.

13.2 Hollow metal waveguides

The simplest type of waveguide we can consider is a hollow metal waveguide, consisting of a tube of perfectly conducting metal that contains no interior structures (such as additional conducting tubes). The interior of the waveguide is taken to be air, i.e. $\epsilon = \epsilon_0$, $\mu = \mu_0$, and $v = c$. For a perfect conductor, the field inside the waveguide must satisfy the boundary conditions $H_\perp = 0$ and $E_\parallel = 0$, where $\perp$ and $\parallel$ indicate field components perpendicular and parallel to the surface of the metal, respectively. Because the boundary conditions do not couple the E- and H-fields, the TE and TM modes can be solved for independently.

We now consider examples of increasing complexity.

13.2.1 Metallic slab waveguide

A metallic slab waveguide is a waveguide that consists of two parallel metal plates with an air gap between them. We start here because the slab waveguide can provide insight into the physical meaning of later results. The configuration is illustrated in figure 13.2. One metal boundary is at $y = 0$ and the other is at $y = d$.

We consider a wave propagating in the z-direction. The air gap is of thickness d in the y-direction. We consider a TE wave with a z-component of **H** of the form

$$H_z(y, z) = H_z(y)e^{ik_l z}. \tag{13.17}$$

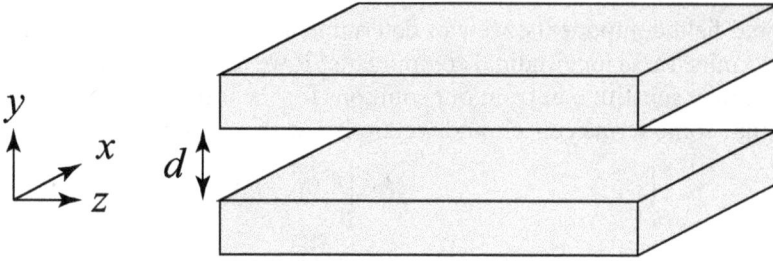

Figure 13.2. Illustration of the geometry for a metallic slab waveguide.

On substituting $H_z(y)$ into equation (13.16), we readily find the general solution

$$H_z(y) = A_\beta \sin(\beta y) + B_\beta \cos(\beta y), \tag{13.18}$$

where

$$\beta = \sqrt{k^2 - k_l^2}, \tag{13.19}$$

where $k = \omega/c$. From equation (13.13), we then find that

$$E_x(y) \propto \partial_y H_z(y) = \beta A_\beta \cos(\beta y) - \beta B_\beta \sin(\beta y). \tag{13.20}$$

We do not worry about the extra constants associated with $E_x(y)$, since we are going to match homogeneous (zero) boundary conditions. The first of these conditions is $E_x(0) = 0$, i.e. the tangential component of the electric field vanishes at the surface of the metal. This immediately leads to $A_\beta = 0$. We also require the tangential component to vanish at the second surface, i.e. $E_x(d) = 0$. This imposes a condition on β, namely that

$$\beta^2 = k^2 - k_l^2 = \left(\frac{\pi n}{d}\right)^2, \tag{13.21}$$

with apparent allowed integers $n = 0, 1, 2, 3, \ldots$. This is a constraint on the allowed values of k_l, which has the form

$$k_l = \sqrt{\left(\frac{\omega}{c}\right)^2 - \left(\frac{\pi n}{d}\right)^2}. \tag{13.22}$$

This is a dispersion relation for modes propagating in a waveguide, relating the frequency ω to the wavenumber k_l for the nth mode. It should be noted that n can be any value for a given frequency, but not all modes are propagating modes.

Let us first consider the explicit form of $H_z(y)$, which is

$$H_z(y) = H_n \cos\left(\frac{\pi n y}{d}\right), \tag{13.23}$$

where the amplitude of the nth mode is labeled H_n. If $n = 0$, H_z is nonzero and constant, which at first glance might appear to be a supported mode. However, from equations (13.11) and (13.12), it can be seen that this implies that all components of the field are zero except H_z. The Poynting vector of the field therefore vanishes, and this solution cannot represent a propagating mode. An alternative view is to note that the solution for $H_z(y, z)$ cannot satisfy $\nabla \cdot \mathbf{H} = 0$. The true allowed values of modes in the slab waveguide are therefore $n = 1, 2, 3, \ldots$.

Furthermore, because n increases without limit, there are only a finite number of modes that result in a real-valued k_l. These are the allowed propagating modes of the waveguide, and the modes corresponding to higher n values exponentially decay.

Let us introduce a quantity

$$\omega_n \equiv \frac{\pi n c}{d}, \tag{13.24}$$

which allows us to write equation (13.22) as

$$k_l = \frac{1}{c}\sqrt{\omega^2 - \omega_n^2}. \tag{13.25}$$

This refers to ω_n, which is the *cutoff frequency* of mode n. For a given waveguide, the nth mode only propagates if the frequency of excitation is above ω_n. Because ω_1 is the lowest cutoff frequency, we say it is the cutoff frequency of the waveguide: electromagnetic waves with frequencies less than $\omega_1 = \pi c/d$ do not propagate at all in the waveguide.

Because we are working with a slab waveguide, the nth mode is, in fact, infinitely degenerate in that it can propagate in any direction in the xz-plane; we chose our wave to propagate along z for convenience.

If we turn to the TM mode, we start with

$$E_z(y, z) = E_z(y)e^{ik_l z}, \tag{13.26}$$

which leads us to

$$E_z(y) = C_\beta \sin(\beta y) + D_\beta \cos(\beta y). \tag{13.27}$$

We can apply our boundary condition $E_\parallel = 0$ directly to this expression, which leads to the $D_\beta = 0$ and k_l given by equation (13.22), again with $n = 1, 2, 3, \ldots$ as allowable solutions. The cutoff frequency for the TM mode is also $\omega_1 = \pi c/d$; TE and TM modes with $n = 1$ have the same cutoff frequencies.

This is not the full story for the slab waveguide, however; there is also a TEM mode that has no cutoff frequency and has both transverse electric and magnetic fields. We determine this mode using a slightly different approach. First, let us assume that $E_x = E_z = 0$ and that only $E_y \neq 0$. By Gauss's law, we have $\partial E_y / \partial y = 0$; in other words, E_y is a constant with respect to y. We have

$$E_y(y, z) = E_0 e^{ik_l z}, \tag{13.28}$$

with E_0 a constant. We find from Faraday's law that

$$H_x(y, z) = -\frac{k_l E_0}{\mu_0 \omega}e^{ik_l z}. \tag{13.29}$$

What is the value of k_l? If we use the Ampère–Maxwell law to determine the electric field again and look for consistency with our original equation, we find that we must have $k_l = \omega/c = k$, i.e. the longitudinal wavenumber is equal to the free-space wavenumber. We can see by inspection that the tangential components of the electric and magnetic fields are continuous at the slab boundaries. We note that the TEM mode can satisfy equations (13.11)–(13.14) because both the numerator and denominator of these equations vanish, yielding an indeterminate result.

Let us now return to some general observations. Because the mode direction is degenerate, a metallic slab waveguide can be used to create quasi-two-dimensional monochromatic waves. This was done in the first experimental test of cloaking devices [3], where a two-dimensional cloak designed for microwaves was sandwiched between metal plates.

For our purposes, the metallic slab waveguide provides insight into the mode structure of waveguides: why are modes with certain special values of k_l the only ones allowed to propagate? Let us consider the picture of a slab waveguide shown in figure 13.3. We look for a solution that consists of an upward-going plane wave and a downward-going plane wave; without being too rigorous, we look at the electric field as being of the form

$$E_x(y, z) = E_0[e^{ik \sin \theta y} - e^{-ik \sin \theta y}]e^{ik \cos \theta z}. \tag{13.30}$$

We include the minus sign between the terms to account for the sign change on reflection for the downward-going wave, and it automatically satisfies our boundary condition at $y = 0$. The field can be written as

$$E_x(y, z) = 2iE_0 \sin(k \sin \theta y)e^{ik \cos \theta z}. \tag{13.31}$$

This superposition only vanishes at the waveguide boundaries if

$$k \sin \theta d = n\pi, \tag{13.32}$$

with $n = 1, 2, 3, \ldots$ again. Using basic trigonometry, this implies that

$$k_l = k \cos \theta = \sqrt{k^2 - (n\pi/d)^2}, \tag{13.33}$$

which is just equation (13.22) again.

This gives us a somewhat intuitive way of thinking about modes in a metallic slab waveguide: modes are plane waves bouncing between surfaces, and higher modes

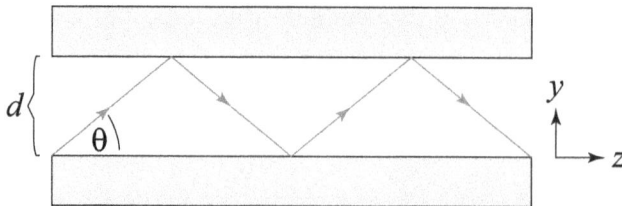

Figure 13.3. Plane wave model of light propagation in a slab waveguide.

are bouncing at higher angles, as equation (13.33) indicates. Only those plane-wave modes that satisfy the boundary conditions are allowed to propagate. We can even determine the angle of propagation for each mode: by noting that $\cos^2\theta = 1 - \sin^2\theta$, we can readily find that

$$\sin\theta = \frac{n\pi}{kd}. \tag{13.34}$$

By making an appropriate choice of kd, we can design our waveguide to support a desired number of modes. The TEM mode is a mode that does not bounce at all but propagates directly down the length of the slab waveguide.

The simple explanation provided here can be extended to waveguides that confine light in x and y, and our dispersion relations for all the metallic waveguides that follow will be directly analogous to equation (13.22).

13.2.2 Metallic rectangular waveguide

The next simplest case to consider is that of a metallic rectangular waveguide, as illustrated in figure 13.4. The waveguide has a width a in the x-direction and b in the y-direction, and we take $a > b$, without loss of generality: the final equations are independent of the particular choice of x and y coordinates.

We first look at TE waves, for which $E_z = 0$, and solve for H_z, which must satisfy equation (13.16). We follow a standard separation of variables approach and look for a particular solution of the form

$$H_z(x, y) = X(x)Y(y). \tag{13.35}$$

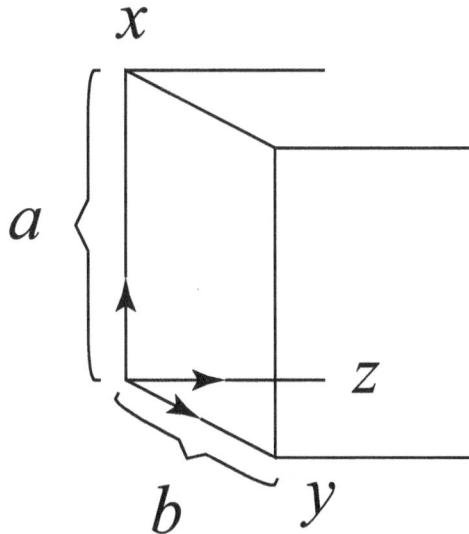

Figure 13.4. Geometry of a metallic rectangular waveguide.

On substitution, this leads to the expression

$$\frac{\partial_x^2 X}{X} + \frac{\partial_y^2 Y}{Y} + [(\omega/c)^2 - k_l^2] = 0. \tag{13.36}$$

The first term of this expression depends only on x, while the second only depends on y. Because this equation must be satisfied for all x, y, and these are independent variables, each term must be equal to a constant. We take those constants to be $-k_x^2$ and $-k_y^2$ (anticipating the form of the solution), and this leads us to three equations:

$$X'' + k_x^2 X = 0, \tag{13.37}$$

$$Y'' + k_y^2 Y = 0, \tag{13.38}$$

$$\left(\frac{\omega}{c}\right)^2 - [k_x^2 + k_y^2 + k_l^2] = 0, \tag{13.39}$$

where the primes indicate derivatives. The third equation tells us how k_l will depend on the calculated values of k_x and k_y; the first and second equations are harmonic oscillator equations and are readily solved. We may write

$$X(x) = A \sin(k_x x) + B \cos(k_x x), \tag{13.40}$$

$$Y(y) = C \sin(k_y y) + D \cos(k_y y), \tag{13.41}$$

where all constants are determined by the boundary conditions. Turning to those boundary conditions, we require that $H_\perp = 0$ on all conducting surfaces; this leads to four conditions,

$$H_x|_{x=0} = H_x|_{x=a} = 0, \tag{13.42}$$

$$H_y|_{y=0} = H_y|_{y=b} = 0. \tag{13.43}$$

To match these conditions, we turn to equations (13.11) and (13.12). We have

$$H_x(x, y) \propto \partial_x H_z(x, y), \tag{13.44}$$

$$H_y(x, y) \propto \partial_y H_z(x, y). \tag{13.45}$$

This indicates that our boundary conditions may be simplified to satisfy

$$X'(0) = X'(a) = 0, \tag{13.46}$$

$$Y'(0) = Y'(b) = 0. \tag{13.47}$$

These conditions can be evaluated quite readily. They indicate that $A = 0$, $C = 0$, and $k_x = m\pi/a$, $k_y = n\pi/a$, where $m = 0, 1, 2, ...$, $n = 0, 1, 2, ...$, but we cannot simultaneously have $m = 0$ and $n = 0$, which would result in a constant H_z that does not satisfy Maxwell's equations. We therefore find that there are a set of modes characterized by the longitudinal component of the **H**-field, given by

$$H_z(x, y) = H_0 \cos(m\pi x/a)\cos(n\pi y/b),\tag{13.48}$$

which we refer to as the TE_{mn} mode. The dispersion relation for the m, nth mode is

$$k_l = \sqrt{\left(\frac{\omega}{c}\right)^2 - \pi^2\left[\left(\frac{m}{a}\right)^2 + \left(\frac{n}{b}\right)^2\right]}.\tag{13.49}$$

We define a cutoff frequency ω_{mn} for the m, nth mode, as follows:

$$\omega_{mn} = c\pi\sqrt{\left(\frac{m}{a}\right)^2 + \left(\frac{n}{b}\right)^2}.\tag{13.50}$$

The m, nth mode does not propagate if the frequency of light is less than ω_{mn}.

The lowest cutoff frequency is ω_{10} because $a > b$ by convention, and so the overall cutoff frequency of the waveguide in TE mode operation is

$$\omega_{10} = \frac{c\pi}{a}.\tag{13.51}$$

It is illustrative to consider the phase and group velocities (recall section 6.4) of each mode, which can readily be derived from equation (13.49). We find that

$$v_p = \frac{\omega}{k} = \frac{c}{\sqrt{1 - \omega_{mn}^2/\omega^2}},\tag{13.52}$$

$$v_g = \frac{1}{dk/d\omega} = c\sqrt{1 - \omega_{mn}^2/\omega^2}.\tag{13.53}$$

We find that the phase velocity is always greater than the vacuum speed of light, and the group velocity is always less than the vacuum speed of light. Though we have noted that the group velocity itself does not generally work as a measure of the speed of light, it works quite well in the case of a metallic waveguide.

We now look at the differences for TM waves, for which $H_z = 0$, and solve for E_z. We follow the same separation of variables approach and get the same forms for $X(x)$ and $Y(y)$ as in equations (13.40) and (13.41). The difference is in the boundary conditions, which are now

$$E_y|_{x=0} = E_y|_{x=a} = 0,\tag{13.54}$$

$$E_x|_{y=0} = E_x|_{y=b} = 0.\tag{13.55}$$

Because we have

$$E_x(x, y) \propto \partial_x E_z(x, y),\tag{13.56}$$

$$E_y(x, y) \propto \partial_y E_z(x, y),\tag{13.57}$$

our boundary conditions become

$$X(0) = X(a) = 0, \qquad (13.58)$$

$$Y(0) = Y(b) = 0. \qquad (13.59)$$

From these, we find that $B = 0$, $D = 0$, and once again $k_x = m\pi/a$, $k_y = n\pi/a$. The longitudinal component of the electric field is of the form

$$E_z(x, y) = E_0 \sin(m\pi x/a)\sin(n\pi y/b), \qquad (13.60)$$

and this is called the TM_{mn} mode. These modes still satisfy equation (13.49) for k_l, with the important exception that both m and n must be nonzero, as is clear from equation (13.60). The lowest TE mode has a cutoff frequency of ω_{10}, and the lowest TM mode has a cutoff frequency of ω_{11}. There are, therefore, frequencies for which the TE mode exists but the TM mode does not. In increasing order, the cutoff frequencies are TE ω_{10}, TE ω_{01}, then the degenerate TE and TM ω_{11} modes, and so on. If $a = b$, then the ω_{10} and ω_{01} modes are also degenerate.

It should be noted that we have focused only on the mode structure of the waveguides and not so much on the detailed structure of the fields of each mode. Generally speaking, it is this mode structure that is most important because it determines the useful range of frequencies of the waveguide for applications. Multiple modes can, of course, propagate simultaneously in a waveguide, and the total field is just the linear superposition of the fields of all the modes. A waveguide might be designed to operate in only a single mode over a range of frequencies or with multiple modes at a single frequency. The mode structure can readily be determined if needed, as all components of the fields are the products of sines and cosines.

We did not concern ourselves with introducing an initial condition, i.e. the field at the entrance to the waveguide, because the coupling of a beam of light to the waveguide entrance is a complicated problem and, in general, not analytically solvable.

13.2.3 Metallic circular waveguides

We now turn to the mode characteristics of a metallic circular waveguide of radius a, as illustrated in figure 13.5.

We start by solving for TM modes; in this case, E_z still satisfies equation (13.15), but now it is more convenient to work in polar coordinates (r, ϕ) instead of Cartesian coordinates (x, y),

$$\partial_r^2 E_z + \frac{1}{r}\partial_r E_z + \frac{1}{r^2}\partial_\phi^2 E_z + [(\omega/c)^2 - k_l^2]E_z = 0. \qquad (13.61)$$

We look for separable solutions of the form $E_z(r, \phi) = R(r)F(\phi)$, which leads to the expression

$$r^2\frac{R''}{R} + r\frac{R'}{R} + \frac{F''}{F} = r^2[k_l^2 - (\omega/c)^2]. \qquad (13.62)$$

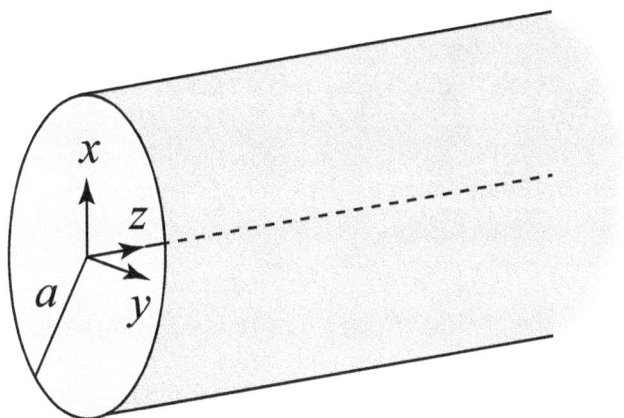

Figure 13.5. Geometry of a metallic circular waveguide.

We note that everything in this equation depends on r alone, with the exception of the term F''/F, which depends on ϕ alone. Because r and ϕ are independent variables, the equation can only be satisfied if F''/F is a constant, which we write as $-m^2$; we then have

$$F'' + m^2 F = 0, \tag{13.63}$$

which is the harmonic oscillator equation again. We write the solution as

$$F(\phi) = Ae^{im\phi} + Be^{-im\phi}, \tag{13.64}$$

where we take $m > 0$ by definition. We may immediately apply a so-called 'hidden' boundary condition, which is that the function $F(\phi)$ must be single-valued: if we go fully around the circle once, so that $\phi \to \phi + 2\pi$, we must arrive at the same value of $F(\phi)$. This implies that m must be an integer, i.e. $m = 0, 1, 2, 3, \ldots$.

Many reference works on metallic circular waveguides write $F(\phi)$ in terms of sines and cosines rather than complex exponentials. We use the exponential notation because the modes then have a spiral phase twist. It is now recognized that waves with a spiral phase twist and an inherent handedness to their phase possess orbital angular momentum (OAM) [4]. These OAM states have been used as independent data channels in fibers to dramatically increase the data transmission rate [5]. Going forward, we will always treat m as a positive integer, but the solutions found are degenerate with phase twists $+m$ and $-m$.

Turning to the equation for $R(r)$, we have

$$R'' + \frac{1}{r}R' + \left[(\omega/c)^2 - k_l^2 - m^2/r^2\right]R = 0. \tag{13.65}$$

For convenience, we define

$$k_r = \sqrt{(\omega/c)^2 - k_l^2}, \tag{13.66}$$

so that our equation becomes

$$R'' + \frac{1}{r}R' + \left[k_r^2 - m^2/r^2\right]R = 0. \tag{13.67}$$

This equation is nontrivial to solve, but it can be identified as Bessel's equation, the solutions of which are Bessel functions $J_m(k_r r)$ and Neumann functions $N_m(k_r r)$, so that we may write

$$R(r) = CJ_m(k_r r) + DN_m(k_r r). \tag{13.68}$$

We refer to chapter 16 of Gbur [6] for a detailed description of the Bessel functions and their properties; here, we simply introduce properties of these functions as needed. Figure 13.6 illustrates the zeroth- and first-order Bessel and Neumann functions. It can be seen that the Neumann functions diverge at the origin, which coincides with the center of the waveguide. We implement another physical 'hidden' boundary condition and discard all the Neumann solutions, i.e. take $D = 0$, as there is no reason to think that light propagating in a finite waveguide results in infinite fields.

We have also labeled the first few zeros of the Bessel functions in the figure; the nth zero of the mth Bessel function is labeled z_{mn}, i.e. $J_m(z_{mn}) = 0$. It can be seen that the zeros are interleaved: each zero of a Bessel function lies between the zeros of a Bessel function of neighboring order. The value of z_{mn} increases with both m and n. The values of z_{mn} must be found numerically; the first few are given in table 13.1.

We require $E_\parallel = 0$ at $r = a$, which implies that $J_m(k_r a) = 0$. Because the Bessel function has an infinite number of zeros, there are an infinite number of solutions that satisfy this boundary condition, with the nth solution given by $k_r = z_{mn}/a$. If we apply this in equation (13.66), we find that k_l is given by

$$k_l = \sqrt{\left(\frac{\omega}{c}\right)^2 - \frac{z_{mn}^2}{a^2}}. \tag{13.69}$$

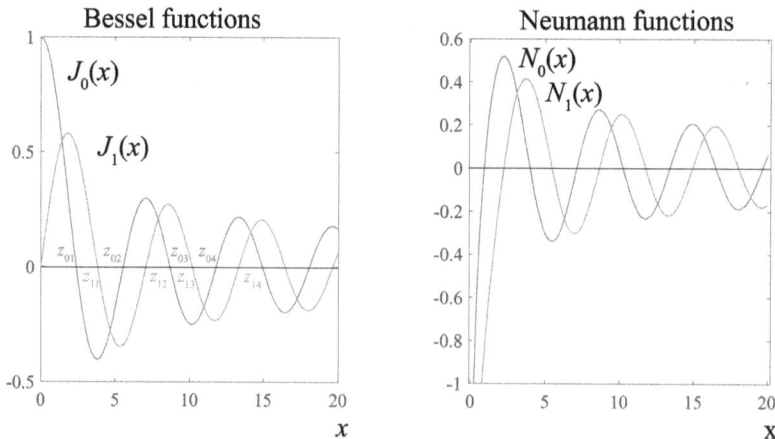

Figure 13.6. Illustration of the zeroth- and first-order Bessel and Neumann functions. The nth zero of the mth Bessel function is labeled z_{mn}.

Table 13.1. Zeros z_{mn} of low-order Bessel functions.

$n\backslash m$	0	1	2
1	2.4048	3.8317	5.1356
2	5.201	7.0156	8.4172
3	8.6537	10.1735	11.6198
4	11.7915	13.3237	14.7960
5	14.9309	16.4706	17.9598

This expression is the dispersion relation for TM modes in a metallic circular waveguide; it should be compared with the analogous relation for rectangular waveguides, equation (13.49). In both cases, we find that distinct modes can be labeled by two integer indices. As in the rectangular case, we find that only a finite number of modes propagate for a given frequency ω; the lowest TM mode therefore has a cutoff frequency of

$$\omega_{01} = \frac{2.40c}{a}. \tag{13.70}$$

We now turn to the TE case and look for separable solutions of equation (13.16) of the form $H_z(r, \phi) = R(r)F(\phi)$. We again find that $F(\phi)$ has the same form as equation (13.64), while $R(r)$ can be written in the form of Bessel and Neumann functions as equation (13.68). We again discard the singular Neumann function contribution and seek solutions described by Bessel functions.

We now need to match the boundary condition $E_\parallel = 0$ on the surface of the circular conductor. In particular, this means that the component E_ϕ of the electric field must vanish. By simple trigonometric reasoning, we obtain

$$E_\phi = -E_x \sin\phi + E_y \cos\phi \propto -\sin\phi\partial_y H_z - \cos\phi\partial_x H_z = -\partial_r H_z. \tag{13.71}$$

Our boundary condition $E_\parallel = 0$ may thus be written explicitly as

$$\partial_r J_m(k_r r)|_{r=a} = 0. \tag{13.72}$$

The allowed *TE* modes are therefore zeros of the *derivatives* of the Bessel functions. Let $\tilde{z}_{mn}$ represent the nth zero of the derivative of the mth Bessel function, i.e. $J_m'(\tilde{z}_{mn}) = 0$; we then have $k_r = \tilde{z}_{mn}/a$. The dispersion relation for TE modes in a metallic circular waveguide is therefore

$$k_l = \sqrt{\left(\frac{\omega}{c}\right)^2 - \frac{\tilde{z}_{mn}^2}{a^2}}. \tag{13.73}$$

Table 13.2 gives some of the zeros of the derivatives of the lowest-order Bessel functions. Of particular note is $\tilde{z}_{01} = 0$, which results in a constant H_z and is also not a propagating mode. Most references leave this zero off the list and start the $m = 0$

Table 13.2. Zeros $\tilde{z}_{mn}$ of the derivative of low-order Bessel functions.

$n\backslash m$	0	1	2
1	0	1.8412	3.0542
2	3.8317	5.3314	6.7061
3	7.0156	8.5363	9.9695
4	10.017 35	11.7060	13.1704
5	13.3237	14.8636	16.3475

column with $\tilde{z}_{02}$ written as $\tilde{z}_{01}$ but I prefer to leave it in the list to clearly show the interleaving of the zeros. It should be noted that this means that my labeling of the TE modes of a metallic circular waveguide differs from the labeling found in other books.

The TE modes therefore follow the same general pattern as the TM modes in a circular waveguide and all modes in a rectangular waveguide: for a given frequency, there are a finite number of propagating modes, whereas for frequencies below the cutoff frequency of the lowest mode, no modes propagate. The lowest TE mode has a cutoff frequency given by

$$\omega_{10} = \frac{1.84c}{a}, \tag{13.74}$$

which is lower than the TM mode. Therefore, there is a frequency range in which only a TE mode can propagate for a given waveguide: in such a case, it is a single-mode waveguide.

It is illuminating to rewrite the TE cutoff as a wavelength; we readily find that modes can only propagate if

$$a > 0.29\lambda. \tag{13.75}$$

This result is relevant to the field of nano-optics [7], in which superresolved imaging is often achieved using light passing through a subwavelength aperture in a metal probe. Such a metal aperture may be considered a very short metallic circular waveguide; equation (13.75) indicates that no propagating modes exist in the waveguide, and the waveguide transmits very little light, once the diameter of the aperture is smaller than roughly half the wavelength. This is, in general, true, and is a significant limitation in nano-optics applications, which explains why the discovery of extraordinary optical transmission via surface plasmons (discussed in chapter 11) drew so much attention.

13.3 Metallic coaxial waveguides

Additional mode possibilities open up when a metal waveguide is not hollow, namely a set of TEM modes in which both E_z and H_z are zero simultaneously. Looking back at equations (13.11)–(13.14), we see this should only be possible if

$k_l = \omega/c$, providing a '0/0' loophole for the fields to be nonzero. In this section we look at the properties of modes in metallic coaxial waveguides, which will also give us additional insight when we finally tackle the dielectric waveguide problem.

13.3.1 Nonexistence of TEM modes in hollow waveguides

We begin by proving that TEM modes cannot exist in a hollow waveguide. Let us focus on the vectors that define the x, y behavior of the field, i.e. $\mathbf{E}_0(x, y)$. We assume that we simultaneously have $E_z = 0$ and $H_z = 0$ and prove that such modes are impossible by contradiction. First, we note that Gauss's law tells us that the divergence of the electric field must be equal to zero, $\nabla \cdot \mathbf{E}_0 = 0$, and from Faraday's law, we find that $\nabla \times \mathbf{E}_0 = 0$, using the fact that $H_z = 0$ and that $\mathbf{E}_0(x, y)$ does not depend on z. We therefore know that the electric field has no divergence or curl and can therefore be written as the gradient of a scalar potential, $\mathbf{E}_0(x, y) = -\nabla U(x, y)$, that satisfies Laplace's equation $\nabla^2 U = 0$.

However, the boundary of the waveguide is a perfect conductor, which must have a constant potential on it; from the properties of Laplace's equation, this leads us to conclude that the potential $U(x, y)$ is constant throughout the hollow waveguide, which, in turn, implies that $\mathbf{E}_0(x, y) = 0$. Then, from Faraday's law, we have $\mathbf{H}_0(x, y) = 0$. No TEM modes can exist in a hollow waveguide.

13.3.2 TEM mode in a metallic coaxial waveguide

We now turn to a circular coaxial waveguide, as illustrated in figure 13.7. The inner radius of the waveguide is a and the outer radius is b.

We again look for a TEM mode, which again means that we may write $\mathbf{E}_0(x, y) = -\nabla U(x, y)$. Because there is an inner surface and an outer surface, however, we may assign different potentials to each surface and get a nonzero result for the potential, in the form

$$U(x, y) = -A \log(r/b) + B, \tag{13.76}$$

with constants A and B determined by the choice of potentials, and thus an electric field

$$\mathbf{E}_0(x, y) = \frac{A}{r}\hat{\mathbf{r}}. \tag{13.77}$$

The boundary condition $E_\parallel = 0$ is satisfied automatically because the electric field is purely radial. We may find the magnetic field by applying Faraday's law to the *complete* expression for the electric field,

$$\mathbf{E}(x, y, z) = \hat{\mathbf{r}}\frac{A}{r}e^{ik_l z}, \tag{13.78}$$

which results in a magnetic field of the form,

$$\mathbf{H}(x, y, z) = \hat{\boldsymbol{\phi}}\frac{ik_l A}{\omega\mu_0 r}e^{ik_l z}. \tag{13.79}$$

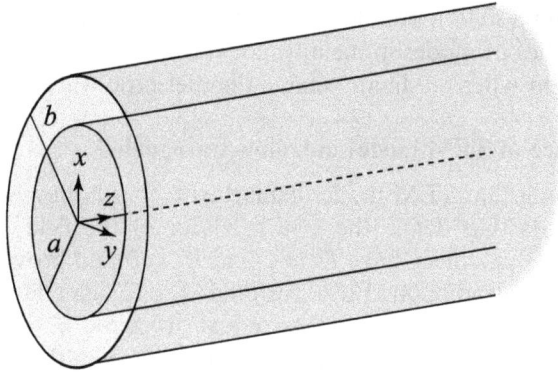

Figure 13.7. Illustration of a metallic coaxial waveguide.

To determine the value of k_l, we apply the Ampère–Maxwell law to the magnetic field and look for consistency with equation (13.78). We have

$$\frac{A}{r}e^{ik_l z}\hat{\mathbf{r}} = \frac{k_l^2 A}{\omega^2 \epsilon_0 \mu_0 r}e^{ik_l z}\hat{\mathbf{r}}. \tag{13.80}$$

This equation is only consistent if $k_l = \omega/c$. The TEM mode, therefore, propagates along the coaxial waveguide with the free-space wavenumber; we may roughly say that it is the only mode we have seen that propagates directly down the axis of the waveguide. If we label the magnitude of the electric field at the inner conductor E_b, we may write

$$\mathbf{E}(x, y, z) = \hat{\mathbf{r}}\frac{bE_b}{r}e^{ik_l z}, \tag{13.81}$$

$$\mathbf{H}(x, y, z) = \hat{\boldsymbol{\phi}}\frac{ik_l bE_b}{\omega \mu_0 r}e^{ik_l z}. \tag{13.82}$$

13.3.3 TE and TM modes in a metallic coaxial waveguide

Higher-order TE and TM modes can also propagate in a coaxial waveguide, though their mathematical form is significantly more complicated. We simply outline the process for deriving these modes and refer the reader to Marcuvitz [8] for quantitative details.

Considering the TM modes first, we return to equation (13.68) and require that $R(a) = 0$ and $R(b) = 0$. This leads us to the pair of equations

$$CJ_m(k_r a) + DN_m(k_r a) = 0, \tag{13.83}$$

$$CJ_m(k_r b) + DN_m(k_r b) = 0. \tag{13.84}$$

This results in an expression for the radial part of E_z,

$$R(r) = C\left[N_m(k_r a)J_m(k_r r) - J_m(k_r a)N_m(k_r r)\right]. \tag{13.85}$$

This automatically satisfies the boundary condition $R(a) = 0$; to satisfy $R(b) = 0$, we must choose $k_r = \alpha_{mn}$ to satisfy

$$N_m(\alpha_{mn}a)J_m(\alpha_{mn}b) - J_m(\alpha_{mn}a)N_m(\alpha_{mn}b) = 0. \tag{13.86}$$

The quantity α_{mn} is the nth solution of the mth-order equation. As one may anticipate at this point, there are an infinite number of discrete solutions, each one representing a distinct waveguide mode.

For the TE case, we use equation (13.68) and require that $R'(a) = 0$ and $R'(b) = 0$. These boundary conditions are written as

$$CJ_m(k_r a) + DN_m(k_r a) = 0, \tag{13.87}$$

$$CJ_m(k_r b) + DN_m(k_r b) = 0, \tag{13.88}$$

which leads to an equation for the radial part of H_z,

$$R(r) = C\left[N'_m(k_r a)J_m(k_r r) - J'_m(k_r a)N_m(k_r r)\right]. \tag{13.89}$$

This satisfies $R'(a) = 0$, and we select $k_r = \beta_{mn}$ that satisfies

$$N'_m(\beta_{mn}a)J_m(\beta_{mn}b) - J'_m(\beta_{mn}a)N_m(\beta_{mn}b) = 0. \tag{13.90}$$

Here, β_{mn} is the nth solution of the mth-order equation. We have a distinct set of mode numbers k_r from these equations. It also follows that we can determine the longitudinal wavenumber for the TM case from the expression

$$k_l = \sqrt{\left(\frac{\omega}{c}\right)^2 - \alpha_{mn}^2}, \tag{13.91}$$

and we can use a similar equation for the TE case. For a fixed frequency, the coaxial waveguide has a finite number of propagating modes, analogous to the metallic circular waveguide.

13.4 Circular dielectric waveguides

We now turn to dielectric waveguides, which confine their modes by TIR. We will consider circular dielectric waveguides with a core of radius a and a core index of n_1 together with a surrounding medium—the cladding—with an index of n_0. Such waveguides are known as *step-index waveguides* due to their step-function-like refractive index profile.

To achieve TIR in the core, we need $n_1 > n_0$. The most common use of such circular dielectric waveguides is for fiber optic communications; we will first consider the general physics of such waveguides, and at the end of our discussion, we will discuss some characteristic values for fiber optics. Our derivation of waveguide modes follows the approach of Okamoto [9], and we refer to his book for additional information about optical waveguides.

We note that equations (13.15) and (13.16) still apply in this case, with the new wrinkle that both the core and the cladding have nontrivial refractive indices. It is more convenient for our purposes to write these equations in cylindrical coordinates:

$$\partial_r^2 E_z + \frac{1}{r}\partial_r E_z + \frac{1}{r^2}\partial_\phi^2 E_z + [k^2 n^2 - k_l^2]E_z = 0, \tag{13.92}$$

$$\partial_r^2 H_z + \frac{1}{r}\partial_r H_z + \frac{1}{r^2}\partial_\phi^2 H_z + [k^2 n^2 - k_l^2]H_z = 0. \tag{13.93}$$

These equations can be applied to the core and the cladding with the appropriate index substituted for n. Furthermore, equations (13.11)–(13.14) still apply, and these can be written in cylindrical coordinates as

$$E_r = \frac{i}{k^2 n^2 - k_l^2}\left[k_l \partial_r E_z + \frac{\omega \mu_0}{r}\partial_\phi H_z\right], \tag{13.94}$$

$$E_\phi = \frac{i}{k^2 n^2 - k_l^2}\left[\frac{k_l}{r}\partial_\phi E_z - \omega \mu_0 \partial_r H_z\right], \tag{13.95}$$

for the electric field components, and

$$H_r = \frac{i}{k^2 n^2 - k_l^2}\left[k_l \partial_r H_z - \frac{\omega \epsilon_0 n^2}{r}\partial_\phi E_z\right], \tag{13.96}$$

$$H_\phi = \frac{i}{k^2 n^2 - k_l^2}\left[\frac{k_l}{r}\partial_\phi H_z + \omega \epsilon_0 n^2 \partial_r E_z\right], \tag{13.97}$$

for the magnetic field components.

The major difference between the mathematics of the dielectric circular waveguide and the mathematics of the metallic circular waveguide discussed previously is the boundary conditions. For the dielectric circular waveguide, the tangential components of the electric and magnetic fields must be continuous between the core and the cladding. These nonzero boundary conditions result in much more complicated dispersion relations for the modes of the waveguide and also allow for the existence of distinct modes for which both E_z and H_z are nonzero. By convention, these hybrid modes are labeled EH and HE, and we will look at their derivation momentarily; in addition, we also have TE and TM modes in the waveguide.

13.4.1 TE modes

Let us begin with a discussion of the TE case, for which $E_z = 0$. If we assume a separable solution for H_z, we already know from prior experience that the azimuthal dependence will have the form $\exp[im\phi]$; in order to obtain an ordinary differential equation for just the radial dependence of H_z, we may therefore consider

$$\partial_r^2 H_z + \frac{1}{r}\partial_r H_z + \left[k^2 n^2 - k_l^2 - \frac{m^2}{r^2} \right] H_z = 0. \tag{13.98}$$

The transverse fields have the simplified forms

$$E_r = \frac{i\omega\mu_0}{k^2 n^2 - k_l^2}\frac{1}{r}\partial_\phi H_z, \tag{13.99}$$

$$E_\phi = \frac{i\omega\mu_0}{k^2 n^2 - k_l^2}\partial_r H_z, \tag{13.100}$$

for the electric field, and

$$H_r = \frac{ik_l}{k^2 n^2 - k_l^2}\partial_r H_z, \tag{13.101}$$

$$H_\phi = \frac{ik_l}{k^2 n^2 - k_l^2}\frac{1}{r}\partial_\phi H_z, \tag{13.102}$$

for the magnetic field.

We consider the boundary condition where $H_\parallel$ must be continuous at $r = a$. This means that H_z and H_ϕ must be continuous at the boundary. We choose H_z to have the form

$$H_z(r,\ \phi) = \begin{cases} g(r)e^{im\phi}, & 0 \leqslant r \leqslant a, \\ h(r)e^{im\phi}, & r > a. \end{cases} \tag{13.103}$$

Our boundary conditions may therefore be written as

$$g(a) = h(a), \qquad H_z \text{ continuous}, \tag{13.104}$$

$$\frac{k_l}{k^2 n_1^2 - k_l^2}\frac{m}{a}g(a) = \frac{k_l}{k^2 n_1^2 - k_l^2}\frac{m}{a}h(a), \quad H_\phi \text{ continuous}. \tag{13.105}$$

The only way that these conditions can be satisfied simultaneously is if $m = 0$, which trivially satisfies the second equation. We thus have $E_r = 0$, $H_\phi = 0$ for the TE case.

We now turn to finding the mode structure of these TE modes, which we will see is a nontrivial exercise. With $m = 0$, equation (13.98) for H_z can be written as

$$r^2\partial_r^2 H_z + r\partial_r H_z + r^2\left[k^2 n^2 - k_l^2 \right] H_z = 0. \tag{13.106}$$

This should be compared to the equation

$$x^2\frac{d^2 y}{dx^2} + x\frac{dy}{dx} \pm x^2 y = 0, \tag{13.107}$$

which represents the Bessel equation of order zero for the positive sign and the modified Bessel equation of order zero for the negative sign. For a TIR guided mode, we expect $k^2 n_1^2 - k_l^2 > 0$ and $k^2 n_0^2 - k_l^2 < 0$, which represents a propagating wave inside the core and an evanescent wave in the cladding. If we define

$$\kappa \equiv \sqrt{k^2 n_1^2 - k_l^2}, \quad \sigma \equiv \sqrt{k_l^2 - k^2 n_0^2}, \tag{13.108}$$

we may write H_z in the two media as

$$H_z(r, \phi) = \begin{cases} A J_0(\kappa r), & 0 \leqslant r \leqslant a, \\ B K_0(\sigma r), & r > a, \end{cases} \tag{13.109}$$

where A and B are constants that must be chosen to satisfy our boundary condition, and J_0 and K_0 are the Bessel function and modified Bessel function of the first kind and order zero, respectively. We have applied 'hidden' boundary conditions again in this choice, by discarding the divergent N_0 in the core and discarding the exponentially growing modified Bessel function I_0 in the cladding; K_0 has the asymptotic behavior $K_0(r) \sim \exp[-\sigma r]/\sigma r$.

We now have two additional boundary conditions to satisfy, corresponding to the case where H_z and E_ϕ are continuous at $r = a$. These conditions become

$$A J_0(\kappa a) = B K_0(\sigma a), \tag{13.110}$$

$$\frac{A}{\kappa} J_0'(\kappa a) = -\frac{B}{\sigma} K_0'(\sigma a), \tag{13.111}$$

where a prime indicates a derivative with respect to the total argument of the Bessel function. It is probably clear that these equations do not have simple solutions! To make things easier, we let $u = \kappa a$, $w = \sigma a$, and take the ratio of the two boundary conditions, which eliminates the constants A and B and gives us a relationship between u and w:

$$\frac{J_0'(u)}{u J_0(u)} = -\frac{K_0'(w)}{w K_0(w)}. \tag{13.112}$$

We may simplify this expression a bit further by using the familiar Bessel identities, $J_0'(u) = -J_1(u)$, $K_0'(w) = -K_1(w)$, which allows us to write

$$\frac{J_1(u)}{u J_0(u)} = -\frac{K_1(w)}{w K_0(w)}, \tag{13.113}$$

and the expression at least does not have derivatives within it. Equation (13.113) is the dispersion relation for TE modes in a circular dielectric waveguide, since it relates the longitudinal wavenumber k_l to the frequency ω. The big question, however, is how to derive the allowed values of k_l for a given frequency ω. To aid in this, we note that

$$u^2 + w^2 = k^2 a^2 (n_1^2 - n_0^2) \equiv v^2, \tag{13.114}$$

which suggests that for a given frequency, allowed solutions lie on a circle of radius v in a u, w-plane. This observation will be true for all modes, and not just the TE ones.

The quantity v is known as the *V-number*, and it is a rough measure of the number of modes that the fiber can sustain, as we will see.

Equations (13.113) and (13.114) trace out two curves in the u, w-plane, and the TE modes allowed at a given frequency are determined by the intersection of these two curves. The radius v of the circle increases linearly with ω, so as the frequency increases, the circle gets bigger and more intersections—more modes—are created.

13.4.2 TM modes

The TM case follows in a very similar manner from the TE case, where now $H_z = 0$. Again, we expect an $\exp[im\phi]$ azimuthal dependence, so we have

$$\partial_r^2 E_z + \frac{1}{r}\partial_r E_z + \left[k^2 n^2 - k_l^2 - \frac{m^2}{r^2} \right] E_z = 0. \tag{13.115}$$

The transverse fields have the simplified forms

$$E_r = \frac{ik_l}{k^2 n^2 - k_l^2}\partial_r E_z, \tag{13.116}$$

$$E_\phi = \frac{ik_l}{k^2 n^2 - k_l^2}\frac{1}{r}\partial_\phi E_z, \tag{13.117}$$

and

$$H_r = \frac{i\omega\epsilon_0 n^2}{k^2 n^2 - k_l^2}\frac{1}{r}\partial_\phi E_z, \tag{13.118}$$

$$H_\phi = -\frac{i\omega\epsilon_0 n^2}{k^2 n^2 - k_l^2}\partial_r E_z. \tag{13.119}$$

Since $E_\parallel = 0$ on the boundary, we follow very much the same reasoning that we used for the *TE* case. We define

$$E_z(r, \phi) = \begin{cases} g(r)e^{im\phi}, & 0 \leqslant r \leqslant a, \\ h(r)e^{im\phi}, & r > a. \end{cases} \tag{13.120}$$

which immediately leads to the condition $m = 0$. We then write

$$E_z(r, \phi) = \begin{cases} AJ_0(\kappa r), & 0 \leqslant r \leqslant a, \\ BK_0(\sigma r), & r > a. \end{cases} \tag{13.121}$$

If we now apply the boundary conditions $E_z = 0$ and $H_\phi = 0$ at $r = a$, we have the following requirements:

$$AJ_0(\kappa a) = BK_0(\sigma a), \tag{13.122}$$

$$\frac{A}{\kappa} J_0'(\kappa a) = -\frac{n_1^2}{n_0^2} \frac{B}{\sigma} K_0'(\sigma a), \tag{13.123}$$

which are structurally similar to the conditions for the TE modes. Taking the ratio and applying Bessel identities again, we arrive at the dispersion relation for the *TM* modes:

$$\frac{J_1(u)}{u J_0(u)} = -\frac{n_0^2}{n_1^2} \frac{K_1(w)}{w K_0(w)}. \tag{13.124}$$

At a given frequency, the allowed modes can be determined from the plots of equations (13.124) and (13.114); the intersection of the curves determines the modes.

Optical fibers typically have a very small refractive index difference between the core and cladding, i.e. $n_0 \approx n_1$, which means that, to a good approximation, we have a dispersion relation of the form

$$\frac{J_1(u)}{u J_0(u)} = -\frac{K_1(w)}{w K_0(w)}. \tag{13.125}$$

This dispersion relation is the same as the dispersion relation of the TE modes, so these modes may be considered degenerate.

13.4.3 Hybrid EH and HE modes

Finally, we consider the possibilities when neither H_z nor E_z is zero. Assuming still an $\exp[im\phi]$ azimuthal dependence, we have the following pair of ordinary differential equations for the radial dependence of the fields:

$$r^2 \partial_r^2 E_z + r \partial_r E_z - m^2 E_z + r^2 \left[k^2 n^2 - k_l^2 \right] E_z = 0, \tag{13.126}$$

$$r^2 \partial_r^2 H_z + r \partial_r H_z - m^2 H_z + r^2 \left[k^2 n^2 - k_l^2 \right] H_z = 0. \tag{13.127}$$

These are simply the Bessel equation of order m or the modified Bessel equation of order m, depending on whether the sign of $k^2 n^2 - k_l^2$ is positive or negative, respectively.

Because we are considering hybrid modes, E_z and H_z are not completely independent, though the equations above indicate that they must have the same mathematical form, namely

$$E_z(r, \phi) = e^{im\phi} \begin{cases} A J_m(\kappa r), & 0 \leqslant r \leqslant a, \\ B K_m(\sigma r), & r > a \end{cases} \tag{13.128}$$

for E_z and

$$H_z(r, \phi) = e^{im\phi} \begin{cases} C J_m(\kappa r), & 0 \leqslant r \leqslant a, \\ D K_m(\sigma r), & r > a \end{cases} \tag{13.129}$$

for H_z. As before, we want to determine a dispersion relation for the modes. We require the tangential components of the electric and magnetic fields to be continuous at the boundary $r = a$, which requires that E_z, H_z, E_ϕ, and H_ϕ must all be continuous. From equations (13.128) and (13.129), we immediately find that

$$AJ_m(u) = BK_m(w),\qquad(13.130)$$

$$CJ_m(u) = DK_m(w),\qquad(13.131)$$

where we have adopted $u = \kappa a$ and $w = \sigma a$ again. For the boundary conditions for E_ϕ and H_ϕ, we turn to equations (13.95) and (13.97) and end up with the pair of complicated expressions

$$Ak_l m\left(\frac{1}{u^2} + \frac{1}{w^2}\right) = -C\omega\mu_0\left[\frac{J_m'(u)}{uJ_m(u)} + \frac{K_m'(w)}{wK_m(w)}\right],\qquad(13.132)$$

$$Ck_l m\left(\frac{1}{u^2} + \frac{1}{w^2}\right) = -A\omega\epsilon_0\left[\frac{J_m'(u)}{uJ_m(u)}n_1^2 + \frac{K_m'(w)}{wK_m(w)}n_0^2\right].\qquad(13.133)$$

These two equations can be solved for a single dispersion relation independent of A and C in the form

$$\left(\frac{k_l m}{k}\right)^2\left(\frac{1}{u^2} + \frac{1}{w^2}\right)^2 = \left[\frac{J_m'(u)}{uJ_m(u)} + \frac{K_m'(w)}{wK_m(w)}\right]\left[\frac{J_m'(u)}{uJ_m(u)}n_1^2 + \frac{K_m'(w)}{wK_m(w)}n_0^2\right].\quad(13.134)$$

This is the dispersion relation for the so-called HE and EH modes, though we have not yet explained how one distinguishes the two classes. This dispersion relation is quite complicated and, of course, requires the use of numerical methods to obtain a solution. It can, however, be simplified somewhat for many cases of interest. From the definitions of u and w, we may find the expression

$$\frac{k_l^2}{k^2}\left(\frac{1}{u^2} + \frac{1}{w^2}\right) = \frac{n_1^2}{u^2} + \frac{n_0^2}{w^2},\qquad(13.135)$$

which leads us to a modified form of the dispersion relation,

$$m^2\left(\frac{1}{u^2} + \left(\frac{n_0}{n_1}\right)^2\frac{1}{w^2}\right)\left(\frac{1}{u^2} + \frac{1}{w^2}\right) = \left[\frac{J_m'(u)}{uJ_m(u)} + \frac{K_m'(w)}{wK_m(w)}\right]\left[\frac{J_m'(u)}{uJ_m(u)} + \left(\frac{n_0}{n_1}\right)^2\frac{K_m'(w)}{wK_m(w)}\right].\quad(13.136)$$

We again take advantage of the fact that optical fibers typically have a very small difference between n_0 and n_1, on the order of 1%. This means that we may, to a good approximation, take $n_0/n_1 \approx 1$, and our dispersion relation becomes a perfect square that can be separated into two different equations,

$$\left[\frac{J_m'(u)}{uJ_m(u)} + \frac{K_m'(w)}{wK_m(w)}\right] = \pm m\left(\frac{1}{u^2} + \frac{1}{w^2}\right).\qquad(13.137)$$

The modes corresponding to the plus sign are called the EH modes, while the modes corresponding to the minus sign are called the HE modes.

These dispersion relations can also be simplified using the Bessel function recurrence relations:

$$\frac{2m}{x} J_m(x) = J_{m-1}(x) + J_{m+1}(x),$$ (13.138)

$$2J'_m(x) = J_{m-1}(x) - J_{m+1}(x),$$ (13.139)

and the modified Bessel function recurrence relations

$$-\frac{2m}{x} K_m(x) = K_{m-1}(x) - K_{m+1}(x),$$ (13.140)

$$-2K'_m(x) = K_{m-1}(x) + K_{m+1}(x).$$ (13.141)

With these relations, we can simplify the EH mode dispersion relation by replacing J_{m-1} and K_{m-1}; the result is of the form

$$\frac{J_{m+1}(u)}{uJ_m(u)} = -\frac{K_{m+1}(w)}{wK_m(w)} \quad \text{(EH modes)}.$$ (13.142)

This relation is structurally similar to the TE/TM dispersion relation, an observation we will use to our advantage momentarily. By instead using the recurrence relations to replace J_{m+1} and K_{m+1} in equation (13.137), we can find a simplified form of the HE mode dispersion relation,

$$\frac{J_{m-1}(u)}{uJ_m(u)} = \frac{K_{m-1}(w)}{wK_m(w)}.$$ (13.143)

This can be simplified even further. We shift down the index in the above equation, $m \rightarrow m - 1$, with the assumption that $m > 0$, and further apply the recurrence relations, equations (13.138) and (13.140); we end up with the expression

$$\frac{J_m(u)}{uJ_{m-1}(u)} = -\frac{K_m(w)}{wK_{m-1}(w)} \quad \text{(HE modes)}.$$ (13.144)

If we compare this equation to equations (13.142) and (13.113), we find that the TE, TM, EH, and HE modes satisfy structurally similar dispersion relations. There is, therefore, a lot of degeneracy between the different modes.

The $m = 0$ HE mode satisfies a slightly different dispersion relation, using $J_{-1}(u) = -J_1(u)$ and $K_{-1}(w) = K_1(w)$, which gives us

$$\frac{J_0(u)}{uJ_1(u)} = \frac{K_0(w)}{wK_1(w)}.$$ (13.145)

13.5 Mode structure of dielectric waveguides

We have pointedly avoided discussing the solutions of the dispersion equations for the TE, TM, HE, and EH modes up to this point, as they are all modes of the same waveguide and possess a significant amount of degeneracy, making it best to discuss them together. In this section, we attempt to gain some insight into the structure of dielectric waveguide modes.

We now have three dispersion relations, corresponding to the TE/TM modes, the EH mode, and the HE mode, listed again here as

$$\frac{J_1(u)}{uJ_0(u)} = -\frac{K_1(w)}{wK_0(w)}, \qquad \text{TE/TM modes}, \tag{13.146}$$

$$\frac{J_{m+1}(u)}{uJ_m(u)} = -\frac{K_{m+1}(w)}{wK_m(w)}, \qquad \text{EH modes}, \tag{13.147}$$

$$\frac{J_m(u)}{uJ_{m-1}(u)} = -\frac{K_m(w)}{wK_{m-1}(w)} \qquad \text{HE modes}. \tag{13.148}$$

(We leave out the $m = 0$ HE mode for the moment.)

Now let us consider the degeneracy of these modes in more detail. We can see that the TE and TM modes are degenerate with HE modes that have $m = 1$. Furthermore, the EH modes of order m are degenerate with the HE modes of order $m + 1$. We expect that for every azimuthal order m, there are an infinite number of discrete modes that we can label by n (though we expect that only a finite number of these are propagating modes for a given frequency).

We label the HE modes using $HE_{m+1,n}$, the EH modes using $EH_{m-1,n}$, and the TE and TM modes using $TE_{0,n}$ and $TM_{0,n}$. These are all unified by their degeneracy under a label of 'LP' (linearly polarized) modes, as summarized in table 13.3.

The term 'linear polarization' may be somewhat confusing, as it is somewhat clear from the definitions of E_z and H_z that all of the TE, TM, HE, and EH modes are, in general, nonuniformly polarized. It can be shown, however, that the degenerate modes, such as the TE_{0n}, TM_{0n}, and HE_{2n} modes, can be combined to form states with a uniformly linear state of polarization: the LP modes. These states are

Table 13.3. Definitions of the LP_{ln} modes.

LP mode	Mode types	Dispersion relation
LP_{0n}	HE_{1n}	$\frac{J_0(u)}{uJ_1(u)} = \frac{K_0(w)}{wK_1(w)}$
LP_{1n}	TE_{0n} TM_{0n} HE_{2n}	$\frac{J_1(u)}{uJ_0(u)} = -\frac{K_1(w)}{wK_0(w)}$
$LP_{ln}, l \geqslant 2$	$EH_{l-1,n}$ $HE_{l+1,n}$	$\frac{J_l(u)}{uJ_{l-1}(u)} = -\frac{K_l(w)}{wK_{l-1}(w)}$

particularly relevant because light that is coupled into fibers is typically linearly polarized and therefore preferentially couples to LP modes. It should be noted, however, that LP modes are only approximately degenerate because the dispersion relations, as derived, rely on the approximation $n_1 \approx n_0$.

With these modes defined, we can now examine the mode structure of a step-index waveguide of the type we have considered. Our first approach is to plot the dispersion relations in (u, w)-space, where they take the form of lines. As we have noted above, the intersection of those lines with the circle defined by equation (13.114), where the radius depends on frequency, determines the modes allowed for a given frequency. The results are shown in figure 13.8, with $v = 5$; it can be seen that the allowed modes at this frequency are the LP_{01}, LP_{11}, LP_{21}, and LP_{02} modes. We can estimate the core diameter of such a waveguide, if we assume that the operating wavelength is 1530 nm; using equation (13.114), we have

$$a^2 = \frac{v^2}{k^2(n_1^2 - n_0^2)}. \tag{13.149}$$

Let us take $n_1 = 1.46$ and $n_0 = 1.45$; we then find that we get $a = 7.1 \ \mu$m.

It should be noted that the lowest-order mode is the LP_{01} mode, i.e. the HE_{11} mode, and this mode has no cutoff. The step-index fiber is a single-mode fiber until it reaches a critical V-number v_c such that

$$v_c \approx 2.405, \tag{13.150}$$

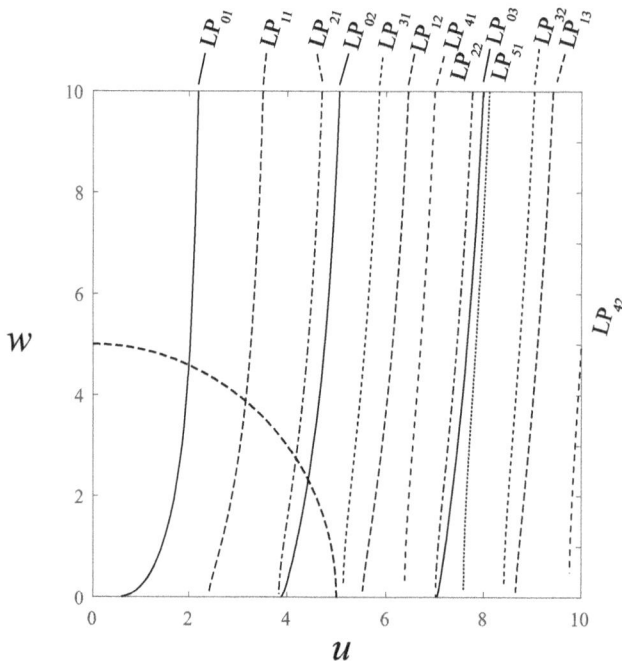

Figure 13.8. The (u, w)-mode structure for a step-index waveguide, with $v = 5$. Modes up to LP_{5n} have been plotted.

which can be determined numerically from the intersection of the LP_{11} line in figure 13.8 with the u-axis.

Because the LP_{01} mode has no cutoff, it is tempting to imagine making single-mode fibers that are as thin as possible. However, the thinner the fiber is made, the more the evanescent waves extend into the cladding of the fiber; we have seen that a similar thing happens for surface plasmons supported by thin films (section 11.7). The wave properties of light prevent us from compressing a mode into an arbitrarily small diameter.

We may also define the numerical aperture (NA) of the fiber as

$$NA = \sqrt{n_1^2 - n_0^2}. \tag{13.151}$$

The NA determines an angle via $NA = \sin\theta$, and beams of light with an angle less than θ can be coupled into a fiber mode.

Though we have referred to the equations in table 13.3 as 'dispersion relations,' they are not in the typical form of a dispersion relation, which relates the propagation constant k_l to the frequency ω. Following the book by Okamoto, we can produce a more conventional set of dispersion curves for the step-index fiber by introducing a normalized propagation constant b, defined as

$$b \equiv \frac{(k_l/k)^2 - n_0^2}{n_1^2 - n_0^2}. \tag{13.152}$$

In these terms, we may write $w = \sqrt{b}\,v$ and $u = \sqrt{1 - b}\,v$. The resulting plots of the mode dispersion are shown in figure 13.9. These plots are still not in exactly the form

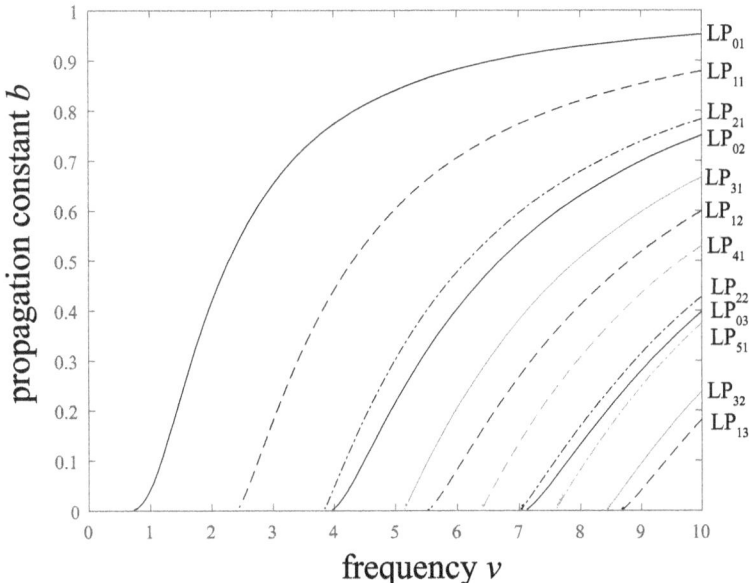

Figure 13.9. The dispersion curves for the step-index fiber as a function of normalized frequency v and normalized propagation constant b.

of the $k - \omega$ dispersion curves we have used in previous chapters, but they never-theless provide an excellent visualization of the mode behavior.

13.6 Optical fibers

We conclude this chapter with a brief discussion of the properties of optical fibers used in practice and typical values for various parameters of the fiber.

We have already noted that the fundamental structure of an optical fiber is a higher-index core surrounded by a lower-index cladding. In practice, additional layers are added beyond the cladding. The next layer, moving outward, is a coating layer around the cladding that provides protection against damage and can also be designed to absorb cladding modes (to be discussed below). Outside this is usually a strength layer made of a material like Kevlar that protects against damage due to stress on the cable. Finally, an outer jacket provides extra protection to the other layers.

For fibers with a low NA, it is difficult to perfectly couple an incident beam of light into a fiber mode. In such cases, much of the light can go into so-called cladding modes that are not confined to the region of the core. For short optical fibers, these cladding modes can propagate to the fiber output and degrade the signal. The coating layer is often made of a polymer that causes high propagation losses for the cladding modes.

Fiber materials can vary widely and depend on practical considerations. For long fibers where absorption is a primary concern, the fibers are commonly made from silica glass such as quartz glass and amorphous SiO_2. Fused silica, for example, has a refractive index of 1.444 at $\lambda = 1530$ nm [10]. The core glass is doped with a dopant such as germania (GeO_2) or alumina (Al_2O_3) to raise its refractive index slightly. As an example, one commercially available single-mode fiber from Thorlabs has a listed core index of $n_1 = 1.452\,59$ and a cladding index of $n_0 = 1.446\,80$ at 1310 nm; we note again that the difference in refractive index is very small. For short propagation distances where signal quality and absorption are not as important, fibers are often made of plastic such as poly(methyl methacrylate) (PMMA).

Fibers may be broadly classified as single mode or multimode. Single-mode fibers possess a V-number less than 2.405 and are used for long-distance propagation because they are not subject to modal dispersion that degrades the signal. It should be noted that a larger V-number corresponds to stronger confinement of the mode within the core, so V-numbers closer to the critical level are preferred. According to the International Telecommunication Union G.652 standard for single-mode fibers designed for an operating wavelength of 1310 nm, the mode field diameter (core plus evanescent width in cladding) should be $8.6 - 9.5$ μm and the cladding diameter should be 125 μm. It is of interest to note that the standard does not specify the material or the refractive index values; any material that provides the appropriate beam profile and loss levels conforms to the standard.

Multimode fibers are used over shorter propagation distances, for example, for interconnects within buildings or between nearby buildings. Because they can carry multiple modes simultaneously, these fibers can transmit at higher data rates, but modal dispersion limits their propagation length. According to the G.651.1 standard for multimode fibers, the nominal core diameter is 50 μm and the cladding diameter is 125 μm.

Multimode fibers are often made with a gradient index core, i.e. a core whose refractive index decreases from its center until it matches the cladding index. This approach reduces the amount of modal dispersion in the fiber. It should be noted that chromatic (frequency) dispersion is, of course, also present in fibers and must be accounted for when using broadband operation.

This is as technical as we will get in discussing the properties of optical fibers, especially considering there are many other applications for them besides sensing, such as in sensors and lasers. For more details, the field guide by Paschotta [11] is a good starting point.

13.7 Exercises

1. An Electronics Industries Alliance (EIA) standard WR-28 rectangular metal waveguide has dimensions of 3.56 mm × 7.11 mm. What TE and TM modes propagate in this waveguide at a driving frequency of 100 GHz? For what range of frequencies is only one TE mode excited in the waveguide?

2. Consider an EIA WR-90 rectangular waveguide with dimensions of 2.28 cm × 1.01 cm. What TE modes propagate in this waveguide if the driving frequency is 17 GHz? Suppose you wanted to excite only one TE mode; what range of frequencies could you use? What are the corresponding wavelengths (in open space)?

3. Let us imagine a long pedestrian tunnel of height 3m and width 3m. Assuming that we treat the walls as a perfect conductor, can AM radio waves propagate in this tunnel? Can FM radio waves propagate in it? Justify your answers.

4. A particular microwave rectangular waveguide (WR-42) is constructed with dimensions 4.32 mm × 10.67 mm. Suppose we wish to propagate only one waveguide mode in this device. What range of frequencies is allowed, and what is the type of mode propagated?

5. A microwave oven uses microwave radiation at a frequency of 2.45 GHz to cook food. The radiation is generated by a magnetron and then conveyed to the cooking area via a rectangular metal waveguide. What is the minimum size of a waveguide that can efficiently convey microwaves of this frequency? Which of the following waveguides would appear to be best for this application: WR-112 (12.62 mm × 28.50 mm), WR-229 (29.08 mm × 58.17 mm), WR-284 (34.03 mm × 72.14 mm), WR-650 (82.55 mm × 165.1 mm), or WR-770 (97.79 mm × 195.58 mm)? (Consider the practical size of the device as well as its operating frequency.)

6. Suppose a hollow metallic rectangular waveguide is filled with a dielectric that has a permittivity $\epsilon_d > \epsilon_0$ instead of air. Derive how the mode frequencies change compared to the case when the waveguide is filled with air.

7. In nano-optics applications, light is often transmitted through a hole of subwavelength size in a metal plate. Let us suppose that light at a frequency

of 500 nm is incident on a square hole of side width 100 nm in a thin silver film, and the thickness of the film is 200 nm. Assuming light is coupled into the hole, which can be treated as a waveguide, estimate how the intensity of the lowest-order TE mode decays upon propagation through the hole. (We neglect resonant effects such as Fabry–Perot resonances.)

8. Consider a resonant cavity produced by closing off the two ends of a rectangular waveguide with side lengths a and b at $z = 0$ and $z = d$, making a perfectly conducting empty box. Following the same strategy as that used for an open waveguide, i.e. solving the wave equation and matching boundary conditions, determine a formula for the resonant frequencies ω_{lmn} of modes that can be excited in this cavity. (The modes are now characterized by three integers: l, m, and n.)

9. We can derive our mode structures for a rectangular waveguide using a simple plane-wave approach, analogous to the one used for the slab waveguide. Assume that a mode in the waveguide consists of four plane waves,

$$E_z(x, y) = E_{\pm,\pm}e^{\pm ikx \sin\theta_x}e^{\pm iky \sin\theta_y}e^{ik_l z},$$

where the plane waves are distinguished by their $\pm$ signs, and θ_x and θ_y represent the angles the plane waves make with the z-axis.

Taking these plane waves to represent the z-component of the electric field, which must vanish at each wall of the rectangular waveguide of sides a and b, show that k_l takes on the expected form.

References

[1] Cregan R F *et al* 1999 Single-mode photonic band gap guidance of light in air *Science* **285** 1537–9

[2] Yeh P, Yariv A and Marom E 1978 Theory of Bragg fiber *J. Opt. Soc. Am.* **68** 1196–201

[3] Schurig D *et al* 2006 Metamaterial electromagnetic cloak at microwave frequencies *Science* **314** 977–80

[4] Allen L, Beijersbergen M W, Spreeuw R J C and Woerdman J P 1992 Orbital angular momentum of light and the transformation of Laguerre-Gaussian laser modes *Phys. Rev.* A **45** 8185–9

[5] Bozinovic N *et al* 2013 Terabit-scale orbital angular momentum mode division multiplexing in fibers *Science* **340** 1545–8

[6] Gbur G J 2011 *Mathematical Methods for Optical Physics and Engineering* (Cambridge: Cambridge University Press)

[7] Novotny L and Hecht B 2012 *Principles of Nano-Optics* 2nd edn (Cambridge: Cambridge University Press)

[8] Marcuvitz N 1951 *Waveguide Handbook* 1st edn (New York: McGraw-Hill)

[9] Okamoto K 2006 *Fundamentals of Optical Waveguides* 2nd edn (Amsterdam: Academic)

[10] Malitson I H 1965 Interspecimen comparison of the refractive index of fused silica *J. Opt. Soc. Am.* **55** 1205–9

[11] Paschotta R 2010 *Field Guide to Optical Fiber Technology* (Bellingham, WA: SPIE Press)

IOP Publishing

Electromagnetic Optics

Gregory J Gbur

Chapter 14

Sources and potentials

Up to this point, we have exclusively focused on the free propagation of electromagnetic waves in a variety of media and have not considered the sources of those waves, i.e. we have let $\rho(\mathbf{r}, t) = 0$ and $\mathbf{J}(\mathbf{r}, t) = 0$. We now turn to investigating the relationship between sources and the electromagnetic fields they produce. We will work in free space, with $\epsilon = \epsilon_0$ and $\mu = \mu_0$, and will primarily work with monochromatic fields at a fixed frequency ω.

A word of caution: though we have focused on electromagnetic waves in the visible spectrum for most of this book, we find that a proper treatment of visible light sources requires additional considerations beyond Maxwell's equations. In most sources of visible light—lightbulbs, stars, the Sun—the source consists of a collection of atoms that radiate largely independently of one another and emit light as discrete photons. To properly characterize ordinary sources of visible light, then, we need to introduce additional physics. At the very least, we need to characterize the correlations between the different parts of the source, which means employing classical coherence theory [1], which applies statistics to the study of random sources and fields. A complete description of electromagnetic wave radiation should also include the quantum physics of the emitters, which follow their own rules outside Maxwell's equations. Indeed, much of the confusion in physics in the early 20th century arose from attempts to apply classical physics to the emission of light from atoms, and this confusion only began to clear when Bohr introduced his model of the atom [2].

In this chapter, we work with monochromatic fields and show how oscillating charges and currents produce electromagnetic waves. This description works well for radio waves and microwaves but is only approximately valid for visible light, due to the omissions mentioned above. Nevertheless, many aspects of this classical radiation theory, such as the use of electromagnetic potentials and multipoles, can be applied directly to the quantum treatment of light emission. Furthermore, the

doi:10.1088/978-0-7503-6064-7ch14

results we derive in this chapter can be directly applied to the problem of electromagnetic scattering, to be discussed in the next chapter.

In this chapter, we follow much of the approach of the classic book by Papas [3] and refer the reader to that book for even more information about radiation problems.

14.1 Sources and potentials

We will be looking at the monochromatic versions of Maxwell's equations in free space with nonzero charge density $\rho(\mathbf{r}, \omega)$ and current density $\mathbf{J}(\mathbf{r}, \omega)$, of the form

$$\nabla \cdot \mathbf{E}(\mathbf{r}, \omega) = \rho(\mathbf{r}, \omega)/\epsilon_0, \tag{14.1}$$

$$\nabla \cdot \mathbf{H}(\mathbf{r}, \omega) = 0, \tag{14.2}$$

$$\nabla \times \mathbf{E}(\mathbf{r}, \omega) = i\omega\mu_0\mathbf{H}(\mathbf{r}, \omega), \tag{14.3}$$

$$\nabla \times \mathbf{H}(\mathbf{r}, \omega) = \mathbf{J}(\mathbf{r}, \omega) - i\omega\epsilon_0\mathbf{E}(\mathbf{r}, \omega). \tag{14.4}$$

Throughout this chapter, we will use the $\mathbf{E}$ and $\mathbf{H}$ fields, which is conventional for radiation problems.

We are now required to solve an inhomogeneous set of vector partial differential equations, a daunting task indeed. However, this problem can be simplified by introducing auxiliary quantities, namely the scalar potential $\phi(\mathbf{r}, \omega)$ and the vector potential $\mathbf{A}(\mathbf{r}, \omega)$, which will allow us to find a general solution much more readily.

It is helpful to first review the use of the scalar potential $\phi(\mathbf{r})$ in electrostatics and the vector potential $\mathbf{A}(\mathbf{r})$ in magnetostatics, which will provide a clear path to introducing them for electromagnetic fields.

In electrostatics, where the charge density $\rho(\mathbf{r})$ is independent of time, we know that the electric field satisfies the equations

$$\nabla \times \mathbf{E}(\mathbf{r}) = 0, \quad \nabla \cdot \mathbf{E}(\mathbf{r}) = \rho(\mathbf{r})/\epsilon_0. \tag{14.5}$$

The vanishing of the curl of $\mathbf{E}$ indicates that $\mathbf{E}$ is an irrotational vector, and it can be written as the gradient of a scalar,

$$\mathbf{E}(\mathbf{r}) = -\nabla\phi(\mathbf{r}). \tag{14.6}$$

This definition automatically satisfies the curl equation, as $\nabla \times \nabla\phi = 0$. Combining this definition with Gauss's law, we can say that the scalar potential $\phi(\mathbf{r})$ satisfies Poisson's equation,

$$\nabla^2\phi(\mathbf{r}) = -\rho(\mathbf{r})/\epsilon_0. \tag{14.7}$$

We are familiar with the expression for the electric field of a point charge q located at the position $\mathbf{r}'$,

$$\mathbf{E}(\mathbf{r}) = \frac{1}{4\pi\epsilon_0} \frac{q\hat{\mathbf{R}}}{R^2}, \tag{14.8}$$

where $R = |\mathbf{r} - \mathbf{r}'|$. The potential difference is determined by taking the integral of the electric field from infinity to R,

$$\Delta\phi = -\frac{1}{4\pi\epsilon_0} \int_\infty^R \frac{q}{R'^2} dR'. \tag{14.9}$$

Because the electric field is defined as the derivative of a scalar function, we are free to choose the zero point of potential without changing the electric field itself. We choose $\phi(\infty) = 0$, which gives us the simple expression

$$\phi(\mathbf{r}) = \frac{1}{4\pi\epsilon_0} \frac{q}{R}. \tag{14.10}$$

If we treat a point charge as the local charge density $\rho(\mathbf{r}')$ multiplied by an infinitesimal volume d^3r', we may write the electric potential of a volume of charge as

$$\phi(\mathbf{r}) = \frac{1}{4\pi\epsilon_0} \int_V \frac{\rho(\mathbf{r}')}{|\mathbf{r} - \mathbf{r}'|} d^3r', \tag{14.11}$$

and the electric field may be found using equation (14.7).

The obvious advantage of solving for the electric potential over the electric field is that we work with a scalar quantity rather than a vector quantity, but it also leads us directly to a solution for the magnetic field in magnetostatics. In magnetostatics, we have

$$\nabla \times \mathbf{H}(\mathbf{r}) = \mathbf{J}(\mathbf{r}), \quad \nabla \cdot \mathbf{H}(\mathbf{r}) = 0. \tag{14.12}$$

Because the divergence of the magnetic field is zero, i.e. it is solenoidal, we may express it in the form of a vector potential $\mathbf{A}$ in the form

$$\mu_0 \mathbf{H}(\mathbf{r}) = \nabla \times \mathbf{A}(\mathbf{r}). \tag{14.13}$$

This automatically satisfies our divergence equation, since the divergence of a curl is always zero, $\nabla \cdot (\nabla \times \mathbf{A}) = 0$. We substitute this definition into Ampère's law, $\nabla \times \mathbf{H} = \mathbf{J}$, and get

$$\nabla \times (\nabla \times \mathbf{A}) = \mu_0 \mathbf{J}. \tag{14.14}$$

We now apply the double curl identity (equation (A.29)) to get

$$\nabla(\nabla \cdot \mathbf{A}) - \nabla^2 \mathbf{A} = \mu_0 \mathbf{J}. \tag{14.15}$$

A solution to this vector partial differential equation is not obvious. However, it turns out that we have significant freedom to choose the form of the vector potential $\mathbf{A}(\mathbf{r})$ and can choose it such that $\nabla \cdot \mathbf{A} = 0$. The freedom to choose the potential in such a way that it satisfies an additional differential equation is called *gauge freedom*, and we will elaborate upon it at the end of the chapter. For now, we simply note that the choice $\nabla \cdot \mathbf{A} = 0$ is known as *working in the Coulomb gauge*. With this choice, our equation takes on the simple form

$$\nabla^2 \mathbf{A} = -\mu_0 \mathbf{J}. \tag{14.16}$$

This expression should be compared with equation (14.7). We see that each component of the vector potential satisfies Poisson's equation, with a source term given by the same component of $\mathbf{J}$, with the only difference being a few simple constants. We may immediately use our results for the scalar potential to write

$$\mathbf{A}(\mathbf{r}) = \frac{\mu_0}{4\pi} \int_V \frac{\mathbf{J}(\mathbf{r}')}{|\mathbf{r} - \mathbf{r}'|} d^3 r'. \tag{14.17}$$

The introduction of potentials into electrostatics and magnetostatics has greatly simplified and unified the solution of these problems, and it is natural to ask whether something similar can be done for Maxwell's equations in general. The answer, of course, is that it can.

We begin with Maxwell's equations in free space, in the time domain, with the following sources:

$$\nabla \cdot \mathbf{E} = \rho / \epsilon_0, \tag{14.18}$$

$$\nabla \cdot \mathbf{H} = 0, \tag{14.19}$$

$$\nabla \times \mathbf{E} = -\mu_0 \frac{\partial \mathbf{H}}{\partial t}, \tag{14.20}$$

$$\nabla \times \mathbf{H} = \mathbf{J} + \epsilon_0 \frac{\partial \mathbf{E}}{\partial t}. \tag{14.21}$$

We immediately note that due to equation (14.19), we may still write the magnetic field as the curl of a vector potential,

$$\mu_0 \mathbf{H}(\mathbf{r}, t) = \nabla \times \mathbf{A}(\mathbf{r}, t). \tag{14.22}$$

We may substitute this expression for $\mathbf{H}$ into Faraday's law. Upon interchanging the order of the derivatives and grouping terms under the curl, we obtain

$$\nabla \times \left(\mathbf{E} + \frac{\partial \mathbf{A}}{\partial t} \right) = 0. \tag{14.23}$$

However, this indicates that the total quantity in parentheses is irrotational and may be written as the gradient of a scalar potential,

$$-\nabla \phi(\mathbf{r}, t) = \mathbf{E}(\mathbf{r}, t) + \frac{\partial \mathbf{A}(\mathbf{r}, t)}{\partial t}. \tag{14.24}$$

We may therefore define a scalar potential and a vector potential for time-varying electromagnetic fields. It should be noted that if the fields are time independent, the time derivative in the equation for ϕ vanishes and our expressions reduce to the static potentials discussed earlier.

If we substitute our expressions for ϕ and $\mathbf{A}$ into the Ampère–Maxwell law, after some rearranging, we get

$$\nabla^2 \mathbf{A} - \mu_0 \epsilon_0 \frac{\partial^2 \mathbf{A}}{\partial t^2} = -\mu_0 \mathbf{J} + \nabla \left(\nabla \cdot \mathbf{A} + \mu_0 \epsilon_0 \frac{\partial \phi}{\partial t} \right). \tag{14.25}$$

If we look at the three leftmost terms of this equation, it would appear that we have a wave equation with a source current. The fourth term causes trouble, however, as it includes nontrivial derivatives of $\mathbf{A}$ as well as the scalar potential ϕ.

We again take advantage of gauge freedom (to be justified later) to choose $\mathbf{A}$ and ϕ such that

$$\nabla \cdot \mathbf{A} + \mu_0 \epsilon_0 \frac{\partial \phi}{\partial t} = 0. \tag{14.26}$$

This choice is known as the *Lorenz gauge*. As a brief aside, this is often misattributed to Lorentz, due to the similar names and the fact that this gauge satisfies relativistic Lorentz invariance.

Using the Lorenz gauge and the familiar $c^2 = 1/\mu_0 \epsilon_0$, we may write

$$\nabla^2 \mathbf{A} - \frac{1}{c^2} \frac{\partial^2 \mathbf{A}}{\partial t^2} = -\mu_0 \mathbf{J}, \tag{14.27}$$

which is exactly a wave equation with a source current.

If we now substitute from equation (14.24) for $\mathbf{E}$ into Gauss's law, we end up with the expression

$$-\nabla^2 \phi - \frac{\partial}{\partial t} (\nabla \cdot \mathbf{A}) = \rho / \epsilon_0. \tag{14.28}$$

We now substitute from equation (14.26) into this expression to eliminate $\nabla \cdot \mathbf{A}$ and are left with

$$\nabla^2 \phi - \frac{1}{c^2} \frac{\partial^2 \phi}{\partial t^2} = -\rho / \epsilon_0. \tag{14.29}$$

We have found that, in the Lorenz gauge, the scalar potential also satisfies a wave equation, in this case with a source charge. The advantage of using the potentials now becomes clearer: though we may not know from memory how to solve a wave equation with a source term, it is a method that is well documented. We have taken a set of eight coupled vector equations in the form of Maxwell's equations and reduced them to a set of four independent scalar wave equations.

You might wonder how we are able to so dramatically reduce the number of equations to solve. This is possible because the scalar and vector potentials have the transversality of the electric and magnetic fields built into them through their definitions. This is significantly different from our solution of Maxwell's equations in free space back in chapter 3, where we found that the electric and magnetic fields themselves satisfy homogeneous wave equations. In that case, after solving the wave equations, we had to go back and impose transversality on our solution; here, transversality is built in.

Let us now turn to the solution of our wave equations. At this point, we restrict ourselves to monochromatic fields, $\mathbf{A}(\mathbf{r}, t) = \mathbf{A}(\mathbf{r}, \omega)\exp[-i\omega t]$ and so forth, which gives us a set of Helmholtz equations with source terms to solve:

$$\nabla^2 \mathbf{A} + k^2 \mathbf{A} = -\mu_0 \mathbf{J}, \tag{14.30}$$

$$\nabla^2 \phi + k^2 \phi = -\rho/\epsilon_0, \tag{14.31}$$

where, as always, $k = \omega/c$.

We approach the solution by the method of Green's functions. We seek a function $G(\mathbf{r}, \mathbf{r}')$ that satisfies the equation

$$(\nabla^2 + k^2)G(\mathbf{r}, \mathbf{r}') = -4\pi\delta^{(3)}(\mathbf{r} - \mathbf{r}'), \tag{14.32}$$

where $\delta^{(3)}(\mathbf{r} - \mathbf{r}')$ is the three-dimensional Dirac delta function. The delta function is zero everywhere except where its argument is equal to zero, where it has an infinite amplitude that lets it satisfy the sifting property, namely

$$\int_V f(\mathbf{r}')\delta^{(3)}(\mathbf{r} - \mathbf{r}')d^3r' = f(\mathbf{r}), \tag{14.33}$$

provided V includes the point $\mathbf{r}$; the integral is zero otherwise.

Let us set $\mathbf{r}' = 0$; this can be done without loss of generality because we can always shift the origin of our coordinate system at the end of the calculation. We are then looking for a solution for the function $G(\mathbf{r})$. The Dirac delta function is basically a localized 'spike' at the origin and is rotationally invariant in spherical coordinates; therefore, we expect that our Green's function will be as well, i.e. $G(\mathbf{r}) = G(r)$. We may then write equation (14.32) entirely in terms of derivatives in r, and we write the derivatives in two different forms for convenience:

$$\frac{1}{r^2}\frac{d}{dr}\left(r^2\frac{dG(r)}{dr}\right) + k^2 G(r) = -4\pi\delta^{(3)}(\mathbf{r}), \tag{14.34}$$

$$\frac{1}{r}\frac{d^2}{dr^2}[rG(r)] + k^2 G(r) = -4\pi\delta^{(3)}(\mathbf{r}). \tag{14.35}$$

It can readily be shown that these two forms are equivalent using the product rule to expand out the derivative terms. We work with equation (14.35) first and introduce a new function $U(r) = rG(r)$; on substitution, we find that $U(r)$ satisfies

$$\frac{d^2 U(r)}{dr^2} + k^2 U(r) = 0, \quad r \neq 0. \tag{14.36}$$

This is just the harmonic oscillator equation away from the origin, so we expect that as long as $r \neq 0$, we may write

$$U(r) = A_+ e^{ikr} + A_- e^{-ikr}, \tag{14.37}$$

with A_+ and A_- as constants to be determined. The solution for $G(r)$ is therefore

$$G(r) = A_+ \frac{e^{ikr}}{r} + A_- \frac{e^{-ikr}}{r}. \tag{14.38}$$

We now turn to some physical intuition to eliminate one of the undetermined constants. The two terms of the solution are spherical waves, one of which is outgoing and one of which is converging. This can be seen clearly by adding the time dependence to the Green's function:

$$G(r)e^{-i\omega t} = A_+ \frac{e^{i(kr-\omega t)}}{r} + A_- \frac{e^{-i(kr+\omega t)}}{r}. \tag{14.39}$$

The first term is clearly one where the surfaces of constant phase go outward as time advances, and the second term is one where the surfaces converge inward. Physically, we are solving a radiation problem, so only the outgoing wave is physically sensible; we set $A_- = 0$. Our Green's function is thus reduced to

$$G(r) = A_+ \frac{e^{ikr}}{r}. \tag{14.40}$$

To determine A_+, we now use equation (14.34) and integrate the entire equation over a sphere of radius r. We will set $r \to 0$ at the end. We have

$$\int_r \frac{1}{r'^2} \frac{d}{dr'}\left(r'^2 \frac{dG(r')}{dr'}\right) r'^2 dr' d\Omega + \int_r k^2 G(r') r'^2 dr' d\Omega = -4\pi \int_r \delta^{(3)}(\mathbf{r}') r'^2 dr' d\Omega, \tag{14.41}$$

where we have used $d^3r' = r'^2 dr' d\Omega$, with $d\Omega$ the infinitesimal unit of solid angle. The integral over the solid angle gives 4π. The integral over the delta function on the right becomes -4π using the sifting property. The middle integral has a finite integrand, due to the $1/r'$ in $G(r')$ and the r'^2 in the numerator, and as $r \to 0$, it will go to zero. The first integral becomes a total derivative with respect to r', which is then evaluated in the limit $r \to 0$. We therefore have the condition

$$r^2 \frac{dG}{dr}\bigg|_{r=0} = -1. \tag{14.42}$$

If we plug in our expression for $G(r)$, we quickly find that $A_+ = 1$. We thus find that the scalar Green's function for the Helmholtz equation is of the form

$$G(\mathbf{r}, \mathbf{r}') = \frac{e^{ik|\mathbf{r}-\mathbf{r}'|}}{|\mathbf{r} - \mathbf{r}'|}, \tag{14.43}$$

where we have taken the liberty of changing the origin of the coordinate system to $\mathbf{r}'$.

We now have a Green's function, but how does that translate back into a solution for the wave equation with a source term? Let us consider a general wave equation with a source, of the form

$$(\nabla'^2 + k^2)u(\mathbf{r}') = -4\pi q(\mathbf{r}'), \tag{14.44}$$

which encompasses both the scalar potential $\phi(\mathbf{r}')$ and each component of the vector potential $\mathbf{A}(\mathbf{r}')$. Here, ∇' represents the derivative with respect to $\mathbf{r}'$. Let us multiply this equation by $G(\mathbf{r}, \mathbf{r}')$ and switch $\mathbf{r}$ and $\mathbf{r}'$ in equation (14.32) and multiply it by $u(\mathbf{r}')$. We then subtract the two resulting equations and integrate the difference over the domain of the problem, in this case, all of three-dimensional space. We get the expression

$$\int [G(\mathbf{r}, \mathbf{r}')\nabla'^2 u(\mathbf{r}') - u(\mathbf{r}')\nabla'^2 G(\mathbf{r}, \mathbf{r}')]d^3r' = -4\pi \int [G(\mathbf{r}, \mathbf{r}')q(\mathbf{r}') - u(\mathbf{r}')\delta^{(3)}(\mathbf{r} - \mathbf{r}')]d^3r'. \quad (14.45)$$

We may use the sifting property of the delta function to simplify the last integral on the right. We also take advantage of the two vector identities,

$$\nabla \cdot [G\nabla u] = \nabla G \cdot \nabla u + G\nabla^2 u, \qquad (14.46)$$

$$\nabla \cdot [u\nabla G] = \nabla G \cdot \nabla u + u\nabla^2 G. \qquad (14.47)$$

Taking the difference between these two equations, we have

$$\nabla \cdot [G\nabla u - u\nabla G] = G\nabla^2 u - u\nabla^2 G. \qquad (14.48)$$

We may use these to write the integral on the left of equation (14.45) in the form

$$\int [G(\mathbf{r}, \mathbf{r}')\nabla'^2 u(\mathbf{r}') - u(\mathbf{r}')\nabla'^2 G(\mathbf{r}, \mathbf{r}')]d^3r' = \int \nabla' \cdot [G(\mathbf{r}, \mathbf{r}')\nabla' u(\mathbf{r}') - u(\mathbf{r}')\nabla' G(\mathbf{r}, \mathbf{r}')]d^3r'. \quad (14.49)$$

We may use Gauss's theorem to write this as

$$\int \nabla' \cdot [G(\mathbf{r}, \mathbf{r}')\nabla' u(\mathbf{r}') - u(\mathbf{r}')\nabla' G(\mathbf{r}, \mathbf{r}')]d^3r' = \int [G(\mathbf{r}, \mathbf{r}')\nabla' u(\mathbf{r}') - u(\mathbf{r}')\nabla' G(\mathbf{r}, \mathbf{r}')] \cdot d\mathbf{a}'. \quad (14.50)$$

Finally, we have

$$u(\mathbf{r}) = \int G(\mathbf{r}, \mathbf{r}')q(\mathbf{r}')d^3r' - \frac{1}{4\pi} \int [G(\mathbf{r}, \mathbf{r}')\nabla' u(\mathbf{r}') - u(\mathbf{r}')\nabla' G(\mathbf{r}, \mathbf{r}')] \cdot d\mathbf{a}'. \quad (14.51)$$

This equation indicates that the field $u(\mathbf{r})$ depends on the source $q(\mathbf{r})$ as well as the properties of the field on the boundary. In our case, the boundary is at infinity, and it can be shown that our choice of Green's function and the behavior of the solution automatically set this boundary term to zero. We are left with

$$u(\mathbf{r}) = \int G(\mathbf{r}, \mathbf{r}')q(\mathbf{r}')d^3r'. \qquad (14.52)$$

At last, we may identify the terms of equation (14.44) with the equations for $\phi(\mathbf{r})$ and $\mathbf{A}(\mathbf{r})$ and find that our solutions for the scalar and vector potential are

$$\phi(\mathbf{r}) = \frac{1}{4\pi\epsilon_0} \int \rho(\mathbf{r}')\frac{e^{ikR}}{R}d^3r', \qquad (14.53)$$

$$\mathbf{A}(\mathbf{r}) = \frac{\mu_0}{4\pi} \int \mathbf{J}(\mathbf{r}')\frac{e^{ikR}}{R}d^3r', \qquad (14.54)$$

where we have used $R \equiv |\mathbf{r} - \mathbf{r}'|$.

We can make one further simplification because we are working with mono-chromatic fields. We refer again to the continuity equation, equation (2.57), which has the form

$$\nabla \cdot \mathbf{J}(\mathbf{r}, \, t) = -\frac{\partial \rho(\mathbf{r}, \, t)}{\partial t}. \tag{14.55}$$

The monochromatic version of this equation is

$$\nabla \cdot \mathbf{J}(\mathbf{r}, \, \omega) = i\omega\rho(\mathbf{r}, \, \omega). \tag{14.56}$$

We may replace $\rho(\mathbf{r})$ in equation (14.53) using the continuity equation and write

$$\phi(\mathbf{r}) = \frac{1}{4\pi i\omega\epsilon_0} \int \nabla' \cdot \mathbf{J}(\mathbf{r}') \frac{e^{ikR}}{R} d^3r'. \tag{14.57}$$

For the monochromatic case, we can therefore determine the scalar and vector potentials, and consequently the electric and magnetic fields, from the form of the current density alone.

Equations (14.54) and (14.57) are exact solutions to Maxwell's equations that relate the structure of the sources to the behavior of the radiated fields. However, the integrals cannot be evaluated analytically except for the simplest source structures. To better understand the physics of electromagnetic radiation, we will need to make some approximations, which we explore in upcoming sections.

14.2 Dyadics

In analyzing the radiation problem, it will be convenient to introduce quantities that are represented by second-rank tensors, or equivalently by 3×3 square matrices. However, the formalisms of tensors and matrices are quite different from the vector formalism that we have used throughout this book, and it is decidedly inconvenient to introduce a radical new notation for a limited set of problems.

Fortunately, we have another option: to write second-rank tensors as combina-tions of the direct products of vectors, quantities known as *dyadics*. We introduce dyadics in this section, as they will be heavily used throughout the rest of this book.

We will explain dyadics by referring back to the more familiar representation of vectors in linear algebra as 3×1 column and 1×3 row vectors. We can represent the unit vector $\hat{\mathbf{x}}$, for example, as

$$|\hat{\mathbf{x}}\rangle = \begin{bmatrix} 1 \\ 0 \\ 0 \end{bmatrix}, \quad \langle\hat{\mathbf{x}}| = \begin{bmatrix} 1 & 0 & 0 \end{bmatrix}. \tag{14.58}$$

The ordinary dot product, or inner product, of this vector with itself can be written in this bra-ket and matrix notation as

$$\hat{\mathbf{x}} \cdot \hat{\mathbf{x}} = \langle\hat{\mathbf{x}}|\hat{\mathbf{x}}\rangle = \begin{bmatrix} 1 & 0 & 0 \end{bmatrix} \begin{bmatrix} 1 \\ 0 \\ 0 \end{bmatrix} = 1. \tag{14.59}$$

We may also, however, take the outer product of the vector with itself, in which we perform a ket-bra multiplication,

$$\hat{\mathbf{x}}\hat{\mathbf{x}} = |\hat{\mathbf{x}}\rangle\langle\hat{\mathbf{x}}| = \begin{bmatrix} 1 \\ 0 \\ 0 \end{bmatrix} \begin{bmatrix} 1 & 0 & 0 \end{bmatrix} = \begin{bmatrix} 1 & 0 & 0 \\ 0 & 0 & 0 \\ 0 & 0 & 0 \end{bmatrix}. \tag{14.60}$$

The result of the outer product of two vectors is a matrix (second-rank tensor). For brevity, we have written the product for ordinary vectors without any symbol, though often the symbol $\otimes$ is used, so we would write $\hat{\mathbf{x}} \otimes \hat{\mathbf{x}}$. It should be noted that the order of the vectors in an outer product is not commutative; the product $\hat{\mathbf{x}}\hat{\mathbf{y}}$ is not the same as $\hat{\mathbf{y}}\hat{\mathbf{x}}$, and so we must take care to preserve the order of vectors when we use them. The outer product of two vectors obtained in this way is known as a *dyad*, and a dyadic is more generally a second-rank tensor written in this notation, as the sum of dyads.

To build on the previous example, the identity matrix may be written in dyadic form as

$$\mathbf{I} = \hat{\mathbf{x}}\hat{\mathbf{x}} + \hat{\mathbf{y}}\hat{\mathbf{y}} + \hat{\mathbf{z}}\hat{\mathbf{z}}. \tag{14.61}$$

In this form, we can see the operation of each dyad when we take the dot product of the identity matrix with a vector: it maps the x-component of the input vector into the x-component of the output vector, and so forth.

A dyadic is simply a way to write a matrix using vector notation. To give another example, we write a simple matrix and its dyadic form,

$$\mathbf{M} = \begin{bmatrix} M_{xx} & M_{xy} & 0 \\ 0 & 0 & 0 \\ M_{zx} & 0 & M_{zz} \end{bmatrix} = M_{xx}\hat{\mathbf{x}}\hat{\mathbf{x}} + M_{xy}\hat{\mathbf{x}}\hat{\mathbf{y}} + M_{zx}\hat{\mathbf{z}}\hat{\mathbf{x}} + M_{zz}\hat{\mathbf{z}}\hat{\mathbf{z}}. \tag{14.62}$$

A dyadic can be premultiplied or postmultiplied by a vector; for example, if we multiply the vector $\mathbf{r} = x\hat{\mathbf{x}} + y\hat{\mathbf{y}} + z\hat{\mathbf{z}}$ by $\mathbf{M}$, we obtain

$$\mathbf{M} \cdot \mathbf{r} = M_{xx}x\hat{\mathbf{x}} + M_{xy}y\hat{\mathbf{x}} + M_{zx}x\hat{\mathbf{z}} + M_{zz}z\hat{\mathbf{z}}, \tag{14.63}$$

$$\mathbf{r} \cdot \mathbf{M} = M_{xx}x\hat{\mathbf{x}} + M_{xy}x\hat{\mathbf{y}} + M_{zx}z\hat{\mathbf{x}} + M_{zz}z\hat{\mathbf{z}}. \tag{14.64}$$

Up to this point, the introduction of dyadics may, in fact, have seemed more cumbersome than simply writing a matrix. However, in the problems we will consider, the dyadics that arise are exceedingly simple and can be written compactly. One of them, for example, is the outer product of the position vector with itself, $\mathbf{rr}$. The del operator, which behaves loosely like a vector, can also be multiplied by itself in an outer product, $\nabla\nabla$, which is a second-rank tensor operator and distinct from $\nabla \cdot \nabla = \nabla^2$, which is a scalar operator.

Finally, we note that the introduction of dyadics allows us to reinterpret familiar vector identities in terms of outer products. For example, the vector triple product can be written as

$$\mathbf{A} \times (\mathbf{B} \times \mathbf{C}) = \mathbf{B}(\mathbf{A} \cdot \mathbf{C}) - \mathbf{C}(\mathbf{A} \cdot \mathbf{B}) = (\mathbf{BA}) \cdot \mathbf{C} - (\mathbf{CA}) \cdot \mathbf{B}, \tag{14.65}$$

where $\mathbf{BA}$ and $\mathbf{CA}$ are now dyads.

14.3 The general radiation problem

In the next section, we introduce the multipole method for characterizing the radiation characteristics of an oscillating charge–current distribution. Before doing so, however, it is worthwhile to demonstrate that we can develop a general formula for the radiation of an electromagnetic source. This formula presents us with some insights into the radiation problem that will serve us well later.

We note again the potentials for a monochromatic charge–current distribution, written entirely in terms of the current density $\mathbf{J}(\mathbf{r}')$,

$$\phi(\mathbf{r}) = \frac{1}{4\pi i\omega\epsilon_0} \int \nabla' \cdot \mathbf{J}(\mathbf{r}')\frac{e^{ikR}}{R}d^3r', \tag{14.66}$$

$$\mathbf{A}(\mathbf{r}) = \frac{\mu_0}{4\pi} \int \mathbf{J}(\mathbf{r}')\frac{e^{ikR}}{R}d^3r'. \tag{14.67}$$

We may make one additional transformation of the scalar potential using the vector identity

$$\nabla' \cdot [\mathbf{J}(\mathbf{r}')\psi(\mathbf{r}')] = \psi(\mathbf{r}')\nabla' \cdot \mathbf{J}(\mathbf{r}') + \nabla'\psi(\mathbf{r}') \cdot \mathbf{J}(\mathbf{r}'), \tag{14.68}$$

where $\psi(\mathbf{r}')$ is a well-behaved scalar function, in this case taken to be the scalar Green's function. We may use this to write the scalar potential as

$$\phi(\mathbf{r}) = \frac{1}{4\pi i\omega\epsilon_0}\left\{ \int \nabla' \cdot \left[\mathbf{J}(\mathbf{r}')\frac{e^{ikR}}{R}\right]d^3r' - \int \mathbf{J}(\mathbf{r}') \cdot \nabla'\frac{e^{ikR}}{R}d^3r'\right\}. \tag{14.69}$$

The first integral is the volume integral of a divergence, which can be converted into a surface integral through the divergence theorem. Since the source is limited to a finite domain, we can choose our surface to completely enclose but lie outside the source domain, and the surface integral then vanishes. We are left with

$$\phi(\mathbf{r}) = -\frac{1}{4\pi i\omega\epsilon_0} \int \mathbf{J}(\mathbf{r}') \cdot \nabla'\frac{e^{ikR}}{R}d^3r'. \tag{14.70}$$

We now turn to evaluating the fields. We know that, by definition,

$$\mathbf{E}(\mathbf{r}) = -\nabla\phi(\mathbf{r}) + i\omega\mathbf{A}(\mathbf{r}), \tag{14.71}$$

$$\mu_0\mathbf{H}(\mathbf{r}) = \nabla \times \mathbf{A}(\mathbf{r}). \tag{14.72}$$

Beginning with the electric field, we substitute the expressions for the potentials into the formula above, with the result

$$\mathbf{E}(\mathbf{r}) = \frac{1}{4\pi i\omega\epsilon_0}\nabla \int \mathbf{J}(\mathbf{r}') \cdot \nabla'\frac{e^{ikR}}{R}d^3r' + \frac{i\omega\mu_0}{4\pi} \int \mathbf{J}(\mathbf{r}')\frac{e^{ikR}}{R}d^3r'. \tag{14.73}$$

Because we are primarily interested in the radiation problem, we restrict our attention to the case where $\mathbf{r}$ lies outside the source domain. In this case, the integrand of the first integral is perfectly well-behaved and differentiable, and we may bring the derivative ∇ within the integral. (Over the next two chapters, we will see the problems that arise when $\mathbf{r}$ lies within the source domain.) Furthermore, we may use

$$\nabla \frac{e^{ikR}}{R} = -\nabla' \frac{e^{ikR}}{R}, \tag{14.74}$$

and through the use of dyadics we may write

$$\mathbf{E}(\mathbf{r}) = \frac{i\omega\mu_0}{4\pi} \int \left\{ \left[\mathbf{I} + \frac{1}{k^2}\nabla'\nabla' \right] \frac{e^{ikR}}{R} \right\} \cdot \mathbf{J}(\mathbf{r}')d^3r', \tag{14.75}$$

where the ∇' only act on the Green's function.

In a similar manner, we may write

$$\mathbf{H}(\mathbf{r}) = -\frac{1}{4\pi} \int \nabla' \left[\frac{e^{ikR}}{R} \right] \times \mathbf{J}(\mathbf{r}')d^3r'. \tag{14.76}$$

Let us now introduce the *far-zone approximation*. We assume that $|\mathbf{r}| \gg |\mathbf{r}'|$ for all source positions $\mathbf{r}'$. In the denominator of the scalar Green's function, we may make the approximation $R \approx r$, and in the exponent, we keep the first two terms of the Taylor series expansion about $\mathbf{r}' = 0$, so that we may write

$$\frac{e^{ikR}}{R} \approx \frac{e^{ikr}}{r}e^{-ik\hat{\mathbf{r}}\cdot\mathbf{r}'}. \tag{14.77}$$

With this approximation, we note that every ∇' in our field expressions simply results in a factor of $-ik\hat{\mathbf{r}}$; we may then simplify the fields to the forms

$$\mathbf{E}(\mathbf{r}) = \frac{i\omega\mu_0}{4\pi} \frac{e^{ikr}}{r}[\mathbf{I} - \hat{\mathbf{r}}\hat{\mathbf{r}}] \cdot \int \mathbf{J}(\mathbf{r}')e^{-ik\hat{\mathbf{r}}\cdot\mathbf{r}'}d^3r', \tag{14.78}$$

$$\mathbf{H}(\mathbf{r}) = \frac{ik}{4\pi} \frac{e^{ikr}}{r}\hat{\mathbf{r}} \times \int \mathbf{J}(\mathbf{r}')e^{-ik\hat{\mathbf{r}}\cdot\mathbf{r}'}d^3r', \tag{14.79}$$

where we have moved the dyadic terms outside the integrals because they do not depend on $\mathbf{r}'$.

The combined vector quantity in the electric field may be rewritten using

$$\hat{\mathbf{r}} \times [\hat{\mathbf{r}} \times \mathbf{J}(\mathbf{r}')] = \hat{\mathbf{r}}\hat{\mathbf{r}} \cdot \mathbf{J}(\mathbf{r}') - \mathbf{J}(\mathbf{r}'), \tag{14.80}$$

so we have

$$\mathbf{E}(\mathbf{r}) = -\frac{i\omega\mu_0}{4\pi} \frac{e^{ikr}}{r}\hat{\mathbf{r}} \times \left\{ \hat{\mathbf{r}} \times \int \mathbf{J}(\mathbf{r}')e^{-ik\hat{\mathbf{r}}\cdot\mathbf{r}'}d^3r' \right\}. \tag{14.81}$$

Now we may turn to the Poynting vector. Let us write the integral over $\mathbf{J}(\mathbf{r}')$ as a vector $\mathbf{F}$; then we can write the fields in shorthand as

$$\mathbf{E}(\mathbf{r}) = -\frac{i\omega\mu_0}{4\pi}\frac{e^{ikr}}{r}\hat{\mathbf{r}} \times [\hat{\mathbf{r}} \times \mathbf{F}], \tag{14.82}$$

$$\mathbf{H}(\mathbf{r}) = \frac{ik}{4\pi}\frac{e^{ikr}}{r}\hat{\mathbf{r}} \times \mathbf{F}. \tag{14.83}$$

With this notation, it becomes straightforward to determine the Poynting vector $\mathbf{S}(\mathbf{r})$, especially if we further write $\mathbf{G} \equiv \hat{\mathbf{r}} \times \mathbf{F}$ to simplify the vector multiplications. The Poynting vector takes the form

$$\mathbf{S}(\mathbf{r}) = \frac{1}{2}\mathbf{E}(\mathbf{r}) \times \mathbf{H}^*(\mathbf{r}) = \sqrt{\frac{\mu_0}{\epsilon_0}}\frac{k^2}{32\pi^2 r^2}\hat{\mathbf{r}}\left| \hat{\mathbf{r}} \times \int \mathbf{J}(\mathbf{r}')e^{-ik\hat{\mathbf{r}}\cdot\mathbf{r}'}d^3 r'\right|^2. \tag{14.84}$$

This is a general expression for the power radiated by an oscillating charge–current distribution into the far zone, and it already provides some worthwhile—though perhaps obvious—physical insights. We note that the Poynting vector points radially outward at every location and that it varies with $1/r^2$. If we integrate the Poynting vector over a sphere of radius r, we find that the power radiated is independent of r. This is an expression of energy conservation—the total power radiated is independent of the surface used to measure it.

The quantity in the absolute square—which we can call the radiation pattern—is the cross product of the unit vector $\hat{\mathbf{r}}$ and the net current of the source (the current integrated over the source domain). This indicates that the radiation in a particular direction only depends on those currents that are oscillating transverse to $\hat{\mathbf{r}}$, which is consistent with our view of light being a transverse wave.

Finally, we note that the radiation pattern depends on the three-dimensional spatial Fourier transform of the current distribution. This indicates that the radiation pattern depends on the mutual interference of the electromagnetic waves that are produced by currents throughout the source. These interference effects can be quite surprising, as illustrated in one exercise at the end of this chapter.

14.4 Multipole sources

We now turn to the multipole method of characterizing the radiation of a general three-dimensional monochromatic charge–current distribution. We will again look at the fields in the far zone of the source. The definition of the far zone is somewhat inexact, but we will assume that we measure the fields at a distance $\mathbf{r}$ such that $|\mathbf{r}'| \ll |\mathbf{r}|$, where $|\mathbf{r}'|$ is the largest distance within the source. Furthermore, we assume that $k|\mathbf{r}| \gg 1$, which indicates that we are sufficiently far from the source that near-field effects, i.e. evanescent waves, do not play a role. We will make this a bit more precise momentarily.

The overall advantage of the multipole method is that the radiation pattern of every source can be divided into different multipoles—electric dipole, magnetic dipole, electric quadrupole, and higher multipoles—where the radiation pattern only depends on the multipole moments of the source, and those multipole moments can be calculated by taking an integral over the source domain. Before we perform the expansion itself, it is convenient to use the continuity equation to derive a number of relationships between integrals of the source components.

As a reminder, the continuity equation for monochromatic fields is of the form

$$i\omega\rho(\mathbf{r}', \omega) = \nabla' \cdot \mathbf{J}(\mathbf{r}', \omega), \tag{14.85}$$

where we will drop the ω argument for brevity moving forward. Let us imagine multiplying this expression by an arbitrary function $f(\mathbf{r}')$ and integrating $\mathbf{r}'$ over the source volume V, which gives us the expression

$$\int_V f(\mathbf{r}')\rho(\mathbf{r}')d^3r' = \frac{1}{i\omega} \int_V f(\mathbf{r}')\nabla' \cdot \mathbf{J}(\mathbf{r}')d^3r'. \tag{14.86}$$

We again take advantage of the vector calculus relation,

$$\nabla' \cdot (f\mathbf{J}) = \nabla'f \cdot \mathbf{J} + f\nabla' \cdot \mathbf{J}, \tag{14.87}$$

which we can substitute into equation (14.86) to get

$$\int_V f(\mathbf{r}')\rho(\mathbf{r}')d^3r' = \frac{1}{i\omega} \int_V \nabla' \cdot [f(\mathbf{r}')\mathbf{J}(\mathbf{r}')]d^3r' - \frac{1}{i\omega} \int_V \mathbf{J}(\mathbf{r}') \cdot \nabla'f(\mathbf{r}')d^3r'. \tag{14.88}$$

The first integral on the right can be converted into a surface integral using Gauss's theorem,

$$\frac{1}{i\omega} \int_V \nabla' \cdot [f(\mathbf{r}')\mathbf{J}(\mathbf{r}')]d^3r' = \int_S f(\mathbf{r}')\mathbf{J}(\mathbf{r}') \cdot d\mathbf{a}', \tag{14.89}$$

where S is the surface bounding the volume. However, the choice of volume and surface is arbitrary, provided it includes the entire source—we can therefore choose a surface much larger than the source itself, and the integral over the surface vanishes. We are left with the relation

$$\int_V f(\mathbf{r}')\rho(\mathbf{r}')d^3r' = -\frac{1}{i\omega} \int_V \mathbf{J}(\mathbf{r}') \cdot \nabla'f(\mathbf{r}')d^3r'. \tag{14.90}$$

This equation indicates that we can find an entire family of integral relations between the charge density and the current density; these will prove incredibly useful in identifying the different multipole moments of a radiation source. We now investigate these integrals for some of the simplest choices of $f(\mathbf{r}, \mathbf{r}')$.

The simplest choice is $f(\mathbf{r}') = 1$. We immediately find that

$$\int_V \rho(\mathbf{r}')d^3r' = 0. \tag{14.91}$$

This is a statement that the *monopole moment* of the source vanishes. There is no monopole ('single pole') radiation of an electromagnetic source; the simplest form of radiation is dipole radiation. We may also interpret this equation as saying that a monochromatic radiation source must always have equal amounts of positive and negative charge within it. If it had a net nonzero charge, it would produce a static electric field and would not strictly be a monochromatic source.

The next choice we can make is $f(\mathbf{r}') = x'_\alpha$, where x'_α is one of the Cartesian components of the position vector, with $\alpha = 1, 2, 3$, i.e. $\mathbf{r}' = x'_1\hat{\mathbf{x}} + x'_2\hat{\mathbf{y}} + x'_3\hat{\mathbf{z}}$. It is easy to see that $\nabla'x'_\alpha = \hat{\mathbf{x}}_\alpha$, with $\hat{\mathbf{x}}_\alpha$ the αth Cartesian unit vector. On substitution into equation (14.90), we obtain

$$\int_V x'_\alpha \rho(\mathbf{r}')d^3r' = \frac{i}{\omega}\int_V J_\alpha(\mathbf{r}')d^3r' \equiv p_\alpha. \tag{14.92}$$

The integral on the left can be identified with the electric dipole moment p_α of the source, as the dipole moment for static systems is a sum over 'charge times distance,' in the form

$$\mathbf{p} = \sum_j q_j\mathbf{r}_j, \tag{14.93}$$

where q_j is the charge at the position $\mathbf{r}_j$. Equation (14.92) therefore shows how the electric dipole moment can be written in terms of both charge density and current density.

We next choose $f(\mathbf{r}') = x'_\alpha x'_\beta$, the product of two components of $\mathbf{r}'$. On substitution, we readily find that

$$\int_V x'_\alpha x'_\beta \rho(\mathbf{r}')d^3r' = \frac{i}{\omega}\int_V \left[J_\alpha x'_\beta + x'_\alpha J_\beta \right]d^3r' \equiv Q_{\alpha\beta}, \tag{14.94}$$

where again, in analogy with statics, we associate the left side of this expression with the electric quadrupole moment, $Q_{\alpha\beta}$, and therefore have an equation that presents the quadrupole moment in terms of the charge density and the current density.

Because this equation is now in second-rank tensor form, we can dispense with the component notation and instead write it as a dyadic,

$$\int_V \mathbf{r}'\mathbf{r}'\rho(\mathbf{r}')d^3r' = \frac{i}{\omega}\int_V [\mathbf{J}\mathbf{r}' + \mathbf{r}'\mathbf{J}]d^3r' \equiv \mathbf{Q}, \tag{14.95}$$

where $\mathbf{r}'\mathbf{r}'$, $\mathbf{J}\mathbf{r}'$, and $\mathbf{r}'\mathbf{J}$ are now three distinct dyads.

There is one final multipole that is worth pursuing here, and that is the magnetic dipole. We approach this by introducing what we call a 'magnetic current' $\mathbf{J}_m(\mathbf{r}')$,

$$\mathbf{J}_m(\mathbf{r}') = \frac{\omega}{2i}\mathbf{r}' \times \mathbf{J}(\mathbf{r}'), \tag{14.96}$$

with a corresponding 'magnetic charge density' $\rho_m(\mathbf{r}')$ defined by the continuity equation,

$$\rho_m(\mathbf{r}') = \frac{1}{i\omega}\nabla' \cdot \mathbf{J}_m(\mathbf{r}'). \tag{14.97}$$

These fictitious densities take advantage of the complementarity between electric and magnetic fields, as we did in section 10.1 in working with stratified media. As we will see, a magnetic charge density effectively produces a magnetic field and a magnetic current density produces an electric field, so that a magnetic dipole acts like an electric dipole except with the fields interchanged.

To simplify things at this point, we choose a vector function $\mathbf{f}(\mathbf{r}') = \mathbf{r}'$, and equation (14.90) takes on the form

$$\int_V \rho_m(\mathbf{r}')\mathbf{r}'d^3r' = \frac{i}{\omega}\int_V \mathbf{J}_m(\mathbf{r}') \cdot [\nabla'\mathbf{r}']d^3r' = \frac{i}{\omega}\int_V \mathbf{J}_m(\mathbf{r}')d^3r', \qquad (14.98)$$

where we have the dyadic $\nabla'\mathbf{r}' = \mathbf{I}$, the identity dyadic. Using our definition of the magnetic current density, we have

$$\int_V \rho_m(\mathbf{r}')\mathbf{r}'d^3r' = \frac{1}{2}\int_V \mathbf{r}' \times \mathbf{J}(\mathbf{r}')d^3r' = \mathbf{m}. \qquad (14.99)$$

which we define as the magnetic dipole moment $\mathbf{m}$. Referring back to equation (5.48), we find that this interpretation is the same as our definition of the magnetic dipole moment for magnetostatics. We can also see that it is possible to think of the magnetic dipole moment as arising from a fictitious magnetic charge in the same way that the electric dipole moment arises from electric charge.

It is possible to continue this process and find even higher multipole moments, e.g. a magnetic quadrupole, an electric octopole, and so on, but the calculations become much more cumbersome and the moments we have described are sufficient for most radiation problems.

14.5 Multipole potentials

We now turn to the multipole expansion of the potentials, which we have seen may be written as

$$\phi(\mathbf{r}) = \frac{1}{4\pi\epsilon_0}\int \rho(\mathbf{r}')\frac{e^{ikR}}{R}d^3r', \qquad (14.100)$$

$$\mathbf{A}(\mathbf{r}) = \frac{\mu_0}{4\pi}\int \mathbf{J}(\mathbf{r}')\frac{e^{ikR}}{R}d^3r'. \qquad (14.101)$$

The basic geometry is provided for convenience in figure 14.1. The origin is taken to lie somewhere within the volume V of the source, $\mathbf{r}'$ specifies a point within the source, and $\mathbf{r}$ specifies a point of observation outside the source.

We now consider the far-zone limit $|\mathbf{r}| \gg |\mathbf{r}'|$ and introduce the Taylor expansion of the Green's function. The three-dimensional Taylor expansion of a function $f(\mathbf{r} - \mathbf{r}')$ about the point $\mathbf{r}$ is of the form

$$f(\mathbf{r} - \mathbf{r}') = \sum_{n=0}^{\infty}\frac{1}{n!}(-\mathbf{r}' \cdot \nabla)^n f(\mathbf{r}), \qquad (14.102)$$

where we stress that ∇ is the derivative with respect to $\mathbf{r}$. Because $\mathbf{r}$ and $\mathbf{r}'$ are independent variables, the higher powers of $(-\mathbf{r}' \cdot \nabla)^n$ can be written in a straightforward way; for example,

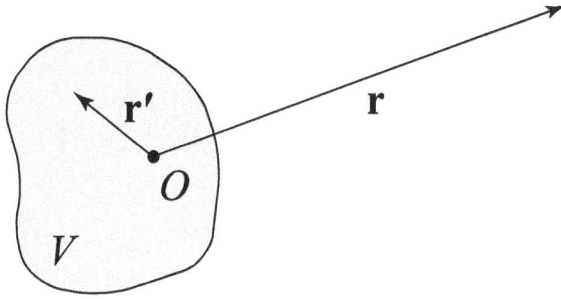

Figure 14.1. Illustration of the basic notation used for the multipole expansion.

$$(-\mathbf{r}' \cdot \nabla)^2 = (x'\partial_x + y'\partial_y)(x'\partial_x + y'\partial_y) = x'^2\partial_x^2 + y'^2\partial_y^2 + 2x'y'\partial_x\partial_y, \quad (14.103)$$

where we have only written the expression in two dimensions for brevity.

The Green's function in each of the potential integrals may be written as

$$\frac{e^{ik|\mathbf{r}-\mathbf{r}'|}}{|\mathbf{r} - \mathbf{r}'|} = \sum_{n=0}^{\infty}(-\mathbf{r}' \cdot \nabla)^n\frac{e^{ikr}}{r}. \quad (14.104)$$

We consider the first three terms of this series substituted into equation (14.100). The $n = 0$ term, which we call $\phi_0(\mathbf{r})$, gives

$$\phi_0(\mathbf{r}) = \frac{1}{4\pi\epsilon_0}\left[\int_V \rho(\mathbf{r}')d^3r'\right]\frac{e^{ikr}}{r} = 0, \quad (14.105)$$

which vanishes following equation (14.91). This is the monopole term, which is trivially zero.

The next $n = 1$ term gives

$$\phi_1(\mathbf{r}) = -\frac{1}{4\pi\epsilon_0}\left[\int_V \mathbf{r}'\rho(\mathbf{r}')d^3r'\right] \cdot \nabla\frac{e^{ikr}}{r} = -\frac{1}{4\pi\epsilon_0}\mathbf{p} \cdot \nabla\frac{e^{ikr}}{r}, \quad (14.106)$$

where we have used equation (14.92) to identify the dipole moment $\mathbf{p}$.

The $n = 2$ term is somewhat more complicated; we have

$$\phi_2(\mathbf{r}) = \frac{1}{4\pi\epsilon_0}\left[\frac{1}{2}\int_V \rho(\mathbf{r}')(\mathbf{r}' \cdot \nabla)(\mathbf{r}' \cdot \nabla)d^3r'\right]\frac{e^{ikr}}{r}. \quad (14.107)$$

To interpret this term, we note that ∇ only acts on $\mathbf{r}$ and can therefore be taken out of the integral. We may write our expression in a dyadic form,

$$\nabla \cdot \left[\frac{1}{2}\int_V \rho(\mathbf{r}')(\mathbf{r}'\mathbf{r}')d^3r'\right] \cdot \nabla\frac{e^{ikr}}{r} = \frac{1}{2}\nabla \cdot \mathbf{Q} \cdot \nabla\frac{e^{ikr}}{r}, \quad (14.108)$$

where we have used equation (14.94) to interpret the expression in terms of the electric quadrupole moment. We thus have

$$\phi_2(\mathbf{r}) = \frac{1}{4\pi\epsilon_0}\frac{1}{2}\left[\nabla \cdot \mathbf{Q} \cdot \nabla\right]\frac{e^{ikr}}{r}. \quad (14.109)$$

In total, the scalar potential may be written, up to the electric quadrupole moment, as

$$\phi(\mathbf{r}) = -\frac{1}{4\pi\epsilon_0}\left\{\mathbf{p}\cdot\nabla\frac{e^{ikr}}{r} - \frac{1}{2}\left[\nabla\cdot\mathbf{Q}\cdot\nabla\right]\frac{e^{ikr}}{r}\right\}. \qquad (14.110)$$

It should be noted that the quadrupole moment is a constant dyadic; the derivatives around it only act on the exp[ikr]/r term. If we need to evaluate this potential explicitly, we can pre- and postmultiply the quadrupole matrix by the quasi-vectors ∇, and the resulting matrix of derivatives can then act on the spherical wave.

So far, the results have been quite straightforward, and one might wonder why we went to all the trouble of writing the multipole moments in terms of charge and current densities. The reason becomes clear as we analogously expand the vector potential, which may be expanded up to the first order as

$$\mathbf{A}(\mathbf{r}) = \frac{\mu_0}{4\pi}\left[\int_V \mathbf{J}(\mathbf{r}')d^3r' - \int_V (\mathbf{J}(\mathbf{r}')\mathbf{r}'\cdot\nabla)d^3r'\right]\frac{e^{ikr}}{r}. \qquad (14.111)$$

From equation (14.92), we see that the first term is simply the electric dipole contribution, or

$$\mathbf{A}_0(\mathbf{r}) = -\frac{i\omega\mu_0}{4\pi}\mathbf{p}\frac{e^{ikr}}{r}. \qquad (14.112)$$

The second term requires a bit more work to interpret. We again note that the ∇ operator can be pulled from the integrand, leaving a dyadic inside,

$$\mathbf{A}_1(\mathbf{r}) = -\frac{\mu_0}{4\pi}\left[\int_V \mathbf{J}(\mathbf{r}')\mathbf{r}'d^3r'\right]\cdot\nabla\frac{e^{ikr}}{r}, \qquad (14.113)$$

where we are integrating over the dyadic $\mathbf{J}\mathbf{r}'$. A dyadic is simply another form of a matrix, however, and a matrix can always be written in terms of pieces that are symmetric and antisymmetric under a transpose, i.e.

$$\mathbf{J}(\mathbf{r}')\mathbf{r}' = \frac{1}{2}[\mathbf{r}'\mathbf{J} + \mathbf{J}\mathbf{r}'] - \frac{1}{2}[\mathbf{r}'\mathbf{J} - \mathbf{J}\mathbf{r}']. \qquad (14.114)$$

The symmetric contribution to the potential, which we label $\mathbf{A}_{1s}$, is therefore

$$\mathbf{A}_{1s}(\mathbf{r}) = -\frac{\mu_0}{4\pi}\frac{1}{2}\left\{\int_V [\mathbf{r}'\mathbf{J} + \mathbf{J}\mathbf{r}']d^3r'\right\}\cdot\nabla\frac{e^{ikr}}{r} = \frac{\mu_0}{4\pi}\frac{i\omega}{2}\mathbf{Q}\cdot\nabla\frac{e^{ikr}}{r}, \qquad (14.115)$$

where we have used equation (14.95) to identify the quadrupole moment.

To identify the final, antisymmetric piece, we resort to some more vector trickery. We consider the product of the magnetic dipole moment $\mathbf{m}$ with a vector $\mathbf{F}$ independent of $\mathbf{r}'$,

$$\mathbf{m}\times\mathbf{F} = \frac{1}{2}\int_V (\mathbf{r}'\times\mathbf{J})d^3r'\times\mathbf{F} = -\frac{1}{2}\mathbf{F}\times\int_V (\mathbf{r}'\times\mathbf{J})d^3r' = \frac{1}{2}\int_V [-\mathbf{r}'(\mathbf{J}\cdot\mathbf{F}) + \mathbf{J}(\mathbf{r}'\cdot\mathbf{F})]d^3r', \qquad (14.116)$$

where we have used equation (14.99) and applied the vector triple product formula in the last step. If we pull out the vector $\mathbf{F}$, we can write this in dyadic form as

$$\frac{1}{2}\int_V [-\mathbf{r}'(\mathbf{J} \cdot \mathbf{F}) + \mathbf{J}(\mathbf{r}' \cdot \mathbf{F})]d^3r' = -\frac{1}{2}\left\{\int_V [\mathbf{r}'\mathbf{J} - \mathbf{J}\mathbf{r}']d^3r'\right\} \cdot \mathbf{F}, \quad (14.117)$$

which is just our antisymmetric dyadic. If we set $\mathbf{F} = \nabla$, we may write the antisymmetric part of the vector potential $\mathbf{A}_{1s}$ as

$$\mathbf{A}_{1s}(\mathbf{r}') = -\frac{\mu_0}{4\pi}\frac{i\omega}{2}\mathbf{m} \times \nabla\frac{e^{ikr}}{r}. \quad (14.118)$$

Therefore, to the first order, the multipole expansion of the vector potential is

$$\mathbf{A}(\mathbf{r}) = \frac{\mu_0}{4\pi}\left\{-i\omega\mathbf{p}\frac{e^{ikr}}{r} + \frac{i\omega}{2}\mathbf{Q} \cdot \nabla\frac{e^{ikr}}{r} - \mathbf{m} \times \nabla\frac{e^{ikr}}{r}\right\}. \quad (14.119)$$

We note that the magnetic dipole makes no contribution to the scalar potential.

Equations (14.110) and (14.119) represent the most significant terms of the multipole expansion of the field. Part of the power of the multipole expansion is that it separates out the detailed physics of the source—the integrals over the charge and current densities—from the properties of the radiation field. This is not only a calculational convenience but also allows us to apply the multipole characterization of the fields to nonclassical problems. The radiation properties of atoms can also be characterized by their dipole moment, quadrupole moment, etc.

It should be noted that all of the multipole potentials have a $1/r$ dependence, which we will see means that they all contribute radiation to the far zone of the source. This is different from the electrostatic and magnetostatic cases, where the potential varies according to $1/r^{n+1}$, where n is the order of the multipole. In most circumstances, we expect the dipole contribution to be the strongest, but it is possible to have sources where the quadrupole or higher multipoles produce significant radiation.

At this point let us note an unexpected twist in our calculation. Referring back to equations (14.81) for the general formula of the electric field, we can see that we expect that the integrals of the current density should be explicitly wavelength-dependent. However, the integrals in equations (14.92), (14.95) and (14.99) for our lowest-order multiple moments are all effectively wavelength independent. In performing the multipole expansion of equation (14.104), we implicitly made a long-wavelength approximation, removing any non-trivial influence of the wavelength on the multipole moments themselves. To avoid making this approximation, we must adopt a more sophisticated albeit complicated approach, which is given in section 14.7.

14.6 Multipole fields and radiation

The potentials themselves are not directly observable, so our next step is to calculate the electric and magnetic fields produced by each of the low-order multipoles we have introduced and, furthermore, to determine the radiation pattern produced by each multipole.

For brevity, we introduce the modified Green's function

$$\mathcal{G}(\mathbf{r}) = \frac{1}{4\pi}\frac{e^{ikr}}{r}. \quad (14.120)$$

14.6.1 Electric dipole fields

We may write the potentials for the electric dipole as

$$\phi_1(\mathbf{r}) = -\frac{1}{\epsilon_0}\mathbf{p} \cdot \nabla \mathcal{G}(\mathbf{r}), \quad \mathbf{A}_0(\mathbf{r}) = -i\omega\mu_0\mathbf{p}\mathcal{G}(\mathbf{r}). \tag{14.121}$$

We can determine the electric dipole fields using the familiar definitions,

$$\mu_0\mathbf{H} = \nabla \times \mathbf{A}, \quad \mathbf{E} = -\nabla\phi + i\omega\mathbf{A}. \tag{14.122}$$

We may then write

$$\mathbf{E}_{ed}(\mathbf{r}) = \frac{1}{\epsilon_0}\{k^2\mathbf{p} + \mathbf{p} \cdot \nabla\}\mathcal{G}(\mathbf{r}), \tag{14.123}$$

where we have written *ed* to refer to 'electric dipole' and $\nabla = \nabla\nabla$ is the dyad constructed from the outer product of the del operator. Noting that $\mathbf{p} \cdot \mathbf{I} = \mathbf{p}$, we can write the electric field entirely in terms of dyadics,

$$\mathbf{E}_{ed}(\mathbf{r}) = \frac{1}{\epsilon_0}\mathbf{p} \cdot \{k^2\mathbf{I} + \nabla\}\mathcal{G}(\mathbf{r}). \tag{14.124}$$

It should be noted again that all the fine details of the source are encapsulated in the dipole moment $\mathbf{p}$, and all the propagation characteristics are carried by the dyadic Green's function. All dipoles, therefore, have a similar behavior, only differing in the magnitude and direction of the dipole moment.

Similarly, we may write the magnetic field as

$$\mathbf{H}_{ed}(\mathbf{r}) = i\omega\mathbf{p} \times \nabla\mathcal{G}(\mathbf{r}). \tag{14.125}$$

These formulas for the electric and magnetic fields, though elegant, are still not simple enough to analyze easily. We now fully consider the fields in the far zone, using an alternative definition of the far zone such that $kr \gg 1$. In this limit, we have

$$\nabla\frac{e^{ikr}}{r} = -\left(\frac{1}{r^2} - \frac{ik}{r}\right)e^{ikr}\hat{\mathbf{r}} \approx ik\hat{\mathbf{r}}\frac{e^{ikr}}{r}. \tag{14.126}$$

We note that this approximation can be used for multiple applications of the del operator, so that in the far zone we may generally use the identity

$$\nabla \leftrightarrow ik\hat{\mathbf{r}}, \tag{14.127}$$

so that, for example, $\nabla \to -k^2\hat{\mathbf{r}}\hat{\mathbf{r}}$.

The electric field in the far zone may therefore be written

$$\mathbf{E}_{ed}(\mathbf{r}) = \frac{k^2}{\epsilon_0}\{\mathbf{p} - \hat{\mathbf{r}}(\hat{\mathbf{r}} \cdot \mathbf{p})\}\mathcal{G}(\mathbf{r}). \tag{14.128}$$

The quantity in the curved brackets can be identified with a vector triple product,

$$\hat{\mathbf{r}} \times (\hat{\mathbf{r}} \times \mathbf{p}) = \hat{\mathbf{r}}(\hat{\mathbf{r}} \cdot \mathbf{p}) - \mathbf{p}(\hat{\mathbf{r}} \cdot \hat{\mathbf{r}}), \tag{14.129}$$

so that we may finally write

$$\mathbf{E}_{ed}(\mathbf{r}) = -\frac{k^2}{\epsilon_0}\hat{\mathbf{r}} \times (\hat{\mathbf{r}} \times \mathbf{p})\mathcal{G}(\mathbf{r}). \tag{14.130}$$

The magnetic field may be similarly found to be

$$\mathbf{H}_{ed}(\mathbf{r}) = k\omega\hat{\mathbf{r}} \times \mathbf{p}\mathcal{G}(\mathbf{r}). \tag{14.131}$$

These expressions for the electric and magnetic fields immediately provide some physical information. From equation (14.131), we can see that the magnetic field is perpendicular both to the direction of propagation $\hat{\mathbf{r}}$ and the dipole moment $\mathbf{p}$. The former observation is a statement that the magnetic field is perpendicular to the direction of propagation; i.e. it is transverse. The latter observation indicates that the field does not propagate in the direction of the dipole, in agreement with the fact that it is transverse.

Equation (14.130) further shows that the electric field is perpendicular to the direction of propagation and the magnetic field, making the field fully transverse in the far zone. The fields of the electric dipole are illustrated in figure 14.2.

We may also calculate the Poynting vector of the electric dipole fields, which can be found in a straightforward manner in the form

$$\mathbf{S} = \frac{1}{2}\text{Re}\{\mathbf{E} \times \mathbf{H}^*\} = \frac{1}{2}\frac{k^3\omega}{(4\pi)^2\epsilon_0}\frac{1}{r^2}[|\mathbf{p}|^2 - |\hat{\mathbf{r}} \cdot \mathbf{p}|^2]\hat{\mathbf{r}}. \tag{14.132}$$

To derive this formula, we apply the vector triple product to the expression for the electric field and then again for the Poynting vector.

To make sense of the Poynting vector, we assume that the dipole moment points along the z-axis; there is no loss of generality because we can always rotate our coordinate system to determine the fields of a dipole pointed in any other direction. We then have $\mathbf{p} \equiv p_0\hat{\mathbf{z}}$, and $\hat{\mathbf{r}} \cdot \mathbf{p} = p_0\cos\theta$; we thus obtain

$$\mathbf{S} = \frac{1}{2}\frac{k^3\omega p_0^2}{(4\pi)^2\epsilon_0 r^2}\sin^2\theta\hat{\mathbf{r}}. \tag{14.133}$$

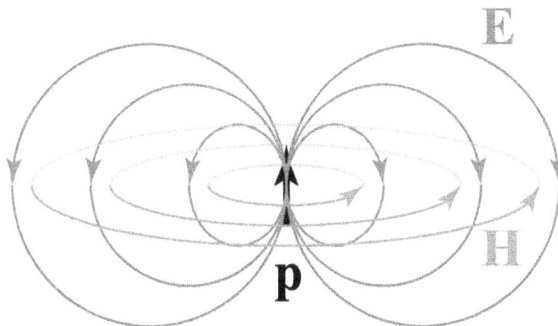

Figure 14.2. Illustration of the fields of the electric dipole.

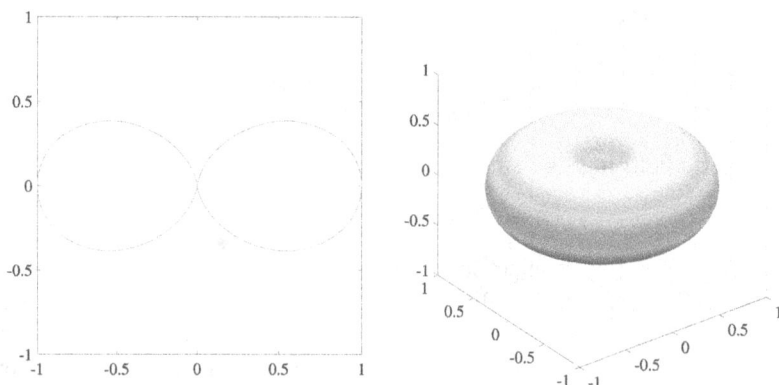

Figure 14.3. The radiation pattern of an electric dipole.

We may deduce several physically significant observations from this formula. First, we note that the power flow is purely radial: in the far zone, all the power flows directly away from the dipole at the origin. Next, we note that the power flow has a $1/r^2$ dependence, which means that the total power P flowing through a sphere centered on the origin is independent of radius, i.e.

$$P = \oint \mathbf{S} \cdot d\mathbf{a} = \text{constant}. \tag{14.134}$$

This is basically a statement of conservation of energy: the total power flowing through a sphere of any radius is the same, and no energy is lost during propagation. Finally, we note that, with $k = \omega/c$, the power radiated by the dipole varies according to ω^4. This will be relevant in our discussion of the scattering of light in the next chapter.

The radiation pattern (power radiated as a function of angle) is illustrated in figure 14.3. The power radiated in a particular direction is found by the intersection between the surface shown and a line drawn from the origin in the direction of propagation. In cross section, the radiation pattern appears as a two-lobed pattern. In three dimensions, however, it is more accurately described as a donut, with no power emitted in the direction of the dipole and the maximum power radiated in the xy-plane, i.e. perpendicular to the dipole.

14.6.2 Magnetic dipole fields

We now turn to the analysis of magnetic dipoles, for which we have the potentials

$$\phi(\mathbf{r}') = 0, \quad \mathbf{A}_{1s}(\mathbf{r}') = -\frac{i\omega\mu_0}{2}\mathbf{m} \times \nabla\mathcal{G}(\mathbf{r}). \tag{14.135}$$

We substitute from these into equation (14.122) and apply our far-zone identity, equation (14.127). We readily find that

$$\mathbf{E}_{md}(\mathbf{r}) = -\omega k\mu_0\hat{\mathbf{r}} \times \mathbf{m}\mathcal{G}(\mathbf{r}), \tag{14.136}$$

$$\mathbf{H}_{md}(\mathbf{r}) = -k^2 \hat{\mathbf{r}} \times (\hat{\mathbf{r}} \times \mathbf{m}) \mathcal{G}(\mathbf{r}). \tag{14.137}$$

Comparing these expressions to equations (14.130) and (14.131), we readily see that the magnetic dipole fields are structurally similar to the electric dipole fields, with the roles of the electric and magnetic fields reversed (to within a sign). The radiation pattern is similarly of identical form to the electric dipole case,

$$\mathbf{S} = \frac{1}{2} \frac{k^3 \omega \mu_0 m_0^2}{(4\pi)^2 r^2} \sin^2 \theta \hat{\mathbf{r}}, \tag{14.138}$$

where we have used $\mathbf{m} = m_0 \hat{\mathbf{z}}$. The observations we made about the electric dipole radiation also apply to the magnetic dipole: the fields are transverse, energy is conserved, and radiation is strongest in directions perpendicular to the dipole moment.

14.6.3 Electric quadrupole fields

Turning to the electric quadrupole, the relevant potentials are

$$\phi_3(\mathbf{r}) = \frac{1}{\epsilon_0} \frac{1}{2} [\nabla \cdot \mathbf{Q} \cdot \nabla] \mathcal{G}(\mathbf{r}), \quad \mathbf{A}_{1s}(\mathbf{r}) = \frac{i\omega \mu_0}{2} \mathbf{Q} \cdot \nabla \mathcal{G}(\mathbf{r}). \tag{14.139}$$

Again using equation (14.122), the corresponding fields are of the form

$$\mathbf{E}_{eq}(\mathbf{r}) = -\frac{1}{2\epsilon_0} [\nabla(\nabla \cdot \mathbf{Q} \cdot \nabla) + k^2 \mathbf{Q} \cdot \nabla] \mathcal{G}(\mathbf{r}), \tag{14.140}$$

$$\mathbf{H}_{eq}(\mathbf{r}) = \frac{i\omega}{2} \nabla \times [\mathbf{Q} \cdot \nabla \mathcal{G}(\mathbf{r})]. \tag{14.141}$$

These expressions look intimidating at first, until we remember that we can replace the derivatives by vectors in the far zone using equation (14.127). We then have

$$\mathbf{E}_{eq}(\mathbf{r}) = -\frac{ik^3}{2\epsilon_0} [-\hat{\mathbf{r}}(\hat{\mathbf{r}} \cdot \mathbf{Q} \cdot \hat{\mathbf{r}}) + \mathbf{Q} \cdot \mathbf{r}] \mathcal{G}(\mathbf{r}), \tag{14.142}$$

$$\mathbf{H}_{eq}(\mathbf{r}) = -\frac{ik^2 \omega}{2} \hat{\mathbf{r}} \times (\mathbf{Q} \cdot \mathbf{r}) \mathcal{G}(\mathbf{r}). \tag{14.143}$$

In order to evaluate the specific form of the fields, we need to specify the form of the quadrupole dyadic, which is a symmetric matrix and has a significant number of degrees of freedom. Instead of worrying about the fields, we turn directly to analyzing the radiation pattern of the quadrupole. To simplify the math, we introduce the vector $\mathbf{q} = \mathbf{Q} \cdot \hat{\mathbf{r}}$, which allows us to write the electric field as

$$\mathbf{E}_{eq}(\mathbf{r}) = -\frac{ik^3}{2\epsilon_0} [-\hat{\mathbf{r}}(\hat{\mathbf{r}} \cdot \mathbf{q}) + \mathbf{q}] \mathcal{G}(\mathbf{r}) = \frac{ik^3}{2\epsilon_0} \hat{\mathbf{r}} \times (\mathbf{q} \times \mathbf{r}) \mathcal{G}(\mathbf{r}), \tag{14.144}$$

where we have taken advantage of the BAC-CAB rule to simplify the formula. The field formulas in terms of $\mathbf{q}$ are roughly analogous to the dipole formulas, and after some vector manipulation, we find that

$$\mathbf{S}(\mathbf{r}) = \frac{k^5\omega}{8(4\pi)^2\epsilon_0 r^2}|\hat{\mathbf{r}} \times (\mathbf{Q}\cdot\hat{\mathbf{r}})|^2\hat{\mathbf{r}}. \qquad (14.145)$$

Again, we find that the power flow in the far zone is entirely radial, and that total power is conserved on propagation due to the $1/r^2$ dependence.

The quadrupole dyadic can take on many forms; we consider a simple example for which

$$\mathbf{Q} = \begin{bmatrix} -Q_0/2 & 0 & 0 \\ 0 & -Q_0/2 & 0 \\ 0 & 0 & Q_0 \end{bmatrix}. \qquad (14.146)$$

In dyadic form, we may write

$$\mathbf{Q}\cdot\hat{\mathbf{r}} = -\frac{Q_0}{2}\hat{\mathbf{x}}(\hat{\mathbf{x}}\cdot\hat{\mathbf{r}}) - \frac{Q_0}{2}\hat{\mathbf{y}}(\hat{\mathbf{y}}\cdot\hat{\mathbf{r}}) + Q_0\hat{\mathbf{z}}(\hat{\mathbf{z}}\cdot\hat{\mathbf{r}}). \qquad (14.147)$$

The dot products may be written in spherical coordinates, which gives us

$$\mathbf{Q}\cdot\hat{\mathbf{r}} = -\frac{Q_0}{2}\hat{\mathbf{x}}\sin\theta\cos\phi - \frac{Q_0}{2}\hat{\mathbf{y}}\sin\theta\sin\phi + Q_0\hat{\mathbf{z}}\cos\theta. \qquad (14.148)$$

We then find that

$$\hat{\mathbf{r}} \times (\mathbf{Q}\cdot\hat{\mathbf{r}}) = \hat{\mathbf{x}}\frac{3Q_0}{2}\cos\theta\sin\theta\sin\phi - \hat{\mathbf{y}}\frac{3Q_0}{2}\cos\theta\sin\theta\cos\phi. \qquad (14.149)$$

Finally, we have

$$\mathbf{S}(\mathbf{r}) = \frac{9k^5\omega}{32(4\pi)^2\epsilon_0 r^2}Q_0^2\cos^2\theta\sin^2\theta\hat{\mathbf{r}}, \qquad (14.150)$$

which can be simplified using trigonometry to

$$\mathbf{S}(\mathbf{r}) = \frac{9k^5\omega}{128(4\pi)^2\epsilon_0 r^2}Q_0^2\sin^2(2\theta). \qquad (14.151)$$

The radiation pattern for this quadrupole is shown in figure 14.4. In cross section, it has a 'four-leaf clover' or four-lobed pattern, though in three dimensions, this pattern is symmetric about the z-axis, making it appear like two donuts squashed together. We specifically chose a symmetric quadrupole dyadic for simplicity; in general, however, it is not required to have any particular axis of symmetry.

With this, we have covered the basic radiation properties of classical currents. In most cases, dipole and quadrupole moments are sufficient for analyzing physical problems. In particular, we will take advantage of our dipole radiation equations in

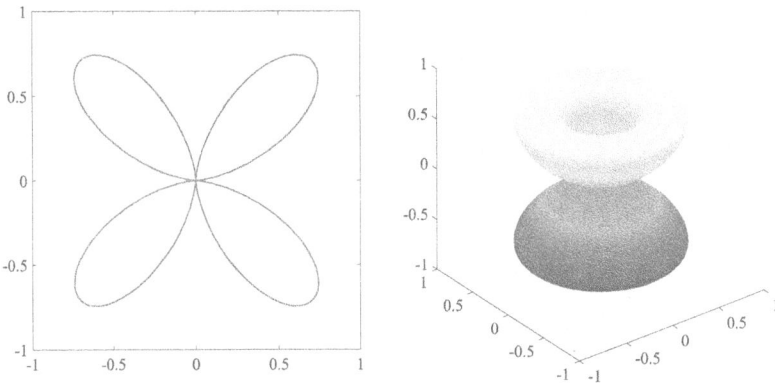

Figure 14.4. The radiation pattern of an electric quadrupole.

the next chapter on scattering. For higher multipole moments, a different, less intuitive but very powerful approach is discussed in the next section.

14.7 Higher-order multipoles and spherical waves

Our approach to multipoles in the past few sections was aimed at giving the reader an intuitive understanding of the structure of multipole fields. Our derivation implicitly used a long wavelength approximation, however, so to get a more rigorous multipole expansion we turn to an alternative method involving what are known as the Debye fields. This approach will also turn out to be incredibly useful in solving the problem of light scattering by a sphere, known as the Mie scattering problem, so it is worth discussing here. The math gets quite involved, so this section can be passed over on a first reading, if desired.

We return to the monochromatic Maxwell's equations in a homogeneous medium with no sources; they take the forms

$$\nabla \cdot \mathbf{E} = 0, \tag{14.152}$$

$$\nabla \cdot \mathbf{H} = 0, \tag{14.153}$$

$$\nabla \times \mathbf{E} = i\omega\mu\mathbf{H}, \tag{14.154}$$

$$\nabla \times \mathbf{H} = -i\omega\epsilon\mathbf{E}, \tag{14.155}$$

where we consider a constant but nonvacuum ϵ and μ, in anticipation of the solution of our scattering problem in the next chapter. These fields also satisfy the vector Helmholtz equations,

$$(\nabla^2 + k^2)\mathbf{E} = 0, \tag{14.156}$$

$$(\nabla^2 + k^2)\mathbf{H} = 0, \tag{14.157}$$

where $k^2 = \epsilon\mu\omega^2$. Because each component of the electric and magnetic fields satisfies a scalar Helmholtz equation, we suspect that we can construct exact electromagnetic waves from solutions of a scalar Helmholtz equation.

We begin by assuming that we have a function $\psi(\mathbf{r})$ that satisfies

$$(\nabla^2 + k^2)\psi = 0. \tag{14.158}$$

Let us introduce functions $\mathbf{M}_\psi$, $\mathbf{N}_\psi$ of the form

$$\mathbf{M}_\psi = \nabla \times (\mathbf{r}\psi), \tag{14.159}$$

$$\mathbf{N}_\psi = \frac{1}{k}\nabla \times \mathbf{M}_\psi. \tag{14.160}$$

It is straightforward to show that

$$\mathbf{M}_\psi = \frac{1}{k}\nabla \times \mathbf{N}_\psi, \tag{14.161}$$

by taking advantage of the fact that ψ satisfies the Helmholtz equation. (At this point in the book, we assume the reader has had sufficient exposure to vector calculus that we do not have to show every step of the calculation!)

For future reference, we explicitly write the components of $\mathbf{M}_\psi$ and $\mathbf{N}_\psi$ in spherical coordinates. This process is simplified by noting that

$$\mathbf{M}_\psi(\mathbf{r}) = \nabla \times [\mathbf{r}\psi(\mathbf{r})] = \psi(\mathbf{r})[\nabla \times \mathbf{r}] - \mathbf{r} \times \nabla\psi(\mathbf{r}) = -\mathbf{r} \times \nabla\psi(\mathbf{r}). \tag{14.162}$$

From this, the components of $\mathbf{M}_\psi$ can straightforwardly be found to be

$$M_r(\mathbf{r}) = 0, \tag{14.163}$$

$$M_\theta(\mathbf{r}) = \frac{1}{\sin\theta}\partial_\phi\psi, \tag{14.164}$$

$$M_\phi(\mathbf{r}) = -\partial_\theta\psi, \tag{14.165}$$

and from equation (14.160) the components of $\mathbf{N}_\psi$ can be then found to be

$$N_r = -\frac{1}{kr\sin^2\theta}\left[\sin\theta\partial_\theta(\sin\theta\partial_\theta\psi) + \partial_\phi^2\psi\right], \tag{14.166}$$

$$N_\theta = \frac{1}{kr}\partial_\theta\partial_r(r\psi), \tag{14.167}$$

$$N_\phi = \frac{1}{kr\sin\theta}\partial_\phi\partial_r(r\psi). \tag{14.168}$$

For particular choices of ψ, we will see that the expression for N_r takes on a particularly simple form.

Looking back at equations (14.154) and (14.155), it becomes clear that we can define electric and magnetic fields using $\mathbf{M}_\psi$ and $\mathbf{N}_\psi$. This is most elegantly done by introducing two independent scalar functions $u(\mathbf{r})$ and $v(\mathbf{r})$ and by defining

$$\mathbf{E}(\mathbf{r}) = \mathbf{N}_u(\mathbf{r}) + i\mathbf{M}_v(\mathbf{r}), \tag{14.169}$$

$$\mathbf{H}(\mathbf{r}) = \sqrt{\frac{\epsilon}{\mu}}\,[\mathbf{N}_v(\mathbf{r}) - i\mathbf{M}_u(\mathbf{r})]. \tag{14.170}$$

The factor of i between $\mathbf{N}_u$ and $\mathbf{M}_v$ is a matter of convenience because it highlights the complementary nature of the electric and magnetic field quantities; it can easily be absorbed into the definition of v if desired.

It can readily be shown that these expressions for the electric and magnetic fields satisfy all of Maxwell's equations. In short, $\mathbf{E}_v = i\mathbf{M}_v$ and $\mathbf{H}_v = \sqrt{\epsilon/\mu}\,\mathbf{N}_v$ form a solution to Maxwell's equations, as do $\mathbf{E}_u = \mathbf{N}_u$ and $\mathbf{H}_u = -i\sqrt{\epsilon/\mu}\,\mathbf{M}_u$.

Why do we have two sets of equations? We can see from the form of $\mathbf{M}_\psi$ and $\mathbf{N}_\psi$ that the fields for u and v are complementary; they form the fields due to electric and magnetic multipoles. In particular, $u(\mathbf{r})$ is associated with electric multipoles, and $v(\mathbf{r})$ is associated with magnetic multipoles. These fields are known as the *Debye potentials*, as Debye was the first to implement them in 1909 [4]. This decomposition is quite general; it has been proven that these potentials can represent the fields within any two concentric spheres in a source-free region [5].

We now have two tasks: (i) to determine the Debye potentials u and v from the source current $\mathbf{J}$, and (ii) to determine the multipole expansion from the Debye potentials. For the first step, we follow a strategy analogous to one we used in chapter 13: we look at the longitudinal fields. It is straightforward to show that

$$\mathbf{r} \cdot \mathbf{E} = \mathbf{r} \cdot \mathbf{N}_u, \tag{14.171}$$

$$\mathbf{r} \cdot \mathbf{H} = \sqrt{\frac{\epsilon}{\mu}}\,\mathbf{r} \cdot \mathbf{N}_v, \tag{14.172}$$

which shows that u is determined by the longitudinal component of the electric field, and v is determined by the longitudinal component of the magnetic field.

We now turn back to Maxwell's equations with sources and take the curl of each curl equation, followed by the dot product with $\mathbf{r}$,

$$\mathbf{r} \cdot \nabla \times (\nabla \times \mathbf{H}) - k^2 \mathbf{r} \cdot \mathbf{H} = \mathbf{r} \cdot \nabla \times \mathbf{J}, \tag{14.173}$$

$$\mathbf{r} \cdot \nabla \times (\nabla \times \mathbf{E}) - k^2 \mathbf{r} \cdot \mathbf{E} = i\omega\mu\mathbf{r} \cdot \mathbf{J}. \tag{14.174}$$

We must now engage in some clever vector calculus to move $\mathbf{r}$ inside the derivatives. Let us consider a general vector $\mathbf{F}$ first; we then have

$$\mathbf{r} \cdot [\nabla \times (\nabla \times \mathbf{F})] = \mathbf{r} \cdot [\nabla(\nabla \cdot \mathbf{F}) - \nabla^2 \mathbf{F}]. \tag{14.175}$$

Let us consider the second term on the right first. For once, it is unwieldy to approach it with dyadics, so we write the dot product as a component sum using the

Einstein summation convention (identical indices summed over). We derive the following expression:

$$\nabla^2(r_i F_i) = r_i \nabla^2 F_i + F_i \nabla^2 r_i + 2\nabla_j F_i \nabla_j r_i. \tag{14.176}$$

The second term on the right vanishes; for the third term, $\nabla_j r_i = \delta_{ij}$. We rearrange the remaining terms to get

$$\mathbf{r} \cdot (\nabla^2 \mathbf{F}) = \nabla^2(\mathbf{r} \cdot \mathbf{F}) - 2\nabla \cdot \mathbf{F}. \tag{14.177}$$

We can substitute from this into equation (14.175) to obtain

$$\mathbf{r} \cdot \nabla \times (\nabla \times \mathbf{F}) = \mathbf{r} \cdot \nabla[\nabla \cdot \mathbf{F}] - \nabla^2(\mathbf{r} \cdot \mathbf{F}) + 2\nabla \cdot \mathbf{F}. \tag{14.178}$$

Let us consider the equation for $\mathbf{H}$ first. Because $\nabla \cdot \mathbf{H} = 0$, equation (14.173) becomes

$$(\nabla^2 + k^2)(\mathbf{r} \cdot \mathbf{H}) = -\mathbf{r} \cdot (\nabla \times \mathbf{J}). \tag{14.179}$$

This is a scalar Helmholtz equation, which we can readily solve, and we will do so in a moment! But first, we consider equation (14.174) for $\mathbf{r} \cdot \mathbf{E}$, which we also want to convert to a Helmholtz equation; this will take a little extra work.

Using equation (14.178) in equation (14.174), we have

$$\mathbf{r} \cdot \nabla[\nabla \cdot \mathbf{E}] - \nabla^2(\mathbf{r} \cdot \mathbf{E}) + 2\nabla \cdot \mathbf{E} - k^2 \mathbf{r} \cdot \mathbf{E} = i\omega\mu \mathbf{r} \cdot \mathbf{J}. \tag{14.180}$$

From the continuity equation, we know that $\nabla \cdot \mathbf{E} = \nabla \cdot \mathbf{J}/(i\omega\epsilon)$; upon using this and rearranging, the expression becomes

$$-(\nabla^2 + k^2)(\mathbf{r} \cdot \mathbf{E}) + \frac{1}{i\omega\epsilon}[\mathbf{r} \cdot \nabla(\nabla \cdot \mathbf{J}) + 2\nabla \cdot \mathbf{J} + k^2(\mathbf{r} \cdot \mathbf{J})] = 0. \tag{14.181}$$

We now use equation (14.178), but with $\mathbf{F} = \mathbf{J}$, to rewrite the first two terms in the brackets above. After some rearranging, we finally have

$$(\nabla^2 + k^2)\left[\mathbf{r} \cdot \mathbf{E} + \frac{i}{\omega\epsilon}\mathbf{r} \cdot \mathbf{J}\right] = -\frac{i}{\omega\epsilon}\mathbf{r} \cdot \nabla \times (\nabla \times \mathbf{J}). \tag{14.182}$$

We therefore have reduced our equation for $\mathbf{r} \cdot \mathbf{E}$ into a scalar Helmholtz equation as well! Following section 14.1, we can immediately write the solutions for $\mathbf{r} \cdot \mathbf{E}$ and $\mathbf{r} \cdot \mathbf{H}$ as

$$\mathbf{r} \cdot \mathbf{H} = \int_D \mathcal{G}(\mathbf{r}, \mathbf{r}')\mathbf{r}' \cdot [\nabla' \times \mathbf{J}(\mathbf{r}')]d^3r', \tag{14.183}$$

$$\mathbf{r} \cdot \mathbf{E} = -\frac{i}{\omega\epsilon}\mathbf{r} \cdot \mathbf{J} + \frac{i}{\omega\epsilon}\int_D \mathcal{G}(\mathbf{r}, \mathbf{r}')\mathbf{r}' \cdot [\nabla' \times [\nabla' \times \mathbf{J}(\mathbf{r}')]]d^3r'. \tag{14.184}$$

where $\mathcal{G}(\mathbf{r}, \mathbf{r}')$ is defined by equation (14.120). Because we are almost always calculating our electric field outside the region where the sources exist, we may simplify $\mathbf{r} \cdot \mathbf{E}$ to

$$\mathbf{r} \cdot \mathbf{E} = \frac{i}{\omega\epsilon}\int_D \mathcal{G}(\mathbf{r}, \mathbf{r}')\mathbf{r}' \cdot [\nabla' \times [\nabla' \times \mathbf{J}(\mathbf{r}')]]d^3r'. \tag{14.185}$$

We have had to do a lot of mathematics, but we are making great progress towards our goal of a general multipole expansion (the amount of math required is a good illustration of why we calculated the lower-order multipoles by a different method).

We now take advantage of a well-known modal expansion of the scalar Helmholtz Green's function, assuming again $r > r'$,

$$\mathcal{G}(\mathbf{r}, \mathbf{r}') = ik\sum_{l=0}^{\infty}\sum_{m=-l}^{l} j_l(kr')h_l^{(1)}(kr)\, Y_{lm}(\theta, \phi)\, Y_{lm}^*(\theta', \phi'), \tag{14.186}$$

where $j_l(x)$ is the spherical Bessel function of the first kind and order l, $h_l^{(1)}(x)$ is the spherical Hankel function of the first kind and order l, and the $Y_{lm}(\theta, \phi)$ are the spherical harmonics of order l, m. For details of the derivation and of the various special functions, see Gbur [6].

The spherical harmonics have the form

$$Y_{lm}(\theta, \phi) = (-1)^m\sqrt{\frac{2l+1}{2}\frac{(l-m)!}{2\pi(l+m)!}}\, P_l^m(\cos\theta)e^{im\phi}, \tag{14.187}$$

where $l = 0, 1, 2, ...,$ $m = -l, -l+1, ... , 0, ... , l-1, l$ and $P_l^m(\cos\theta)$ are the associated Legendre functions. The spherical harmonics are defined to be ortho-normal over the solid angle of a sphere, i.e.

$$\int_0^{\pi}\int_0^{2\pi} Y_{lm}(\theta, \phi)\, Y_{l'm'}(\theta, \phi)\sin\theta d\theta d\phi = \delta_{ll'}\delta_{mm'}. \tag{14.188}$$

The functions $\mathbf{M}_{lm}$ and $\mathbf{N}_{lm}$ derived from the set of spherical harmonic basis functions are often referred to as vector spherical harmonics.

For brevity, let us rewrite the Green's function expansion as

$$\mathcal{G}(\mathbf{r}, \mathbf{r}') = ik\sum_{l=0}^{\infty}\sum_{m=-l}^{l} \psi_{lm}^*(\mathbf{r}')\eta_{lm}(\mathbf{r}), \tag{14.189}$$

where

$$\psi_{lm}(\mathbf{r}') = j_l(kr')\, Y_{lm}(\theta', \phi'), \tag{14.190}$$

$$\eta_{lm}(\mathbf{r}) = h_l^{(1)}(kr)\, Y_{lm}(\theta, \phi). \tag{14.191}$$

These definitions will make the mathematical expressions cleaner as we continue. We are finally approaching a multipole expansion of the field, as substitution from equation (14.189) into equations (14.183) and (14.184) gives us the expressions

$$\mathbf{r}\cdot\mathbf{H} = ik\sum_{l=0}^{\infty}\sum_{m=-l}^{l} \eta_{lm}(\mathbf{r})\int_D \psi_{lm}^*(\mathbf{r}')\mathbf{r}'\cdot[\nabla'\times\mathbf{J}(\mathbf{r}')]d^3r', \tag{14.192}$$

$$\mathbf{r}\cdot\mathbf{E} = -\sqrt{\frac{\mu}{\epsilon}}\sum_{l=0}^{\infty}\sum_{m=-l}^{l} \eta_{lm}(\mathbf{r})\int_D \psi_{lm}^*(\mathbf{r}')\mathbf{r}'\cdot[\nabla'\times[\nabla'\times\mathbf{J}(\mathbf{r}')]]d^3r'. \tag{14.193}$$

It is more convenient to write these expressions directly in terms of $\mathbf{J}$ instead of the derivatives of $\mathbf{J}$. We may take advantage of the modified form of Gauss's theorem,

$$\int_D \nabla' \cdot (\mathbf{F} \times \mathbf{G}) d^3 r' = \int_D [\mathbf{G} \cdot (\nabla' \times \mathbf{F}) - \mathbf{F} \cdot (\nabla' \times \mathbf{G})] d^3 r' = \int_S (\mathbf{F} \times \mathbf{G}) \cdot d\mathbf{a}', \quad (14.194)$$

to rewrite $\mathbf{r} \cdot \mathbf{H}$ and $\mathbf{r} \cdot \mathbf{E}$. Because the currents are finite, the surface integrals vanish; we apply this form of Gauss's theorem twice to $\mathbf{r} \cdot \mathbf{E}$ to get the final results,

$$\mathbf{r} \cdot \mathbf{H} = ik \sum_{l=0}^{\infty} \sum_{m=-l}^{l} \eta_{lm}(\mathbf{r}) \int_D \mathbf{J}(\mathbf{r}') \cdot [\nabla' \times (\mathbf{r}' \psi_{lm}^*(\mathbf{r}'))] d^3 r', \quad (14.195)$$

$$\mathbf{r} \cdot \mathbf{E} = -\sqrt{\frac{\mu}{\epsilon}} \sum_{l=0}^{\infty} \sum_{m=-l}^{l} \eta_{lm}(\mathbf{r}) \int_D \mathbf{J}(\mathbf{r}') \cdot \left\{ \nabla' \times [\nabla' \times (\mathbf{r}' \psi_{lm}^*(\mathbf{r}'))] \right\} d^3 r'. \quad (14.196)$$

These expressions form the basis of our multipole expansion, where l represents the order of the multipole and m effectively represents the components of the multipole. For example, for $l = 1$, we have the dipole moments, which potentially have three components determined by $m = -1, 0, 1$. For $l = 2$, we have the quadrupole moments, which have five components determined by $m = -2, -1, 0, 1, 2$. Why five? The quadrupole tensor is symmetric, which means it has six independent components, but one can also show that the trace of the tensor is zero, providing one constraint.

We now turn to deriving the other field components from the radial components. We first note that

$$\mathbf{r} \cdot \nabla \times [\nabla \times (\mathbf{r}\eta_{lm})] = (r^2 \partial_r^2 + 2r\partial_r k^2 r^2)\eta_{lm} = l(l+1)\eta_{lm}, \quad (14.197)$$

the latter step following from the radial differential equation that the spherical Bessel functions satisfy. This suggests that we can expand our radial field components in the series

$$\mathbf{r} \cdot \mathbf{E} = \sum_{l=0}^{\infty} \sum_{m=-l}^{l} l(l+1)a_{lm}\eta_{lm}, \quad (14.198)$$

$$\mathbf{r} \cdot \mathbf{H} = \sum_{l=0}^{\infty} \sum_{m=-l}^{l} l(l+1)b_{lm}\eta_{lm}, \quad (14.199)$$

with the coefficients a_{lm} and b_{lm} to be determined. The coefficients a_{lm} are associated with the electric multipoles, and the coefficients b_{lm} are associated with the magnetic multipoles.

Let us consider the electric multipole case first, for which $H_r = 0$. We know the general form of the electric field from equation (14.171); this and equation (14.197) suggest that we can introduce the electric field of the lmth multipole as

$$\mathbf{E}_{lm} = \nabla \times [\nabla \times (\mathbf{r}\eta_{lm})]. \quad (14.200)$$

We may use the expression

$$\nabla \times (\mathbf{r}\eta_{lm}) = \eta_{lm}\nabla \times \mathbf{r} - \mathbf{r} \times \nabla\eta_{lm} = -\mathbf{r} \times \nabla\eta_{lm} \qquad (14.201)$$

to write the electric field as

$$\mathbf{E}_{lm} = -\nabla \times (\mathbf{r} \times \nabla\eta_{lm}). \qquad (14.202)$$

We can now directly calculate the curls to find the transverse field components; the complete electric field components may be written as

$$E_r = \frac{l(l+1)}{r}\eta_{lm}, \qquad (14.203)$$

$$E_\theta = \frac{1}{r}\partial_r\partial_\theta(r\eta_{lm}), \qquad (14.204)$$

$$E_\phi = \frac{1}{r\sin\theta}\partial_r\partial_\phi(r\eta_{lm}). \qquad (14.205)$$

We may then determine the coefficients of $\mathbf{H}$ by noting that the magnetic field is of the form

$$\mathbf{H}_{lm} = -i\sqrt{\frac{\epsilon}{\mu}}k\mathbf{M}_u = i\omega\epsilon\,\mathbf{r} \times \nabla\eta_{lm}, \qquad (14.206)$$

and we can readily find that

$$H_r = 0, \qquad (14.207)$$

$$H_\theta = -\frac{i\omega\epsilon}{\sin\theta}\partial_\phi\eta_{lm}, \qquad (14.208)$$

$$H_\phi = i\omega\epsilon\,\partial_\theta\eta_{lm}. \qquad (14.209)$$

We may follow similar steps to determine the magnetic multipole fields from the expression for H_r and $E_r = 0$; we simply present the results here. First, we note that $\mathbf{H}_{lm}$ must be of the form

$$\mathbf{H}_{lm} = \nabla \times [\nabla \times (\mathbf{r}\eta_{lm})] = k\mathbf{N}_v, \qquad (14.210)$$

with a corresponding electric field of the form

$$\mathbf{E}_{lm} = i\sqrt{\frac{\mu}{\epsilon}}k\mathbf{M}_v = -i\omega\mu\mathbf{r} \times \nabla\eta_{lm}. \qquad (14.211)$$

We then follow the same approach as that used earlier for the electric multipoles and the magnetic dipoles. For $\mathbf{H}$, we have

$$H_r = \frac{l(l+1)}{r}\eta_{lm}, \qquad (14.212)$$

$$H_\theta = \frac{1}{r}\partial_r\partial_\theta(r\eta_{lm}), \tag{14.213}$$

$$H_\phi = \frac{1}{r\sin\theta}\partial_r\partial_\phi(r\eta_{lm}), \tag{14.214}$$

and for **E**, we have

$$E_r = 0, \tag{14.215}$$

$$E_\theta = \frac{i\omega\mu}{\sin\theta}\partial_\phi\eta_{lm}, \tag{14.216}$$

$$E_\phi = -i\omega\mu\partial_\theta\eta_{lm}. \tag{14.217}$$

Finally, we note that the expressions for a_{lm} and b_{lm} can be found from equations (14.195) and (14.196); the results are

$$a_{lm} = -\sqrt{\frac{\mu}{\epsilon}}\frac{1}{l(l+1)}\int_D \mathbf{J}(\mathbf{r}') \cdot \left\{\nabla' \times [\nabla' \times (\mathbf{r}'\psi_{lm}^*(\mathbf{r}'))]\right\}d^3r', \tag{14.218}$$

$$b_{lm} = \frac{ik}{l(l+1)}\int_D \mathbf{J}(\mathbf{r}') \cdot [\nabla' \times (\mathbf{r}'\psi_{lm}^*(\mathbf{r}'))]d^3r'. \tag{14.219}$$

In the far zone, we may write the fields of each multipole moment in an even more intuitive form. Let us introduce an angular momentum operator **L**, as follows:

$$\mathbf{L} \equiv \frac{1}{i}\mathbf{r} \times \nabla. \tag{14.220}$$

This operator will only include derivatives of θ and ϕ, because the cross product of two vectors is perpendicular to both.

We now consider the electric multipoles again. We may then write

$$\mathbf{E}_{lm} = i\nabla \times (\mathbf{L}\eta_{lm}), \tag{14.221}$$

$$\mathbf{H}_{lm} = -\omega\epsilon\mathbf{L}\eta_{lm}. \tag{14.222}$$

We now focus in particular on propagation in the far zone ($kr \to \infty$). In this limit, it can be shown that the asymptotic expression for the spherical Hankel function is of the form

$$h_l^{(1)}(kr) \sim (-i)^{l+1}\frac{e^{ikr}}{kr}. \tag{14.223}$$

We may apply the rule $\partial_r \leftrightarrow ik$, as we have done in previous far-zone calculations; we can then show that

$$\mathbf{E}_{lm} = -i(-i)^l\frac{e^{ikr}}{r}\hat{\mathbf{r}} \times (\mathbf{L}Y_{lm}), \tag{14.224}$$

$$\mathbf{H}_{lm} = i(-i)^l \sqrt{\frac{\epsilon}{\mu}} \frac{e^{ikr}}{r} \mathbf{L} Y_{lm}. \tag{14.225}$$

A similar approach for the magnetic multipoles gives the fields as

$$\mathbf{E}_{lm} = -i(-i)^l \sqrt{\frac{\mu}{\epsilon}} \frac{e^{ikr}}{r} \mathbf{L} Y_{lm}, \tag{14.226}$$

$$\mathbf{H}_{lm} = -i(-i)^l \frac{e^{ikr}}{r} \hat{\mathbf{r}} \times (\mathbf{L} Y_{lm}). \tag{14.227}$$

Thus, in principle, we can perform a complex expansion of a radiation field into multipoles based on spherical harmonics. The total electric field is thus of the form,

$$\mathbf{E}(\mathbf{r}) = \left\{ \sum_{l=0}^{\infty} \sum_{m=-l}^{l} \left[a_{lm} \mathbf{E}_{lm}^{(e)}(\mathbf{r}) + b_{lm} \mathbf{E}_{lm}^{(m)}(\mathbf{r}) \right] \right\}. \tag{14.228}$$

It should be noted that this expansion is distinct from the Taylor series-based expansion we used in earlier sections, as the coefficients a_{lm} and b_{lm} are integrals over spherical Bessel functions, where the spherical Bessel functions are nowhere to be found in, for example, equations (14.92) and (14.99) for the electric and magnetic dipole moments derived from the Taylor series expansion. This is the result of implicitly adopting a long-wavelength approximation in the earlier derivation.

The situation can be considered analogous to the observation that a well-behaved function $f(x)$ on the interval $-1 \leqslant x \leqslant 1$ can be written in terms of its Taylor series expansion, but it can also be written using a variety of distinct orthogonal polynomial sets, such as the Chebyshev polynomials and the Legendre polynomials. The first few terms of the Taylor series expansion will only be an accurate representation near the origin of the expansion, while the same number of terms of the polynomial expansion gives roughly the same accuracy over the entire domain.

We can see that the Taylor series and Bessel series expansions qualitatively give the same behaviors for the electric dipole terms. We first note that, according to the Taylor series calculation, the electric field of an electric dipole in the far zone is given by equation (14.131). Let us consider an electric dipole aligned along the $\hat{\mathbf{z}}$-axis, i.e. $\mathbf{p} = p_0 \hat{\mathbf{z}}$; the electric field then becomes

$$\mathbf{E}_{ed}(\mathbf{r}) = -\frac{k^2 p_0}{\epsilon_0} \frac{e^{ikr}}{r} \sin\theta \hat{\boldsymbol{\theta}}. \tag{14.229}$$

Let us turn to the Bessel series multipoles, which feature the spherical harmonics; a list of the lowest-order spherical harmonics is given in table 14.1 for convenience.

It should be clear that the three $l=1$ spherical harmonics are related, albeit in a nontrivial way, to the three independent components of the electric dipole moment.

Table 14.1. Table of the lowest-order spherical harmonics.

$m\backslash l$	0	1	2
-2			$\frac{1}{4}\sqrt{\frac{15}{2\pi}}\sin^2\theta e^{-2i\phi}$
-1		$\frac{1}{2}\sqrt{\frac{3}{2\pi}}\sin\theta e^{-i\phi}$	$\frac{1}{4}\sqrt{\frac{15}{2\pi}}\cos\theta\sin\theta e^{-i\phi}$
0	$\frac{1}{2}\sqrt{\frac{1}{\pi}}$	$\frac{1}{2}\sqrt{\frac{3}{\pi}}\cos\theta$	$\frac{1}{4}\sqrt{\frac{5}{\pi}}(3\cos^2\theta - 1)$
1		$-\frac{1}{2}\sqrt{\frac{3}{2\pi}}\sin\theta e^{i\phi}$	$-\frac{1}{4}\sqrt{\frac{15}{2\pi}}\cos\theta\sin\theta e^{i\phi}$
2			$\frac{1}{4}\sqrt{\frac{15}{2\pi}}\sin^2\theta e^{2i\phi}$

For our choice $\mathbf{p} = p_0\hat{\mathbf{z}}$, the system is independent of the angle ϕ and thus we expect that only the Y_{10} spherical harmonic contributes to the field. The electric field of the electric dipole should thus be of the form

$$\mathbf{E}_{10}(\mathbf{r}) = a_{10}\nabla \times [\nabla \times (\mathbf{r}\eta_{10}(\mathbf{r}))]. \tag{14.230}$$

These curls can be calculated relatively easily, and the result is of the form

$$\mathbf{E}_{10}(\mathbf{r}) = a_{10}\left\{ \hat{\mathbf{r}}\frac{l(l+1)}{r}h_1(kr)Y_{10}(\theta,\phi) + \hat{\boldsymbol{\theta}}\frac{\partial_r[rh_1(kr)]}{r}\partial_\theta Y_{10}(\theta,\phi) + \hat{\boldsymbol{\phi}}\frac{\partial_r[rh_1(kr)]}{r\sin\theta}\partial_\phi Y_{10}(\theta,\phi) \right\}. \tag{14.231}$$

We now turn to the far zone, and immediately note that the $\hat{\mathbf{r}}$ component of the field is negligible compared to the others. We also note that $\partial_\phi Y_{10} = 0$ because this spherical harmonic has no ϕ-dependence. If we substitute the expression for the spherical harmonic into the remaining $\hat{\boldsymbol{\theta}}$ term, we obtain

$$\mathbf{E}_{10}(\mathbf{r}) = a_{10}\frac{i}{2}\sqrt{\frac{3}{\pi}}\frac{e^{ikr}}{r}\sin\theta\hat{\boldsymbol{\theta}}. \tag{14.232}$$

A comparison of this equation and equation (14.229) shows that the functional dependence is the same in both cases, showing that the $l = 1$ terms indeed relate to the dipole contribution to the field. A similar calculation can be done to show that the magnetic fields in the two cases have the same functional dependence.

14.8 Gauge transformations

To conclude the chapter, we return to the concept of gauge freedom and gauge transformation: the change from one gauge to another. We have argued, for example, that we have the freedom to require that the vector and scalar potentials satisfy the Lorenz gauge,

$$\nabla \cdot \mathbf{A} + \mu_0\epsilon_0\frac{\partial\phi}{\partial t} = 0. \tag{14.233}$$

But why do we have this freedom, and how can we determine whether a gauge is valid or not? To explore the first question, we note that the fields are defined as the derivatives of potentials,

$$\mathbf{H}(\mathbf{r},\ t) = \frac{1}{\mu_0}\nabla \times \mathbf{A}(\mathbf{r},\ t), \tag{14.234}$$

$$\mathbf{E}(\mathbf{r},\ t) = -\nabla\phi(\mathbf{r},\ t) - \frac{\partial \mathbf{A}(\mathbf{r},\ t)}{\partial t}, \tag{14.235}$$

which indicates that, at the very least, the potentials can be redefined up to a constant. But we can go even further than this: suppose that $\mathbf{A}$ and ϕ represent valid potentials that produce the desired electric and magnetic fields, and let

$$\mathbf{A}'(\mathbf{r},\ t) = \mathbf{A}(\mathbf{r},\ t) + \nabla\Lambda(\mathbf{r},\ t), \tag{14.236}$$

$$\phi'(\mathbf{r},\ t) = \phi(\mathbf{r},\ t) - \frac{\partial\Lambda(\mathbf{r},\ t)}{\partial t}, \tag{14.237}$$

where $\Lambda(\mathbf{r},\ t)$ is a scalar field. It is immediately obvious that the magnetic field is the same for these potentials, as $\nabla \times \nabla\Lambda = 0$. If we evaluate the electric field for the new potentials, we find

$$\mathbf{E}'(\mathbf{r},\ t) = -\nabla\phi(\mathbf{r},\ t) + \nabla\frac{\partial\Lambda(\mathbf{r},\ t)}{\partial t} - \frac{\partial\mathbf{A}(\mathbf{r},\ t)}{\partial t} - \frac{\partial}{\partial t}\nabla\Lambda(\mathbf{r},\ t) = \mathbf{E}(\mathbf{r},\ t), \tag{14.238}$$

where the last step comes from the assumption that the spatial and temporal derivatives of $\Lambda(\mathbf{r},\ t)$ can be interchanged, which is a reasonable assumption provided $\Lambda(\mathbf{r},\ t)$ is taken to be a well-behaved function.

The nontrivial change of the vector and scalar potentials according to equations (14.236) and (14.237) is what we call a gauge transformation. In a gauge transformation, the potentials are chosen to satisfy an auxiliary condition in order to simplify the calculation process. We have seen two of the most familiar gauge choices: the Lorenz gauge given by equation (14.26) and the Coulomb gauge that satisfies

$$\nabla \cdot \mathbf{A}(\mathbf{r},\ t) = 0. \tag{14.239}$$

There are many other gauges that have been used in various situations to simplify electromagnetic calculations; we refer to Jackson and Okun [7] for a detailed historical overview.

The other important question to answer about a particular gauge is whether the gauge condition is valid, i.e. is it consistent with Maxwell's equations, and can it be described by well-behaved functions? We consider this question in the context of the Lorenz gauge. Let us suppose that we have already found $\mathbf{A}(\mathbf{r},\ t)$ and $\phi(\mathbf{r},\ t)$ for a particular problem but that these potentials *do not* satisfy the Lorenz gauge condition. We now introduce new potentials $\mathbf{A}'(\mathbf{r},\ t)$ and $\phi'(\mathbf{r},\ t)$ as in equations (14.236) and (14.237) and require them to satisfy

$$\nabla \cdot \mathbf{A}' + \mu_0 \epsilon_0 \frac{\partial \phi'}{\partial t} = 0. \qquad (14.240)$$

The question now becomes: what sort of condition must the function $\Lambda(\mathbf{r}, t)$ satisfy in order to satisfy the Lorenz gauge? We substitute our new potentials into equation (14.240) and rearrange the resulting terms as

$$\nabla^2 \Lambda - \mu_0 \epsilon_0 \frac{\partial^2 \Lambda}{\partial t^2} = -\nabla \cdot \mathbf{A} - \mu_0 \epsilon_0 \frac{\partial \phi}{\partial t}. \qquad (14.241)$$

This equation suggests that $\Lambda(\mathbf{r}, t)$ must satisfy a wave equation with a source term, where the source term is dictated by the original potentials $\mathbf{A}(\mathbf{r}, t)$ and $\phi(\mathbf{r}, t)$. This is, in principle, possible, which indicates that the Lorenz gauge condition is achievable. It should be noted that if $\mathbf{A}(\mathbf{r}, t)$ and $\phi(\mathbf{r}, t)$ already satisfy the Lorenz gauge, then the solution for Λ is $\Lambda = 0$.

It is important to note, however, that we almost never actually calculate the function $\Lambda(\mathbf{r}, t)$. Once we are aware that a gauge is achievable, we simply use the gauge condition directly to solve a particular problem, as we did in the case of electromagnetic radiation.

The earliest application of gauge theory was in electromagnetism, but it has been found to be important in the study of all fundamental particle interactions, as well as hypothetical 'theories of everything' such as string theory. We refer the curious reader to the review by O'Raifeartaigh and Straumann [8].

14.9 The Aharonov–Bohm experiments

Throughout this chapter, and back in chapter 5, we have treated the potentials $\mathbf{A}$ and ϕ entirely as calculational tools without any physical significance of their own. Indeed, the fact that the potentials are themselves nonunique would seem to indicate that only the electric and magnetic fields have any measurable effect on the behavior of charged particles. We would be remiss if we did not mention that this is evidently untrue when quantum physics is taken into account. In 1959, Aharonov and Bohm [9] argued that there are situations where a potential can influence the behavior of a quantum particle, even when there are no fields present. This result has been confirmed in experiments and has led to a longstanding debate over what, exactly, the so-called Aharonov–Bohm (AB) effect means.

Though we will not delve deeply into the quantum aspects of this phenomenon, we introduce a few key concepts to provide context. For an electron propagating in a classical magnetic field $\mathbf{B}(\mathbf{r}, t)$ and an electric field $\mathbf{E}(\mathbf{r}, t)$, the Hamiltonian H is given by

$$H = \frac{[\mathbf{P} - q\mathbf{A}(\mathbf{r}, t)]^2}{2m} + q\phi(\mathbf{r}, t), \qquad (14.242)$$

where $\mathbf{P}$ is the momentum operator, q is the charge of the electron, and m is its mass. The function $\mathbf{A}(\mathbf{r}, t)$ is the vector potential associated with the fields, and $\phi(\mathbf{r}, t)$ is the scalar potential. This Hamiltonian is discussed and justified in numerous textbooks, such as Grynberg, Aspect, and Fabre [10]. The Hamiltonian determines the dynamics of the electron, and it is striking that it depends upon the potentials rather than the fields. Until Aharonov and Bohm came along, however, it was more or less assumed that any observable results could be traced back in some way to the fields themselves rather than depending explicitly on the potentials.

However, let us calculate the vector potential of an infinite solenoid of radius a with N turns per unit length and a constant current I, whose magnetic field we calculated back in section 2.4.3. We may do so by taking advantage of a method analogous to Ampère's law. We note that

$$\oint_C \mathbf{A}(\mathbf{r}) \cdot d\mathbf{l} = \int_S [\nabla \times \mathbf{A}(\mathbf{r})] \cdot d\mathbf{a} = \int_S \mathbf{B} \cdot d\mathbf{a}. \tag{14.243}$$

If we choose a circular loop centered on the z-axis in the $+\hat{\phi}$-direction as our integration path, we can use our knowledge of the $\mathbf{B}$-field to determine $\mathbf{A}$. We recall that the electric field is $\mathbf{B} = \mu_0 N I \hat{\mathbf{z}}$ inside the solenoid and $\mathbf{B} = 0$ outside. We may then readily determine that the vector potential has the form

$$\mathbf{A}(\mathbf{r}) = \begin{cases} \dfrac{\mu_0 N I r}{2} \hat{\phi}, & r < a, \\ \dfrac{\mu_0 N I a^2}{2r} \hat{\phi}, & r > a. \end{cases} \tag{14.244}$$

Here, we have the unusual result that the magnetic field is zero outside the solenoid but the vector potential is nonzero. We can readily demonstrate that $\nabla \times \mathbf{A} = 0$ outside the solenoid to show that this result is not inconsistent with the math.

Now consider the system shown in figure 14.5. A beam of coherent electrons is incident upon a Young's double-slit interferometer, and the diffracted beams passing through the slits form an interference pattern on an observation screen some distance away. We now imagine placing an infinite solenoid vertically (out of the

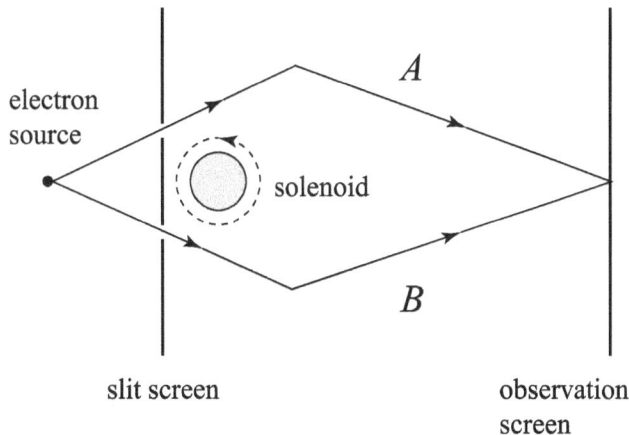

Figure 14.5. Thought experiment used to demonstrate the Aharonov–Bohm effect.

page) between the two slits, in the shadow of the electron beam. If the system is designed carefully, the electrons pass through a region of nonzero vector potential but do not penetrate into the solenoid itself and therefore do not experience a magnetic field directly.

The electrons passing through the two slits experience vector potentials that are effectively opposite to each other, due to the $\hat{\phi}$ dependence of the potential. In quantum physics, the phase accumulated by the electron wave function in traversing a path through a vector potential is given by

$$\varphi = -\frac{q}{h} \int_P \mathbf{A}(\mathbf{r}) \cdot d\mathbf{l}, \tag{14.245}$$

where P symbolizes the path and h is Planck's constant. We label the paths above and below the solenoid as A and B; then, according to equation (14.243), we may write

$$\varphi_{A-B} = -\frac{q}{h} \int_{A-B} \mathbf{A}(\mathbf{r}) \cdot d\mathbf{l} = \frac{q\mu_0 NI}{h}, \tag{14.246}$$

where '$A - B$' denotes the closed path that follows A and then the opposite of B. However, φ_{A-B} is simply the difference in phase $\Delta\varphi$ that the electron wave function experiences on passing on either side of the solenoid, so that

$$\Delta\varphi = \frac{q\mu_0 NI}{h}. \tag{14.247}$$

A phase difference in the wave function emanating from the two slits causes a shift in the interference pattern. As we turn on the current in the solenoid, we expect to see the interference fringes in Young's experiment shift on the screen along the line parallel to the one connecting the two slits. We have a measurable experimental effect that apparently only arises from the electron's interaction with the vector potential.

The experimental setup shown in figure 14.5 is for illustrative purposes only. In their original paper, Aharonov and Bohm proposed the use of electron diffraction to create parallel beams of coherent electrons that could be passed around a thin solenoid and then recombined; such electron interference experiments had already been demonstrated years earlier [11]. The first experimental test of the AB effect was reported by Chambers in 1960 [12], and this led to the first big debate about the effect. Chambers used a magnetized iron whisker as his solenoid that possessed strong fringe magnetic fields; he argued that the fringe fields produced an effect distinct from the AB effect and that one could be distinguished from the other. Other researchers, however, claimed the fringe fields were mimicking the AB effect, and some even went so far as to state that the AB effect does not exist. An additional argument against the early AB experiments was that it was unclear whether the electrons were blocked from the interior of the solenoid or its experimental equivalent: any interaction between the electron wave function and the magnetic field could invalidate the interpretation in terms of potentials.

These debates appear to have been largely resolved in Aharonov and Bohm's favor in the 1986 experiment of Tonomura *et al* [13]. They used a toroidal ferromagnet as their magnetic source—it acted effectively as a solenoid bent to match its two open ends together. To prevent magnetic fields from escaping the toroid, they coated it with a superconducting film and followed this with a copper coating to prevent electrons from penetrating the interior. For this experiment, the 'left' and 'right' sides of the infinite straight solenoid are replaced by the 'inside' and 'outside' of the toroid. Below the superconducting critical temperature, the magnetic flux is quantized, which results in a quantized phase difference that is an integer multiple of π; one experimental result is shown in figure 14.6.

Another argument continues, however: exactly what is the Aharonov–Bohm effect telling us about physics? The biggest debate is whether the AB effect demonstrates the physical reality of potentials or not, though authors such as Vaidman [15] have argued that the effect can be demonstrated without the use of potentials at all by considering the influence of the electron field on the source of the vector potential. For my part, I can only borrow a quote from the late, great Douglas Adams [16]: 'It was hard to avoid the feeling that somebody, somewhere, was missing the point. I couldn't even be sure it wasn't me.'

Aharonov and Bohm also predicted an electric version of their effect, where an electron wave propagates through regions of different scalar potentials but zero electric field. Though attempts have been made, an experimental test of their original idea has yet to be verified.

There is also a parallel phenomenon to the AB effect, known as the Aharonov–Casher effect, introduced in 1984 [17]. In this effect, the quantum phase of an electrically neutral particle with a magnetic dipole can be manipulated when it passes by an electric line charge.

There is much more that can be said about the AB effect; we refer the reader to the review by Batelaan and Tonomura [14]. For our purposes, it is a reminder that the role of potentials in electromagnetics is much more subtle than it first appears.

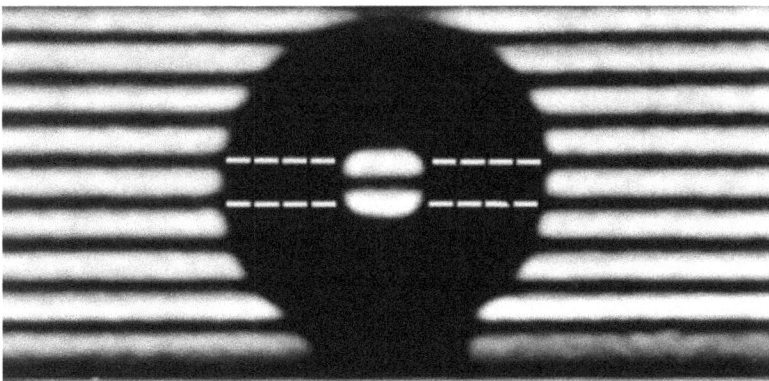

Figure 14.6. Interference fringes obtained during the experiment of Tonomura *et al*. The dashed lines illustrate that the interference pattern viewed through the hole of the toroid is π out of phase with the interference pattern viewed outside the toroid. Reprinted from [14], with the permission of AIP Publishing.

14.10 Exercises

1. Write the following matrix **M** in dyadic notation, i.e. in terms of unit vectors:

$$\mathbf{M} = \begin{bmatrix} 5 & 2 & 0 \\ 0 & 1 & 3 \\ 4 & 6 & 0 \end{bmatrix}.$$

2. Write the following dyadic **N** in matrix form:

$$\mathbf{N} = \hat{\mathbf{x}}\hat{\mathbf{x}} + 4\hat{\mathbf{x}}\hat{\mathbf{y}} + 3\hat{\mathbf{x}}\hat{\mathbf{z}} + 2\hat{\mathbf{y}}\hat{\mathbf{y}} + 5\hat{\mathbf{y}}\hat{\mathbf{z}} + 6\hat{\mathbf{z}}\hat{\mathbf{x}}.$$

3. Figure 14.7 illustrates two distributions of point charges that oscillate with a harmonic time dependence, i.e. $Q = Qe^{-i\omega t}$. Calculate the electric dipole and quadrupole moments of the distributions, and sketch the leading contribution to the radiation pattern. The point charges may be taken to have a delta function spatial density, i.e. $Q(\mathbf{r}) = q_0\delta^{(3)}(\mathbf{r} - \mathbf{r}_0)$ for a charge q_0 at $\mathbf{r}_0$.

4. Let us suppose we have a time-harmonic source which has the form $\mathbf{J} = J_0\hat{\mathbf{z}}$ within a sphere of radius a centered on the origin and is zero elsewhere. By using the far-zone form of the vector potential, with the far-zone form of the Green's function,

$$\frac{e^{ikR}}{R} \approx \frac{e^{ikr}}{r}e^{-ik\hat{\mathbf{r}}\cdot\mathbf{r}'},$$

show that radii a exist such that the source does not radiate at all. Give the first three values of these radii as a function of the wavenumber k. (Hint: You will need to look up and use the integral form of the spherical Bessel function $j_0(x)$.)

5. The other useful gauge for electromagnetic problems is the Coulomb gauge, for which $\nabla \cdot \mathbf{A} = 0$, though the derivation is more subtle. Derive integral relations for $\phi(\mathbf{r}, t)$ and $\mathbf{A}(\mathbf{r}, t)$ through the following steps:

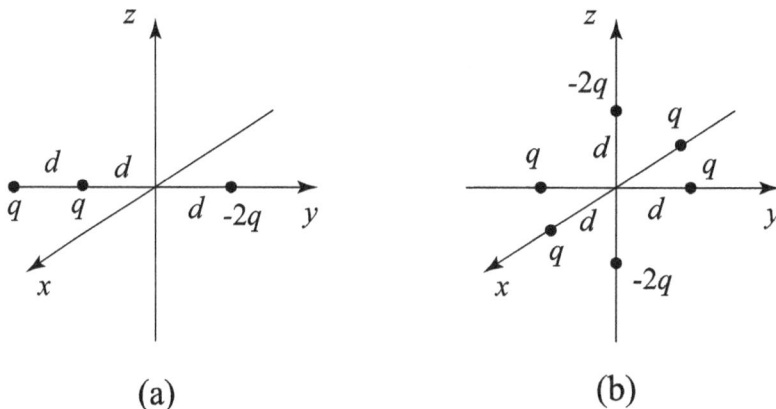

(a)   (b)

Figure 14.7. Arrangement of point charges for time-harmonic point charge distributions.

(a) Insert the potential definition for $\mathbf{E}(\mathbf{r}, t)$ into Gauss's law to get a differential equation for $\phi(\mathbf{r}, t)$, and write the integral solution of that equation.

(b) Insert the definitions of the potentials into the Ampère–Maxwell law to get a differential equation for $\mathbf{A}(\mathbf{r}, t)$ that also includes $\phi(\mathbf{r}, t)$.

(c) Using the integral,

$$\mathbf{K}(\mathbf{r}) = \int \frac{\mathbf{J}(\mathbf{r}')}{|\mathbf{r} - \mathbf{r}'|} d^3 r',$$

and the vector identity $\nabla \times (\nabla \times \mathbf{K}) = \nabla(\nabla \cdot \mathbf{K}) - \nabla^2 \mathbf{K}$, show that the current density $\mathbf{J}(\mathbf{r})$ can be decomposed into a transverse part and a longitudinal part, $\mathbf{J} = \mathbf{J}_t + \mathbf{J}_l$, where the transverse part has zero divergence and the longitudinal part has zero curl. (You will need the identity $\nabla^2(1/|\mathbf{r} - \mathbf{r}'|) = \delta^{(3)}(\mathbf{r} - \mathbf{r}')$.)

(d) Using these results, and considering the monochromatic case, write an integral expression for the vector potential in terms of the transverse current.

6. Calculate the electric dipole, electric quadrupole, and magnetic dipole moments for the following current distributions:

(a) A straight thin wire of length L aligned along the z-axis and centered on the origin, with $\mathbf{J}(\mathbf{r}) = I_0 \delta(x)\delta(y)\hat{\mathbf{z}}$.

(b) A thin wire loop of radius a lying in the xy-plane and centered on the origin, with $\mathbf{J}(\mathbf{r}) = I_0 \delta(r - a)\delta(z)\hat{\boldsymbol{\phi}}$.

(c) Two thin straight wires of length L aligned along the z-axis, one centered on $x = +a$ and the other on $x = -a$, with $\mathbf{J}(\mathbf{r}) = I_0 \delta(x - a)\delta(y)\hat{\mathbf{z}} - I_0 \delta(x + a)\delta(y)\hat{\mathbf{z}}$.

7. For very simple problems, we can calculate the radiation pattern exactly, without multipoles. Consider a simple line antenna, with current density

$$\mathbf{J}(\mathbf{r}) = I_0 \delta(x)\delta(y)\hat{\mathbf{z}}, \quad -L/2 \leqslant z \leqslant L/2.$$

Calculate the radiation pattern of the field, i.e. the quantity

$$R \equiv \frac{1}{\lambda^2} \left| \hat{\mathbf{r}} \times \int \mathbf{J}(\mathbf{r}') e^{-ik\hat{\mathbf{r}}\cdot\mathbf{r}'} d^3 r' \right|^2,$$

and describe the behavior of this pattern as a function of the polar angle θ for the cases $l \ll \lambda$, $l < \lambda$, and $\lambda < l < 2\lambda$.

References

[1] Wolf E 2007 *Introduction to the Theory of Coherence and Polarization of Light* (Cambridge: Cambridge University Press)
[2] Bohr N 1913 On the constitution of atoms and molecules *Phil. Mag.* **26** 1–25
[3] Papas C H 1988 *Theory of Electromagnetic Wave Propagation* (New York: Dover)
[4] Debye P 1909 Der Lichtdruck auf Kugeln von beliebigem Material *Ann. Phys., Lpz.* **335** 57–136

[5] Wilcox C H 1957 Debye potentials *J. Math. Mech.* **6** 167–201

[6] Gbur G J 2011 *Mathematical Methods for Optical Physics and Engineering* (Cambridge: Cambridge University Press)

[7] Jackson J D and Okun L B 2001 Historical roots of gauge invariance *Rev. Mod. Phys.* **73** 663–80

[8] O'Raifeartaigh L and Straumann N 2000 Gauge theory: historical origins and some modern developments *Rev. Mod. Phys.* **72** 1–23

[9] Aharonov Y and Bohm D 1959 Significance of electromagnetic potentials in the quantum theory *Phys. Rev.* **115** 485–91

[10] Grynberg G, Aspect A and Fabre C 2010 *Introduction to Quantum Optics* (Cambridge: Cambridge University Press)

[11] Marton L 1952 Electron interferometer *Phys. Rev.* **85** 1057–8

[12] Chambers R G 1960 Shift of an electron interference pattern by enclosed magnetic flux *Phys. Rev. Lett.* **5** 3–5

[13] Tonomura A *et al* 1986 Evidence for Aharonov-Bohm effect with magnetic field completely shielded from electron wave *Phys. Rev. Lett.* **56** 792–5

[14] Batelaan H and Tonomura A 2009 The Aharonov-Bohm effects: variations on a subtle theme *Phys. Today* **62** 38–43

[15] Vaidman L 2012 Role of potentials in the Aharonov-Bohm effect *Phys. Rev. A* **86** 040101

[16] Adams D and Carwardine M 1990 *Last Chance to See* (New York: Ballantine Books)

[17] Aharonov Y and Casher A 1984 Topological quantum effects for neutral particles *Phys. Rev. Lett.* **53** 319–21

IOP Publishing

Electromagnetic Optics

Gregory J Gbur

Chapter 15

Electromagnetic scattering

When light changes direction as a result of moving from one homogeneous medium of constant optical properties to another, we typically refer to such a change of direction as refraction. When light changes direction while traveling in a medium with inhomogeneous optical properties, i.e. permittivity and/or permeability that depend on position, we refer to it as *scattering*. Alternatively, we may say that scattering refers to the distortion of a light wave by spatial inhomogeneities that cannot be characterized by the law of refraction. What constitutes scattering versus refraction is often a matter of the scale of the wavelength. In early attempts to understand the nature of x-rays, researchers were puzzled by the fact that x-rays reflected diffusely from a flat metal surface, rather than specularly according to the law of reflection. This turned out to be due to the fact that x-rays have a significantly smaller wavelength and thus are sensitive to smaller surface irregularities and even to the spacing of atoms.

Scattering theory represents one of the most difficult mathematical topics in optics, and scattering problems cannot be solved analytically in most cases. Approximations and special cases are often considered in order to derive some intuition for the physics and to make the calculations tractable. In this chapter, we introduce some of the general formalism of electromagnetic scattering theory and present some of the most important special cases.

Before we delve into the mathematics, however, we note that there is a close relationship between scattering theory and the radiation theory we discussed in the previous chapter. This relationship is illustrated in figure 15.1. We may roughly view the scattering problem as follows: 1. an electromagnetic wave is incident upon an object with inhomogeneous optical properties; 2. the incident wave excites oscillating currents in the object; 3. the oscillating currents produce their own electromagnetic radiation, which we call the scattered field. When dealing with a radiation problem, we start at step 2, with the assumption of oscillating currents, and calculate the radiation of those currents in step 3.

doi:10.1088/978-0-7503-6064-7ch15

Radiation problem

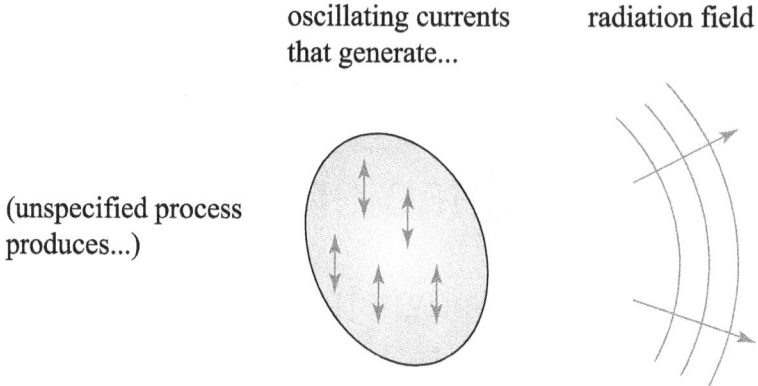

oscillating currents   radiation field
that generate...

(unspecified process
produces...)

Scattering problem

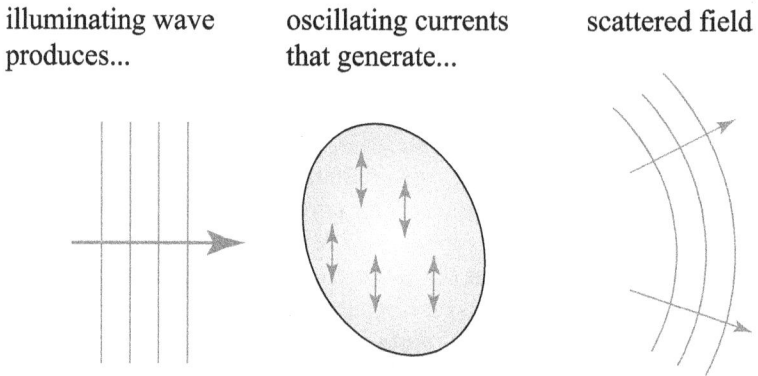

illuminating wave   oscillating currents   scattered field
produces...   that generate...

Figure 15.1. Illustration of the conceptual relationship between radiation theory and scattering theory.

In a sense, the scattering problem may be considered a special case of the radiation problem. In a radiation problem, we simply assume that a set of oscillating currents exists, without asking how they were set into motion. In a scattering problem, we assume that the currents were set into motion by an external field interacting with an inhomogeneous object. In general, determining the form of those induced currents will turn out to be the key challenge in scattering theory.

15.1 The electromagnetic Green's dyadics

In order to properly determine how an electromagnetic wave is scattered from an inhomogeneous object, we will need to be able to calculate the fields inside and outside the domain D of the object. The best way to solve this problem is to first determine the free-space Green's dyadic, which represents the fields produced by a point dipole.

For the first time, we use Maxwell's equations in their most general mono-chromatic form,

$$\nabla \cdot \mathbf{D}(\mathbf{r}) = \rho(\mathbf{r}), \tag{15.1}$$

$$\nabla \cdot \mathbf{B}(\mathbf{r}) = 0, \tag{15.2}$$

$$\nabla \times \mathbf{E}(\mathbf{r}) = i\omega \mathbf{B}(\mathbf{r}), \tag{15.3}$$

$$\nabla \times \mathbf{H}(\mathbf{r}) = \mathbf{J}(\mathbf{r}) - i\omega \mathbf{D}(\mathbf{r}). \tag{15.4}$$

If we use

$$\mathbf{D}(\mathbf{r}) = \varepsilon_0 \mathbf{E}(\mathbf{r}) + \mathbf{P}(\mathbf{r}), \tag{15.5}$$

$$\mathbf{B}(\mathbf{r}) = \mu_0[\mathbf{H}(\mathbf{r}) + \mathbf{M}(\mathbf{r})], \tag{15.6}$$

we may instead write Maxwell's equations in terms of $\mathbf{E}$ and $\mathbf{H}$ as

$$\nabla \cdot \mathbf{E}(\mathbf{r}) = \frac{1}{\varepsilon_0}[\rho(\mathbf{r}) - \nabla \cdot \mathbf{P}(\mathbf{r})], \tag{15.7}$$

$$\nabla \cdot \mathbf{H}(\mathbf{r}) = -\nabla \cdot \mathbf{M}(\mathbf{r}), \tag{15.8}$$

$$\nabla \times \mathbf{E}(\mathbf{r}) - i\omega\mu_0 \mathbf{H}(\mathbf{r}) = i\omega\mu_0 \mathbf{M}(\mathbf{r}), \tag{15.9}$$

$$\nabla \times \mathbf{H}(\mathbf{r}) + i\omega\varepsilon_0 \mathbf{E}(\mathbf{r}) = \mathbf{J}(\mathbf{r}) - i\omega \mathbf{P}(\mathbf{r}). \tag{15.10}$$

We now have a striking symmetry between the $\mathbf{E}$ and $\mathbf{H}$ fields, which we may fully reveal if we define

$$\rho_e \equiv \rho - \nabla \cdot \mathbf{P}, \quad \mathbf{J}_e = \mathbf{J} - i\omega \mathbf{P}, \tag{15.11}$$

$$\rho_m \equiv -\mu_0 \nabla \cdot \mathbf{M}, \quad \mathbf{J}_m = -i\omega\mu_0 \mathbf{M}. \tag{15.12}$$

These definitions are related to the ones used in section 14.4 to define the magnetic dipole moment. We stress that ρ_m and J_m are not real magnetic charges and currents but objects of mathematical convenience; we still have $\nabla \cdot \mathbf{B} = 0$, even though $\nabla \cdot \mathbf{H} \neq 0$.

Maxwell's equations may now be written as

$$\nabla \cdot \mathbf{E}(\mathbf{r}) = \rho_e(\mathbf{r})/\varepsilon_0, \tag{15.13}$$

$$\nabla \cdot \mathbf{H}(\mathbf{r}) = \rho_m(\mathbf{r})/\mu_0, \tag{15.14}$$

$$\nabla \times \mathbf{E}(\mathbf{r}) - i\omega\mu_0 \mathbf{H}(\mathbf{r}) = -\mathbf{J}_m(\mathbf{r}), \tag{15.15}$$

$$\nabla \times \mathbf{H}(\mathbf{r}) + i\omega\varepsilon_0 \mathbf{E}(\mathbf{r}) = \mathbf{J}_e(\mathbf{r}). \tag{15.16}$$

The magnetic charges and currents are due to the magnetic properties of matter, if any, e.g. a nontrivial permeability $\mu(\mathbf{r})$. Let us ignore those properties for the moment.

Let us follow our usual strategy for deriving a wave equation: we take the curl of Faraday's law and use the Ampère–Maxwell law to write everything in terms of the electric field,

$$\nabla \times (\nabla \times \mathbf{E}) = i\omega\mu_0 \nabla \times \mathbf{H} = k^2\mathbf{E} + i\omega\mu_0\mathbf{J}_e. \tag{15.17}$$

We thus have the equation for $\mathbf{E}$,

$$\nabla \times (\nabla \times \mathbf{E}) - k^2\mathbf{E} = i\omega\mu_0\mathbf{J}_e. \tag{15.18}$$

This equation shows that there is a linear relationship between the electric field $\mathbf{E}$ and the electric current $\mathbf{J}_e$ and indicates that there must be a linear integral equation relating the two, of the form

$$\mathbf{E}(\mathbf{r}) = \int_D \boldsymbol{\Gamma}_e(\mathbf{r}, \mathbf{r}') \cdot \mathbf{J}_e(\mathbf{r}')d^3r', \tag{15.19}$$

where $\boldsymbol{\Gamma}_e$ is the *electric Green's dyadic*. Our goal is now to find the form of this dyadic.

You might wonder about the difference between this Green's dyadic and the scalar Green's function that we used in section 14.1. In the previous chapter, we derived a scalar Green's function that allowed us to calculate the potentials from the electric current via an integral. Once we had the potentials, we could perform a second mathematical step to derive the fields. Here, we are looking for a dyadic Green's function that will allow us to directly calculate the electric and magnetic fields from the electric current via an integral. This will lead to an additional complication, however, as we shall soon see.

To determine our Green's dyadic, we begin by substituting from equation (15.19) into equation (15.18). We also write the electric current in integral form using the identity dyadic $\mathbf{I}$ and the Dirac delta function,

$$\mathbf{J}_e(\mathbf{r}) = \int_D \delta^{(3)}(\mathbf{r} - \mathbf{r}')\mathbf{I} \cdot \mathbf{J}_e(\mathbf{r}')d^3r'. \tag{15.20}$$

We then have

$$\nabla \times \left[\nabla \times \int_D \boldsymbol{\Gamma}_e(\mathbf{r}, \mathbf{r}') \cdot \mathbf{J}_e(\mathbf{r}')d^3r'\right] - k^2\int_D \boldsymbol{\Gamma}_e(\mathbf{r}, \mathbf{r}') \cdot \mathbf{J}_e(\mathbf{r}')d^3r' = i\omega\mu_0\int_D \delta^{(3)}(\mathbf{r} - \mathbf{r}')\mathbf{I} \cdot \mathbf{J}_e(\mathbf{r}')d^3r'. \tag{15.21}$$

We would like to have three integrals of the same form, which will allow us to argue that the integrands must satisfy an equation independently. To do this, we must bring the double cross product of the first term inside the integral. At first glance, this seems quite reasonable, which leads us to the expression

$$\int_D \{\nabla \times [\nabla \times \boldsymbol{\Gamma}_e(\mathbf{r}, \mathbf{r}') \cdot \mathbf{J}_e(\mathbf{r}')] - k^2\boldsymbol{\Gamma}_e(\mathbf{r}, \mathbf{r}') \cdot \mathbf{J}_e(\mathbf{r}') - i\omega\mu_0\delta^{(3)}(\mathbf{r} - \mathbf{r}')\mathbf{I} \cdot \mathbf{J}_e(\mathbf{r}')\}d^3r' = 0. \tag{15.22}$$

From this, we can make the argument that we must have

$$\nabla \times [\nabla \times \boldsymbol{\Gamma}_e(\mathbf{r}, \mathbf{r}')] - k^2\boldsymbol{\Gamma}_e(\mathbf{r}, \mathbf{r}') - i\omega\mu_0\delta^{(3)}(\mathbf{r} - \mathbf{r}')\mathbf{I} = 0, \tag{15.23}$$

since our integral is evidently satisfied for any choice of domain D and current $\mathbf{J}$.

However, there is a significant problem with our interchange of the order of integration and differentiation. Keep in mind that we will physically identify the Green's dyadic with the field of a point dipole. From the previous chapter, we expect the electric field of a dipole to decay as $1/R$. This singularity in the dyadic is integrated out because the volume element in spherical coordinates disappears when $d^3r = r^2dr \sin\theta d\theta d\phi$, leaving the result of the integral in equation (15.21) well-behaved and differentiable. When we bring the double curl inside the integral, however, we are now taking two r derivatives of $1/R$, which creates a singularity of the form $1/R^3$. This singular behavior does not cancel out in the integral, and if we attempt to calculate the field at a point $\mathbf{r}$ within the volume D, the integral diverges. Our interchange of integration and differentiation is only valid, then, for points $\mathbf{r}$ not belonging to D, and the resulting Green's dyadic is only valid under the same conditions.

All is not lost inside the domain D, however; it turns out that in computational problems, there are methods to fix the singularity of the Green's dyadic in order to calculate the field inside and outside the domain. We will discuss those corrections in the next chapter; here, we simply continue our calculation for the dyadic and assume that there is a method for making it valid for all values of $\mathbf{r}$.

Returning to equation (15.23), we expand the double curl in the usual way and write

$$-\nabla^2\mathbf{\Gamma}_e(\mathbf{r}, \mathbf{r}') + \nabla[\nabla \cdot \mathbf{\Gamma}_e(\mathbf{r}, \mathbf{r}')] - k^2\mathbf{\Gamma}_e(\mathbf{r}, \mathbf{r}') = i\omega\mu_0\delta^{(3)}(\mathbf{r} - \mathbf{r}')\mathbf{I}. \qquad (15.24)$$

It is not immediately obvious how to solve this vector equation, but we may perform a devious trick: if we take the divergence of both sides of equation (15.23) and note that the divergence of a curl is zero, we get

$$-k^2\nabla \cdot \mathbf{\Gamma}_e(\mathbf{r}, \mathbf{r}') = i\omega\mu_0\nabla\delta^{(3)}(\mathbf{r} - \mathbf{r}'), \qquad (15.25)$$

where we have also used $\nabla \cdot \mathbf{I} = \nabla$. We may substitute this expression for the divergence of the Green's dyadic into equation (15.24), which results in the expression

$$(\nabla^2 + k^2)\mathbf{\Gamma}_e(\mathbf{r}, \mathbf{r}') = -i\omega\mu_0\left[\mathbf{I} + \frac{1}{k^2}\nabla\right]\delta^{(3)}(\mathbf{r} - \mathbf{r}'). \qquad (15.26)$$

We are getting closer, because now we have a Helmholtz equation with a source term to solve for the Green's dyadic; the dyadic terms on the right of this equation make it nontrivial to solve, however. We may use another clever trick, though, and assume that $\mathbf{\Gamma}_e(\mathbf{r}, \mathbf{r}')$ is of the form

$$\mathbf{\Gamma}_e(\mathbf{r}, \mathbf{r}') = i\omega\mu_0\left[\mathbf{I} + \frac{1}{k^2}\nabla\right]\mathcal{G}(\mathbf{r}, \mathbf{r}'). \qquad (15.27)$$

We substitute from this into equation (15.26) and, assuming that we can interchange the orders of differentiation, we find that $\mathcal{G}(\mathbf{r}, \mathbf{r}')$ must satisfy

$$[\nabla^2 + k^2]\mathcal{G}(\mathbf{r}, \mathbf{r}') = -\delta^{(3)}(\mathbf{r} - \mathbf{r}'). \qquad (15.28)$$

But referring back to equation (14.32), this is within a constant factor equal to the Helmholtz equation for the scalar Green's function. We are thus able to write

$$\mathcal{G}(\mathbf{r}, \mathbf{r}') = \frac{1}{4\pi} \frac{e^{ik|\mathbf{r}-\mathbf{r}'|}}{|\mathbf{r} - \mathbf{r}'|}. \tag{15.29}$$

Then, finally, we are able to write our electric Green's dyadic as

$$\mathbf{\Gamma}_e(\mathbf{r}, \mathbf{r}') = \frac{i\omega\mu_0}{4\pi} \left[\mathbf{I} + \frac{1}{k^2}\boldsymbol{\nabla} \right] \frac{e^{ik|\mathbf{r}-\mathbf{r}'|}}{|\mathbf{r} - \mathbf{r}'|}. \tag{15.30}$$

It should be noted that, because of the presence of $\boldsymbol{\nabla}$, this Green's dyadic does indeed include a $1/R^3$ contribution, which results in a singularity if $\mathbf{r}$ lies within the domain of integration of $\mathbf{r}'$. Moving forward, however, we will assume that we have properly corrected this singular behavior in any use of $\mathbf{\Gamma}_e(\mathbf{r}, \mathbf{r}')$. As we will see, we will, for the most part, not be able to use the exact form of the Green's dyadic in any analytic calculations anyway.

What does this Green's dyadic tell us? Referring back to equation (15.19), we note that $\mathbf{j} \equiv \mathbf{J}(\mathbf{r}')d^3r'$ represents an elementary element of electric current. The product $\mathbf{\Gamma}_e(\mathbf{r}, \mathbf{r}') \cdot \mathbf{j}$ therefore tells us the electric field at location $\mathbf{r}$ that arises from an elementary current element $\mathbf{j}$ at location $\mathbf{r}'$. We need a dyadic Green's function for this problem because the vector field produced by a particular current element not only depends on the magnitude of the current but also its direction.

We may also calculate the magnetic field produced by an electric current directly from Faraday's law,

$$\mathbf{H}(\mathbf{r}) = \frac{1}{i\omega\mu_0}\boldsymbol{\nabla} \times \mathbf{E}(\mathbf{r}, \mathbf{r}') = \frac{1}{i\omega\mu_0}\boldsymbol{\nabla} \times \int_D \mathbf{\Gamma}_e(\mathbf{r}, \mathbf{r}') \cdot \mathbf{J}_e(\mathbf{r}')d^3r'. \tag{15.31}$$

We again make our dubious assumption that we can switch the integration and the differentiation. We may also calculate the magnetic field produced by an electric current directly from Faraday's law,

$$\mathbf{H}(\mathbf{r}) = \frac{1}{i\omega\mu_0} \int_D [\boldsymbol{\nabla} \times \mathbf{\Gamma}_e(\mathbf{r}, \mathbf{r}')] \cdot \mathbf{J}_e(\mathbf{r}')d^3r'. \tag{15.32}$$

Again, we expect that this formula will not be valid for $\mathbf{r}$ within D. But what is $\boldsymbol{\nabla} \times \mathbf{\Gamma}_e$? If we write it out explicitly, we obtain

$$\boldsymbol{\nabla} \times \mathbf{\Gamma}_e(\mathbf{r}, \mathbf{r}') = i\omega\mu_0 \left[\boldsymbol{\nabla} \times \mathbf{I} + \frac{1}{k^2}(\boldsymbol{\nabla} \times \boldsymbol{\nabla})\boldsymbol{\nabla} \right] \mathcal{G}(\mathbf{r}, \mathbf{r}'). \tag{15.33}$$

The $(\boldsymbol{\nabla} \times \boldsymbol{\nabla})$ on the right of this expression is, in essence, the curl of a gradient, and it vanishes. Furthermore, $\boldsymbol{\nabla} \times \mathbf{I}\mathcal{G} = \boldsymbol{\nabla}\mathcal{G}\times$, with the cross product acting on whatever is to the right of it in the final expression. We have

$$\mathbf{H}(\mathbf{r}) = \int_D \boldsymbol{\nabla}\mathcal{G}(\mathbf{r}, \mathbf{r}') \times \mathbf{J}_e(\mathbf{r}')d^3r'. \tag{15.34}$$

We therefore have expressions that tell us the electric and magnetic fields produced by a distribution of electric current.

This process can be repeated for the magnetic current $\mathbf{J}_m$. If we assume that $\rho_e = 0$ and $\mathbf{J}_e = 0$, we can readily find the wave equation,

$$\nabla \times (\nabla \times \mathbf{H}) - k^2 \mathbf{H} = i\omega\varepsilon_0 \mathbf{J}_m. \tag{15.35}$$

If we compare this equation to equation (15.18), we see that we can take the steps used to find the electric Green's dyadic and use analogous steps to find the magnetic Green's dyadic, with the result

$$\mathbf{\Gamma}_m(\mathbf{r}, \mathbf{r}') = \frac{i\omega\varepsilon_0}{4\pi}\left[\mathbf{I} + \frac{1}{k^2}\nabla\right]\frac{e^{ik|\mathbf{r}-\mathbf{r}'|}}{|\mathbf{r}-\mathbf{r}'|}, \tag{15.36}$$

where this dyadic calculates the radiated magnetic field, i.e.

$$\mathbf{H}(\mathbf{r}) = \int_D \mathbf{\Gamma}_m(\mathbf{r}, \mathbf{r}') \cdot \mathbf{J}_m(\mathbf{r}')d^3r'. \tag{15.37}$$

Using the Ampère–Maxwell law, we can find the corresponding electric field,

$$\mathbf{E}(\mathbf{r}) = -\int_D \nabla\mathcal{G}(\mathbf{r}, \mathbf{r}') \times \mathbf{J}_m(\mathbf{r}')d^3r'. \tag{15.38}$$

If we compare these two equations with equations (15.19) and (15.31), we can again see the symmetry in Maxwell's equations, in which a magnetic dipole produces a magnetic field that matches the electric field produced by an electric dipole.

These solutions for electric and magnetic dipoles are, in principle, independent, in that we are considering a linear radiation problem. In scattering, however, there is usually a nontrivial relationship between the electric and magnetic dipoles induced by an incident field. For simplicity, we will restrict the rest of the chapter to electric materials only, for which $\mathbf{J}_m = 0$.

15.2 Scattering theory

We now look to formally solve a general scattering problem. As we will see, we will not be able to get very far in general, but we will introduce some important definitions and approximations along the way.

We return to the monochromatic Maxwell's equations and consider a scattering object restricted to a domain D with no free charges or currents, vacuum permeability $\mu = \mu_0$, and an inhomogeneous permittivity $\varepsilon(\mathbf{r})$. We imagine a monochromatic wave $\mathbf{E}_0(\mathbf{r})$ incident upon the object, and the interaction produces a scattered wave $\mathbf{E}_s(\mathbf{r})$. We will clarify the definitions of the incident and scattered fields momentarily. The overall geometry is shown in figure 15.2.

Maxwell's curl equations for this system are of the form

$$\nabla \times \mathbf{E} = i\omega\mu_0\mathbf{H}, \tag{15.39}$$

$$\nabla \times \mathbf{H} = -i\omega\varepsilon\mathbf{E}. \tag{15.40}$$

$$\mathbf{E}_0(\mathbf{r}) \qquad\qquad\qquad \mathbf{E}_s(\mathbf{r})$$

$$D$$

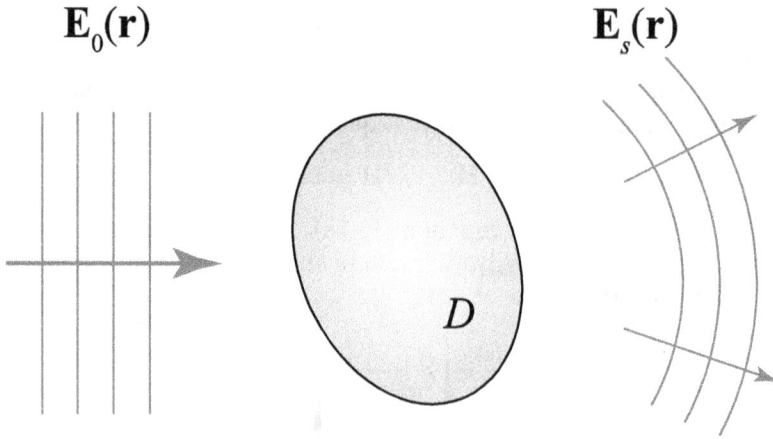

Figure 15.2. Illustration showing the geometry of our scattering system.

From these, we can readily find the wave equation,

$$\nabla \times (\nabla \times \mathbf{E}) - \omega^2 \mu_0 \varepsilon \mathbf{E} = 0. \tag{15.41}$$

We note that the electric field $\mathbf{E}(\mathbf{r})$ represents the total field in the region of the scatterer, i.e. the combination of the incident field and the scattered field. Because of the spatially varying $\varepsilon(\mathbf{r})$, this equation is not solved as readily as the wave equation in free space. We can, however, simplify things by adding $-\omega^2 \mu_0 (\varepsilon_0 - \varepsilon)\mathbf{E}$ to both sides of the equation. We have

$$\nabla \times (\nabla \times \mathbf{E}) - k^2 \mathbf{E} = -\omega^2 \mu_0 (\varepsilon_0 - \varepsilon)\mathbf{E}. \tag{15.42}$$

Let us immediately compare this equation to equation (15.18) for the electric field generated by an electric current. We see that they are structurally the same, which means that we can write the electric field $\mathbf{E}(\mathbf{r})$ for our scattering problem using the electric Green's dyadic $\boldsymbol{\Gamma}_e(\mathbf{r}, \mathbf{r}')$. Before we do so, however, we define

$$F(\mathbf{r}) \equiv \frac{\omega^2 \mu_0}{4\pi} [\varepsilon(\mathbf{r}) - \varepsilon_0]. \tag{15.43}$$

Scattering theory was first studied for quantum wave functions scattered by static potentials and later adapted to optics. In analogy with the quantum case, we refer to $F(\mathbf{r})$ as the *scattering potential*. It should be noted that the scattering potential is localized in space if the object is localized in space, because $\varepsilon(\mathbf{r}) - \varepsilon_0 = 0$ outside the object.

We thus have

$$\nabla \times (\nabla \times \mathbf{E}) - k^2 \mathbf{E} = 4\pi F(\mathbf{r}) \mathbf{E}(\mathbf{r}). \tag{15.44}$$

We now must be careful because we are particularly interested in calculating the field scattered from our object and must come up with a sensible definition for it. We write the total field as

$$\mathbf{E}(\mathbf{r}) \equiv \mathbf{E}_0(\mathbf{r}) + \mathbf{E}_s(\mathbf{r}), \tag{15.45}$$

where $\mathbf{E}_0(\mathbf{r})$ represents the illuminating field *in the absence of the scatterer*, and $\mathbf{E}_s(\mathbf{r})$ represents the scattered field. Because of this definition of the illuminating field, the scattered field must partially cancel out the illuminating field to satisfy energy conservation. This leads to some interesting physical consequences, as we will note later.

The incident field, which is the field propagating in vacuum, must satisfy the homogeneous wave equation,

$$\nabla \times [\nabla \times \mathbf{E}_0] - k^2 \mathbf{E}_0 = 0, \tag{15.46}$$

and this means that the incident field falls out of the left-hand side of equation (15.44). We are left with

$$\nabla \times (\nabla \times \mathbf{E}_s) - k^2 \mathbf{E}_s = 4\pi F(\mathbf{r})\mathbf{E}(\mathbf{r}). \tag{15.47}$$

We may now make the association with the electric current problem we discussed earlier. If we introduce the identity

$$\mathbf{J}_e(\mathbf{r}) = \frac{4\pi}{i\omega\mu_0}F(\mathbf{r})\mathbf{E}(\mathbf{r}), \tag{15.48}$$

we may write an integral expression for the scattered field, following equation (15.19), of the form

$$\mathbf{E}_s(\mathbf{r}) = \int_D \mathbf{G}(\mathbf{r}, \mathbf{r}') \cdot \mathbf{E}(\mathbf{r}')F(\mathbf{r}')d^3r', \tag{15.49}$$

where, for convenience, we have introduced a Green's dyadic with a different constant factor,

$$\mathbf{G}(\mathbf{r}, \mathbf{r}') = \frac{4\pi}{i\omega\varepsilon_0}\mathbf{\Gamma}_e(\mathbf{r}, \mathbf{r}') = \left[\mathbf{I} + \frac{1}{k^2}\nabla\right]\frac{e^{ik|\mathbf{r}-\mathbf{r}'|}}{|\mathbf{r} - \mathbf{r}'|}. \tag{15.50}$$

At first glance, it might appear that the scattering problem was not, in fact, so hard to solve at all! A closer look at equation (15.49), however, shows that the scattered field appears on both sides of the equation, hidden in the integral in the definition of $\mathbf{E}(\mathbf{r})$. We have not solved the scattering problem, but rather replaced a partial differential equation that we did not know how to solve with an integral equation that we do not know how to solve. This integral form can be used to find approximate solutions to the scattering problem that provide physical insight, which we discuss in the next section.

Equation (15.49) may also be used to find computational solutions to the scattering problem, which we will discuss in detail in the next chapter.

15.3 The Born series

Now that we know an exact analytic solution of equation (15.49) is out of the question, we look for an approximate solution. The natural choice is to assume that the object is only weakly scattering, i.e. $|\mathbf{E}_s| \ll |\mathbf{E}_0|$, which implies both that the permittivity is close to that of free space, $\varepsilon(\mathbf{r}) \approx \varepsilon_0$, and also that the scattering object is small, probably with dimensions comparable to the wavelength. This definition of weak scattering is unavoidably vague, as it is not possible to come up with a quantitative universal definition.

Under conditions of weak scattering, we can say $\mathbf{E}(\mathbf{r}) \approx \mathbf{E}_0(\mathbf{r})$, and if we use this approximation in equation (15.49), we get the simplified expression

$$\mathbf{E}_s(\mathbf{r}) = \int_D \mathbf{G}(\mathbf{r}, \mathbf{r}') \cdot \mathbf{E}_0(\mathbf{r}')F(\mathbf{r}')d^3r'. \tag{15.51}$$

The equation looks nearly the same as before, but the change is significant. The scattered field now only appears on the left-hand side of the equation, so equation (15.51) is a direct integral solution for the scattered field. This integral can only be evaluated for very rare special cases, but having a direct solution allows us to understand the physics of the scattering process.

Equation (15.51) is known as the *first Born approximation*, or simply the *Born approximation*, after Max Born, who first formulated a perturbation expansion of the scattered field in quantum mechanics. Let us write this approximation as $\mathbf{E}_s^{(1)}(\mathbf{r})$; we may now develop a higher-order approximation by substituting $\mathbf{E} = \mathbf{E}_0 + \mathbf{E}_s^{(1)}$ into equation (15.49), and we find the second Born approximation,

$$\mathbf{E}_s^{(2)}(\mathbf{r}) = \int_D \mathbf{G}(\mathbf{r}, \mathbf{r}') \cdot \mathbf{E}_0(\mathbf{r}')F(\mathbf{r}')d^3r' + \int_D \int_D \mathbf{G}(\mathbf{r}, \mathbf{r}') \cdot \mathbf{G}(\mathbf{r}', \mathbf{r}'') \cdot \mathbf{E}_0(\mathbf{r}'')F(\mathbf{r}')F(\mathbf{r}'')d^3r'd^3r''. \tag{15.52}$$

We can then take $\mathbf{E} = \mathbf{E}_0 + \mathbf{E}_s^{(2)}$ and substitute it into equation (15.49) to derive the third Born approximation, and so on. We can see the trend: the Nth order Born approximation involves N integrals, N Green's dyadics, and N scattering potentials. It is cumbersome to write the exact formula for orders beyond the second, but we can write the entire series in a symbolic form, with the integrations implied, as the *Born series*,

$$\mathbf{E}_s = \sum_{n=1}^{\infty} (F\mathbf{G})^n \cdot \mathbf{E}_0. \tag{15.53}$$

How do we interpret the Born series? Each integral represents an elementary scattering event, so the Nth-order term of the Born series represents the process in which the incident field scatters N times within the object before emerging as the scattered field. This is illustrated in figure 15.3.

The 'Born approximation' became much more famous than the entire series that Born developed, much to his irritation. He once told Emil Wolf [1], 'I developed in that paper the whole perturbation expansion for the scattered field, valid to all orders, yet I am only given credit for the first term in that series!'

It is important to note, however, that the Born series does not converge and therefore does not represent a valid solution to a scattering problem in many cases. This can be traced to the built-in assumption that light always scatters a finite number of times within the scattering object, and this assumption fails when the scattering object can support bound states. An illustrative example of a quasi-bound state is a whispering gallery mode [2], in which a wave creeps along the boundary of

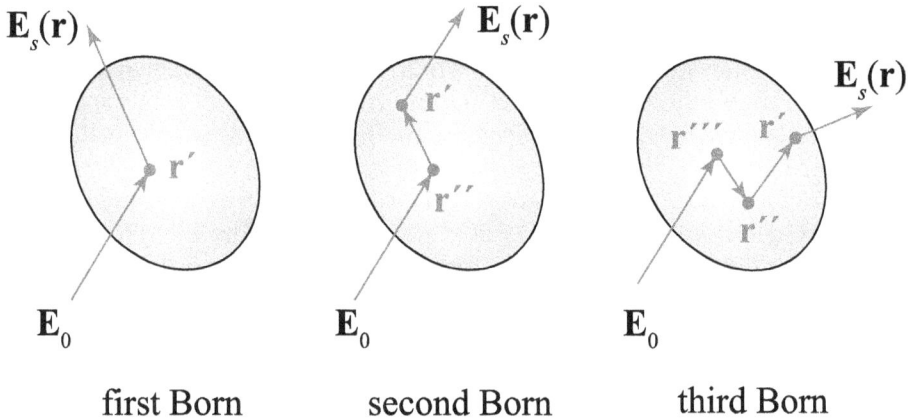

Figure 15.3. Conceptual illustration of the Born series. The Nth-order term of the series represents the incident field scattering N times within the object.

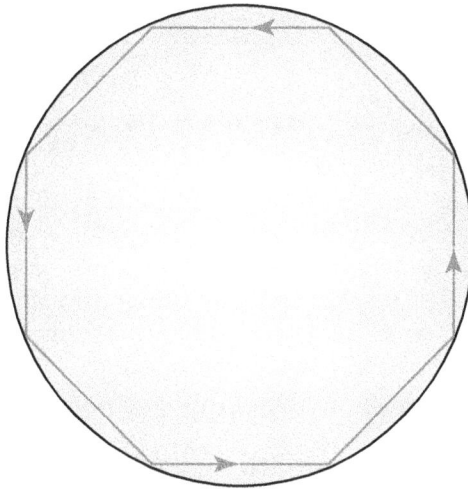

Figure 15.4. Ray picture of a whispering gallery mode, where the light ray is confined to a periodic trajectory in a sphere by total internal reflection.

a circular or cylindrical region by total internal reflection; this is roughly illustrated in figure 15.4. Davies [3] investigated the convergence of the Born series for a limited class of potentials and concluded that the series diverges or converges in the presence or absence of bound states, respectively. One significant implication of this result is that even the full Born series is only valid for objects that scatter relatively weakly.

The Born series highlights the difficulties in solving scattering problems, especially the difficulties encountered while trying to solve them with any generality. More success and physically useful expressions have been found by looking at special cases, as we will soon consider.

15.4 The optical theorem

Despite the fact that scattering problems typically cannot be solved in analytic form, a number of important and even practical results can be drawn from the theory of scattering, and these have surprising generality. One of these is known as the *optical theorem*, or more specifically the *optical cross-section theorem*, which we derive in this section for electromagnetic waves. This will also give us the opportunity to introduce some common terminology used in scattering theory.

Let us consider a general scattering system in which a monochromatic electric field $\mathbf{E}_i$ illuminates a scattering object, thus producing a scattered field $\mathbf{E}_s$. As always, the total field is the sum of these two. We may also introduce the corresponding magnetic fields $\mathbf{H}_i$ and $\mathbf{H}_s$. The incident field is taken to be a plane wave propagating in the direction $\mathbf{s}_0$, satisfying the relations

$$\mathbf{E}_i(\mathbf{r}) = \mathbf{E}_0 e^{ik\mathbf{s}_0 \cdot \mathbf{r}}, \quad \mathbf{H}_i(\mathbf{r}) = \mathbf{s}_0 \times \mathbf{E}_0 \frac{k}{\omega\mu_0} e^{ik\mathbf{s}_0 \cdot \mathbf{r}}. \tag{15.54}$$

We consider the scattered field far from the scattering object, where a far-zone approximation may be applied. In that limit, we expect the fields to be spherical waves of the form

$$\mathbf{E}_s(\mathbf{r}) = \mathbf{A}(\mathbf{s}, \mathbf{s}_0) \frac{e^{ikr}}{r}, \quad \mathbf{H}_s(\mathbf{r}) = \mathbf{s} \times \mathbf{A}(\mathbf{s}, \mathbf{s}_0) \frac{k}{\omega\mu_0} \frac{e^{ikr}}{r}, \tag{15.55}$$

where $\mathbf{r} = r\mathbf{s}$, with $\mathbf{s}$ the unit vector in the direction of observation. The quantity $\mathbf{A}(\mathbf{s}, \mathbf{s}_0)$ is the vector *scattering amplitude*, which depends upon both the direction of illumination $\mathbf{s}_0$ and the direction of scattering $\mathbf{s}$. It should be noted that we make no attempt to solve the scattering problem; we simply assume a broad form of the scattered field in the far zone. Let us also assume, for simplicity, that $\mathbf{E}_0$ is a real-valued vector, i.e. that the field is linearly polarized.

Let us now calculate the time-averaged Poynting vector for the total field, i.e.

$$\mathbf{S}(\mathbf{r}) = \frac{1}{2}\mathrm{Re}\{\mathbf{E}(\mathbf{r}) \times \mathbf{H}^*(\mathbf{r})\} = \frac{1}{2}\mathrm{Re}\{[\mathbf{E}_i(\mathbf{r}) + \mathbf{E}_s(\mathbf{r})] \times [\mathbf{H}_i(\mathbf{r}) + \mathbf{H}_s(\mathbf{r})]^*\}. \tag{15.56}$$

There are three distinct contributions to the total Poynting vector, and we may write the total as

$$\mathbf{S}(\mathbf{r}) = \mathbf{S}_i(\mathbf{r}) + \mathbf{S}_s(\mathbf{r}) + \mathbf{S}'(\mathbf{r}), \tag{15.57}$$

with

$$\mathbf{S}_i(\mathbf{r}) = \frac{1}{2}\mathrm{Re}\left\{\mathbf{E}_i(\mathbf{r}) \times \mathbf{H}_i^*(\mathbf{r})\right\}, \tag{15.58}$$

$$\mathbf{S}_s(\mathbf{r}) = \frac{1}{2}\mathrm{Re}\left\{\mathbf{E}_s(\mathbf{r}) \times \mathbf{H}_s^*(\mathbf{r})\right\}, \tag{15.59}$$

$$\mathbf{S}'(\mathbf{r}) = \frac{1}{2}\text{Re}\Big\{\mathbf{E}_i(\mathbf{r}) \times \mathbf{H}_s^*(\mathbf{r}) + \mathbf{E}_s(\mathbf{r}) \times \mathbf{H}_i^*(\mathbf{r})\Big\}. \tag{15.60}$$

Clearly, $\mathbf{S}_i$ represents the power flux for the incident field by itself, $\mathbf{S}_s$ represents the power flux for the scattered field by itself, and $\mathbf{S}'$ represents the modification of the flux due to interference between these fields.

Next, we integrate the flux of each term over a large spherical surface of radius r, taken to be the far zone, to get the net power flow through the surface associated with each. For example, we have

$$W_i = \int_S \mathbf{S}_i(r\mathbf{s}) \cdot \mathbf{s}\,da, \tag{15.61}$$

where $da = r^2 d\Omega$ and $d\Omega$ is again the differential of the solid angle. If the object is nonabsorbing, we expect energy to be conserved; in which case, we would expect that $W_i + W_s + W' = 0$. If the scatterer is also absorbing light, we introduce W_a as the net power absorbed, and we may then write the relation

$$W_i + W_s + W' = -W_a. \tag{15.62}$$

This expression indicates that the amount of power absorbed must equal the power lost from the combination of the other terms, i.e. there must be a net flux of power into the sphere.

We have implicitly taken the incident field to be propagating in vacuum, where there is no absorption; we must therefore have $W_i = 0$. (This can be proven explicitly by integration, if desired.) We may then rewrite equation (15.62) as

$$W_a + W_s = -W' = -\frac{1}{2}\text{Re}\Big\{\int_S [\mathbf{E}_i(\mathbf{r}) \times \mathbf{H}_s^*(\mathbf{r}) + \mathbf{E}_s(\mathbf{r}) \times \mathbf{H}_i^*(\mathbf{r})] \cdot \mathbf{s}r^2 d\Omega\Big\}. \tag{15.63}$$

The quantity on the left is the total power *extinguished* from the incident field, either through absorption or scattering. This equation demonstrates that the total extinguished power, which we label W_e, can be derived entirely from the interference term of the Poynting vector.

Now let us attempt to evaluate equation (15.63) using our definitions of the incident and scattered fields. We have

$$W_e = -\frac{r}{2\mu_0 c}\text{Re}\int_S \{\mathbf{E}_0 \times [\mathbf{s} \times \mathbf{A}^*(\mathbf{s}, \mathbf{s}_0)]e^{-ikr(1-\mathbf{s}_0 \cdot \mathbf{s})} + \mathbf{A}(\mathbf{s}, \mathbf{s}_0) \times [\mathbf{s}_0 \times \mathbf{E}_0]e^{ikr(1-\mathbf{s}_0 \cdot \mathbf{s})}\} \cdot \mathbf{s}\,d\Omega. \tag{15.64}$$

Our next step is to simplify the vector products in the formula; here, we take advantage of the transverse nature of plane waves and spherical waves, i.e. $\mathbf{s}_0 \cdot \mathbf{E}_0 = 0$ and $\mathbf{s} \cdot \mathbf{A} = 0$. With this, we have

$$W_e = -\frac{r}{2\mu_0 c}\text{Re}\int_S \{\mathbf{E}_0 \cdot \mathbf{A}^*(\mathbf{s}, \mathbf{s}_0)e^{-ikr(1-\mathbf{s}_0 \cdot \mathbf{s})} + [(\mathbf{s}_0 \cdot \mathbf{s})(\mathbf{E}_0 \cdot \mathbf{A}(\mathbf{s}, \mathbf{s}_0)) - (\mathbf{E}_0 \cdot \mathbf{s})(\mathbf{A} \cdot \mathbf{s}_0)]e^{ikr(1-\mathbf{s}_0 \cdot \mathbf{s})}\}d\Omega. \tag{15.65}$$

For convenience, let us integrate in spherical coordinates (θ, ϕ) with $\mathbf{s}_0$ taken to be the z-axis. Our integrand is then independent of ϕ, which may be integrated out, and our expression may be simplified to

$$W_e = -\frac{\pi r}{\mu_0 c} \text{Re} \int_0^\pi \{\mathbf{E}_0 \cdot \mathbf{A}^* e^{-ikr(1-\cos\theta)} + [(\mathbf{s}_0 \cdot \mathbf{s})(\mathbf{E}_0 \cdot \mathbf{A}) - (\mathbf{E}_0 \cdot \mathbf{s})(\mathbf{A} \cdot \mathbf{s}_0)]e^{ikr(1-\cos\theta)}\} \sin\theta \, d\theta. \quad (15.66)$$

Because we are in the far zone, the quantity $kr \gg 1$. The exponents in the integrand oscillate rapidly with θ, and this is a perfect opportunity to use the *method of stationary phase* (see Gbur, chapter 21 [4]). In short, let us assume we have an integral of the form

$$F(\alpha) = \int_a^b f(\theta) e^{i\alpha g(\theta)} d\theta, \quad (15.67)$$

where $\alpha \gg 1$. The exponent oscillates rapidly for most values of θ, and integrals over small regions around these points tend to vanish due to equal parts of positive and negative contributions. The only exceptions are points where the function $g(\theta)$ is stationary, i.e. points θ_0 where $g'(\theta_0) = 0$ and endpoints of the domain of integration. In our case, $\alpha = kr$, $g(\theta) = 1 - \cos\theta$, and the endpoints of the domain ($\theta = 0, \pi$) are the only stationary points. The value of the integral based on these endpoint contributions can be shown to be

$$F(\alpha) = \frac{1}{i\alpha} \left[\frac{f(b)}{g'(b)} e^{i\alpha g(b)} - \frac{f(a)}{g'(a)} e^{i\alpha g(a)} \right]. \quad (15.68)$$

We use this result to evaluate equation (15.66), with $a = 0$ and $b = \pi$. We note that $g'(\theta) = \sin\theta$, which cancels the $\sin\theta$ in the integrand. We are left with

$$W_e = -\frac{\pi r}{\mu_0 c} \text{Re} \left\{ -\frac{1}{ikr} \mathbf{E}_0 \cdot \mathbf{A}^*(-\mathbf{s}_0, \mathbf{s}_0) e^{-2ikr} + \frac{1}{ikr} \mathbf{E}_0 \cdot \mathbf{A}^*(\mathbf{s}_0, \mathbf{s}_0) \right.$$

$$+ \frac{1}{ikr}[(-\mathbf{s}_0 \cdot \mathbf{s}_0)(\mathbf{E}_0 \cdot \mathbf{A}(-\mathbf{s}_0, \mathbf{s}_0)) + (\mathbf{E}_0 \cdot \mathbf{s}_0)(\mathbf{A}(-\mathbf{s}_0, \mathbf{s}_0) \cdot \mathbf{s}_0)]e^{2ikr} \quad (15.69)$$

$$\left. - \frac{1}{ikr}[(\mathbf{s}_0 \cdot \mathbf{s}_0)(\mathbf{E}_0 \cdot \mathbf{A}(\mathbf{s}_0, \mathbf{s}_0)) - (\mathbf{E}_0 \cdot \mathbf{s}_0)(\mathbf{A}(\mathbf{s}_0, \mathbf{s}_0) \cdot \mathbf{s}_0)] \right\}.$$

This can be immediately simplified using $\mathbf{s}_0 \cdot \mathbf{s}_0 = 1$ and $\mathbf{E}_0 \cdot \mathbf{s}_0 = 0$. We then have

$$W_e = -\frac{\pi r}{\mu_0 c} \text{Re} \left\{ -\frac{1}{ikr} \mathbf{E}_0 \cdot \mathbf{A}^*(-\mathbf{s}_0, \mathbf{s}_0) e^{-2ikr} + \frac{1}{ikr} \mathbf{E}_0 \cdot \mathbf{A}^*(\mathbf{s}_0, \mathbf{s}_0) \right.$$

$$\left. + \frac{1}{ikr}(-\mathbf{s}_0 \cdot \mathbf{s}_0)(\mathbf{E}_0 \cdot \mathbf{A}(-\mathbf{s}_0, \mathbf{s}_0))e^{2ikr} - \frac{1}{ikr} \mathbf{E}_0 \cdot \mathbf{A}(\mathbf{s}_0, \mathbf{s}_0) \right\}. \quad (15.70)$$

The terms that contain the exponents combine to make a purely imaginary piece that is eliminated when we take the real part. We are left with

$$W_e = \frac{2\pi}{\mu_0 \omega} \text{Im}\{\mathbf{E}_0 \cdot \mathbf{A}(\mathbf{s}_0, \mathbf{s}_0)\}. \quad (15.71)$$

This is the basic statement of the optical theorem. It shows that the power extinguished from the incident field may be directly related to the component of the scattering amplitude in the forward direction $\mathbf{s}_0$ that has the same polarization $\mathbf{E}_0$ as the incident field.

The optical theorem may be considered a statement of energy conservation in the context of scattering theory. Recall that the incident field $\mathbf{E}_i$ was defined as the field that exists in the absence of the scatterer, and the total field with the scatterer present is simply $\mathbf{E} = \mathbf{E}_i + \mathbf{E}_s$. As light is scattered, it must be depleted from the incident field in the shadow of the object, and this is achieved by complete destructive interference between the incident field and the scattered field in the $\mathbf{s}_0$ direction. The amount of power depleted from the incident field is therefore related to the scattering amplitude in the forward direction; the optical theorem quantifies this statement.

The quantity W_e has dimensions of power; let us divide this by the magnitude of the incident Poynting vector, which has dimensions of power per unit area. The result is a quantity that we label σ_e, which has dimensions of area; it is known as the *extinction cross section* of the scatterer and has the form

$$\sigma_e = \frac{W_e}{|\mathbf{S}_i|} = \frac{4\pi}{k} \frac{1}{|\mathbf{E}_0|^2} \text{Im}\{\mathbf{E}_0 \cdot \mathbf{A}(\mathbf{s}_0, \mathbf{s}_0)\}. \tag{15.72}$$

(We note that $\mathbf{A}$ has dimensions of electric field times length, as can be seen from equation (15.55).)

The cross section represents the effective size of the object with respect to the process in question; a larger cross section indicates a stronger effect. Cross sections are used in all branches of physics that involve scattering phenomena, including nuclear and particle physics. Because the optical theorem is broadly based on conservation law arguments, it exists in some form in all these fields. In particle physics, the units of cross section are *barns*, as in 'you couldn't hit the broadside of a barn.' A barn is 10^{-28} m^2; in optics, it is more reasonable to measure cross sections in mm^2 or μm^2.

We may also look individually at the cross sections for scattering and absorption by normalizing W_s and W_a to the incident Poynting vector; these quantities depend on the specific properties of the scattering object.

The optical theorem is of practical as well as physical importance. It allows the total extinguished power in a scattering process to be determined directly from the forward scattering amplitude, which can simplify experimental measurements. In 1999, Carney, Wolf, and Agarwal showed how the optical theorem could be used to solve inverse scattering problems without the need to determine the phase of the scattered field [5]; we will discuss inverse problems in an upcoming section.

We have determined the optical theorem for electromagnetic waves, in fitting with the topic of the text. It can also be derived for scalar waves; chapter 13 of Born and Wolf [6] contains information on this and more on scattering theory.

15.5 Rayleigh scattering

Because analytic solutions to the scattering problem do not exist in general, we turn to special cases that are physically relevant and simple enough to be analytically tractable. The simplest possible case we can imagine is a spherical particle of subwavelength radius $a \ll \lambda$, with a constant permittivity ε. This is an extremely limited model, but fortunately, one that works well for a variety of physical problems, most notably for explaining the scattering of sunlight by the atmosphere and the blue color of the sky.

Because the particle is taken to be much smaller than the wavelength, the electric field is effectively constant over the entire volume of the particle and oscillates with an $\exp[-i\omega t]$ time dependence. In this case, we can treat the system locally as an electrostatics problem and calculate the electrostatic potential that a particle of permittivity ε produces in response to an applied electric field $\mathbf{E}_0 = E_0 \hat{\mathbf{z}}$. The configuration and relevant parameters are illustrated in figure 15.5.

We therefore seek solutions of Laplace's equation, $\nabla^2 \phi(\mathbf{r}) = 0$. We can solve for the potential $\phi_{in}(\mathbf{r})$ inside the sphere and $\phi_{out}(\mathbf{r})$ outside the sphere while reconciling these solutions with the appropriate boundary conditions. The system is subject to three boundary conditions,

$$\phi_{in}(a) = \phi_{out}(a), \text{ (continuity)}, \tag{15.73}$$

$$\varepsilon \frac{\partial \phi_{in}(a)}{\partial r} = \varepsilon_0 \frac{\partial \phi_{in}(a)}{\partial r}, \quad \text{(normal electric field)}, \tag{15.74}$$

$$\lim_{r \to \infty} \phi_{out}(\mathbf{r}) = -E_0 r \cos\theta. \tag{15.75}$$

The first condition, continuity of the potential, ensures that the electric field does not diverge at the boundary, as $\mathbf{E} = -\nabla\phi$. The second condition is the boundary condition for the normal component of the electric field at an interface, equation (9.33), converted to potential form. The third equation is the requirement that, far from the sphere, the electric field should take on the value $\mathbf{E}_0 = E_0 \hat{\mathbf{z}}$.

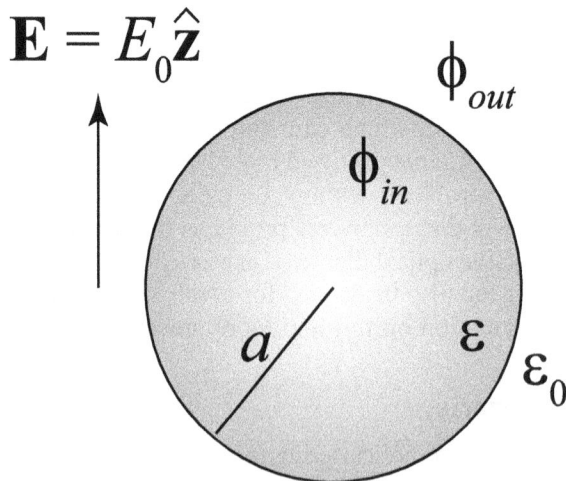

Figure 15.5. Illustration of the geometry and parameters related to Rayleigh scattering.

We now borrow results from electrostatics, which can be found in any elementary electromagnetism textbook. In particular, for an isotropic spherical particle, the potentials inside and outside the sphere may be written in the form

$$\phi_{in}(\mathbf{r}) = \sum_{l=0}^{\infty} A_l r^l P_l(\cos\theta) + \sum_{l=0}^{\infty} \frac{B_l}{r^{l+1}} P_l(\cos\theta), \tag{15.76}$$

$$\phi_{out}(\mathbf{r}) = \sum_{l=0}^{\infty} C_l r^l P_l(\cos\theta) + \sum_{l=0}^{\infty} \frac{D_l}{r^{l+1}} P_l(\cos\theta), \tag{15.77}$$

where $P_l(x)$ are the Legendre polynomials.

We now apply some hidden boundary conditions. Because the potential and field within the sphere must be finite, we find that $B_l = 0$ for all l. The potential generated by a finite sphere must also decrease far away from the sphere, so all $C_l = 0$ except in the case $l = 1$, for which we can match our third boundary condition by letting $C_1 = -E_0$.

We next consider the first two boundary conditions. These conditions are homogeneous, and for any $l \neq 1$ they can be satisfied by the choice $A_l = D_l = 0$. The only nontrivial case is the $l = 1$ case, and our two boundary conditions become

$$A_1 a = -E_0 a + \frac{D_1}{a^2}, \tag{15.78}$$

$$\varepsilon A_1 = -\varepsilon_0 E_0 - 2\frac{\varepsilon_0 D_1}{a^3}. \tag{15.79}$$

These equations can readily be solved for A_1 and D_1. Using these results, and knowing that $P_1(\cos\theta) = \cos\theta$, we find the following expressions for the potentials inside and outside the sphere:

$$\phi_{in}(\mathbf{r}) = -\frac{3\varepsilon_0}{\varepsilon + 2\varepsilon_0} E_0 r \cos\theta, \tag{15.80}$$

$$\phi_{out}(\mathbf{r}) = \frac{\varepsilon - \varepsilon_0}{\varepsilon + 2\varepsilon_0} \frac{a^3 E_0 \cos\theta}{r^2} - E_0 r \cos\theta. \tag{15.81}$$

In ϕ_{out}, the second term on the right is the potential of the incident field; the first term is therefore the potential $\phi_s(\mathbf{r})$ of the scattered field,

$$\phi_s(\mathbf{r}) = \frac{\varepsilon - \varepsilon_0}{\varepsilon + 2\varepsilon_0} \frac{a^3 E_0 \cos\theta}{r^2}. \tag{15.82}$$

It is worth noting that the structure of the permittivities on the right-hand side of this equation is in agreement with the Clausius–Mossotti equation, equation (6.21).

We now refer all the way back to chapter 5, when we noted that the electrostatic potential of an ideal dipole pointing in the z-direction is of the form

$$\phi_{dip}(\mathbf{r}) = \frac{1}{4\pi\varepsilon_0} \frac{p_0 \cos\theta}{r^2}, \tag{15.83}$$

where p_0 is the magnitude of the dipole moment. We conclude from this that our sphere scatters exactly like an ideal dipole with a dipole moment given by

$$\mathbf{p} = 4\pi\varepsilon_0 E_0 a^3 \frac{\varepsilon - \varepsilon_0}{\varepsilon + 2\varepsilon_0}\hat{\mathbf{z}}. \tag{15.84}$$

The field scattered from the sphere should then be a pure dipole field. We may also define the polarizability α of the particle, according to the relation $\mathbf{p} = \alpha\mathbf{E}_0$, as

$$\alpha = 4\pi\varepsilon_0 a^3 \frac{\varepsilon - \varepsilon_0}{\varepsilon + 2\varepsilon_0}. \tag{15.85}$$

We note two significant observations here, which make intuitive sense: the polarizability increases as the size of the particle increases and as the permittivity of the particle increases.

Before considering the scattered field properties, we note that the electric field inside the particle may be written as

$$\mathbf{E}_{in}(\mathbf{r}) = \frac{3\varepsilon_0}{\varepsilon + 2\varepsilon_0}E_0\hat{\mathbf{z}}. \tag{15.86}$$

Because we know that $\mathbf{P} = (\varepsilon - \varepsilon_0)\mathbf{E}$, we may write the polarization density within the sphere as

$$\mathbf{P} = 3\varepsilon_0 \frac{\varepsilon - \varepsilon_0}{\varepsilon + 2\varepsilon_0}\mathbf{E}_0. \tag{15.87}$$

We will find a natural use for this in the next section.

At this point, the astute reader may say, 'Wait a moment—the electrostatic potential of a dipole decays as $1/r^2$, while the potential of a radiation field goes as $1/r$—aren't these formulas fundamentally inconsistent?' This is true, but what we are really arguing is that the dipole moment of a Rayleigh sphere excited by a monochromatic wave is the same as the dipole moment of a sphere in a static field. We will see that this argument is correct through a rigorous argument when we derive the theory of Mie scattering in an upcoming section. For now, we assume that equation (15.84) accurately represents the dipole moment of a subwavelength-sized sphere excited by a monochromatic wave.

We now recall from equation (14.133) that the radiation emitted by an electric dipole is given by

$$\mathbf{S} = \frac{1}{2}\frac{k^3\omega p_0^2}{(4\pi)^2\varepsilon_0 r^2}\sin^2\theta\hat{\mathbf{r}}, \tag{15.88}$$

where θ is the angle between the dipole moment and the direction of radiation. Substituting from equation (15.84) into this expression and using $k = 2\pi/\lambda$ and $\omega = 2\pi c/\lambda$, we find after some manipulation that the radiation pattern of the scattered field of a Rayleigh particle is

$$\mathbf{S} = \frac{1}{2}\frac{\pi^2}{\lambda^4}\frac{c}{\varepsilon_0 r^2}|\alpha|^2|E_0|^2\sin^2\theta\hat{\mathbf{r}}. \tag{15.89}$$

Of particular interest in this expression is the $1/\lambda^4$ dependence of the scattered power on the wavelength, λ. This indicates that smaller wavelengths are scattered more strongly, i.e. blue light is scattered more strongly than red light. Lord Rayleigh developed a theory of small-particle scattering over a number of years, starting in 1871, to account for why the sky is blue [7]; this, of course, is why it is called Rayleigh scattering. The molecules of the atmosphere act as Rayleigh scatterers, and thus all the light in the sky during the day is the blue light preferentially scattered in accordance with the theory. At sunset, the Sun and the sky near it appear red because all of the blue light has been preferentially scattered away from the line of sight.

Violet light has an even shorter wavelength, so one naturally wonders why we do not see a violet daytime sky. Our eyes are significantly less sensitive to violet light than to blue light, and the Sun emits a lower intensity of violet light than blue; the net result is a blue sky.

It should be noted that θ is the angle between the radiation direction and the polarization of light; a much more natural angle to choose is the angle γ between the radiation direction and the direction of propagation of the incident light. We further note that the scattering is independent of the azimuthal angle ϕ.

Sunlight is, in general, unpolarized, and the radiation pattern of the scattered field is polarization dependent. We define the plane of scattering as that plane that includes both the incident propagation vector and the radiation direction; we can then introduce transverse electric (TE) and transverse magnetic (TM) cases where the polarization is out-of-plane or in-plane, respectively. The geometry is illustrated in figure 15.6.

For TM polarization, we use equation (15.89) above and write

$$\mathbf{S}_{TM} = \frac{1}{2}\frac{\pi^2}{\lambda^4}\frac{c}{\varepsilon_0 r^2}|\alpha|^2|E_0|^2\cos^2\gamma\,\hat{\mathbf{r}}. \tag{15.90}$$

For TE polarization, the radiation direction is always perpendicular to the polarization of the incident electric field, so the radiation pattern is independent of γ. We have

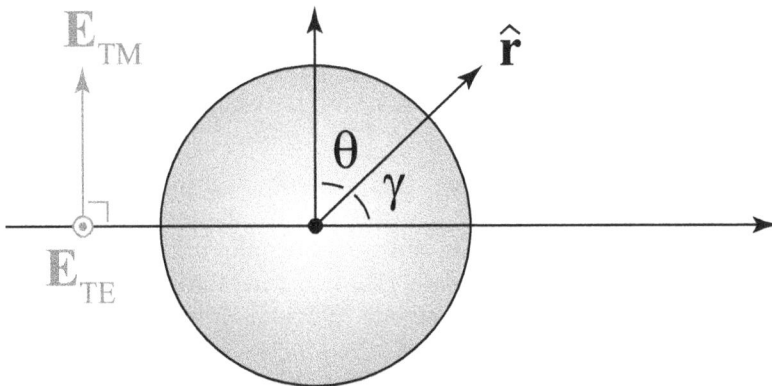

Figure 15.6. The notation related to TE and TM scattering and the relation between θ and γ.

$$\mathbf{S}_{TE} = \frac{1}{2} \frac{\pi^2}{\lambda^4} \frac{c}{\varepsilon_0 r^2} |\alpha|^2 |E_0|^2 \hat{\mathbf{r}}. \tag{15.91}$$

If we want to consider the total radiation scattered by a Rayleigh particle when the incident field is unpolarized, we simply add the two contributions, noting that the two states are mutually incoherent and therefore there are no interference terms present,

$$\mathbf{S}_{tot} = \frac{1}{2} \frac{\pi^2}{\lambda^4} \frac{c}{\varepsilon_0 r^2} |\alpha|^2 |E_0|^2 [1 + \cos^2 \gamma] \hat{\mathbf{r}}. \tag{15.92}$$

The scattered light intensity therefore has a dependence on the angle of observation. Because the two polarization states have different intensities in general, we may say that daytime skylight is partially polarized. This can be confirmed by looking through the sky—away from the Sun—with any polarizer and rotating the polarizer. One will see that the amount of light transmitted depends on the orientation of the polarizer. This polarization dependence is even maintained to some extent on cloudy days; it has been suggested that Vikings used Iceland spar (optical calcite) as a polarization sensor in navigating from Norway to North America [8].

Rayleigh scattering also appears as the dominant loss mechanism in fiber optics: small defects and inhomogeneities in the glass can act as Rayleigh scatterers, scattering light out of the fiber [9].

15.6 Rayleigh–Gans scattering

Rayleigh scattering has proven very useful for understanding a number of physical problems, but it is limited to small particles of uniform refractive index and high symmetry (spheres, ellipsoids). In 1925, Richard Gans came up with an alternative approximation [10] that relaxes some conditions while making others more strict. The result is now known as the *Rayleigh–Gans approximation*, and we briefly discuss it here.

We consider a particle with refractive index n that possesses a longest spatial dimension d. Rayleigh–Gans theory is considered valid when these two quantities satisfy the conditions

$$|n - 1| \ll 1, \tag{15.93}$$

$$2kd|n - 1| \ll 1. \tag{15.94}$$

The first condition is the requirement that the refractive index must not be very different from that of free space. This means that the scattering is generally weak. The left side of the second condition represents the amount of phase shift a light wave experiences relative to vacuum when propagating through the longest side of the particle. The second condition amounts to assuming that the particle imposes very small phase shifts on the light wave.

We now imagine breaking the general scatterer into a collection of infinitesimal point scatterers of volume $d^3 r'$. We now make the very strong assumption that *the scattered field of each point scatterer does not interact with the fields of any of the other point scatterers.* This is, in essence, assuming that the scattered field of each point scatterer is negligible compared to the incident wave. Let us assume that the incident field is given by $E_0(r')$; treating each point scatterer as a Rayleigh particle, we may then use equation (15.87) to write the polarization density as

$$\mathbf{P}(\mathbf{r}') = 3\varepsilon_0 \frac{\varepsilon(\mathbf{r}') - \varepsilon_0}{\varepsilon(\mathbf{r}') + 2\varepsilon_0} \mathbf{E}_0(\mathbf{r}'). \tag{15.95}$$

Assuming that $n \approx 1$ leads us to write the denominator as $3\varepsilon_0$; we then have

$$\mathbf{P}(\mathbf{r}') \approx [\varepsilon(\mathbf{r}') - \varepsilon_0]\mathbf{E}_0(\mathbf{r}'). \tag{15.96}$$

The dipole moment $\mathbf{p}$ at a point $\mathbf{r}'$ may then be written as

$$\mathbf{p}(\mathbf{r}') \approx [\varepsilon(\mathbf{r}') - \varepsilon_0]\mathbf{E}_0(\mathbf{r}')d^3r'. \tag{15.97}$$

Let us consider fields far from the scatterer, so that we may use the far-zone approximation. From equation (14.130), we then have

$$\mathbf{E}_s(\mathbf{r}) = -\frac{k^2}{\varepsilon_0}\hat{\mathbf{R}} \times (\hat{\mathbf{R}} \times \mathbf{p})\mathcal{G}(\mathbf{R}), \tag{15.98}$$

where $\mathbf{R} = \mathbf{r} - \mathbf{r}'$. We may further simplify this by noting that in the far zone, $\hat{\mathbf{R}} \approx \hat{\mathbf{r}}$. We further simplify this to

$$\mathbf{E}_s(\mathbf{r}) = -\frac{k^2}{\varepsilon_0}\hat{\mathbf{r}} \times (\hat{\mathbf{r}} \times \mathbf{p})\mathcal{G}(\mathbf{R}). \tag{15.99}$$

We substitute in the definition for $\mathcal{G}$ as well as $\mathbf{p}$ and have

$$\mathbf{E}_s(\mathbf{r}) = -\int_D \frac{\omega^2\mu_0(\varepsilon - \varepsilon_0)}{4\pi}\hat{\mathbf{r}} \times [\hat{\mathbf{r}} \times \mathbf{E}_0(\mathbf{r}')]\frac{e^{ik|\mathbf{r}-\mathbf{r}'|}}{|\mathbf{r} - \mathbf{r}'|}d^3r'. \tag{15.100}$$

We now use our definition for the scattering potential, equation (15.43), and integrate over $\mathbf{r}'$. We finally get an expression for the scattered field,

$$\mathbf{E}_s(\mathbf{r}) = -\int_D F(\mathbf{r}')\hat{\mathbf{r}} \times [\hat{\mathbf{r}} \times \mathbf{E}_0(\mathbf{r}')]\frac{e^{ik|\mathbf{r}-\mathbf{r}'|}}{|\mathbf{r} - \mathbf{r}'|}d^3r'. \tag{15.101}$$

However, let us compare this to the first Born approximation, equation (15.51),

$$\mathbf{E}_s^{(1)}(\mathbf{r}) = \int_D \mathbf{G}(\mathbf{r}, \mathbf{r}') \cdot \mathbf{E}_0(\mathbf{r}')F(\mathbf{r}')d^3r'. \tag{15.102}$$

Using the definition of $\mathbf{G}(\mathbf{r}, \mathbf{r}')$ and the far-zone identity $\nabla \leftrightarrow \hat{\mathbf{r}}$, we find that the Born approximation may be written as

$$\mathbf{E}_s^{(1)}(\mathbf{r}) = -\int_D \hat{\mathbf{r}} \times [\hat{\mathbf{r}} \times \mathbf{E}_0(\mathbf{r}')]F(\mathbf{r}')\frac{e^{ik|\mathbf{r}-\mathbf{r}'|}}{|\mathbf{r} - \mathbf{r}'|}d^3r', \tag{15.103}$$

which is exactly the Rayleigh–Gans approximation! The first Born approximation and the Rayleigh–Gans approximation are equivalent. The Rayleigh–Gans formulation, however, gives us some intuition for the conditions under which the first Born approximation is valid. It is clear that the particle has to have a refractive index very close to unity, and its size can be larger than a wavelength but not significantly so.

15.7 Mie scattering

We have seen that significant restrictions and approximations are typically needed in order to derive any analytical solution to a scattering problem. There is one case, however, where an exact solution to the electromagnetic scattering problem can be found: a sphere of radius a and constant permittivity ε_1, permeability μ_1, and refractive index n_1. We will assume that the background medium is vacuum, though it can be any nonabsorbent medium with constant permittivity and permeability. The solution to this problem was first found by Mie in 1908 and is therefore known as *Mie scattering* [11]. We have already developed most of the tools needed to solve the Mie problem in the course of developing the multipole expansion of the previous chapter, so we consider the derivation here.

Our approach will be somewhat similar to the approach used to derive the Rayleigh scattering solution: we determine a series expansion of the fields inside and outside of the sphere and then match our tangential boundary conditions, namely

$$\hat{\mathbf{r}} \times (\mathbf{H}_{out} - \mathbf{H}_{in}) = 0, \tag{15.104}$$

$$\hat{\mathbf{r}} \times (\mathbf{E}_{out} - \mathbf{E}_{in}) = 0. \tag{15.105}$$

We will write these fields as we did in section 14.7, as a combination of vector functions $\mathbf{M}_\psi(\mathbf{r})$ and $\mathbf{N}_\psi(\mathbf{r})$, with

$$\mathbf{M}_\psi = \nabla \times (\mathbf{r}\psi), \tag{15.106}$$

$$\mathbf{N}_\psi = \frac{1}{k} \nabla \times \mathbf{M}_\psi, \tag{15.107}$$

but with k in this case depending on the refractive index of the particular region, i.e. inside or outside the sphere.

We assume the field incident upon the sphere is a monochromatic plane wave traveling in the z-direction and polarized in the x-direction, i.e.

$$\mathbf{E}_0(\mathbf{r}) = E_0 \hat{\mathbf{x}} e^{ik_0 z} = E_0 [\hat{\mathbf{r}} \sin\theta\cos\phi + \hat{\boldsymbol{\phi}}\cos\theta\cos\phi - \hat{\boldsymbol{\phi}}\sin\phi] e^{ik_0 r \cos\theta}. \tag{15.108}$$

Let us draw some intuition from the Rayleigh scattering problem. In that case, we found that the scattered fields inside and outside the sphere had to match the angular dependence of the incident field, which was itself written in spherical coordinates. We apply a similar approach here: we want to find an expansion of the incident field in terms of an infinite sum of spherical waves; the form of those spherical waves will determine the form of the scattered fields we choose.

We have already seen in section 14.7 that all free-propagating fields in a constant medium can be written in terms of spherical harmonics and spherical Bessel functions; here, we choose a slightly different basis of even and odd azimuthal functions and drop the normalization constants,

$$\psi_{elm}(k\mathbf{r}) = \cos(m\phi)P_l^m(\cos\theta)z_l(kr), \tag{15.109}$$

$$\psi_{olm}(k\mathbf{r}) = \sin(m\phi)P_l^m(\cos\theta)z_l(kr), \tag{15.110}$$

where $z_l(kr)$ represents some combination of spherical Bessel functions $j_l(kr)$ and spherical Neumann functions $n_l(kr)$ to be determined. We may then introduce a set of vectors $\mathbf{M}_{elm}(k\mathbf{r})$, $\mathbf{N}_{elm}(k\mathbf{r})$, $\mathbf{M}_{olm}(k\mathbf{r})$, and $\mathbf{N}_{olm}(k\mathbf{r})$, constructed using the scalars $\psi_{elm}(k\mathbf{r})$ and $\psi_{olm}(k\mathbf{r})$. We note that k will be different inside and outside the sphere, and so we have modified the arguments of our functions to include k explicitly.

The actual derivation of the incident fields in terms of these functions is quite involved; we will simply present the results here, as we only need them once! A full derivation can be found in the book by Bohren and Huffman [12]. The electric and magnetic fields of the incident wave are given by

$$\mathbf{E}_0(\mathbf{r}) = E_0\sum_{n=1}^{\infty}i^n\frac{2n+1}{n(n+1)}\left[\mathbf{M}_{o1n}^{(j)}(k_0\mathbf{r}) - i\mathbf{N}_{e1n}^{(j)}(k_0\mathbf{r})\right], \tag{15.111}$$

$$\mathbf{H}_0(\mathbf{r}) = -\frac{k_0E_0}{\omega\mu_0}\sum_{n=1}^{\infty}i^n\frac{2n+1}{n(n+1)}\left[\mathbf{M}_{e1n}^{(j)}(k_0\mathbf{r}) + i\mathbf{N}_{o1n}^{(j)}(k_0\mathbf{r})\right], \tag{15.112}$$

where the superscript '(j)' is used to indicate that $z_l(k_0r)$ is taken to be the spherical Bessel function $j_l(kr)$. Though we do not derive these results, we can see signs of their plausibility. For example, it should be noted that only the $m=1$ term arises in the expansion; referring back to equation (15.108), we can see that this is because our plane wave only depends on $\cos\phi$ and $\sin\phi$.

We now introduce similar expansions for the scattered field outside the sphere and the internal field. Outside the sphere, we write

$$\mathbf{E}_s(\mathbf{r}) = \sum_{n=1}^{\infty}E_n\left[ia_n\mathbf{N}_{e1n}^{(h)}(k_0\mathbf{r}) - b_n\mathbf{M}_{o1n}^{(h)}(k_0\mathbf{r})\right], \tag{15.113}$$

$$\mathbf{H}_s(\mathbf{r}) = \frac{k_0}{\omega\mu_0}\sum_{n=1}^{\infty}E_n\left[a_n\mathbf{M}_{e1n}^{(h)}(k_0\mathbf{r}) + ib_n\mathbf{N}_{o1n}^{(h)}(k_0\mathbf{r})\right], \tag{15.114}$$

where '(h)' now refers to the spherical Hankel function of the first kind $h_l^{(1)}(k_0r)$, and

$$E_n = E_0i^n\frac{2n+1}{n(n+1)}. \tag{15.115}$$

This latter definition allows us to more easily match the boundary conditions of the incident, scattered, and internal fields. The choice of the Hankel function guarantees that the scattered field is an outgoing spherical wave.

The internal fields may be written in a similar manner as

$$\mathbf{E}_1(\mathbf{r}) = \sum_{n=1}^{\infty} E_n \left[-i d_n \mathbf{N}_{e1n}^{(j)}(k_1 \mathbf{r}) + c_n \mathbf{M}_{o1n}^{(j)}(k_1 \mathbf{r}) \right], \tag{15.116}$$

$$\mathbf{H}_1(\mathbf{r}) = -\frac{k_1}{\omega \mu_0} \sum_{n=1}^{\infty} E_n \left[d_n \mathbf{M}_{e1n}^{(j)}(k_1 \mathbf{r}) + i c_n \mathbf{N}_{o1n}^{(j)}(k_1 \mathbf{r}) \right], \tag{15.117}$$

where inside the sphere, the fields are determined by $j_l(k_1 r)$ because they must be finite at the origin. The positions of the i's in the expressions are again chosen to match the incident field.

Let us now turn to matching the boundary conditions. We require the tangential components of the fields to be continuous at the surface. Referring back to equations (14.163)–(14.168), we can explicitly write the boundary conditions associated with the electric and magnetic fields at $r = a$, which are the conditions that the $\hat{\phi}$ and $\hat{\theta}$ of each field are continuous. A little thought, however, indicates that we can also just require the tangential components of $\mathbf{M}_e$, $\mathbf{M}_o$, $\mathbf{N}_e$, and $\mathbf{N}_o$ to be individually continuous. This would suggest that eight conditions need to be satisfied, but one can readily find that four of these conditions are redundant. We therefore choose to enforce continuity on the $\hat{\phi}$ components of $\mathbf{M}$ and $\mathbf{N}$. This amounts to requiring continuity of

$$\psi_{elm}, \quad \psi_{olm}, \quad \frac{1}{kr} \partial_r \left[r \psi_{elm} \right], \quad \frac{1}{kr} \partial_r \left[r \psi_{olm} \right]. \tag{15.118}$$

One other simplification will help greatly here. We introduce the Riccati–Bessel functions,

$$\psi_l(z) = z j_l(z), \quad \zeta_l(z) = z h_l^{(1)}(z). \tag{15.119}$$

Applying the boundary conditions results in a system of four equations for a_n, b_n, c_n, and d_n. Solving this system is easier than it appears because the equations for a_n and d_n are decoupled from the equations for b_n and c_n. Let us introduce $\beta_0 \equiv k_0 a$ and $\beta_1 \equiv k_1 a$. Then, for the scattered field amplitudes, we have

$$a_n = \frac{\psi_n(\beta_0)\psi_n'(\beta_1) - \dfrac{n_1}{\tilde{\mu}_1}\psi_n(\beta_1)\psi_n'(\beta_0)}{\psi_n'(\beta_1)\zeta_n(\beta_0) - \dfrac{n_1}{\tilde{\mu}_1}\psi_n(\beta_1)\zeta_n'(\beta_0)}, \tag{15.120}$$

$$b_n = \frac{\psi_n'(\beta_0)\psi_n(\beta_1) - \dfrac{n_1}{\tilde{\mu}_1}\psi_n'(\beta_1)\psi_n(\beta_0)}{\psi_n(\beta_1)\zeta_n'(\beta_0) - \dfrac{n_1}{\tilde{\mu}_1}\psi_n'(\beta_1)\zeta_n(\beta_0)}, \tag{15.121}$$

where the prime represents the derivative with respect to the argument of the function, and $\tilde{\mu}_1$ represents the relative permeability.

The coefficients for the interior fields are less commonly needed but are readily derived as well. After some effort, we obtain

$$c_n = \frac{n_1\left[\psi_n(\beta_0)\zeta_n'(\beta_0) - \psi_n'(\beta_0)\zeta_n(\beta_0)\right]}{\psi_n(\beta_1)\zeta_n'(\beta_0) - \dfrac{n_1}{\tilde{\mu}_1}\psi_n'(\beta_1)\zeta_n(\beta_0)}, \tag{15.122}$$

$$d_n = \frac{n_1\left[\psi_n'(\beta_0)\zeta_n(\beta_0) - \psi_n(\beta_0)\zeta_n'(\beta_0)\right]}{\psi_n'(\beta_1)\zeta_n(\beta_0) - \dfrac{n_1}{\tilde{\mu}_1}\psi_n(\beta_1)\zeta_n'(\beta_0)}. \tag{15.123}$$

It is worth noting that there is no completely standard notation for solving the Mie problem. As we have already noted, we made certain choices for the placement of the i's to make our solution more elegant; other authors make different choices. In the end, the results for the fields are all the same.

The Mie solution is, in principle, an exact analytic result but has a summation over an infinite number of terms. It was shown early on by Debye [13] that the scattering coefficients go rapidly to zero for $n > ka$; this has become the guideline for determining how many coefficients are needed for an accurate result. Furthermore, we note that this condition implies that even the Mie theory is impractical for spheres significantly greater than the wavelength; if we take $\lambda = 500$ nm, for example, and a modest sphere of $a = 10\ \mu$m, we are already at $n > 125$. For larger spheres, one must incorporate tools from geometrical optics; see, for example, Grandy [14].

Because our fields are written in the form of the functions $\mathbf{N}$ and $\mathbf{M}$, we can roughly connect them to the higher-order multipoles of section 14.7. The coefficients a_n appear to be associated with electric multipoles, while b_n are associated with magnetic multipoles. In the limit of a small sphere, i.e $ka \ll 1$, we therefore expect only the a_1 term to be significant and for its form to mirror the results from Rayleigh scattering in section 15.5. We can see this by looking at the small-argument forms of the spherical Bessel functions, i.e.

$$\psi_1(z) \approx \frac{z^2}{3}, \tag{15.124}$$

$$\zeta_1(z) \approx \frac{z^2}{3} - \frac{i}{z}. \tag{15.125}$$

These results follow from the elementary properties of spherical Bessel functions. On substitution into equation (15.120), and using the approximate limit $ka \to 0$, we can find that

$$a_n \approx 2(ka)^3\frac{n^2 - 1}{n^2 + 2}. \tag{15.126}$$

If we compare this to the Rayleigh polarizability, equation (15.85), we can see a direct correspondence.

Mie theory is often used for calculating the total scattering cross section and extinction cross section of a spherical particle. The scattering cross section is found to be of the form

$$\sigma_s = \frac{2\pi}{k^2} \sum_{n=1}^{\infty} (2n+1)[|a_n|^2 + |b_n|^2], \qquad (15.127)$$

and the extinction cross section is found to be

$$\sigma_e = \frac{2\pi}{k^2} \sum_{n=1}^{\infty} (2n+1)\mathrm{Re}\{a_n + b_n\}. \qquad (15.128)$$

These formulas take some effort to derive, so we only provide a starting point to the calculation for the scattering cross section, which involves applying properties of associated Legendre polynomials in the end. Because $\mathbf{M} = -\mathbf{r} \times \nabla\psi$, it follows that the field $\mathbf{M}$ has no radial component. If we look at the form of the corresponding $\mathbf{N}$, we find it has components in all three orthogonal directions; however, the radial component of $\mathbf{N}$ does not contribute to an outgoing radiation field, so we ignore it. If we use the far-zone approximation for $h_n^{(1)}(k_0 r)$ and neglect any terms that decay at a rate greater than $1/r$, we can find that

$$\mathbf{M} = \hat{\boldsymbol{\theta}} M_\theta + \hat{\boldsymbol{\phi}} M_\phi, \qquad (15.129)$$

$$\mathbf{N} = i[-\hat{\boldsymbol{\theta}} M_\phi + \hat{\boldsymbol{\phi}} M_\theta], \qquad (15.130)$$

where we have included a general expression for $\mathbf{M}$ as well. In the far zone, then, the Poynting vector only has a radial component that can be integrated over. From here, we must apply equations (15.109) and (15.110) and the orthogonality properties of these functions to derive the result.

The Mie solution highlights the richness and complexity of scattering problems, even for the 'simplest' scatterers possible. The coefficients a_n and b_n depend on frequency, and resonance effects associated with the different multipole moments can be excited at special frequencies. One advantage of Mie theory is that the different multipoles, represented by a_n and b_n, can be plotted separately to isolate the physical origin of a particular resonance in the total scattered field. Collections of Mie-type particles are often used for modeling meta-atoms, which we have seen require strong electric and magnetic resonances to create novel optical effects.

An example of a Mie scattering calculation is shown in figure 15.7. The sphere is taken to have a 100 nm radius and an index of $n_1 = 4.0$. Instead of showing the cross sections, the figure shows the scattering efficiency Q_s, which is defined as the scattering cross section divided by the geometrical cross section πa^2 of the sphere. Also shown are the individual contributions of the lowest-order multipole moments to the scattered power. One can see that the sphere exhibits multiple resonant features, each associated with a distinct multipole. Strong resonances, as shown in

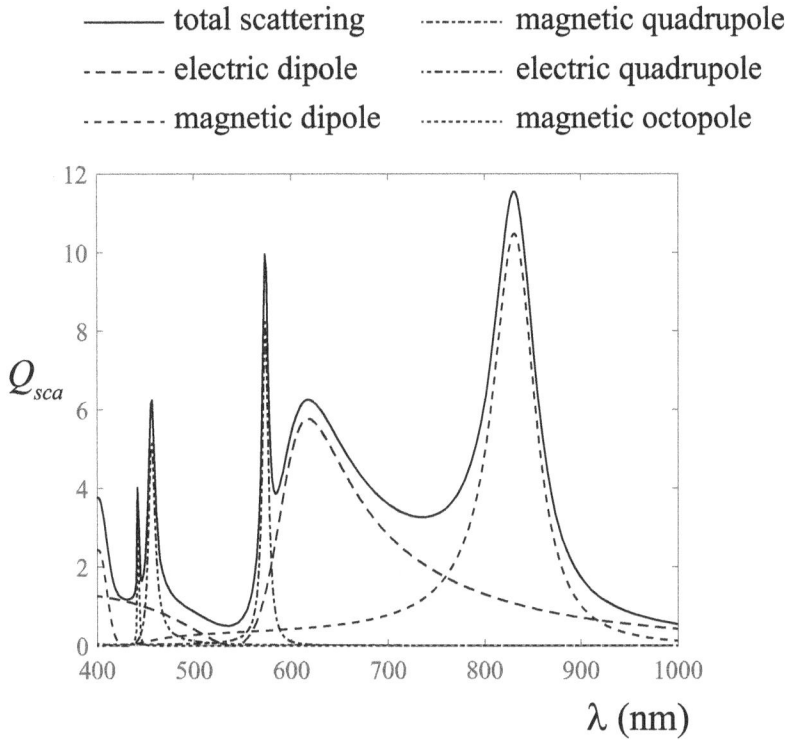

Figure 15.7. Scattering efficiency of Mie scattering, with $a = 100$ nm and $n_1 = 4.0$, as a function of wavelength. Parameters taken from the excellent Wikipedia article on Mie scattering [15] because I could not find a source that more clearly showed the different multipole resonances.

the figure, typically only arise when the refractive index of the sphere is large; for more reasonable refractive index values, the Mie scattering spectrum is much smoother.

The Mie solution is also useful for testing new computational models for electromagnetic scattering and also for testing inverse scattering techniques. It is worth noting in this context that the Mie formalism can be extended to a multilayer sphere with a constant refractive index in each layer. In 2003, Xu *et al* [16] introduced a transfer matrix formalism that might be considered the spherical analogy to the matrix approach used in chapter 10. In particular, they investigated so-called onion resonators, spheres coated with a periodic multilayer film that can have extremely high quality factors [17] for cavity quantum electrodynamics. We also refer the reader to an earlier iterative approach for calculating multilayer sphere scattering by Wu and Wang [18].

Mie scattering is commonly used to model the interaction of light with particulate matter in the atmosphere, such as dust, water droplets, smoke, or pollen. In such cases, the particle size can be comparable to the wavelength of light, and Mie theory gives an excellent description of the light scattering.

15.8 Inverse problems

Historically, physics has concerned itself with finding an 'effect' due to a 'cause.' In first-year physics classes, for example, one calculates the trajectory of a ball (the effect) from the knowledge of how it was thrown (the cause). In scattering theory, we have considered problems where the scattered field (the effect) is derived from knowledge of the scattering object and the illuminating field (the cause).

It is also possible to consider the problem in reverse, i.e. to deduce the cause of an experimentally measured effect. In the case of the ball, one might attempt to determine the initial speed and position of the ball from knowledge of its trajectory; in scattering theory, one might attempt to determine the structure of the scattering object (the cause) from knowledge of the scattered field (the effect). Such problems are now referred to as *inverse problems*, and they became a major field of study with the advent of medical imaging techniques such as computed tomography (CT) and magnetic resonance imaging (MRI). In CT, x-rays are passed through a body from different directions, and the absorption as a function of position is measured for each of these projections. From the collected data, one can use a computer to mathematically construct an image of the interior of the body. Hounsfield published his first results of medical imaging via CT in 1973 [19], and it revolutionized both medicine and imaging science.

Since Hounsfield's publication, many different inverse problems have been developed for imaging the interior of objects for a variety of wavelengths and wave types, including electromagnetic waves, acoustic waves, and even electron beams. Many of these can be referred to as *inverse scattering problems*. Every inverse scattering problem is the reverse of a *direct scattering problem* that follows the usual cause–effect relationship. In this section, we consider the mathematical solution of one of the simplest inverse scattering problems: determining the structure of a scattering object within the context of the first Born approximation.

The problem we want to solve may be formulated as follows: we illuminate the scattering object with plane waves from multiple directions, and for each of these illuminating directions, we measure the far-zone scattered field in multiple directions. From this data, can we determine the structure of the scattering potential $F(\mathbf{r})$?

We consider again the equation for the scattered field in the far zone under the first Born approximation,

$$\mathbf{E}_s(\mathbf{r}) = -\int_D \hat{\mathbf{r}} \times [\hat{\mathbf{r}} \times \mathbf{E}_0(\mathbf{r}')] F(\mathbf{r}') \frac{e^{ik|\mathbf{r}-\mathbf{r}'|}}{|\mathbf{r} - \mathbf{r}'|} d^3r', \qquad (15.131)$$

and we would like to see whether it is possible to mathematically determine the form of $F(\mathbf{r})$ from measurements of the scattered field $\mathbf{E}_s(\mathbf{r})$. First, let us assume that the incident field is a plane wave propagating in the direction $\mathbf{s}_0$,

$$\mathbf{E}_0(\mathbf{r}) = \mathbf{E}_{\mathbf{s}_0} e^{ik\mathbf{s}_0 \cdot \mathbf{r}}, \qquad (15.132)$$

where $\mathbf{E}_{\mathbf{s}_0}$ is a constant electric field vector perpendicular to $\mathbf{s}_0$.

We now make one more far-zone approximation for the Green's function. We note that

$$|\mathbf{r} - \mathbf{r}'| = \sqrt{r^2 - 2\mathbf{r} \cdot \mathbf{r}' + r'^2} = r\sqrt{1 - 2\mathbf{r} \cdot \mathbf{r}'/r^2 + r'^2/r^2}. \qquad (15.133)$$

In the limit $r \gg r'$, we can neglect the third term in the square root. We may then use the binomial expansion to approximate this expression as

$$|\mathbf{r} - \mathbf{r}'| \approx r - \mathbf{r} \cdot \mathbf{r}'/r. \qquad (15.134)$$

It is convenient to write the unit vector $\hat{\mathbf{r}}$ as $\mathbf{s}$, with $\mathbf{r} = r\mathbf{s}$, so that we have

$$|\mathbf{r} - \mathbf{r}'| \approx r - \mathbf{s} \cdot \mathbf{r}'. \qquad (15.135)$$

In the exponent of the Green's function, we use this approximation. In the denominator of the Green's function, we use $|\mathbf{r} - \mathbf{r}'| \approx r$. We may then write equation (15.131) as

$$\mathbf{E}_s(\mathbf{r}) = -\mathbf{s} \times [\mathbf{s} \times \mathbf{E}_{\mathbf{s}_0}] \frac{e^{ikr}}{r} \int_D F(\mathbf{r}') e^{-ik(\mathbf{s} - \mathbf{s}_0) \cdot \mathbf{r}'} d^3 r'. \qquad (15.136)$$

The integral itself in this expression represents the three-dimensional spatial Fourier transform of the scattering potential. We may write this transform as

$$\tilde{F}(\mathbf{K}) = \frac{1}{(2\pi)^3} \int F(\mathbf{r}') e^{-i\mathbf{K} \cdot \mathbf{r}'} d^3 r', \qquad (15.137)$$

so that the scattered field may be written as

$$\mathbf{E}_s(\mathbf{r}) = -(2\pi)^3 \mathbf{s} \times [\mathbf{s} \times \mathbf{E}_{\mathbf{s}_0}] \frac{e^{ikr}}{r} \tilde{F}[k(\mathbf{s} - \mathbf{s}_0)]. \qquad (15.138)$$

We can see from this expression that every measurement of the scattered field for a given $(\mathbf{s}, \mathbf{s}_0)$ pair gives us the value of one component of $\tilde{F}[\mathbf{K}]$. If we measure the scattered field for all directions of scattering $\mathbf{s}$ for a given direction of incidence $\mathbf{s}_0$, we find that we get information about $\tilde{F}(\mathbf{K})$ on a spherical shell centered on $k\mathbf{s}_0$, as illustrated in figure 15.8. These spheres are known as *Ewald spheres* after Paul Peter Ewald, who introduced the concept in the context of x-ray crystallography [20]. If we measure the scattered field on Ewald spheres for all directions of illumination $\mathbf{s}_0$, we have the values of $\tilde{F}(\mathbf{K})$ within the volume of a sphere of radius $2k$, often called the *Ewald limiting sphere* Σ_L.

It follows from a theorem of Fourier analysis that a unique band-limited reconstruction of the scattering potential can be determined from values of the Fourier transform of the potential within a volume. This band-limited version may be written as

$$F_{BL}(\mathbf{r}) = \int_{|\mathbf{K}| \leqslant 2k} \tilde{F}(\mathbf{K}) e^{i\mathbf{K} \cdot \mathbf{r}} d^3 K, \qquad (15.139)$$

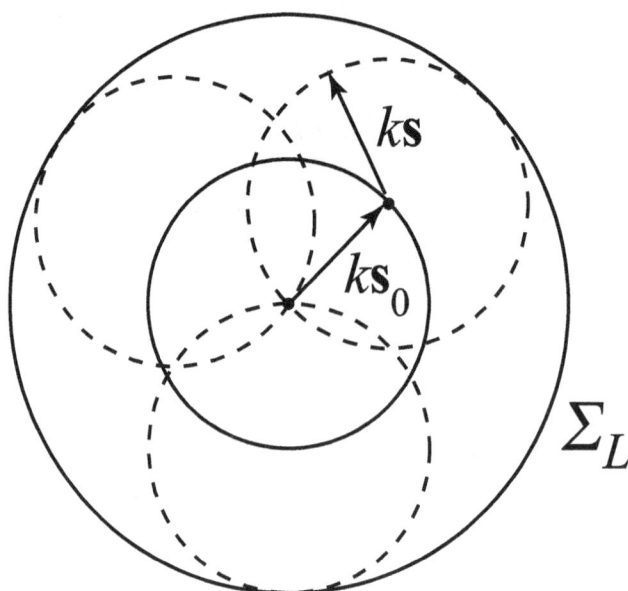

Figure 15.8. Illustration of Ewald spheres and the Ewald limiting sphere Σ_L.

and it is effectively a blurred image of the scattering potential, lacking any features that involve spatial frequencies higher than $2k$. This result was proven by Wolf and Habashy [21].

It should be noted that the vector properties of the scattered field, given by the triple product in equation (15.138), are easily accounted for in the analysis of data, assuming the direction of scattering and the direction of $\mathbf{E}_{s_0}$ are known. This vector is generally nonzero everywhere except for isolated directions of scattering.

One additional limitation of our inverse method is that it requires measurement of the scattered electric field, which includes both phase and amplitude measurements. Applying this method therefore requires some method such as interferometry to determine the phase.

Inverse problems reverse the usual cause–effect way of thinking, and as such, they can suffer from limitations that the corresponding direct problems do not possess. One of these is *noncontinuity*: small errors in the data can lead to large errors in the reconstructed image. This is a significant problem that can be addressed at least in part with mathematical regularization methods. A more serious problem is *nonuniqueness*: an inverse problem may, in fact, not have a unique solution at all, at least as it is posed.

As an example, consider the inverse scattering problem discussed above, but let us consider the case where we only collect scattering data for a finite number of directions of illumination. As was shown by Devaney [22], it is possible to construct objects that are nonscattering, i.e. invisible, for a finite number of illumination directions. These invisible objects cannot be reconstructed at all, and this also implies that any object we attempt to reconstruct might have unseen features. It is only in the ideal limit when a continuous range of illumination and scattering directions is considered that we can get a unique reconstruction. In general, the nonuniqueness of an inverse problem is related to the existence of invisible objects for that problem [23]. Proving uniqueness is a priority in the development of any new inverse problem.

The problem we have solved here is very limited in application because it applies only to weakly scattering objects, i.e. extremely small and transparent objects. Most inverse scattering problems that have a formal analytic solution involve some strong limitation in their scope. Attempting to solve the problem for objects that exhibit strong scattering invariably requires computational methods.

For more information on the basics of inverse problems, I recommend the book by Bertero and Boccacci [24]. For more information specifically about electromagnetic inverse problems, see Colton and Kress [25].

15.9 Anapoles

Referring back to equations (15.120) and (15.121) that provide the scattered field amplitudes in Mie theory, there is one subtle but striking implication of these formulas: the numerators can potentially go identically to zero, meaning that there is no scattered field for that multipole. However, the corresponding amplitudes c_n and d_n for the fields within the scatterer, and the corresponding induced polarization, will in general not be zero, resulting in a scattered electromagnetic field localized to the scattering object itself. This sort of excitation is known as an *anapole* excitation, from the Greek 'ana' meaning 'without,' i.e. 'without poles.'

To understand the origin of an anapole, let us return to the spherical multipole expansions of section 14.7 and consider the electric dipole moments. We will assume that we have a current distribution rotationally symmetric around the z-axis; in this case, we expect only the a_{10} moment to be of significance,

$$a_{10} = -\sqrt{\frac{\mu}{\epsilon}} \frac{1}{2} \int_D \mathbf{J}(\mathbf{r}') \cdot \left\{ \nabla' \times \left[\nabla' \times \left(\mathbf{r}' \psi_{10}^*(\mathbf{r}') \right) \right] \right\} d^3 r', \tag{15.140}$$

where we can write $\psi_{10}(\mathbf{r})$ as

$$\mathbf{r}' \psi_{10}(\mathbf{r}') = \sqrt{\frac{3}{4\pi}} r' j_1(kr') \cos\theta' \hat{\mathbf{r}}'. \tag{15.141}$$

Let us further consider only the $\hat{\mathbf{z}}$-component of $\mathbf{J}(\mathbf{r}')$; the other components can be non-zero, but by the assumed symmetry of the problem, only the $\hat{\mathbf{z}}$-component will contribute to the final dipole. We write $\mathbf{J}(\mathbf{r}') = \hat{\mathbf{z}} J_z(\mathbf{r}')$, and we can evaluate the curls explicitly in equation (15.140). In doing so, we will take advantage of the following spherical Bessel function identities,

$$\frac{2n+1}{x} j_n(x) = j_{n-1}(x) + j_{n+1}(x), \tag{15.142}$$

$$(2n+1) j_n'(x) = n j_{n-1}(x) - (n+1) j_{n+1}(x), \tag{15.143}$$

as well as basic trigonometric identities. With some straightforward manipulation, we finally arrive at

$$a_{10} = -\sqrt{\frac{\mu}{\epsilon}}\frac{k}{3}\sqrt{\frac{3}{4\pi}}\int J_z(\mathbf{r'})\left\{j_0(kr') + \frac{1}{2}\left[3\cos^2\theta' - 1\right]j_2(kr')\right\}d^3r'. \quad (15.144)$$

It is to be noted that this expression is quite different from the simple electric dipole formula of equation (14.92), which we now recognize as being exact only in the long wavelength limit. In that limit, in which $k \to 0$, the Bessel functions in the curved brackets can be expanded in terms of the lowest terms of their Taylor series approximations; the leading term of this expansion is

$$a_{10} \approx -\sqrt{\frac{\mu}{\epsilon}}\frac{k}{3}\sqrt{\frac{3}{4\pi}}\int J_z(\mathbf{r'})d^3r'. \quad (15.145)$$

If this is substituted into equation (14.232) for the field of a dipole in the spherical multipole expansion, one finds that it agrees perfectly with equation (14.229), the long wavelength approximation. disc So how can we interpret the higher-order Taylor series terms that arise in equation (15.144)? These represent more complicated distributions of current that still produce an electric dipole field. The next piece, of order r'^2 in the integrand, is of the form

$$a_{10}^{(2)} \approx -\sqrt{\frac{\mu}{\epsilon}}\frac{k}{15}\sqrt{\frac{3}{4\pi}}\int J_z(\mathbf{r'})\left[\cos^2\theta' - 2\right](kr')^2 d^3r'. \quad (15.146)$$

This is what is known as the toroidal dipole moment. The simplest form of a toroidal dipole would be a current that circulates along a torus oriented with its normal to the $\hat{\mathbf{z}}$-direction, with the current following a circle that lies along a line from the origin.

Beyond this, higher-order currents can be constructed that will also result in a dipole field. And a similar process can be taken with higher-order multipoles, as well. Our calculation was done only for the simplest source of high symmetry; the general case is considered by Fernandez-Corbaton et al. [26] and Alaee et al [27].

The basic anapole, illustrated in figure 15.9, consists of a linear dipole whose field cancels out exactly the field of a toroidal dipole. In this way, an anapole becomes a particular class of venerable objects known as *nonradiating sources*: sources with oscillating charges and currents that produce no radiation due to complete destructive interference [28]. These nonradiating sources have been studied on and off for more than a century, and are of particular relevance to the inverse source problem, in that they form 'invisible objects' in the context of the problem, as described in the previous section. A nonradiating source can be considered an extreme form of anapole, in which all multipole moments of a source vanish to all orders.

The anapole was first introduced by Zeldovich in 1957 in particle physics as an object that would naturally arise in weak force interactions [29]. In the context of elementary particles, they have also been considered as candidates for cosmic dark matter [30]; their effectively zero dipole moment would make them weakly

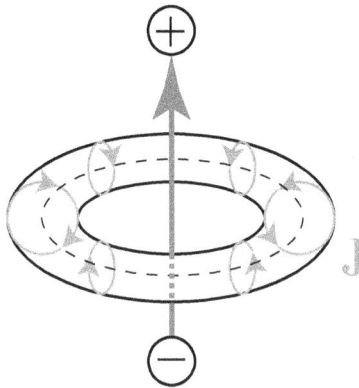

Figure 15.9. Illustration of the basic anapole: a linear dipole whose field cancels out the field of a toroidal dipole.

interacting with electromagnetic waves. In optics, interest in anapoles was sparked in 2015 with the observation that anapole excitations could be seen in dielectric nanoparticles [31]. Since then, anapoles have been studied extensively as another phenomenon that can be applied in metamaterial design [32, 33].

For our purposes, it suffices to say that anapoles demonstrate that very interesting phenomena can lie hidden in even very familiar physical problems like radiation and scattering.

15.10 Exercises

Note that these problems will feature a lot of Bessel function and spherical Bessel function manipulation, which is common in scattering. They also involve significantly more intensive math than the exercises of previous chapters; scattering theory is like that.

1. Calculate the three-dimensional Fourier transform of the scattering potential of a circular cylinder of length L and radius a that is centered and aligned on the z-axis with a uniform refractive index n. Evaluate the form of $\tilde{F}[k(\mathbf{s} - \mathbf{s}_0)]$, where $\mathbf{s}_0 = \hat{\mathbf{x}}$, using cylindrical coordinates. This is the Fourier transform that will appear in the first Born approximation.

2. Let us consider a weak scatterer that is a sphere of radius a and uniform refractive index n_0. Using the first Born approximation, write an expression for the scattered field, assuming the incident field is aligned in the $\hat{\mathbf{z}}$-direction and is $\hat{\mathbf{x}}$-polarized. Plot the radiation pattern of the scattered field,

$$R \equiv r^2 |\mathbf{E}_s(r\mathbf{s})|^2,$$

for the cases $a = 4\lambda$, $a = \lambda$, $a = \lambda/4$. Does the radiation pattern approach an expected dipole form as the sphere gets smaller?

3. Write an expression for the integrated far-zone scattered power entirely in terms of the scattered electric field vector, instead of the electric and magnetic

fields. (This is a handy simplification—no need to always calculate the **H**-field for every problem!)

4. Consider a spherical particle of radius a and uniform real-valued refractive index n_0 under the first Born approximation. The incident field is aligned in the $\hat{\mathbf{z}}$-direction and polarized in the $\hat{\mathbf{x}}$-direction. With the help of the results of exercise 2, use the optical theorem to determine the extinguished power. Is the result what you expect?

5. Let us study the Mie scattering of a sphere of radius 200 nm in vacuum. Write your own code for Mie scattering and plot the scattering efficiency as a function of wavelength for the cases $n = 1.33$ and $n = 2.4$ over the range $400\,\text{nm} \leqslant \lambda \leqslant 100\,\text{nm}$.

6. Consider a spherical particle of constant refractive index and radius a. In the limit that $a \ll \lambda$, show that the Mie scattering cross section approaches the Rayleigh scattering cross section.

7. Here, we take a brief look at the resolution limits that can appear in an inverse scattering problem. Let us consider the scattering of light from a sphere of uniform refractive index n and radius a. Analytically calculate the three-dimensional Fourier transform of the scattering potential. Now, assuming that the wavelength is $\lambda = 500\,\text{nm}$, numerically calculate and plot the inverse three-dimensional Fourier transform of the scattering potential for spheres of radius (a) $a = 1000\,\text{nm}$, (b) $a = 500\,\text{nm}$, (c) $a = 250\,\text{nm}$, (d) $a = 100\,\text{nm}$. (Because the scatterer is rotationally symmetric, you only need to plot the reconstruction as a function of radius.)

References

[1] Wolf E 1983 Recollections of Max Born *Opt. News.* **9** 10–6
[2] Oraevsky A N 2002 Whispering-gallery waves *Quantum Electron.* **32** 377
[3] Davies H 1960 On the convergence of the Born approximation *Nucl. Phys.* **14** 465–71
[4] Gbur G J 2011 *Mathematical Methods for Optical Physics and Engineering* (Cambridge: Cambridge University Press)
[5] Carney P S, Wolf E and Agarwal G S 1999 Diffraction tomography using power extinction measurements *J. Opt. Soc. Am.* A **16** 2643–8
[6] Born M and Wolf E 1999 *Principles of Optics* 7th edn (Cambridge: Cambridge University Press)
[7] Strutt J W 1871 XV. On the light from the sky, its polarization and colour *London, Edinburgh Dublin Phil. Mag. J. Sci.* **41** 107–20
[8] Ropars G, Gorre G, Floch A L, Enoch J and Lakshminarayanan V 2012 A depolarizer as a possible precise sunstone for Viking navigation by polarized skylight *Proc. R. Soc.* A **468** 671–84
[9] Wang Z, Wu H, Hu X, Zhao N, Mo Q and Li G 2016 Rayleigh scattering in few-mode optical fibers *Sci. Rep.* **6** 35844
[10] Gans R 1925 Strahlungsdiagramme ultramikroskopischer Teilchen *Ann. Phys., Lpz.* **381** 29–38
[11] Mie G 1908 Beiträge zur Optik trüber Medien, speziell kolloidaler Metallösungen *Ann. Phys., Lpz.* **330** 377–445

[12] Bohren C F and Huffman D R 1983 *Absorption and Scattering of Light by Small Particles* (New York: Wiley)

[13] Debye P 1909 Der Lichtdruck auf Kugeln von beliebigem Material *Ann. Phys., Lpz.* **335** 57–136

[14] Grandy W T Jr 2000 *Scattering of Waves from Large Spheres* (Cambridge: Cambridge University Press)

[15] Wikipedia contributors. Mie scattering—Wikipedia, The Free Encyclopedia; 2024. [Online] Available from: https://en.wikipedia.org/w/index.php?title=Mie_scattering&oldid=1225594381 (accessed 5 July 2024)

[16] Xu Y, Liang W, Yariv A, Fleming J G and Lin S Y 2003 High-quality-factor Bragg onion resonators with omnidirectional reflector cladding *Opt. Lett.* **28** 2144–6

[17] Liang W, Xu Y, Huang Y, Yariv A, Fleming J G and Lin S Y 2004 Mie scattering analysis of spherical Bragg 'onion' resonators *Opt. Express* **12** 657–69

[18] Wu Z S and Wang Y P 1991 Electromagnetic scattering for multilayered sphere: recursive algorithms *Radio Sci.* **26** 1393–401

[19] Hounsfield G N 1973 Computerized transverse axial scanning (tomography): part I. Description of system *Br. J. Radiol.* **46** 1016–22

[20] Ewald P P 1921 Die Berechnung optischer und elektrostatischer Gitterpotentiale *Ann. Phys., Lpz.* **369** 253–87

[21] Wolf E and Habashy T 1993 Invisible bodies and uniqueness of the inverse scattering problem *J. Mod. Opt.* **40** 785–92

[22] Devaney A J 1978 Nonuniqueness in the inverse scattering problem *J. Math. Phys.* **19** 1526–31

[23] Gbur G 2013 Invisibility physics: past, present, and future *Progress in Optics* vol 58 ed E Wolf (Amsterdam: Elsevier) 65–114

[24] Bertero M and Boccacci P 1998 *Introduction to Inverse Problems in Imaging* (Bristol and Philadelphia, PA: Institute of Physics Publishing)

[25] Colton D and Kress R 2019 *Inverse Acoustic and Electromagnetic Scattering Theory* (Switzerland: Springer)

[26] Fernandez-Corbaton Ivan, Nanz Stefan, Alaee Rasoul and Rockstuhl Carsten 2015 Exact dipolar moments of a localized electric current distribution *Opt. Express* **23** 33044–64

[27] Alaee Rasoul, Rockstuhl Carsten and Fernandez-Corbaton Ivan 2019 Exact multipolar decompositions with applications in nanophotonics *Advanced Optical Materials* **7** 1800783

[28] Greg Gbur 2003 Chapter 5—nonradiating sources and other 'invisible' objects *Progress in Optics* **vol 45** (Amsterdam: Elsevier) 273–315 pp

[29] Zel'dovich I B 1958 Electromagnetic interaction with parity violation *Soviet Phys. JETP* **6** 06

[30] Ho C M and Scherrer R J 2013 Anapole dark matter *Physics Letters B* **722** 341–6

[31] Miroshnichenko Andrey E *et al* 2015 Nonradiating anapole modes in dielectric nanoparticles *Nat. Commun.* **6** 8069

[32] Baryshnikova K V, Smirnova D A, Luk'yanchuk B S and Kivshar Y S 2019 Optical anapoles: Concepts and applications *Advanced Optical Materials* **7** 1801350

[33] Yang Y and Bozhevolnyi S I 2019 Nonradiating anapole states in nanophotonics: from fundamentals to applications *Nanotechnology* **30** 204001

IOP Publishing

Electromagnetic Optics

Gregory J Gbur

Chapter 16

Computational methods for Maxwell's equations

Throughout this book, we have endeavored to find analytical ('pen and paper') solutions to Maxwell's equations for a wide variety of systems. The advantage of this approach is that the exact solutions provide a lot of physical intuition. Often, these solutions use some sort of simplifying assumptions to make the problem tractable, for example, by assuming that the object has a permittivity close to the vacuum value in scattering problems or assuming that absorption is negligible in dispersion problems.

In many problems of interest, there are no analytical solutions available, or the analytical solutions use approximations that are inappropriate for the given problem. In these cases, we must turn to computational solutions in which the system or Maxwell's equations themselves are discretized.

There are many specialized techniques for solving very specific problems involving Maxwell's equations, and we cannot do justice to all of them here. We focus on three techniques that represent the solution of Maxwell's equations with some generality to give the reader some intuition as to the advantages and limitations of various techniques. A discussion of the proper implementation of these methods in practice would—and has—filled many books, so we explain the fundamental principles here and refer the reader to detailed references for further information.

Because the chapter is entirely computational, we have provided no exercises. The reader is encouraged to test out the computational techniques themselves.

16.1 The Foldy–Lax method

The simplest case we may consider is the scattering of monochromatic light from a finite collection of N Rayleigh-like particles of the type discussed in section 15.5. We label the positions of the particles r_i, where $i = 1, \ldots, N$, and the polarizability of each particle is denoted by α_i. The dipole moment of each particle is therefore

doi:10.1088/978-0-7503-6064-7ch16

$\mathbf{p}_i = \alpha_i \mathbf{E}(\mathbf{r}_i)$, where $\mathbf{E}(\mathbf{r}_i)$ is the *total* field experienced by the particle, which includes not only the incident field but also the scattered fields of every other particle.

What is the scattered field of each particle? Here, we borrow our formula for the field of an electric dipole from equation (14.124) and may write

$$\mathbf{E}(\mathbf{r}) = \frac{1}{\varepsilon_0}\mathbf{p}_i \cdot \{k^2\mathbf{I} + \boldsymbol{\nabla}\}\frac{1}{4\pi}\frac{e^{ikR}}{R} = \mathbf{p}_i \cdot \mathbf{G}(\mathbf{r}, \mathbf{r}_i), \tag{16.1}$$

where $R = |\mathbf{r}_i - \mathbf{r}|$ and we have introduced a Green's dyadic expression

$$\mathbf{G}(\mathbf{r}, \mathbf{r}_i) \equiv \frac{1}{4\pi\varepsilon_0}\{k^2\mathbf{I} + \boldsymbol{\nabla}\}\frac{e^{ikR}}{R}. \tag{16.2}$$

Assuming that we have an incident field of the form $\mathbf{E}_0(\mathbf{r})$, we may write an expression for the total field of the form

$$\mathbf{E}(\mathbf{r}) = \mathbf{E}_0(\mathbf{r}) + \sum_j \alpha_j \mathbf{E}(\mathbf{r}_j) \cdot \mathbf{G}(\mathbf{r}, \mathbf{r}_j). \tag{16.3}$$

We note that this expression fails if $\mathbf{r} = \mathbf{r}_j$, the position of one of the particles. Finally, we consider the case where $\mathbf{r} = \mathbf{r}_i$ and exclude the $i = j$ case from the sum. We then have

$$\mathbf{E}(\mathbf{r}_i) = \mathbf{E}_0(\mathbf{r}_i) + \sum_{i \neq j} \alpha_j \mathbf{E}(\mathbf{r}_j) \cdot \mathbf{G}(\mathbf{r}_i, \mathbf{r}_j). \tag{16.4}$$

We may now view this as a matrix equation, where we introduce

$$\mathbf{E}_{0i} = \mathbf{E}_0(\mathbf{r}_i), \quad \mathbf{E}_i = \mathbf{E}(\mathbf{r}_i), \quad \mathbf{G}_{ij} = \mathbf{G}(\mathbf{r}_i, \mathbf{r}_j), \tag{16.5}$$

and we define $\mathbf{G}_{ii} \equiv 0$. The resulting matrix equation becomes

$$\sum_j \delta_{ij}\mathbf{I} \cdot \mathbf{E}_j = \mathbf{E}_{0i} + \sum_j \alpha_j \mathbf{E}_j \cdot \mathbf{G}_{ij}, \tag{16.6}$$

and we have added the Kronecker delta δ_{ij}—the identity matrix—to the left of the expression, as well as the identity dyadic. We may rewrite this as

$$\mathbf{E}_{0i} = \sum_j [\mathbf{I}\delta_{ij} - \alpha_j \mathbf{G}_{ij}] \cdot \mathbf{E}_j. \tag{16.7}$$

Our result is a matrix equation and a matrix that is $3N \times 3N$ in size. The factor of three is due to the fact that we not only have N particles but also three components of polarization. We can then solve this matrix equation by a variety of methods for $\mathbf{E}_j$, the scattered field at each dipole. Direct matrix inversion is impractical for all but the smallest systems of particles, but there are many computationally efficient linear equation solvers available.

Once the scattered field is found at each of the particles, we may use equation (16.3) to find the scattered field at any other position. Because of the singular behavior of the field at each of the particles, this can only be done to find the fields outside the particle domain. However, we can 'regularize' our solution by defining a maximum cutoff amplitude for the Green's function around the singularities. This is not as difficult as it might appear. We have the explicit form of the field inside a Rayleigh scatterer, given by equation (15.86), and can use this as the maximum.

This approach for solving the field scattered by a system of small particles is often referred to as the Foldy–Lax method because of the independent work of Leslie Foldy [1] and Melvin Lax [2, 3]. Their original work predates modern computer simulation, and they used additional statistical methods to calculate the average scattering properties of a random system of particles. This is still one of the main uses of this method. It is also another method which can be used to test inverse scattering algorithms, as it involves a computationally 'exact' solution to Maxwell's equations.

Our choice $\mathbf{G}_{ii} \equiv 0$ amounts to ignoring any multiple scattering of light within individual particles. As long as we are considering small, Rayleigh-like particles, this appears to be a reasonable approximation.

To actually perform calculations using this method, equation (16.2) for the Green's function is insufficient due to the presence of the derivative operations. We can evaluate these derivatives explicitly in Cartesian coordinates, with the result in dyadic form given by

$$\mathbf{G}(\mathbf{r}, \mathbf{r}_i) \equiv \frac{1}{4\pi\varepsilon_0} \left\{ k^2[\mathbf{I} - \hat{\mathbf{R}}\hat{\mathbf{R}}] + \frac{ikR - 1}{R^2}[\mathbf{I} - 3\hat{\mathbf{R}}\hat{\mathbf{R}}] \right\} \frac{e^{ikR}}{R}. \qquad (16.8)$$

For much more on the multiple scattering of light by particles, see the book by Mishchenko, Travis, and Lacis [4].

16.2 Integral equation solutions and the dyadic Green's function

In the next section, we will look at methods for computationally solving the integral scattering equation,

$$\mathbf{E}(\mathbf{r}) = \mathbf{E}_0(\mathbf{r}) + \int_D \mathbf{G}(\mathbf{r}, \mathbf{r}') \cdot \mathbf{E}(\mathbf{r}')F(\mathbf{r}')d^3r', \qquad (16.9)$$

where we have formally taken $\mathbf{G}(\mathbf{r}, \mathbf{r}')$ to be of the form

$$\mathbf{G}(\mathbf{r}, \mathbf{r}') = \frac{4\pi}{i\omega\varepsilon_0}\Gamma_e(\mathbf{r}, \mathbf{r}') = \left[\mathbf{I} + \frac{1}{k^2}\nabla\right]\frac{e^{ik|\mathbf{r}-\mathbf{r}'|}}{|\mathbf{r} - \mathbf{r}'|}. \qquad (16.10)$$

In order to solve the integral equation, however, we will need to evaluate the electric field within the scatterer itself, which means that we will need to address the nonintegrable singularity in the Green's dyadic. Let us use the version given by equation (15.30), which provides the electric field due to an infinitesimal current density. As you may have noted by now, this dyadic only differs from the other forms we have introduced in scattering theory by a constant factor. The dyadic is given by

$$\Gamma_e(\mathbf{r}, \mathbf{r}') = \frac{i\omega\mu_0}{4\pi}\left[\mathbf{I} + \frac{1}{k^2}\nabla\right]\frac{e^{ik|\mathbf{r}-\mathbf{r}'|}}{|\mathbf{r} - \mathbf{r}'|}, \qquad (16.11)$$

and the electric field produced by a current is (formally) given by equation (15.19),

$$\mathbf{E}(\mathbf{r}) = \int_D \mathbf{\Gamma}_e(\mathbf{r}, \mathbf{r}') \cdot \mathbf{J}_e(\mathbf{r}') d^3 r'. \tag{16.12}$$

We have seen that the singularity of this dyadic is nonintegrable, as it is of the $1/R^3$ form, where only singularities of the $1/R^2$ form can be integrated out (which we can roughly attribute to the r'^2 in the volume element $d^3 r'$ in spherical coordinates). The singularity is not a problem for a collection of point particles, because we do not perform any integrals where the field point is within the scatterers, but it is a significant issue for bulk scatterers.

To deal with the singularity, we recall that we have another expression for the electric field of an infinitesimal current based on the potentials. From equation (14.24), we have

$$\mathbf{E}(\mathbf{r}) = -\nabla \phi(\mathbf{r}) + i\omega \mathbf{A}(\mathbf{r}), \tag{16.13}$$

and from equations (14.54) and (14.57), we have

$$\mathbf{A}(\mathbf{r}) = \frac{\mu_0}{4\pi} \int_D \mathbf{J}(\mathbf{r}') G(R) d^3 r', \tag{16.14}$$

$$\phi(\mathbf{r}) = \frac{1}{4\pi i \omega \varepsilon_0} \int_D \nabla' \cdot \mathbf{J}(\mathbf{r}') G(R) d^3 r', \tag{16.15}$$

where we have written $G(R) = \exp[ikR]/R$.

The first thing we note is that we may perform an integration by parts on $\phi(\mathbf{r})$ and discard the resulting surface integral, as the surface S may be taken to fully enclose the current. We may then write

$$\phi(\mathbf{r}) = -\frac{1}{4\pi i \omega \varepsilon_0} \int_D \nabla' G(R) \cdot \mathbf{J}(\mathbf{r}') d^3 r'. \tag{16.16}$$

The total electric field can then be written in the form

$$\mathbf{E}(\mathbf{r}) = \frac{i\omega \mu_0}{4\pi} \left[\mathbf{I} \cdot \int_D \mathbf{J}(\mathbf{r}') G(R) d^3 r' - \frac{1}{k^2} \nabla \int_D \nabla' G(R) \cdot \mathbf{J}(\mathbf{r}') d^3 r' \right]. \tag{16.17}$$

This expression for the electric field, unlike our Green's function expression, is perfectly well-behaved for points $\mathbf{r}$ within D. On comparison with equation (16.12), we can see that the singularity problem arises from attempting to bring the second 'del' into the integral in equation (16.17). We must therefore determine a way to do so that (mostly) does not result in a nonintegrable singularity.

Before continuing, I should note that our arguments will be rather nonrigorous here, in an attempt to make the reasoning clear. Rigorous derivations of what follows can be found throughout the work of Van Bladel [5], Fikioris [6], Lee et al [7], Yaghjian [8] and Wang [9]. These articles can be hard to follow; hence, we attempt a more intuitive derivation.

Our first step is to divide the volume of integration D into a small volume D' that includes the point $\mathbf{r}$ and the remaining volume $D - D'$. In the region $D - D'$, the

Green's function is well-behaved, and we may immediately bring the derivative into the integral and use $\nabla' G(R) = -\nabla G(R)$. We then have

$$\mathbf{E}(\mathbf{r}) = \int_{D-D'} \mathbf{\Gamma}_e(\mathbf{r}, \mathbf{r}') \cdot \mathbf{J}_e(\mathbf{r}') d^3 r' + \frac{i\omega\mu_0}{4\pi}\left[\mathbf{I} \cdot \int_{D'} \mathbf{J}(\mathbf{r}') G(R) d^3 r' - \frac{1}{k^2}\nabla \int_{D'} \nabla' G(R) \cdot \mathbf{J}(\mathbf{r}') d^3 r'\right]. \quad (16.18)$$

The trick is now to write the contribution to the electric field due to D', which we label $\mathbf{E}_{D'}(\mathbf{r})$, in a form that is straightforward to evaluate and is well-behaved. Let us introduce the static Green's function,

$$G_0(\mathbf{r}, \mathbf{r}') = \frac{1}{|\mathbf{r} - \mathbf{r}'|}, \quad (16.19)$$

and let us add (and subtract) $\nabla' G_0(R) \cdot \mathbf{J}(\mathbf{r})$ to (from) the second integrand on the right of equation (16.18). We then have

$$\mathbf{E}_{D'}(\mathbf{r}) = \frac{i\omega\mu_0}{4\pi}\left[\mathbf{I} \cdot \int_{D'} \mathbf{J}(\mathbf{r}') G(R) d^3 r' - \frac{1}{k^2}\nabla\int_{D'}[\nabla' G(R) \cdot \mathbf{J}(\mathbf{r}') - \nabla' G_0(R) \cdot \mathbf{J}(\mathbf{r})]d^3 r'\right.$$
$$\left. - \frac{1}{k^2}\nabla\int_{D'}\nabla' G_0(R) \cdot \mathbf{J}(\mathbf{r})d^3 r'\right]. \quad (16.20)$$

Let us consider the integrand of the second term on the right. In the limit $\mathbf{r}' \to 0$, the most singular parts of the integrand—the ones that behave as $1/R^2$—cancel out between the $G(R)$ and $G_0(R)$ terms. This means that the integrand is sufficiently well-behaved to bring the second ∇ inside. This is usually described in terms of the so-called Hölder condition, which says that the interchange may be done provided $|\mathbf{J}(\mathbf{r}') - \mathbf{J}(\mathbf{r})| \sim R^\alpha$, with $\alpha > 0$. We may also use $\nabla G(R) = -\nabla' G(R)$. But then the first two integrals can be combined, and we get the abbreviated form

$$\mathbf{E}_{D'}(\mathbf{r}) = \int_{D'}[\mathbf{\Gamma}_e(R) \cdot \mathbf{J}(\mathbf{r}') - \mathbf{\Gamma}_0(R) \cdot \mathbf{J}(\mathbf{r})]d^3 r' - \frac{i\omega\mu_0}{4\pi k^2}\nabla\int_{D'}\nabla' G_0(R) \cdot \mathbf{J}(\mathbf{r})d^3 r', \quad (16.21)$$

where

$$\mathbf{\Gamma}_0(\mathbf{r}, \mathbf{r}') = \frac{i\omega\mu_0}{4\pi k^2}\nabla G_0(R). \quad (16.22)$$

Now let us examine the final integral in this expression, which we label $\mathbf{E}_S(\mathbf{r})$ for convenience. The integrand may be rewritten using

$$\nabla' \cdot [G_0(R)\mathbf{J}(\mathbf{r})] = \nabla' G_0(R) \cdot \mathbf{J}(\mathbf{r}). \quad (16.23)$$

(This is straightforward because the expression uses $\mathbf{J}(\mathbf{r})$, not $\mathbf{J}(\mathbf{r}')$.) So we can write

$$\mathbf{E}_S(\mathbf{r}) = -\frac{i\omega\mu_0}{4\pi k^2}\nabla\int_{D'}\nabla' \cdot [G_0(R)\mathbf{J}(\mathbf{r})]d^3 r'. \quad (16.24)$$

We can use the divergence theorem to write this in terms of an integral over the surface S' of D',

$$\mathbf{E}_S(\mathbf{r}) = -\frac{i\omega\mu_0}{4\pi k^2}\nabla\int_{S'} G_0(R)\hat{\mathbf{n}}' \cdot \mathbf{J}(\mathbf{r})da', \quad (16.25)$$

where $\hat{\mathbf{n}}'$ represents the normal to the surface. We are now not even integrating over the singularity, which lies within D', so we bring the ∇ in to operate on the integrand; we neglect the derivative of $\mathbf{J}(\mathbf{r})$, which makes a vanishing contribution in the limit of a small surface. We then have

$$\mathbf{E}_S(\mathbf{r}) = -\frac{i\omega\mu_0}{4\pi k^2}\left[\int_{S'} \frac{\hat{\mathbf{R}}\hat{\mathbf{n}}'}{R^2}da'\right] \cdot \mathbf{J}(\mathbf{r}). \tag{16.26}$$

where $\hat{\mathbf{R}}$ is the unit vector pointing in the direction of $\mathbf{r} - \mathbf{r}'$.

Let us now put all the pieces back together, multiplying out some constant factors, to write

$$\int_D \mathbf{G}(\mathbf{r}, \mathbf{r}') \cdot \mathbf{K}(\mathbf{r}')d^3r' = \int_{D-D'} \mathbf{G}(\mathbf{r}, \mathbf{r}') \cdot \mathbf{K}(\mathbf{r}')d^3r' + \left[\mathbf{M}(\mathbf{r}) - \frac{1}{k^2}\mathbf{L}(\mathbf{r})\right] \cdot \mathbf{K}(\mathbf{r}), \tag{16.27}$$

where we have introduced $\mathbf{K}(\mathbf{r})$ as an arbitrary vector field, related either to the current density (in source problems) or the induced polarization (in scattering problems). In this expression, we have defined

$$\mathbf{M}(\mathbf{r}) \equiv \int_{D'} [\mathbf{G}(R) - \mathbf{G}_0(R)]d^3r', \tag{16.28}$$

where we have taken advantage of the fact that $\mathbf{J}(\mathbf{r}') \approx \mathbf{J}(\mathbf{r})$ in D', and

$$\mathbf{L}(\mathbf{r}) \equiv \int_{S'} \frac{\hat{\mathbf{R}}\hat{\mathbf{n}}'}{R^2}da'. \tag{16.29}$$

In the expression for $\mathbf{M}(\mathbf{r})$, we have defined $\mathbf{G}(R)$ using equation (16.10) and $\mathbf{G}_0(R)$ using

$$\mathbf{G}_0(R) \equiv \frac{1}{k^2}\nabla\frac{1}{R}. \tag{16.30}$$

The dyadic $\mathbf{M}(\mathbf{r})$ vanishes in the limit that D' becomes vanishingly small; the dyadic $\mathbf{L}(\mathbf{r})$, however, remains finite, as the $1/R^2$ in the denominator is effectively canceled out by the R^2 in the area element in spherical coordinates. In computational integral equation solutions of Maxwell's equations, where the volume D' is finite, the dyadic $\mathbf{M}(\mathbf{r})$ is often taken to be zero, though it can have a significant influence on the results. Lakhtakia [10] has classified integral equation methods as 'strong' or 'weak' depending on the presence or absence of $\mathbf{M}(\mathbf{r})$ in the calculation.

The dyadic $\mathbf{L}(\mathbf{r})$ can be evaluated analytically for simple geometries. It is straightforward to determine that, for a sphere centered on $\mathbf{r}$, the source dyadic has the form

$$\mathbf{L} = \frac{4\pi}{3}\mathbf{I}, \tag{16.31}$$

where $\mathbf{I}$ is the identity dyadic. For a cube centered on $\mathbf{r}$, $\mathbf{L}$ has the same form, regardless of orientation. For other cases, such as cylinders and rectangles, see Yaghjian [8].

16.3 The method of moments and the discrete dipole approximation

If it is possible to computationally solve the multiple scattering of a system of discrete particles, it is not a great conceptual leap to imagine treating a continuous scattering medium as a finite system of closely spaced particles and to look for a solution in a manner similar to the Foldy–Lax method. This is the basis of two integral-equation-based approaches for calculating the scattering of bulk objects, known as the *method of moments* and the *discrete dipole approximation*. These methods are very similar, and in fact, in their most advanced forms, are effectively equivalent; we examine them in this section.

In both approaches, the volume D of the scattering object is broken into a number of discrete subvolumes D_i, where i is an index that ranges over the total number of subvolumes N. The basic idea is illustrated in figure 16.1. Within each subvolume, the permittivity ε_i is typically assumed to be constant, as is any electric field passing through it; these approximations are valid provided the size of the subvolume is taken to be sufficiently small. Usually, the wavelength sets the needed size of the subvolume—in order to provide a good approximation, the side of a subvolume must be significantly smaller than the wavelength, at least $\lambda/10$, or ideally $\lambda/20$. It is important to note that the subvolumes must be small compared to the wavelength of the wave in the given medium; i.e. if the refractive index of the medium is n, the volume must be small compared to λ/n. For simplicity, we will take the subvolumes to be cubes of side d, though there is, in principle, no restriction on the chosen shape.

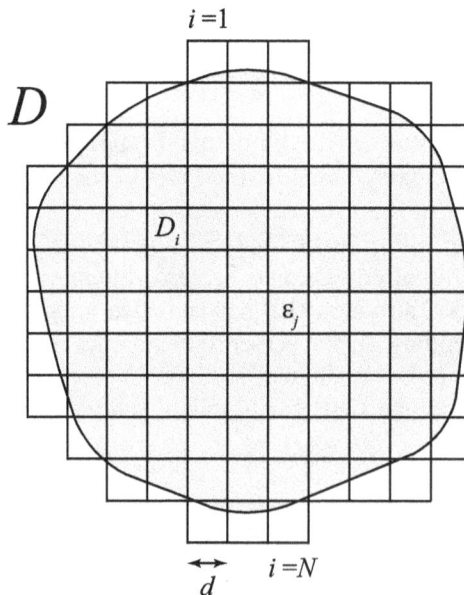

Figure 16.1. Illustration of the discrete subvolumes used in integral-equation-based computational scattering techniques.

The center of the cube will be taken as the point at which to measure and calculate any field properties in our computations.

The discrete dipole approximation (DDA) was first introduced in 1964 by DeVoe [11] to study the scattering properties of molecular aggregates. DeVoe ignored retardation effects in the Green's function, dropping the $\exp[ikR]$ in the expression; this imposed a strong restriction on the size of the aggregates that could be studied. The first complete form of the DDA was introduced by Purcell and Pennypacker in 1973 [12], and it has been a standard method ever since.

In the earliest implementations of the DDA, each subvolume was treated as a discrete dipole, with a polarizability given by the Clausius–Mossotti relation of equation (6.21), i.e.

$$\alpha = 3d^3 \frac{\varepsilon/\varepsilon_0 - 1}{\varepsilon/\varepsilon_0 + 2}. \tag{16.32}$$

In this case, we are almost directly following the Foldy–Lax approach, treating the continuous medium as a set of Rayleigh-type particles. We have, in a sense, already seen this sort of logic in approximate form in deriving the Rayleigh–Gans scattering formula in section 15.6, where we treated a bulk object as a collection of Rayleigh-like scatterers. This approach to the DDA, however, is inexact, and a number of corrections to the Clausius–Mossotti relation have been introduced over the years to improve the DDA; see the early review by Draine [13].

One noteworthy but subtle implication of using the Clausius–Mossotti polarizability is that it already includes any self-interaction of the particle with its own electric field. The given polarizability is only due to the external, 'exciting' fields. This will be a significant observation momentarily.

Let us now consider the integral equation approach more rigorously and return to equation (15.49), which, when written in terms of the total field, has the form

$$\mathbf{E}(\mathbf{r}) = \mathbf{E}_0(\mathbf{r}) + \int_D \mathbf{G}(\mathbf{r}, \mathbf{r}') \cdot \mathbf{E}(\mathbf{r}')F(\mathbf{r}')d^3r', \tag{16.33}$$

where the integral over the Green's dyadic has been regularized as discussed in the previous section, i.e.

$$\int_D \mathbf{G}(\mathbf{r}, \mathbf{r}') \cdot \mathbf{K}(\mathbf{r}')d^3r' = \int_{D-D'} \mathbf{G}(\mathbf{r}, \mathbf{r}') \cdot \mathbf{K}(\mathbf{r}')d^3r' + \left[\mathbf{M}(\mathbf{r}) - \frac{1}{k^2}\mathbf{L}(\mathbf{r})\right] \cdot \mathbf{K}(\mathbf{r}). \tag{16.34}$$

We turn this into a self-consistent integral equation by dividing up the domain D into N subvolumes and considering the value of the field within each subvolume. We therefore write

$$\mathbf{E}(\mathbf{r}_i) = \mathbf{E}_0(\mathbf{r}_i) + \sum_{j \neq i}\int_{D_j} \mathbf{G}(\mathbf{r}_i, \mathbf{r}') \cdot \mathbf{E}(\mathbf{r}')F(\mathbf{r}')d^3r' + \left[\mathbf{M}(\mathbf{r}_i) - \frac{1}{k^2}\mathbf{L}(\mathbf{r}_i)\right] \cdot \mathbf{E}(\mathbf{r}_i)F(\mathbf{r}_i), \tag{16.35}$$

where we have taken the exclusion volume D' of the Green's dyadic to match the ith subvolume. The sum is therefore over all subvolumes $j \neq i$. If the subvolume is small enough, we may treat all quantities in the integral as constant and write

$$\mathbf{E}(\mathbf{r}_i) = \mathbf{E}_0(\mathbf{r}_i) + \sum_{j \neq i} d^3 \mathbf{G}(\mathbf{r}_i, \mathbf{r}_j) \cdot \mathbf{E}(\mathbf{r}_j)F(\mathbf{r}_j) + \left[\mathbf{M}(\mathbf{r}_i) - \frac{1}{k^2}\mathbf{L}(\mathbf{r}_i)\right] \cdot \mathbf{E}(\mathbf{r}_i)F(\mathbf{r}_i), \qquad (16.36)$$

We now have what amounts to a vector-matrix equation that can be solved for the internal fields, and there are two approaches to doing so. In what is now referred to as the method of moments, we rearrange this equation to solve for $\mathbf{E}(\mathbf{r}_i)$ directly, i.e.

$$\mathbf{A}_{ij} \cdot \mathbf{E}(\mathbf{r}_j) = \mathbf{E}_0(\mathbf{r}_i), \qquad (16.37)$$

where we define

$$\mathbf{A}_{ij} = \begin{cases} [\mathbf{I}\delta_{ij} - d^3\mathbf{G}(\mathbf{r}_i, \mathbf{r}_j)F(\mathbf{r}_j)], & i \neq j, \\ \left\{\mathbf{I} - \left[\mathbf{M}(\mathbf{r}_i) - \dfrac{1}{k^2}\mathbf{L}(\mathbf{r}_i)\right]\right\}F(\mathbf{r}_i), & i = j. \end{cases} \qquad (16.38)$$

For a system of N subvolumes, it should be noted that this matrix will be $3N \times 3N$ in size, due to the inclusion of polarization effects.

We are left with a system of equations that can be solved for $\mathbf{E}(\mathbf{r}_i)$ by any method desired, such as Gaussian elimination or the conjugate gradient method. Once the fields inside the scatterer have been found, we can calculate the field at any point external to D by substituting back into the integral equation, i.e.

$$\mathbf{E}(\mathbf{r}) = \mathbf{E}_0(\mathbf{r}_i) + \sum_j d^3\mathbf{G}(\mathbf{r}, \mathbf{r}_j) \cdot \mathbf{E}(\mathbf{r}_j)F(\mathbf{r}_j). \qquad (16.39)$$

This is the solution approach for the method of moments (MoM). We note that the MoM is a quite general technique for solving integral equations for fields, popularized by Harrington in a classic 1968 book [14]. The general method of moments involves decomposing the field in the domain D into a set of orthonormal functions, which transforms the continuous integral equation into a discrete matrix equation. In our case, the orthonormal functions are a set of nonoverlapping functions that are constant within their respective cubic volumes; however, for very symmetric problems, the functions could be polynomials that are orthogonal over the domain of the scatterer.

It should be noted that the dyadics $\mathbf{M}(\mathbf{r})$ and $\mathbf{L}(\mathbf{r})$ do not directly depend upon the fields generated at other locations and may therefore be said to represent the self-interaction of the particle with its own field. With this in mind, the discrete dipole approximation is performed by defining an *exciting electric field* $\mathbf{E}_{exc}(\mathbf{r}_i)$ as

$$\mathbf{E}_{exc}(\mathbf{r}_i) = \mathbf{E}_0(\mathbf{r}_i) + \sum_{j \neq i} d^3\mathbf{G}(\mathbf{r}_i, \mathbf{r}_j) \cdot \mathbf{E}(\mathbf{r}_j)F(\mathbf{r}_j). \qquad (16.40)$$

This exciting field can be written in terms of the total field as

$$\mathbf{E}_{exc}(\mathbf{r}_i) = \mathbf{E}(\mathbf{r}_i) - \left[\mathbf{M}(\mathbf{r}_i) - \frac{1}{k^2}\mathbf{L}(\mathbf{r}_i)\right] \cdot \mathbf{E}(\mathbf{r}_i)F(\mathbf{r}_i) = \mathbf{A}_{ii} \cdot \mathbf{E}(\mathbf{r}_i). \qquad (16.41)$$

We can invert this expression to write

$$\mathbf{E}(\mathbf{r}_i) = \mathbf{A}_{ii}^{-1} \cdot \mathbf{E}_{exc}(\mathbf{r}_i), \tag{16.42}$$

and this can be substituted into equation (16.40) to write an equation for the exciting electric field,

$$\mathbf{E}_{exc}(\mathbf{r}_i) - \sum_{j \neq i} d^3 \mathbf{G}(\mathbf{r}_i, \mathbf{r}_j) \cdot \mathbf{A}_{jj}^{-1} \cdot \mathbf{E}_{exc}(\mathbf{r}_j) F(\mathbf{r}_j) = \mathbf{E}_0(\mathbf{r}_i). \tag{16.43}$$

Finally, we can compress this into a compact matrix equation,

$$\mathbf{B}_{ij} \cdot \mathbf{E}_{exc}(\mathbf{r}_j) = \mathbf{E}_0(\mathbf{r}_i), \tag{16.44}$$

where

$$\mathbf{B}_{ij} = \begin{cases} \left[\mathbf{I}\delta_{ij} - d^3 \mathbf{G}(\mathbf{r}_i, \mathbf{r}_j) \cdot \mathbf{A}_{jj}^{-1} F(\mathbf{r}_j) \right], & i \neq j, \\ 0, & i = j. \end{cases} \tag{16.45}$$

Again, this system of equations can be solved by a variety of numerical techniques. Because we are only solving for the exciting fields, the diagonal elements of this matrix all vanish. Once we have the excited field, we can determine the total field by substituting back into equation (16.42), if desired.

The MoM and DDA methods can be seen to be very closely related, with the difference being the choice of field to solve for. In the method of moments, the total field within each subvolume is calculated, while in the discrete dipole approximation, only the exciting field is calculated. The DDA, as mentioned earlier, solves the problem by building the self-interaction of the dipole with the field into the dipole model. In principle, the two methods should be equivalent, though numerical inaccuracies could result in slight differences in the calculated fields.

There are a number of advantages and disadvantages to integral equation solutions of Maxwell's equations; the greatest disadvantage is the computational intensiveness of the method. We have already noted that it is necessary to use 10–20 subvolumes per wavelength to get accurate results using integral equations. If we imagine roughly attempting to model even a single particle of wavelength size, we would require 10^3 subvolumes, and with three polarization components to calculate, we would need a matrix that has $3 \times 10^3 \times 3 \times 10^3$ elements, or a matrix with roughly 9 million elements! The solution of such problems requires significant time and computational resources. Whenever possible, researchers often solve problems in a two-dimensional geometry, where the number of needed elements drops significantly.

Another limitation of integral equation solutions is that Green's functions can only be calculated for linear systems, which means that the MoM and the DDA cannot be applied to problems involving nonlinear optics. In such cases, one must turn to the finite-difference time-domain (FDTD) method, discussed in the next section.

Integral equation solutions also have a number of advantages that arise directly from the mathematics of the Green's dyadic. First, it should be noted that the only regions we must discretize to solve our matrix equation are those where the system deviates from the background system, and we do not need to discretize the whole domain in which the fields are calculated. This contrasts with, for example, the FDTD method to be discussed in the next section. Furthermore, the Green's dyadic can be calculated for a wide variety of background systems, including a stratified medium consisting of an arbitrary number of layers. Light propagation through a multilayer system with a small number of defects can then be calculated relatively efficiently. For example, figure 11.9(b), showing the plasmonic contribution of extraordinary optical transmission, was calculated using the method of moments (also referred to in this case as the domain integral equation solution). The Green's dyadic for an air–metal–air thin-film system was derived, and the two slits were treated as scatterers. All of the complicated physics of thin-film surface plasmonics, as well as reflection and refraction from the metal film, were built into the Green's dyadic.

Finally, it should be noted that integral equation solutions naturally incorporate any boundary conditions in the Green's dyadic. For the free-space case, the dyadic automatically has the far-zone radiation conditions built into it, removing any worry about effects that might be caused by an artificial boundary in the calculations. This, again, contrasts with the FDTD method, which requires careful consideration to properly avoid spurious boundary effects.

There are a number of excellent references that discuss the method of moments and the discrete dipole approximation in more detail. We refer the reader to a general article on numerical methods from 2003 by Kahnert [15] and a 2007 review of the DDA by Yurkin and Hoekstra [16]. For an analysis of the Green's dyadic for multilayer films, see the 2005 dissertation by Schouten [17].

16.4 Finite-difference time-domain (FDTD) method

To conclude this chapter and the book, we turn to the most general approach for solving Maxwell's equations computationally: a direct discretization of the equations themselves. As a set of coupled vector partial differential equations, this is a nontrivial task, but it is one that is much more straightforward than the previous techniques discussed. The method is known as the finite-difference time-domain (FDTD) method and was first introduced by Yee in 1966 [18].

There are a number of excellent books that focus on the implementation of the FDTD method. An ideal one for beginners is by Sullivan [19], as it walks the reader through the process from one-dimensional to three-dimensional problems, with practical code given at every step of the way; we follow much of their approach in this section. More detailed books on FDTD include the one by Inan and Marshall [20] and the hefty tome by Taflove and Hagness [21]. In this section, we restrict our attention to the full three-dimensional case.

We return one last time to the free-space form of Maxwell's equations, in particular the curl equations,

$$\nabla \times \mathbf{E}(\mathbf{r},\,t) = -\mu_0 \frac{\partial \mathbf{H}(\mathbf{r},\,t)}{\partial t},$$ (16.46)

$$\nabla \times \mathbf{H}(\mathbf{r}, t) = \varepsilon_0 \frac{\partial \mathbf{E}(\mathbf{r}, t)}{\partial t}. \tag{16.47}$$

(We know that the divergence equations are implicitly built into the curl equations, as can be shown by taking the divergence of either.)

As a first step, we note that the magnitudes of $\mathbf{E}$ and $\mathbf{B}$ are significantly different from each other, which can be an issue with computational accuracy. In monochromatic plane waves in free space, we have seen that the amplitudes of the magnetic and electric fields are related by

$$H_0 = \frac{1}{\mu_0 c} E_0 = \sqrt{\frac{\mu_0}{\varepsilon_0}} \, E_0. \tag{16.48}$$

To make their scales comparable, it therefore makes sense to introduce a modified electric field $\tilde{\mathbf{E}}(\mathbf{r}, t)$ of the form

$$\tilde{\mathbf{E}}(\mathbf{r}, t) = \sqrt{\frac{\varepsilon_0}{\mu_0}} \, \mathbf{E}(\mathbf{r}, t). \tag{16.49}$$

With this change, we can write

$$\nabla \times \tilde{\mathbf{E}}(\mathbf{r}, t) = \frac{1}{c} \frac{\partial \mathbf{H}(\mathbf{r}, t)}{\partial t}, \tag{16.50}$$

$$\nabla \times \mathbf{H}(\mathbf{r}, t) = \frac{1}{c} \frac{\partial \tilde{\mathbf{E}}(\mathbf{r}, t)}{\partial t}, \tag{16.51}$$

where we have, of course, used $1/\sqrt{\varepsilon_0 \mu_0} = c$. (This formulation is similar to, but not quite like, using CGS units, for which the equations look similar but with $\mathbf{B}$ instead of $\mathbf{H}$.)

We now have the daunting task of not only discretizing these equations but also discretizing them in such a way that they can be readily solved. For every derivative, we will be using a central difference approximation; for example, if we wish to take a time derivative of $\tilde{\mathbf{E}}(\mathbf{r}, t)$, we will use

$$\frac{\partial \tilde{\mathbf{E}}(\mathbf{r}, t)}{\partial t} \approx \frac{\tilde{\mathbf{E}}(\mathbf{r}, t + \Delta t/2) - \tilde{\mathbf{E}}(\mathbf{r}, t - \Delta t/2)}{\Delta t}. \tag{16.52}$$

The fundamental idea for discretizing Maxwell's equations, introduced by Yee, is that we interleave the electric and magnetic fields in space and time on a discrete mesh. In space, the unit cell of this interleaved mesh is known as the Yee cell, illustrated in figure 16.2. All of space is therefore discretized into an array of Yee cells, with the corners of the cells representing integer values. The integers (i, j, k) are used to represent the Cartesian positions of the corners.

The conceptual elegance of this arrangement can be seen by writing the different components of Maxwell's curl equations explicitly:

$$\frac{\partial \tilde{E}_x}{\partial t} = \frac{1}{c} \left(\frac{\partial H_z}{\partial y} - \frac{\partial H_y}{\partial z} \right), \tag{16.53}$$

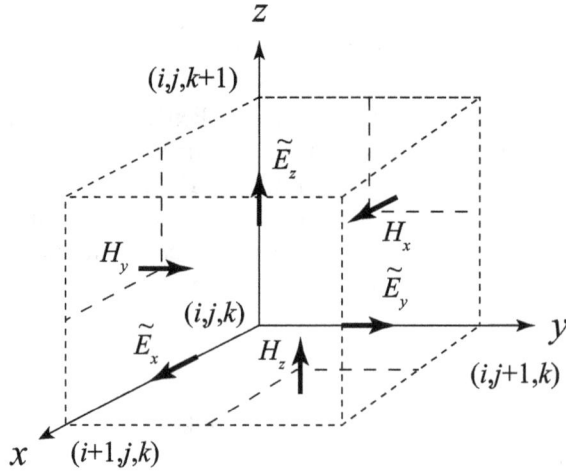

Figure 16.2. Illustration of the Yee cell used to interleave electric and magnetic fields.

$$\frac{\partial \tilde{E}_y}{\partial t} = \frac{1}{c}\left(\frac{\partial H_x}{\partial z} - \frac{\partial H_z}{\partial x}\right), \tag{16.54}$$

$$\frac{\partial \tilde{E}_z}{\partial t} = \frac{1}{c}\left(\frac{\partial H_y}{\partial x} - \frac{\partial H_x}{\partial y}\right), \tag{16.55}$$

$$\frac{\partial H_x}{\partial t} = \frac{1}{c}\left(\frac{\partial \tilde{E}_y}{\partial z} - \frac{\partial \tilde{E}_z}{\partial y}\right), \tag{16.56}$$

$$\frac{\partial H_y}{\partial t} = \frac{1}{c}\left(\frac{\partial \tilde{E}_z}{\partial x} - \frac{\partial \tilde{E}_x}{\partial z}\right), \tag{16.57}$$

$$\frac{\partial H_z}{\partial t} = \frac{1}{c}\left(\frac{\partial \tilde{E}_x}{\partial y} - \frac{\partial \tilde{E}_y}{\partial x}\right). \tag{16.58}$$

The change of $\tilde{E}_x$, for example, depends on the nearby values of H_z in the y-direction and the nearby values of H_y in the z-direction. Every field component is placed between the values of those field components that influence it.

Because the corners of the cell represent integer values of the discrete spatial mesh, the field components are evaluated at some locations that are half-integer values. For example, in figure 16.2, H_y is measured at the position $(i + 1/2, j, k + 1/2)$. Each **H**-field component is evaluated at the integer value of its parallel component and half-integer values of its orthogonal components, and the opposite is true for **E**-field components. Of course, when we implement these arrays on a computer, we do not use half-integer values: symbolically, this representation shows how the arrays of each field component are spatially related to each other.

Similarly, we will evaluate the electric field components on half-integer time steps and magnetic field components on integer time steps. This represents the observation that we will derive the change in $\tilde{\mathbf{E}}$ from $\mathbf{H}$ and vice versa.

Let us consider a discrete algorithm with a time step Δt and a spatial mesh size Δx. We may then apply a finite-difference approximation for each derivative in Maxwell's curl equations to get equations of evolution for the fields. The equations are lengthy, so we write just two components, one of $\tilde{\mathbf{E}}$ and one of $\mathbf{H}$,

$$\tilde{E}_z^{n+1/2}(i, j, k + 1/2) = \tilde{E}_z^{n-1/2}(i, j, k + 1/2)$$
$$+ \frac{c\Delta t}{\Delta x}\left[H_y^n(i + 1/2, j, k + 1/2) - H_y^n(i - 1/2, j, k + 1/2)\right] \quad (16.59)$$
$$- \frac{c\Delta t}{\Delta x}[H_x^n(i, j + 1/2, k + 1/2) - H_x^n(i, j - 1/2, k + 1/2)],$$

and

$$H_z^{n+1}(i + 1/2, j + 1/2, k) = H_z^n(i + 1/2, j + 1/2, k)$$
$$- \frac{c\Delta t}{\Delta x}\left[\tilde{E}_y^{n+1/2}(i + 1, j + 1/2, k) - \tilde{E}_y^{n+1/2}(i, j + 1/2, k)\right] \quad (16.60)$$
$$+ \frac{c\Delta t}{\Delta x}\left[\tilde{E}_x^{n+1/2}(i + 1/2, j + 1, k) - \tilde{E}_x^{n+1/2}(i + 1/2, j, k)\right].$$

The approach is then, in principle, straightforward: we loop over the entire mesh and calculate the new values of $\tilde{\mathbf{E}}$ from the current values of $\tilde{\mathbf{E}}$ and $\mathbf{H}$; we then loop over the mesh again and calculate the new values of $\mathbf{H}$ from the new current values of $\mathbf{H}$ and $\tilde{\mathbf{E}}$. The initial values are typically set at zero, so we have a well-defined starting point. It should be noted that the order in which we loop over the spatial mesh makes no difference, as the new value of $\tilde{\mathbf{E}}$ at a point, for example, depends only on the previous value and the values of the $\mathbf{H}$-field.

Of course, the devil is in the details, as they say, and there are a lot of details here. First, we need to determine appropriate spatial and temporal steps with which to accurately calculate the field. Let us look at the time step first. For a field traveling at the vacuum speed of light, propagation along an axis of the cell a distance Δx takes a time of $\Delta t = \Delta x/c$. To properly capture the time evolution of the field, we must choose a Δt that is equal to or smaller than this value. In three dimensions, the argument is more subtle but amounts to the requirement that $\Delta t < \Delta x/\sqrt{3}\,c$; we refer to Taflove and Brodwin for the derivation based on computational stability [22]. In fact, Yee gave an incorrect estimate for Δt in his original paper! Many authors use $\Delta t = \Delta x/2c$; it exceeds the minimum requirement, and the factors in the FDTD equations simply reduce to $1/2$.

For the spatial step, we apply reasoning analogous to that used in the integral equation method: we need a spatial step that is significantly smaller than the shortest relevant wavelength in the problem. This depends on both the highest frequency in the spectrum of waves considered as well as the highest refractive index that appears. If we consider n_{max} to be the largest index and λ_{min} the shortest wavelength, a good guideline is

$$\Delta x < \frac{\lambda_{min}}{10 n_{max}}. \tag{16.61}$$

The factor of ten represents the guideline that we need approximately ten points per wavelength to get a good measure of the field.

Next, we note that we will not have much of a simulation if we do not have a source of electromagnetic waves. The simplest approach for including a source in the algorithm is to impose an electric field at a single mesh location; the magnetic field is then taken care of by the algorithm. For example, if we want to drive a z-pointing electric field in the vicinity of the point (i_0, j_0, k_0), we could implement the definition

$$\tilde{E}_z^{n+1/2}(i_0, j_0, k_0 + 1/2) = E_S(n + 1/2), \tag{16.62}$$

where $E_S(n + 1/2)$ is the desired time oscillation of the imposed field, such as a Gaussian pulse or monochromatic oscillation. This is an example of a 'hard' source, so named because the strict imposition of the field value causes other impinging fields to be blocked and reflected.

As an alternative, we can construct a 'soft' source by introducing a point-like electric current into Maxwell's equations. Returning to the Ampère–Maxwell law, let us define a current density in the z-direction, which gives us

$$\frac{\partial \tilde{E}_z}{\partial t} = \frac{1}{c}\left(\frac{\partial H_y}{\partial x} - \frac{\partial H_x}{\partial y}\right) - J_z. \tag{16.63}$$

If we assume that we have a current $I_0(n + 1/2)$ in the z-direction that is extended over the unit cell, we may write $J_z(n + 1/2) = I_0(n + 1/2)/(\Delta x)^2$. If we place this dipole at $(i_0, j_0, k_0 + 1/2)$, we can then implement the definition

$$\tilde{E}_z^{n+1/2}(i_0, j_0, k_0 + 1/2) = \tilde{E}_z^{n-1/2}(i_0, j_0, k_0 + 1/2) + \frac{I_0(n + 1/2)}{(\Delta x)^2} + [\text{H–terms}], \tag{16.64}$$

where we reuse the magnetic field terms from equation (16.59). Because it is an additive relation, fields that impinge upon the point $(i_0, j_0, k_0 + 1/2)$ propagate through it, which is why it is a 'soft' source. For more on such sources, see Costen [23].

This approach of simply adding a fixed source follows the reasoning of section 5.4, where we noted that we often artificially distinguish between external currents that are generated by physics outside our calculation and induced currents that are created by the fields within the calculation. Here, we have sources that act as external currents.

It is also possible to implement a plane wave source in a system in a relatively straightforward way. The plane wave is added as a soft source at one end of the simulation space and is subtracted at the other end of the space. Following the reasoning of section 15.2, any field that remains at the boundaries of the system must be the scattered field. This approach allows us to study and calculate the scattering and extinction cross sections of objects.

So far, we have looked at FDTD solely in free space. When we introduce scatterers, we run into a major complication: in the time domain, dispersion effects must be taken into account, i.e. we have to incorporate

$$\mathbf{D}(\mathbf{r}, \omega) = \varepsilon(\mathbf{r}, \omega)\mathbf{E}(\mathbf{r}, \omega). \tag{16.65}$$

Let us first introduce a normalized version $\tilde{\mathbf{D}}$ of the displacement field that satisfies the relation

$$\tilde{\mathbf{D}} = \frac{1}{\sqrt{\varepsilon_0 \mu_0}} \mathbf{D}. \tag{16.66}$$

This leads to the expression

$$\tilde{\mathbf{D}}(\mathbf{r}, \omega) = \varepsilon_r(\mathbf{r}, \omega)\tilde{\mathbf{E}}(\mathbf{r}, \omega), \tag{16.67}$$

where $\varepsilon_r(\mathbf{r}, \omega)$ is now the relative permittivity, normalized to unity in vacuum. We now use the Ampère–Maxwell law with $\tilde{\mathbf{D}}$ instead of $\tilde{\mathbf{E}}$,

$$\nabla \times \mathbf{H}(\mathbf{r}, t) = \frac{1}{c} \frac{\partial \tilde{\mathbf{D}}(\mathbf{r}, t)}{\partial t}. \tag{16.68}$$

In the time domain, equation (16.67) may be written as a convolution. Recalling the analogous formula we used in our discussion of the Kramers–Kronig relations, equation (6.87), we may write $\tilde{\mathbf{D}}(\mathbf{r}, t)$ as

$$\tilde{\mathbf{D}}(\mathbf{r}, t) = \frac{1}{2\pi} \int_0^\infty \varepsilon_r(\mathbf{r}, t')\tilde{\mathbf{E}}(\mathbf{r}, t - t')dt', \tag{16.69}$$

where causality indicates that the lower limit of integration is zero. If we consider a discrete version of this formula in time, we obtain the discrete convolution

$$\tilde{\mathbf{D}}(\mathbf{r}, n) = \frac{1}{2\pi} \sum_{n'=0}^{n} \varepsilon_r(\mathbf{r}, n')\tilde{\mathbf{E}}(\mathbf{r}, n - n')\Delta t, \tag{16.70}$$

where we have also assumed that the signal begins at time step $n' = n$, which gives us an upper limit to the summation. Formally, however, we will extend these summations from $-\infty$ to ∞. (Note that we have dropped the half-integer time-step notation for the moment for clarity.)

It would be extremely tedious to evaluate $\varepsilon_r(\mathbf{r}, t)$ for every medium we encounter (Debye, Drude, Lorentz, and more) and perform the discretization each time, so we look for a simplifying process.

Here, we follow the approach championed by Sullivan [19, 24] and introduce the z-transform of a discrete signal $x(n)$ in time, defined as

$$Z[x(n)] = X(z) = \sum_{n=-\infty}^{\infty} x(n)z^{-n}. \tag{16.71}$$

The z-transform is, in essence, an analogue to the Fourier transform for discrete signals. For those familiar with complex analysis, we are, in effect, taking our discrete time-domain signal and converting it into a Laurent series in the complex z-plane.

If we take the z-transform of equation (16.70), it is straightforward to show that

$$\tilde{\mathbf{D}}(\mathbf{r}, z) = \frac{1}{2\pi}\varepsilon_r(\mathbf{r}, z)\tilde{\mathbf{E}}(\mathbf{r}, z)\Delta t. \tag{16.72}$$

A convolution in the discrete time domain becomes a product in the z-domain.

We will need one other important property of the z-transform to continue. We note that we may write

$$z^{-1}Z[x(n)] = \sum_{n=-\infty}^{\infty} x(n)z^{-n-1} = \sum_{n'=-\infty}^{\infty} x(n'-1)z^{-n'} = Z[x(n-1)]. \tag{16.73}$$

Every power of z^{-1} multiplying a transform corresponds to a delay operator or a shift of the signal one step lower.

The final piece of the puzzle is to note that we may construct a table of common z-transforms, just as we do for common Fourier transforms. This table is given as table 16.1, which connects the continuous time-domain and frequency-domain signals to their sampled time-domain and z-transform results. In this table, $S(t)$ represents the step function, which has a value of zero for $t < 0$ and a value of one for $t \geqslant 0$.

We now have the tools to modify our FDTD algorithm to account for dispersive materials. As an example, let us consider the case of waves propagating in a simple conductor, for which we may write the following using equation (5.85):

$$\varepsilon_r(\mathbf{r}, \omega) = \varepsilon_c(\mathbf{r}) + i\frac{\sigma(\mathbf{r})}{\omega}, \tag{16.74}$$

where we have ignored any dispersive effects in the normalized conductivity σ or normalized background permittivity ε_c. From table 16.1, we may associate this with a z-transform

$$\varepsilon_r(\mathbf{r}, z) = 2\pi\varepsilon_c(\mathbf{r}) - \frac{2\pi\sigma(\mathbf{r})}{1 - z^{-1}}. \tag{16.75}$$

Table 16.1. Table of common z-transforms used in dispersion calculations.

$f(t)$	$\tilde{f}(\omega)$	$f[n]$	$F(z)$
$\delta(t)$	$\frac{1}{2\pi}$	$\delta[n]$	1
$S(t)$	$\frac{1}{2\pi i\omega}$	$S[n]$	$\frac{1}{1-z^{-1}}$
$tS(t)$	$\frac{1}{2\pi(i\omega)^2}$	$nTS[n]$	$\frac{Tz^{-1}}{(1-z^{-1})^2}$
$\exp[-\alpha t]S(t)$	$\frac{1}{2\pi}\frac{1}{\alpha + i\omega}$	$\exp[-\alpha nT]S[n]$	$\frac{1}{1-\exp[-\alpha T]z^{-1}}$
$\exp[\pm i\beta t]\exp[-\alpha t]S(t)$	$\frac{1}{2\pi}\frac{1}{\alpha + i\omega \mp i\beta}$	$\exp[-\alpha nT]\exp[\pm i\beta nT]S[n]$	$\frac{1}{1-z^{-1}\exp[\alpha T]\exp[\mp i\beta T]}$

We now substitute this into our relation between $\tilde{\mathbf{D}}$ and $\tilde{\mathbf{E}}$. We have

$$\tilde{\mathbf{D}}(\mathbf{r},\, z) = 2\pi\varepsilon_c(\mathbf{r})\tilde{\mathbf{E}}(\mathbf{r},\, z) - \frac{2\pi\sigma(\mathbf{r})}{1 - z^{-1}}\tilde{\mathbf{E}}(\mathbf{r},\, z). \tag{16.76}$$

We may expand the term on the right in a geometric series,

$$\frac{1}{1 - z^{-1}} = \sum_{m=0}^{\infty} z^{-m}, \tag{16.77}$$

and then write

$$\tilde{\mathbf{D}}(\mathbf{r},\, z) = 2\pi\varepsilon_c(\mathbf{r})\tilde{\mathbf{E}}(\mathbf{r},\, z) - 2\pi\sigma(\mathbf{r})\sum_{m=0}^{\infty} z^{-m}\tilde{\mathbf{E}}(\mathbf{r},\, z). \tag{16.78}$$

Using the delay operator property of the z-transform, we may immediately convert this expression into a discrete time-domain relationship,

$$\tilde{\mathbf{D}}(\mathbf{r},\, n) = 2\pi\varepsilon_c(\mathbf{r})\tilde{\mathbf{E}}(\mathbf{r},\, n) - 2\pi\sigma(\mathbf{r})\sum_{m=0}^{\infty} \tilde{\mathbf{E}}(\mathbf{r},\, n - m). \tag{16.79}$$

Next, we extract the $m = 0$ term of the series, truncate the series at $m = n$ (which implies that our signal starts at time step zero), and write

$$\tilde{\mathbf{D}}(\mathbf{r},\, n) = [2\pi\varepsilon_c(\mathbf{r}) - 2\pi\sigma(\mathbf{r})]\tilde{\mathbf{E}}(\mathbf{r},\, n) - 2\pi\sigma(\mathbf{r})\sum_{m=0}^{n-1} \tilde{\mathbf{E}}(\mathbf{r},\, m), \tag{16.80}$$

where we have taken the liberty of reordering the series in the last term for convenience. Finally, we write this as

$$\tilde{\mathbf{E}}(\mathbf{r},\, n) = \frac{1}{2\pi\varepsilon_c(\mathbf{r}) - 2\pi\sigma(\mathbf{r})}\left[\tilde{\mathbf{D}}(\mathbf{r},\, n) + 2\pi\sigma(\mathbf{r})\sum_{m=0}^{n-1} \tilde{\mathbf{E}}(\mathbf{r},\, m)\right]. \tag{16.81}$$

We now have a new step in our FDTD algorithm, which proceeds as follows. We calculate the new values of $\tilde{\mathbf{D}}$ from the current values of $\tilde{\mathbf{D}}$ and the previous value of $\mathbf{H}$. We then determine the new values of $\tilde{\mathbf{E}}$ from equation (16.81). In this, it is helpful to introduce an auxiliary function $\mathbf{I}(\mathbf{r},\, n)$, where

$$\mathbf{I}(\mathbf{r},\, n) = \mathbf{I}(\mathbf{r},\, n - 1) + 2\pi\sigma(\mathbf{r})\tilde{\mathbf{E}}(\mathbf{r},\, n - 1). \tag{16.82}$$

We therefore do not need to save all past values of $\tilde{\mathbf{E}}$ to calculate the $\tilde{\mathbf{D}}$-field, only the net sum that is characterized by $\mathbf{I}(\mathbf{r},\, n)$. We can use these new values of $\tilde{\mathbf{E}}$ to calculate the new values of $\mathbf{H}$ via Faraday's law. From there, the process repeats.

A similar approach can be taken for more complicated media: table 16.1 provides enough information to evaluate the response of a Lorentz-type medium or one with multiple Lorentz resonances.

With dispersion reasonably handled, there is one other significant issue that must be addressed: effects due to the boundary of our calculation space. We are often interested in simulating light interactions in unbounded space, for example, in the study of light scattering. In FDTD calculations, however, we must discretize the entirety of our simulation space, which is always limited to a finite domain. After enough time steps, waves generated in the simulation space reach the bounds of the domain, and if nothing is done, they are reflected at the boundary and generate spurious results. This is especially problematic if we are looking at simulations of monochromatic waves, where we must allow the simulation to run sufficiently long to reach a steady state. We therefore need to incorporate some sort of boundary conditions in our simulation that effectively absorb any waves that reach them, simulating waves escaping to infinity.

The standard approach used today is to introduce what is known as a *perfectly matched layer* (PML) at the bounds of the simulation space, as first described by Berenger [25]. This idea is based on the impedance matching that can be done for magnetic materials, as discussed in section 9.8. It turns out that if we construct a boundary layer from an appropriately designed anisotropic magnetic material, we can, in fact, make it impedance matched for all directions of incidence. Furthermore, we take the permittivity and permeability to have the behavior of conductors, which makes them highly absorbing. Waves that hit the interface of the boundary layer pass through without reflection and are mostly absorbed. The layer needs to have a finite thickness, looking much like a frame around the simulation space in two dimensions or padding around a rectangular volume in three dimensions.

Here, we only outline an appropriate method for constructing such a layer, again following Sullivan [26]. The permittivity and permeability of the layer in the x-direction are taken to be

$$\varepsilon_{Fx}(\mathbf{r}, \omega) = \varepsilon_{Dx}(\mathbf{r}) + i\frac{\sigma_{Dx}(\mathbf{r})}{\omega}, \tag{16.83}$$

$$\mu_{Fx}(\mathbf{r}, \omega) = \mu_{Hx}(\mathbf{r}) + i\frac{\sigma_{Hx}(\mathbf{r})}{\omega}, \tag{16.84}$$

where the 'F' indicates a fictitious medium, and there are similar definitions for the y- and z-directions. Again, we use normalized values of permittivity, permeability, and conductivity. It was shown by Sacks [27] that the layer is, in principle, perfectly matched if the impedance going from the background medium to the PML is constant, i.e.

$$\eta_0 = \eta_x \equiv \sqrt{\frac{\mu_{Fx}}{\varepsilon_{Fx}}}, \tag{16.85}$$

with similar relations for y and z, and if the permittivity and permeability normal to each surface is the inverse of the tangential components, i.e. for the x-boundary, we have

$$\frac{1}{\varepsilon_{Fx}} = \varepsilon_{Fy} = \varepsilon_{Fz}, \tag{16.86}$$

$$\frac{1}{\mu_{Fx}} = \mu_{Fy} = \mu_{Fz}. \tag{16.87}$$

These constraints can be satisfied if we take

$$\varepsilon_{Dm} = \mu_{Hm} = 1, \quad \text{for all } m = x, y, z, \tag{16.88}$$

$$\sigma_{Dm} = \sigma_{Hm} = \sigma_0, \quad \text{for all } m = x, y, z. \tag{16.89}$$

The conductivity σ_0 is still a function of position and is taken to gradually increase as one propagates further into the PML. It is important to note that the above conditions imply that, for a rectangular simulation space, the optical properties of the PML are different on each face of the space, with transition regions needed on the edges and corners of each face.

The preceding discussion presents all of the fundamental pieces required to construct a three-dimensional FDTD simulation. It should be noted that three-dimensional simulations are still extremely computationally expensive, and researchers tend to study two-dimensional problems whenever possible, as they can be run with reasonable efficiency on desktop computers. It should also be noted that the size of the domain studied is strongly limited by the operating wavelength—since roughly 10 points are needed per wavelength, the simulation space is itself likely to be limited in practical cases to a small number of wavelengths. At optical frequencies, this means simulations are restricted to the study of nano-optical systems; for microwaves and radio waves, however, much larger domains can be investigated.

16.5 So long, and thanks for all the physics

If you have read this far, you have completed a study of the fundamentals of electromagnetic optics—along with some advanced topics along the way! Thank you for joining me on this expedition, and I hope this book and the lessons learned will serve you well in your future studies of electromagnetic waves.

References

[1] Foldy L L 1945 The multiple scattering of waves *Phys. Rev.* **67** 107–19
[2] Lax M 1951 Multiple scattering of waves *Rev. Mod. Phys.* **23** 287–310
[3] Lax M 1952 Multiple scattering of waves. II. The effective field in dense systems *Phys. Rev.* **85** 621–9
[4] Mishchenko M I, Travis L D and Lacis A A 2006 *Multiple Scattering of Light by Particles* (Cambridge: Cambridge University Press)
[5] Van Bladel J 1961 Some remarks on Green's dyadic for infinite space *IRE Trans. Antennas Propag.* **9** 563–6
[6] Fikioris J G 1965 Electromagnetic field inside a current-carrying region *J. Math. Phys.* **6** 1617–20
[7] Lee S W, Boersma J, Law C L and Deschamps G 1980 Singularity in Green's function and its numerical evaluation *IEEE Trans. Antennas Propag.* **28** 311–7
[8] Yaghjian A D 1980 Electric dyadic Green's functions in the source region *Proc. IEEE* **68** 248–63
[9] Wang J 1982 A unified and consistent view on the singularities of the electric dyadic Green's function in the source region *IEEE Trans. Antennas Propag.* **30** 463–8

[10] Lakhtakia A 1992 Strong and weak forms of the method of moments and the coupled dipole method for scattering of time-harmonic electromagnetic fields *Int. J. Mod. Phys. C—IJMPC* **3** 583–603

[11] DeVoe H 1964 Optical properties of molecular aggregates. I. Classical model of electronic absorption and refraction *J. Chem. Phys.* **41** 393–400

[12] Purcell E M and Pennypacker C R 1973 Scattering and absorption of light by nonspherical dielectric grains *Astrophys. J.* **186** 705–14

[13] Draine B T and Flatau P J 1994 Discrete-dipole approximation for scattering calculations *J. Opt. Soc. Am.* A **11** 1491–9

[14] Harrington R F 1968 *Field Computation by Moment Methods* reprint edn (Malabar, FL: Robert E. Krieger Publishing Company)

[15] Kahnert F M 2003 Numerical methods in electromagnetic scattering theory *J. Quant. Spectrosc. Radiat. Transfer* **79–80** 775–824 Electromagnetic and Light Scattering by Non-Spherical Particles

[16] Yurkin M A and Hoekstra A G 2007 The discrete dipole approximation: an overview and recent developments *J. Quant. Spectrosc. Radiat. Transfer* **106** 558–89 *IX Conf. on Electromagnetic and Light Scattering by Non-Spherical Particles*

[17] Schouten H F 2005 Light transmission through sub-wavelength apertures *Dissertation (external)* (Amsterdam: Vrije Universiteit)

[18] Yee K 1966 Numerical solution of initial boundary value problems involving Maxwell's equations in isotropic media *IEEE Trans. Antennas Propag.* **14** 302–7

[19] Sullivan D M 2000 *Electromagnetic Simulation Using the FDTD Method* (New Jersey: Wiley-IEEE)

[20] Inan U S and Marshall R A 2011 *Numerical Electromagnetics* (Cambridge: Cambridge University Press)

[21] Taflove A and Hagness S C 2005 *Computational Electrodynamics* 3rd edn (Norwood, MA: Artech House)

[22] Taflove A and Brodwin M E 1975 Numerical solution of steady-state electromagnetic scattering problems using the time-dependent Maxwell's equations *IEEE Trans. Microw. Theory Tech.* **23** 623–30

[23] Costen F, Berenger J P and Brown A K 2009 Comparison of FDTD hard source with FDTD soft source and accuracy assessment in Debye media *IEEE Trans. Antennas Propag.* **57** 2014–22

[24] Sullivan D M 1996 Z-transform theory and the FDTD method *IEEE Trans. Antennas Propag.* **44** 28–34

[25] Berenger J P 1994 A perfectly matched layer for the absorption of electromagnetic waves *J. Comput. Phys.* **114** 185–200

[26] Sullivan D M 1997 An unsplit step 3-D PML for use with the FDTD method *IEEE Microw. Guided Wave Lett.* **7** 184–6

[27] Sacks Z S, Kingsland D M, Lee R and Lee J F 1995 A perfectly matched anisotropic absorber for use as an absorbing boundary condition *IEEE Trans. Antennas Propag.* **43** 1460–3

IOP Publishing

Electromagnetic Optics

Gregory J Gbur

Appendix A

A brief vector calculus review

This book assumes that the reader has taken a basic course in vector calculus and knows the basic vector integration and vector differentiation operations. This appendix is presented, however, as a small refresher of the minimum 'need to know' concepts. Since we will be working with vector fields throughout the book, vector calculus tools and identities will be used again and again. We primarily state results without proof and refer the reader to vector calculus texts for more information.

A.1 Vector algebra

We begin with a discussion of vector algebra. A vector $\mathbf{A}$, in the simplest definition, is a quantity with magnitude and direction. It can be represented by an arrow, where the length of the arrow indicates the magnitude and the direction of the arrow represents the vector direction. Examples of vectors in physics include the velocity, momentum, and angular momentum vectors of an object, the force vector acting on an object, and the position vector $\mathbf{r}$ of an object. Of these, the position vector is anomalous because it depends on the choice of the origin of the coordinate system; the position vector $\mathbf{r}$ represents the vector distance from the chosen origin to the object.

Vectors can be added together geometrically using the parallelogram rule, which indicates that the sum of two vectors can be found by placing them 'tip to tail,' as illustrated in figure A.1(a); the resultant represents the sum. From the parallelogram rule, we can deduce that vector addition is commutative, i.e. it does not depend on the order of summation. If $\mathbf{A}$ and $\mathbf{B}$ are vectors, then we have

$$\mathbf{A} + \mathbf{B} = \mathbf{B} + \mathbf{A}. \tag{A.1}$$

From the parallelogram rule, we can deduce the trapezoid rule of figure A.1(b), which indicates that vector addition is also associative, i.e.

$$\mathbf{A} + (\mathbf{B} + \mathbf{C}) = (\mathbf{A} + \mathbf{B}) + \mathbf{C}. \tag{A.2}$$

doi:10.1088/978-0-7503-6064-7ch17   A-1   © IOP Publishing Ltd 2025. All rights,

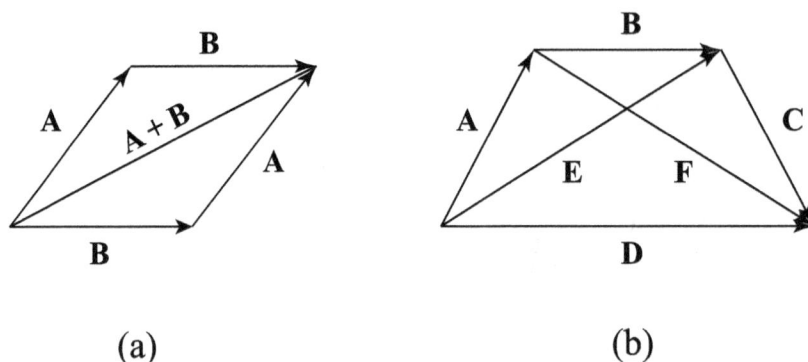

(a)   (b)

Figure A.1. Illustration of (a) the sum of two vectors **A** and **B** according to the parallelogram rule and (b) the associative property of vectors according to the trapezoid rule. We can see that $\mathbf{E} + \mathbf{C} = \mathbf{A} + \mathbf{F}$.

A vector should be thought of as a physical quantity whose properties are independent of the particular choice of a static (nonmoving) coordinate system. We may, however, express a vector in terms of a particular coordinate system in component form. If we choose a set of orthogonal axes x, y, z in three-dimensional space, a vector **A** may be written simply as an (ordered) set of numbers, $\mathbf{A} = (A_x, A_y, A_z)$, where the components represent the projection of the vector along the respective axes. Such a coordinate system, where the axes are fixed and point in the same direction at every point in space, is known as a Cartesian coordinate system.

Let us write the magnitude, or length, of the vector **A** as $|\mathbf{A}|$ and refer to θ as the angle between the vector and the z-axis, as illustrated in figure A.2. Then $A_z = |\mathbf{A}|\cos\theta$, and $|\mathbf{A}|\sin\theta$ represents the projection of the vector into the xy-plane. We then further introduce ϕ as the angle between this projection and the x-axis and obtain $A_x = |\mathbf{A}|\sin\theta\cos\phi$, $A_y = |\mathbf{A}|\sin\theta\sin\phi$. The angles θ and ϕ introduced in this way are also the angles used in a spherical coordinate system. In this system, the position vector **r** is characterized by the ordered numbers (r, θ, ϕ), where r is the length of the position vector.

We may also naturally introduce the unit vectors $\hat{\mathbf{x}}$, $\hat{\mathbf{y}}$, and $\hat{\mathbf{z}}$. The unit vector $\hat{\mathbf{x}}$, for example, is a vector of unit length that points directly along the x-axis. In terms of the unit vectors, a general vector may be written as

$$\mathbf{A} = A_x\hat{\mathbf{x}} + A_y\hat{\mathbf{y}} + A_z\hat{\mathbf{z}}. \tag{A.3}$$

This notation will be very convenient in discussing vector multiplication, which we consider next.

In vector algebra, there are three fundamental forms of multiplication that we will commonly use, called scalar multiplication, the dot product (inner product), and the cross product. Why are there only three fundamental forms when it appears that there are many ways to multiply the three components of one vector with another? We recall that a vector is supposed to represent a physical quantity whose properties are independent of any specific coordinate representation, and we require the same behavior for vector multiplication. For example, the dot product, which is a scalar,

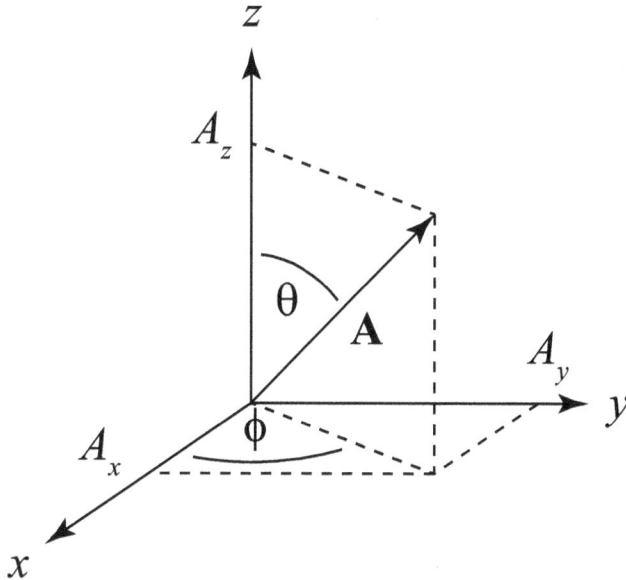

Figure A.2. Definitions of A_x, A_y, and A_z in terms of the length of the vector $|\mathbf{A}|$ and the angles θ, ϕ.

has the same scalar value regardless of the coordinate system we choose to represent the vectors. Under this condition, there are three fundamental types of vector multiplication (well, really, there are four, but the fourth—the outer product—is used much less often and discussed separately in chapter 14).

Scalar multiplication of a vector is the simplest type of vector multiplication and is the multiplication of a vector $\mathbf{A}$ by a scalar α, producing a resultant vector $\alpha\mathbf{A}$. The result of scalar multiplication is to scale the length of the vector $\mathbf{A}$ by the factor α, only changing its direction by at most a sign; if $\alpha < 0$, the resultant vector points in the opposite direction. In component form, scalar multiplication simply scales each component of the vector by the same multiplier α,

$$\alpha\mathbf{A} = \alpha A_x \hat{\mathbf{x}} + \alpha A_y \hat{\mathbf{y}} + \alpha A_z \hat{\mathbf{z}}. \tag{A.4}$$

The *dot product* is the product of two vectors $\mathbf{A}$ and $\mathbf{B}$, and the result is a scalar. If θ is the angle between two vectors, then the dot product may be written as

$$\mathbf{A} \cdot \mathbf{B} = |\mathbf{A}||\mathbf{B}|\cos\theta. \tag{A.5}$$

It can immediately be seen that this quantity is independent of the choice of coordinate axes, as the lengths of the vectors and the angle between them do not change under an axis rotation. From this definition, we can immediately determine that the dot product of two different unit vectors vanishes, i.e. $\hat{\mathbf{x}} \cdot \hat{\mathbf{y}} = 0$, and the dot product of two identical unit vectors is unity, i.e. $\hat{\mathbf{x}} \cdot \hat{\mathbf{x}} = 1$. From this, the dot product of two vectors in a component representation can readily be found,

$$\mathbf{A} \cdot \mathbf{B} = A_x B_x + A_y B_y + A_z B_z = \sum_{i=1}^{3} A_i B_i, \tag{A.6}$$

where we have also written it in a summation form, where $A_1 = A_x$ and so forth. It should be noted that, from either expression, we can see that $\mathbf{A} \cdot \mathbf{B} = \mathbf{B} \cdot \mathbf{A}$.

The dot product gives us a convenient way to write the magnitude, or length, of a real-valued vector. The dot product of a vector with itself is

$$\mathbf{A} \cdot \mathbf{A} = A_x^2 + A_y^2 + A_z^2. \tag{A.7}$$

The magnitude of the vector, or its length, is simply

$$|\mathbf{A}| = \sqrt{A_x^2 + A_y^2 + A_z^2}. \tag{A.8}$$

We will have need of the magnitude of complex vectors as well, which comes from the straightforward modification

$$|\mathbf{A}|^2 = \mathbf{A}^* \cdot \mathbf{A} = |A_x|^2 + |A_y|^2 + |A_z|^2. \tag{A.9}$$

where an asterisk represents the complex conjugate of a complex quantity.

Once we have a definition of length, or a metric, for our vectors, we may prove the triangle inequality, which indicates that for any two vectors $\mathbf{A}$ and $\mathbf{B}$, we have

$$|\mathbf{A} + \mathbf{B}| \leqslant |\mathbf{A}| + |\mathbf{B}|, \tag{A.10}$$

where equality only occurs when the vectors point in the same direction. We do not prove the triangle inequality here, though it follows quite readily from the definition of the dot product. Geometrically, the triangle inequality is the mathematical statement that the sum of any two sides of a triangle is larger than the third side. More generally, the triangle inequality is often used as the definition of a norm in a vector space: any function of vectors that satisfies the triangle inequality is a valid norm for the vector space.

The cross product of two vectors is itself a vector, and we must therefore characterize both its magnitude and direction. The magnitude of $\mathbf{C} \equiv \mathbf{A} \times \mathbf{B}$ is complementary to the dot product, i.e.

$$|\mathbf{C}| \equiv |\mathbf{A} \times \mathbf{B}| = |\mathbf{A}||\mathbf{B}|\sin\theta, \tag{A.11}$$

which is, in fact, the area of the parallelogram formed by the two vectors. The vectors $\mathbf{A}$ and $\mathbf{B}$ form a plane, and we may define the direction of the cross product as being perpendicular to this plane; in fact, with some reflection, one can be convinced that this is the only way that the direction can be defined where it does not depend on the choice of coordinate system. But a plane has two sides, like a sheet of paper, and therefore has two perpendiculars, pointing in opposite directions; which do we use? Here, we follow a right-handed convention for the definition of the cross product, in which one imagines pointing the right-hand index finger along $\mathbf{A}$, the right-hand middle finger along $\mathbf{B}$, and then the resultant vector $\mathbf{C}$ points in the direction of the thumb. For example, if $\mathbf{A}$ points along $\hat{\mathbf{x}}$ and $\mathbf{B}$ points along $\hat{\mathbf{y}}$, then $\mathbf{C}$ will point along $\hat{\mathbf{z}}$. Conversely, if $\mathbf{A}$ points along $\hat{\mathbf{y}}$ and $\mathbf{B}$ points along $\hat{\mathbf{x}}$, then $\mathbf{C}$ will point along $-\hat{\mathbf{z}}$.

Using the right-hand rule and the unit length of the unit vectors, we can take the cross product of two vectors in Cartesian components, and with some effort we can find that

$$\mathbf{C} = \mathbf{A} \times \mathbf{B} = \hat{\mathbf{x}}(A_y B_z - A_z B_y) + \hat{\mathbf{y}}(A_z B_x - A_x B_z) + \hat{\mathbf{z}}(A_x B_y - A_y B_x). \quad \text{(A.12)}$$

This is a seemingly tricky expression to remember, so for those familiar with linear algebra, the cross product can be written in determinant form as

$$\mathbf{C} = \begin{vmatrix} \hat{\mathbf{x}} & \hat{\mathbf{y}} & \hat{\mathbf{z}} \\ A_x & A_y & A_z \\ B_x & B_y & B_z \end{vmatrix}. \quad \text{(A.13)}$$

It should also be noted, however, that in summation form, the ith component of the cross product can also be written as

$$C_i = \sum_{j,k=1}^{3} \epsilon_{ijk} A_j B_k, \quad \text{(A.14)}$$

where ϵ_{ijk} is the Levi-Civita tensor that takes on values of $+1$ if ijk are cyclic, e.g. 123, 231, 312; -1 if ijk are anticyclic, e.g. 321, 132, 213; and 0 for any other case. From this, we can see that equation (A.12) includes every cyclic and anticyclic combination of a unit vector, component of $\mathbf{A}$, and component of $\mathbf{B}$, which allows us to easily write the cross product without needing to memorize any complicated equation.

It is important to note (and turns out to be surprisingly useful at times) that the cross product of a vector with itself is zero, i.e. $\mathbf{A} \times \mathbf{A} = 0$. Furthermore, we note that the cross product is not a commutative operation, as it depends on the order of multiplication; in particular, $\mathbf{A} \times \mathbf{B} = -\mathbf{B} \times \mathbf{A}$.

Two other significant results from vector algebra are worth noting here, as they will be used constantly throughout the book. These are the vector triple product formula, also known as the *BAC–CAB* rule, which is the following relation,

$$\mathbf{A} \times (\mathbf{B} \times \mathbf{C}) = \mathbf{B}(\mathbf{A} \cdot \mathbf{C}) - \mathbf{C}(\mathbf{A} \cdot \mathbf{B}). \quad \text{(A.15)}$$

The name '*BAC–CAB* rule' originates from the arrangement of vectors on the right of the above equation and is a handy mnemonic.

There is also a scalar triple product formula, which is of the form

$$\mathbf{A} \cdot (\mathbf{B} \times \mathbf{C}) = \mathbf{C} \cdot (\mathbf{A} \times \mathbf{B}) = \mathbf{B} \cdot (\mathbf{C} \times \mathbf{A}). \quad \text{(A.16)}$$

This expression says that the scalar triple product is unchanged by a cyclic permutation of the vectors.

A.2 Vector fields

So far, we have focused on vectors that represent the properties of a single, localized object: the velocity of a car, the force of gravity acting on the car, and the vector position of the car. However, many phenomena in physics are extended disturbances where we can associate a vector quantity with every point in the disturbance. For example, in a flowing fluid, the velocity **v** of the fluid depends on the position **r** within the fluid, even if the system as a whole is time independent. We introduce a vector field **v(r)**, a vector that depends on position, to characterize the behavior of the disturbance as a whole. Most relevant for our purposes are the electric field **E(r**, t) and the magnetic field **H(r**, t) of light waves, which also represent extended disturbances throughout space.

We may also introduce scalar fields, which are scalar quantities associated with an extended phenomenon whose value depends upon position. In fluids, the mass density $\rho_m(\mathbf{r})$ can vary with position. In electromagnetism, the electric charge density $\rho(\mathbf{r})$ and the scalar potential $\phi(\mathbf{r})$ are good examples of scalar fields.

Once we have scalar and vector fields, it is natural to characterize their properties and study their evolution through the use of calculus. Derivatives of fields can tell us about the local behavior of the fields, and integrals of fields can tell us about the effect of the fields as a whole. Vector calculus, of course, is therefore our next step.

A.3 Vector differentiation

In single-variable calculus, the derivative of a function tells us about its local rate of change. In vector calculus, derivatives have similar interpretations, albeit more complicated ones due to the vector nature of the fields and the multivariable dependence of the functions.

All of our vector derivatives are based on the 'del' operator ∇, which may be written as

$$\nabla \equiv \hat{\mathbf{x}}\frac{\partial}{\partial x} + \hat{\mathbf{y}}\frac{\partial}{\partial y} + \hat{\mathbf{z}}\frac{\partial}{\partial z}. \tag{A.17}$$

This quantity is an operator and not, strictly speaking, a vector, but because it has an analogous component structure, we can often treat it as one in deriving vector derivative identities, as we will see.

Let us consider a scalar field $\phi(\mathbf{r})$ first. A scalar field associates a single number $\phi(\mathbf{r})$ with every position **r** in space. We can introduce the *gradient* of the field as

$$\nabla\phi(\mathbf{r}) \equiv \hat{\mathbf{x}}\frac{\partial\phi(\mathbf{r})}{\partial x} + \hat{\mathbf{y}}\frac{\partial\phi(\mathbf{r})}{\partial y} + \hat{\mathbf{z}}\frac{\partial\phi(\mathbf{r})}{\partial z}. \tag{A.18}$$

The gradient is the three-dimensional analogy of the derivative in single-variable calculus. The magnitude of the gradient represents the largest rate of change of $\phi(\mathbf{r})$ at the point **r**, while the direction of the gradient represents the direction of the largest rate of change. We may express these statements formally using the differential relation

$$\nabla\phi \cdot d\mathbf{r} = \frac{\partial\phi}{\partial x}dx + \frac{\partial\phi}{\partial y}dy + \frac{\partial\phi}{\partial z}dz = d\phi, \tag{A.19}$$

where we have introduced the infinitesimal displacement

$$d\mathbf{r} = \hat{\mathbf{x}}dx + \hat{\mathbf{y}}dy + \hat{\mathbf{z}}dz. \tag{A.20}$$

Equation (A.19) illustrates that the gradient relates an infinitesimal change in position to the corresponding infinitesimal change in ϕ.

It is well-known that a conservative force field $\mathbf{F}(\mathbf{r})$ can be expressed in terms of the gradient of a scalar potential $\phi(\mathbf{r})$ in the form

$$\mathbf{F}(\mathbf{r}) = -\nabla\phi(\mathbf{r}), \tag{A.21}$$

where the minus sign is a convention that makes the force point 'downhill' in $\phi(\mathbf{r})$. Examples include a classical gravitational potential and the electric potential in electrostatic problems. This form of the force $\mathbf{F}$ is equivalent to saying that the work done by the force over a path is path independent, which comes from integrating equation (A.19) from a point A to a point B in space,

$$\int_A^B \mathbf{F}(\mathbf{r}) \cdot d\mathbf{r} = -\int_A^B \nabla\phi(\mathbf{r}) \cdot d\mathbf{r} = -\int_{\phi_A}^{\phi_B} d\phi = \phi_B - \phi_A, \tag{A.22}$$

where ϕ_A and ϕ_B are the values of the potential at the endpoints.

Let us now consider a vector field $\mathbf{v}(\mathbf{r})$ and the relevant derivative operations that may be performed upon it. Thinking of ∇ as a quasi-vector, we may immediately see that we can take the dot product and cross product of ∇ with $\mathbf{v}(\mathbf{r})$, which are the operations known as *divergence* and *curl*, respectively. It is straightforward to determine that the form of the divergence in Cartesian components is given by

$$\nabla \cdot \mathbf{v}(\mathbf{r}) = \frac{\partial v_x(\mathbf{r})}{\partial x} + \frac{\partial v_y(\mathbf{r})}{\partial y} + \frac{\partial v_z(\mathbf{r})}{\partial z}, \tag{A.23}$$

where v_x is the x component of $\mathbf{v}$, and so forth.

The divergence may be shown to quantify the net flow of the vector field into or out of an infinitesimal volume at the point $\mathbf{r}$; positive divergence represents an outward flow. In other words, the divergence characterizes the local 'sources' or 'sinks' of the vector field. For example, Gauss's law in differential form $\nabla \cdot \mathbf{E} = \rho/\epsilon_0$ indicates that electric charges represent locations where electric field lines begin or end; the 'no magnetic monopoles' law for the magnetic field $\nabla \cdot \mathbf{B} = 0$ indicates that magnetic field lines must form closed loops or extend to infinity.

The form of the curl in Cartesian components is given as

$$\nabla \times \mathbf{v}(\mathbf{r}) = \hat{\mathbf{x}}\left[\frac{\partial v_z}{\partial y} - \frac{\partial v_y}{\partial z}\right] + \hat{\mathbf{y}}\left[\frac{\partial v_x}{\partial z} - \frac{\partial v_z}{\partial x}\right] + \hat{\mathbf{z}}\left[\frac{\partial v_y}{\partial x} - \frac{\partial v_x}{\partial y}\right]. \tag{A.24}$$

Like the cross product, the curl may be written and remembered in a compact determinant form,

$$\nabla \times \mathbf{v} = \begin{vmatrix} \hat{\mathbf{x}} & \hat{\mathbf{y}} & \hat{\mathbf{z}} \\ \frac{\partial}{\partial x} & \frac{\partial}{\partial y} & \frac{\partial}{\partial z} \\ v_x & v_y & v_z \end{vmatrix}. \tag{A.25}$$

The curl may be shown to quantify the local circulation of the vector field, i.e. how much it goes around in circles. Each component of the curl may be interpreted as circulation around the component axis in a right-handed sense: if one's right thumb points in the direction of the curl, then one's fingers curl in the direction of circulation.

We will regularly come across second derivatives in the course of studying Maxwell's equations, and there are some general identities that will arise again and again in calculations. The first two of these may seem trivial but have significant physical implications down the line. By calculating in Cartesian components, one can show that the curl of a gradient is always zero,

$$\nabla \times (\nabla \phi) = 0. \tag{A.26}$$

This indicates that there is no circulation in a vector field derived from the gradient of a scalar; such fields are called *irrotational*.

One can also show that the divergence of a curl is always zero, i.e.

$$\nabla \cdot (\nabla \times \mathbf{v}) = 0. \tag{A.27}$$

A field that can be expressed as the curl of another vector field is referred to as *solenoidal* because it acts similarly to the zero-divergence magnetic field.

Next, we note that we may take the divergence of a gradient, producing an operator known as the Laplacian, which is written as ∇^2; in Cartesian components, the Laplacian of a scalar function is given by

$$\nabla^2 \phi(\mathbf{r}) = \nabla \cdot [\nabla \phi(\mathbf{r})] = \frac{\partial^2 \phi(\mathbf{r})}{\partial x^2} + \frac{\partial^2 \phi(\mathbf{r})}{\partial y^2} + \frac{\partial^2 \phi(\mathbf{r})}{\partial z^2}. \tag{A.28}$$

There is no special rule associated with the Laplacian, but we define it explicitly here because we will encounter it often.

Finally, we note that the curl of a curl satisfies the expression

$$\nabla \times [\nabla \times \mathbf{v}(\mathbf{r})] = \nabla[\nabla \cdot \mathbf{v}(\mathbf{r})] - \nabla^2 \mathbf{v}(\mathbf{r}). \tag{A.29}$$

This result can be proven in a straightforward manner by expanding both sides in Cartesian components, but it is a cumbersome calculation. We may also take advantage of the vector-like nature of the del operator and apply the *BAC–CAB* rule; naively, this produces a formula

$$\nabla \times [\nabla \times \mathbf{v}(\mathbf{r})] = \nabla[\nabla \cdot \mathbf{v}(\mathbf{r})] - \mathbf{v}(\mathbf{r})\nabla^2. \quad \text{(WARNING: not complete!)} \tag{A.30}$$

We note, however, that any derivative operations must act on $\mathbf{v}$, so we immediately move the ∇^2 in the second term to the left of $\mathbf{v}$, and the double curl formula follows directly.

Many vector calculus identities can be evaluated using this sort of approach. The vector multiplications are independent of the derivative operations, so when faced with a complicated expression involving multiple vector derivatives that we need to simplify, the best approach is to first use the product rule of derivatives to write the expression explicitly in terms of the derivatives of specific components and then apply vector multiplication rules to find the final result. As a simple example, let us consider the expression

$$\nabla \times (\mathbf{A} \times \mathbf{B}), \tag{A.31}$$

where $\mathbf{A}$ and $\mathbf{B}$ are both vector fields. We note that by the product rule, we expect one term where the derivative only acts on $\mathbf{A}$ and another term where it acts only on $\mathbf{B}$,

$$\nabla \times (\mathbf{A} \times \mathbf{B}) = \nabla_A \times (\mathbf{A} \times \mathbf{B}) + \nabla_B \times (\mathbf{A} \times \mathbf{B}), \tag{A.32}$$

where ∇_A represents the derivatives with respect to $\mathbf{A}$ only, and so forth. We now apply the *BAC–CAB* rule to each term:

$$\nabla \times (\mathbf{A} \times \mathbf{B}) = (\mathbf{B} \cdot \nabla_A)\mathbf{A} - \mathbf{B}(\nabla_A \cdot \mathbf{A}) + \mathbf{A}(\nabla_B \cdot \mathbf{B}) - (\mathbf{A} \cdot \nabla_B)\mathbf{B}, \tag{A.33}$$

where we have reorganized terms as allowed in order to place the derivatives to the left of their targets. Because everything is in its proper place, we can drop the subscripts to obtain

$$\nabla \times (\mathbf{A} \times \mathbf{B}) = (\mathbf{B} \cdot \nabla)\mathbf{A} - \mathbf{B}(\nabla \cdot \mathbf{A}) + \mathbf{A}(\nabla \cdot \mathbf{B}) - (\mathbf{A} \cdot \nabla)\mathbf{B}, \tag{A.34}$$

which is our final result.

This method of solving vector derivative identities is not always the best approach, but it is an excellent starting point in many cases; if it fails, one can fall back on Cartesian components if needed.

A.4 Vector integration

We have briefly alluded to vector integration in previous sections. Here, we elaborate on the types of vector integration and their interpretation. In three-dimensional space, we have three broad classes of integrals: path integrals (one parameter), surface integrals (two parameters), and volume integrals (three parameters).

The most common path integral is an integral of the form

$$\int_C \mathbf{v}(\mathbf{r}) \cdot d\mathbf{r}, \tag{A.35}$$

where $d\mathbf{r}$ is *formally* given by equation (A.20) and C represents a one-dimensional path in three-dimensional space with a starting point labeled A and an ending point labeled B. This formalism and the method used to evaluate the integral are illustrated in figure A.3. We consider a Riemannian integration, breaking the path into N straight line segments, each of which is represented by a vector length $\Delta\mathbf{r}_i$, where i is the index of the ith segment. We introduce $\mathbf{r}_i$ as the midpoint of this segment and then define the path integral as

$$\int_C \mathbf{v}(\mathbf{r}) \cdot d\mathbf{r} = \lim_{N \to \infty} \sum_{i=1}^{N} \mathbf{v}(\mathbf{r}_i) \cdot \Delta\mathbf{r}_i. \tag{A.36}$$

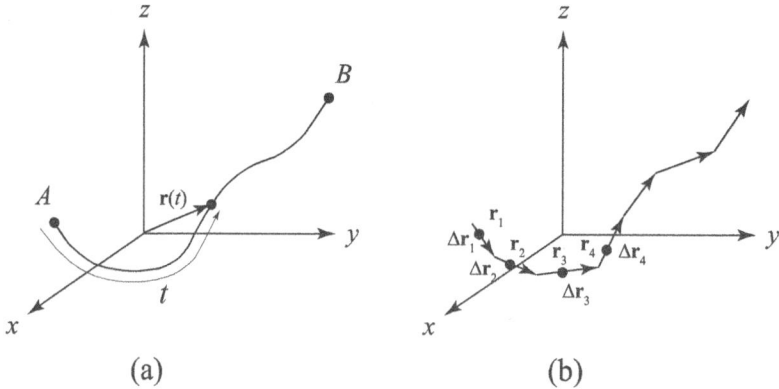

Figure A.3. Illustration of (a) the parameterization of a path integral, and (b) the Riemannian integration method used to evaluate the integral.

This is a fair formal definition and also demonstrates how one can evaluate such an integral computationally by choosing a large (but finite) value of N.

If we wish to attempt an analytical evaluation of the integral, we must parameterize the path. We introduce a function $\mathbf{r}(t)$, where $0 \leqslant t \leqslant 1$, such that $\mathbf{r}(t)$ traces out the path as t ranges from zero to one. Here, t is a dimensionless parameter that serves as a measure of how far along the path we are. We may then write

$$d\mathbf{r} = \frac{d\mathbf{r}}{dt}dt, \tag{A.37}$$

which allows us to replace our integration over $d\mathbf{r}$ with an integration over t. We may also write

$$\frac{d\mathbf{r}}{dt} = \hat{\mathbf{x}}\frac{dx}{dt} + \hat{\mathbf{y}}\frac{dy}{dt} + \hat{\mathbf{z}}\frac{dz}{dt}. \tag{A.38}$$

The integral becomes

$$\int_C \mathbf{v}(\mathbf{r}) \cdot d\mathbf{r} = \int_0^1 \left\{ v_x[\mathbf{r}(t)]\frac{dx}{dt} + v_y[\mathbf{r}(t)]\frac{dy}{dt} + v_z[\mathbf{r}(t)]\frac{dz}{dt} \right\} dt. \tag{A.39}$$

Now, everything in the integrand is expressed in terms of the variable t, and we may evaluate the integral according to the rules of single-variable calculus, if possible (as always, many integrals simply cannot be evaluated analytically).

It should be noted that the most difficult part of the process is finding the parameterizing function $\mathbf{r}(t)$, as, in principle, the choice is not a unique one. We also note that the path is often taken to be closed, i.e. $A = B$, and in such a case, the integral sign is often written as $\oint_C$, with the circle indicating a closed path.

The most common surface integral is of the form

$$\int_S \mathbf{v}(\mathbf{r}) \cdot d\mathbf{a}, \tag{A.40}$$

where S indicates an integral over a two-dimensional surface and the vector area element $d\mathbf{a}$ has a magnitude equal to the area of the element and a direction normal to the surface. However, any surface has two sides (think of a sheet of paper), so we must introduce conventions that define which of the two normals is used for $d\mathbf{a}$.

If the surface is closed, like a sphere with no openings, we choose the outward normal, i.e. the direction that points outside the surface. For an open surface, we note that most integrals over an open surface are associated with an integral over a closed path following the edge of the surface. We introduce another right-hand rule: we let the fingers of our right hand curve in the direction of the path, and then our thumb points in the direction of the appropriate surface normal.

Integrals of the form of equation (A.40) are a measure of the total flux of the vector field flowing through the surface; only the component of the vector field pointing through the surface contributes to the integral. Flux integrals are usually taken over closed surfaces, and $\oint_S$ is used to represent the closed surface.

Formally, we also evaluate a surface integral in a Riemannian manner, as illustrated in figure A.4. We imagine dividing the surface into N flat surface elements of vector area $\Delta\mathbf{a}_i$ and let $\mathbf{r}_i$ represent the central point of the ith surface element. The integral can then be written as

$$\int_S \mathbf{v}(\mathbf{r}) \cdot d\mathbf{a} = \lim_{N\to\infty} \sum_{i=1}^{N} \mathbf{v}(\mathbf{r}_i) \cdot \Delta\mathbf{a}_i. \tag{A.41}$$

We may also parameterize the integral in an attempt to evaluate it analytically. In this case, we note that a point on a surface is typically indicated by two parameters. For example, a point on the surface $z=0$, which is the xy-plane, can be parameterized by the values of x and y; a point on a sphere can be parameterized by the polar angle θ (latitude) and the azimuthal angle ϕ (longitude). Let us call these parameters $0 \leqslant s \leqslant 1$ and $0 \leqslant t \leqslant 1$; we may then introduce a function $\mathbf{r}(s, t)$ that represents the position vector of all points on the surface and can parameterize our surface integral as a two-variable integral over s and t. This can usually be done analytically only for the simplest cases of high symmetry, such as a flat surface or a spherical surface, so we do not elaborate on it here.

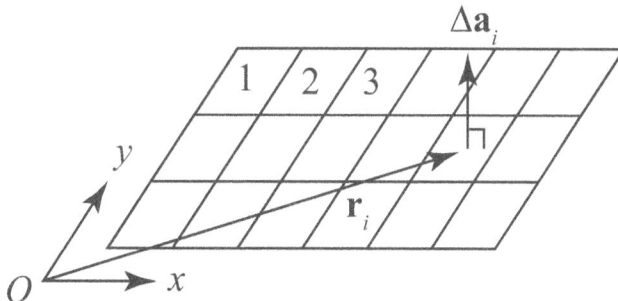

Figure A.4. Illustration of the notation related to a surface integral and the Riemannian integration method used to evaluate the integral.

The attentive reader might wonder why we have to parameterize a surface with two variables but were able to write the sum in equation (A.41) in terms of a single summation index. If there is a clear parameterization, we could also define the surface integral in terms of a double sum, but for complicated surfaces, a good parameterization is often unclear, and the single sum is easiest to implement; this is not a problem as long as we are considering finite sums. Is there a contradiction here between having a double integral and a single sum as we take the limit to infinity? We draw the reader's attention to the concept of space-filling curves [1], first discovered by Giuseppe Peano in 1890. A space-filling curve is a path—a one-dimensional manifold—that can be constructed to fill a two-dimensional region *completely* and continuously. It is like taking a pen with an infinitely thin tip and, starting from one corner of a square, drawing a line that completely fills the square without lifting the pen from the paper. In fact, any finite region of a finite-dimensional space can be parameterized by a single path that runs through and completely fills it, which indicates that there is at least nothing mathematically inconsistent about imagining a surface integral as the limit of a single sum of pieces.

Volume integrals are typically of a scalar function $u(\mathbf{r})$ and have the form

$$\int_V u(\mathbf{r})d^3r, \tag{A.42}$$

where V represents the volume of integration and d^3r indicates an infinitesimal volume. In Cartesian coordinates, this infinitesimal volume can be written as

$$d^3r = dxdydz. \tag{A.43}$$

In spherical coordinates, where the variables are r, θ, and ϕ, the volume element can be written as

$$d^3r = r^2dr \sin \theta d\theta d\phi. \tag{A.44}$$

A.5 Integral theorems

There are two integral theorems of fundamental importance to electromagnetic theory that we will have cause to use again and again throughout the book. Again, we will not prove these theorems here, but we state the theorems and provide an intuitive argument for their validity.

Gauss's theorem, also known as the divergence theorem, states that the integral of the divergence of a vector field over a volume is equal to the flux of the vector field through the surface bounding the volume,

$$\int_V \nabla \cdot \mathbf{v}(\mathbf{r})d^3r = \oint_S \mathbf{v}(\mathbf{r}) \cdot d\mathbf{a}. \tag{A.45}$$

We already noted in section A.3 that the divergence at a point in space represents the net flow of a vector field out of that point; the integral on the left of Gauss's theorem, therefore, is the net sum of all sources/sinks of the vector field within the volume. The integral on the right represents the flux of the vector field through the surface. Gauss's theorem, in words, thus says: *the flux of the vector field through the surface bounding a volume is equal to the net sum of all the sources and sinks within the volume.* Gauss's theorem is, in a sense, a mathematical conservation law; if there is a net nonzero flux of vector field through a closed surface, then there must be points within the volume where the vector field is being created or destroyed.

Gauss's theorem may be said to hold for 'well-behaved' vector fields $\mathbf{v}(\mathbf{r})$. To be well-behaved, the field must be continuous and have continuous first partial derivatives; this includes many, if not most, cases of interest in physics. The need for these conditions is clear from the presence of the divergence in equation (A.45).

Stokes's theorem, which curiously rarely seems to be called the curl theorem, states that the integral of the curl of a vector field over an open surface is equal to the path integral of the vector field along the curve bounding the open surface,

$$\int_S \nabla \times \mathbf{v}(\mathbf{r}) \cdot d\mathbf{a} = \oint_C \mathbf{v}(\mathbf{r}) \cdot d\mathbf{r}. \qquad (A.46)$$

Stokes's theorem may be considered a 'conservation of circulation' law for mathematics. We have noted previously that the curl of a vector field at a point quantifies the net circulation of the field at that point; the integral over the open surface, therefore, determines the net circulation within the surface. The path integral on the right of Stokes's theorem is a measure of the overall circulation around the entire boundary; for example, if $d\mathbf{r}$ is parallel to $\mathbf{v}$ over the entire closed path, then the path integral has a large positive number and the vector field 'goes in a circle' around the boundary.

Stokes's theorem also holds only for 'well-behaved' vector fields that are continuous and have continuous first partial derivatives.

It is worth noting that we can develop modified versions of Stokes's theorem and Gauss's theorem by clever choices of the form of $\mathbf{v}(\mathbf{r})$. For example, let us choose $\mathbf{v} = \mathbf{c} \times \mathbf{u}$, where $\mathbf{c}$ is an arbitrary constant vector. Then Gauss's theorem has the form

$$\int_V \nabla \cdot [\mathbf{c} \times \mathbf{u}(\mathbf{r})] d^3 r = \oint_S [\mathbf{c} \times \mathbf{u}(\mathbf{r})] \cdot d\mathbf{a}. \qquad (A.47)$$

We use the scalar triple product formula of equation (A.16) to write

$$-\int_V \mathbf{c} \cdot [\nabla \times \mathbf{u}(\mathbf{r})] d^3 r = -\oint_S \mathbf{c} \cdot [d\mathbf{a} \times \mathbf{u}(\mathbf{r})]. \qquad (A.48)$$

Because $\mathbf{c}$ is a constant vector, it may be removed from the integrals to obtain

$$\mathbf{c} \cdot \int_V \nabla \times \mathbf{u}(\mathbf{r}) d^3 r = \mathbf{c} \cdot \oint_S d\mathbf{a} \times \mathbf{u}(\mathbf{r}). \qquad (A.49)$$

We now note that this expression must hold for any choice of $\mathbf{c}$; we can, in turn, choose it to be each of the unit vectors $\hat{\mathbf{x}}$, $\hat{\mathbf{y}}$, and $\hat{\mathbf{z}}$. Because these unit vectors cover all of three-dimensional space, this implies that the related vector integrals must be equal to each other, or

$$\int_V \nabla \times \mathbf{u}(\mathbf{r}) d^3r = \oint_S d\mathbf{a} \times \mathbf{u}(\mathbf{r}). \tag{A.50}$$

We have therefore derived a form of Gauss's theorem where a curl appears in the volume integral.

Gauss's theorem and Stokes's theorem may themselves be shown to be different manifestations of the same general integral theorem, as discussed in the mathematical theory of forms [2]. We will not consider this relationship further, but it is worthwhile to be aware of it.

References

[1] Sagan H 1994 *Space-Filling Curves* (New York: Springer)
[2] Bachman D 2012 *A Geometric Approach to Differential Forms* (New York: Springer)